Quality

Second Edition

Donna C. S. Summers
University of Dayton

Prentice Hall
Upper Saddle River, New Jersey Columbus, Ohio

Library of Congress Cataloging-in-Publication Data
Summers, Donna C. S.
Quality / Donna C. S. Summers.
p. cm.
Includes bibliographical references and index.
ISBN 0-13-099924-5
1. Quality assurance. 2. Process control—Statistical methods.
I. Title.
TS156.6.S86 2000 98-51773
658.5'62—dc21 CIP

Editor: Stephen Helba
Production Editor: Rachel Besen
Design Coordinator: Karrie Converse-Jones
Cover Designer: Ceri Fitzgerald
Production Manager: Deidra M. Schwartz
Marketing Manager: Chris Bracken

This book was set in Goudy and Helvetica by York Graphic Services, Inc. and was printed and bound by R. R. Donnelley & Sons Company. The cover was printed by Phoenix Color Corp.

Printed in the United States of America

10 9 8 7 6 5 4 3 2 1

ISBN: 0-13-099924-5

Prentice-Hall International (UK) Limited, *London*
Prentice-Hall of Australia Pty. Limited, *Sydney*
Prentice-Hall of Canada, Inc., *Toronto*
Prentice-Hall Hispanoamericana, S. A., *Mexico*
Prentice-Hall of India Private Limited, *New Delhi*
Prentice-Hall of Japan, Inc., *Tokyo*
Prentice-Hall (Singapore) Pte. Ltd., *Singapore*
Editora Prentice-Hall do Brasil, Ltda., *Rio de Janeiro*

To my loving family,
Karl,
and all my boys

Preface

In the time that has elapsed since the first edition of *Quality* appeared, quality has woven itself into the very fabric of business. This second edition represents a continuous improvement effort on my part to enhance the text using input from students, professors, readers, and reviewers nationwide. I appreciate all the helpful comments, information, and ideas.

The underlying theme throughout the text is one of achieving total quality assurance by improving processes through the reduction of variation. This text introduces fundamental quality concepts of statistical process control and the application of these concepts to issues arising in industry.

Quality encompasses a wide variety of topics. In making the decision about which topics to cover in this text, I asked the industries who hire the individuals who would take this course. Another significant source of information was the article "What Higher Education Should Be Teaching About Quality, But Is Not" from the April 1998 issue of *Quality Progress*. This text covers 28 of the 30 topic areas listed in the article's table entitled "Baldridge Award Winners' Assessments of Importance of Quality Concepts and Skills for Entry Level College Graduates." These topics include creating a process focus, understanding variation, process control charts, root-cause analysis, the seven tools of quality, systems thinking, costs of quality, and the voice of the customer.

The changes users of the first edition will notice the most are the restructuring of several chapters, the use of integrated example problems within the chapters, ten additional cases, and more end-of-chapter problems.

KEY FEATURES OF THE TEXT

Many of the unique features of this text have been designed to develop a greater understanding of the complexities of quality improvement efforts. The use of comprehensive examples and case studies is the book's strongest feature. Taken from real-life situations, these examples and cases support learning by asking readers to consider the application and interpretation of the statistical techniques they have studied.

There are five key features of this text:

1. Emphasis is placed on the practical application of quality principles. Overall quality assurance serves as the underlying theme.
2. Emphasis is also placed on the interpretation, understanding, and use of quality principles and concepts throughout the problem-solving process.
3. Computer software is provided with the text to support the extensive example problems, end-of-chapter problems, and case studies.

4. Lengthy examples and case studies based on real-life situations provide the student with an understanding of the knowledge and effort necessary to solve quality problems.
5. Within each chapter, continuity in student learning is maintained through the use of certain elements—a list of learning opportunities, italicized key words and concepts, integrated examples, a summary of lessons learned, formula summaries, and finally case studies.

To expand on these features:

1. *Emphasis is placed on the practical application of quality principles. Overall quality assurance serves as the underlying theme.* The purpose of this text is to expose the reader to a wide variety of topics within quality assurance. The basic statistical analysis and control-chart principles are enhanced through emphasis on the interpretation and use of information to solve quality problems. Throughout the text, example problems and case studies, based on realistic situations, encourage the student to consider the complexities of improving quality.

2. *Emphasis is also placed on the interpretation, understanding, and use of quality principles and concepts throughout the problem-solving process.* In the work world, statistical analysis and control-chart information must be used to make decisions concerning the manufacture of products and the provision of services. To make the best decisions, those trained in quality concepts must be able to synthesize and interpret the calculations, graphs, and control charts they create. The example problems, end-of-chapter problems, and case studies are constructed to support the use and interpretation of the calculations, charts, and graphs in dealing with quality-assurance issues. This is especially true in the case studies, where the student will need to consider all aspects of a situation—as well as the calculations, charts, and graphs—in order to arrive at a solution.

3. *Computer software is provided with the text to support the extensive example problems, end-of-chapter problems, and case studies.* The analysis and interpretation of control charts is as key to improving quality as is the creation of the charts themselves. To allow the students the opportunity to analyze and interpret realistic charts created with a significant amount of data, a comprehensive computer software program should be utilized to enable the student to work the problems without bogging down in the calculations. The software provided with the text, a student version of the industrial software package sold by PQ Systems, gives the user the ability to create X, R, p, u, and c charts.

4. *Lengthy examples and case studies based on real-life situations provide the student with an understanding of the knowledge and effort necessary to solve quality problems.* The use of case studies brings insight into the complexities of real-world quality assurance. The cases have been taken from a wide variety of industries, including manufacturing, environmental and chemical processing, packaging, service, education, and government. The cases cover the analysis and use of quality-assurance techniques, integrating the material covered in the chapter. Some of the cases are open-ended; some are narrowly focused and have specific answers. Because

problem solving in industry links a variety of techniques, many cases merge information from previous chapters. For example, the case study for Chapter 11, Costs of Quality, uses Pareto diagrams and flowcharts as well as cost-of-quality concepts.

Throughout the course, the case studies can be used as assignments to supplement shorter homework problems or as portions of take-home tests. Using the cases in parallel with class discussions and discussing case results in class will enhance the student's understanding of the course material.

5. *Within each chapter, continuity in student learning is maintained through the use of certain elements—a list of learning opportunities, italicized key words and concepts, integrated examples, a summary of lessons learned, formula summaries, and finally case studies.* Each chapter begins with a set of learning opportunities and an opening paragraph discussing the applicability of the chapter material to quality. Throughout the chapter, key words and concepts are italicized to draw the student's attention to critical ideas. Each chapter concludes with Lessons Learned, essentially a summary of the vital material presented in the chapter. When applicable, the formulas introduced in the chapter are also summarized at chapter's end in a table format. This approach has been adopted to help the student see at a glance the calculations necessary for statistical analysis and the creation of quality-control charts. Example problems and case studies created from actual experiences expose students to a variety of quality-improvement situations.

ORGANIZATION OF THE TEXT

The text is divided into four parts: Setting the Stage, Control Charts for Variables, Control Charts for Attributes, and Expanding the Scope of Quality. The first part, Setting the Stage, includes Chapters 1, 2, and 3. Chapter 1, Quality Basics, introduces quality definitions and concepts. Quality Advocates and Total Quality Management, Chapter 2, presents information about total quality and the individuals who shaped the quality movement in the United States and abroad. Chapter 3, Quality Improvement: Problem Solving, provides the student with a foundation in quality problem solving.

The second part, Control Charts for Variables, begins with Chapter 4 on statistics. This presentation prepares the reader for Chapters 5, 6, and 7 on control charts for variables, process capability, and other variable control charts.

The third part, Control Charts for Attributes, begins with Chapter 8, a review of probability. Chapter 9, Quality Control Charts for Attributes, presents p, c, and u charts.

Extending the discussion of basic quality concepts, the fourth part encourages the reader to examine other areas beyond traditional control charts. Expanding the Scope of Quality comprises Chapters 10 through 14. In these chapters, the concepts of reliability, quality costs, product liability, quality systems (ISO 9000, QS 9000) and the Malcolm Baldridge Award), benchmarking, and auditing are all introduced.

The book concludes with five appendices—dealing with values for the area under the normal curve, factors for computing $\overline{X}$, s, and R charts, binomial and Poisson probability distributions, and PQ Systems software, respectively—a comprehensive glossary, and a bibliography.

INSTRUCTOR'S MANUAL

The instructor's manual is divided into two sections. The first includes solutions to the end-of-chapter problems; the second discusses the case studies and provides sample answers for them.

SOFTWARE AND THE USE OF COMPUTERS

In the chapters dealing with control charts, several of the end-of-chapter problems are designed to be solved on computer since they contain a large amount of data. Once the data are entered, students will see how the computer enhances their ability to interpret control charts.

SQCpack for Windows Student Version

SQCpack for Windows Student Version, the software provided here, is a Windows 3.1 application that combines powerful SPC techniques with the flexibility and ease of use of the Windows environment. Minimum hardware requirements are a PC compatible with Windows 3.1, a 4-MB RAM, and 5 MB of hard-disk space. The software supports all printers supported by Windows, including PostScript.

SQCpack for Windows Student Version enables students to create SPC charts and reports with their data. They can enter limited amounts of new data or can use the data stored in ASCII files or on the Clipboard. After selecting the data they want to use, students can create and draw charts. The software includes on-line help files that describe the application's commands and windows. These files appear in response to a Help command or when F1 is pressed. In addition, the SQC Quality Advisor provides an overview for using SPC tools. Opened from the Help menu, the Advisor explains terminology and answers frequently asked questions.

For further information on PQ Systems, see Appendix 5.

ACKNOWLEDGMENTS

This book has been created with the help and insight of some very important people—my students and the people who have hired them. I would like to express my sincere appreciation to both those groups. Their input has been invaluable. My thanks to the reviewers of this second edition whose suggestions added polish to the text: O. G. Peterson, Ph.D., Milwaukee School of Engineering; Matthew P. Stephens, Ph.D., Purdue University; Bruce A. Feodoroff, New England Institute of Technology; and William H. McPherson, East Carolina University. I would like to extend special thanks to the many individuals who helped create the case studies: David Sweeney, Kim Horrox, Greg Sesso, Joe McGravey, Tami Duncan, and Kwesi Amoa-Awuah. To those who listened to me throughout the creation of this second edition—Karl Summers, Carol Sweeney, Frannie Gracey Rye, Max, Rachel Besen, and Steve Helba—my heartfelt thanks!

Contents

List of Case Studies

I

Setting the Stage

1

Quality Basics

■ *Learning Opportunities:*

1. To understand the complexities of defining quality
2. To develop a definition for quality
3. To become familiar with differing definitions of quality from such sources as the American Society for Quality Control, Dr. W. Edwards Deming, Philip Crosby, and Armand Feigenbaum
4. To gain insight into the evolution of total quality management concepts
5. To become familiar with the definitions of specifications, tolerance limits, inspection, prevention, quality, quality control, statistical quality control, statistical process control, total quality management, and process improvement
6. To understand the differences between the philosophies of inspection, quality control, statistical quality control, statistical process control, total quality management, and continuous improvement
7. To understand the differences between actions necessary in inspection, quality control, statistical quality control, statistical process control, total quality management, and continuous improvement ■

SOURCE: Don Kite, *Parts Pups,* Nov. 1971, and *Reader's Digest,* October 1973.

The cartoon above is meant to make us chuckle a little about the difficulties consumers can experience in communicating what they want. But when you take a closer look, it isn't so funny. How can a company expect to stay in business if no connection is made between what the customer wants and what the company provides? This chapter begins the exploration into using process-improvement concepts to fulfill customers' needs, requirements, and expectations.

IDENTIFYING QUALITY

Portable computers, e-mail, DNA testing, cellular phones, gene therapy, compact discs, CNN (Cable News Network), hand-held video games, MTV, the Internet, solar-cell power plates—what do they all have in common? They are all innovations developed since 1980. Imagine what innovations await us in the next 20 years. All these products or services have been designed, developed, tested and produced, or provided for consumers. Are they quality products or services? How do we know? How do their manufacturers or providers know they are providing a quality product or service? How will their customers define quality? You have probably used some of these products or services. How do you view their quality? What does the word *quality* mean to you? What does quality represent? Think about it: Is there one simple definition or is it a case of knowing what quality is when you see it?

The American Society for Quality Control defines ***quality*** as *a subjective term for which each person has his or her own definition. In technical usage, quality can have two meanings:* (1) *the characteristics of a product or service that bear on its ability to satisfy stated or implied needs and* (2) *a product or service free of deficiencies*. Is this formal definition enough? Is it how you would define quality? Let's see what others have to say about quality.

Definitions have been developed by many prominent professionals in the field. Dr. W. Edwards Deming, well-known consultant and author on the subject of quality, describes quality as *nonfaulty systems*. In *Out of the Crisis*, Dr. Deming stresses that quality efforts should be directed at the present and future needs of the consumer. Deming is careful to point out that future needs of the customer may not be identified *by* the customer but rather *for* the customer. The need for some of the products we use in our day-to-day lives (VCRs, microwaves, electronic mail, deodorants) was developed by the companies designing, manufacturing, and advertising the products. In other words, consumers do not necessarily know what they want until they have seen the product or received the service. From there, consumers may refine the attributes they desire in a product or service.

Dr. Joseph M. Juran, in his book *Juran's Quality Control Handbook*, describes quality as *fitness for use*. In his text, *Quality Is Free*, Philip Crosby discusses quality as *conformance to requirements* and nonquality as *nonconformance*.

Quality can take many forms. The above definitions mention three types: quality of design, quality of conformance, and quality of performance. *Quality of design* means that the product has been designed to successfully fill a consumer need, real or perceived. While the flat swing seat seen in the opening cartoon is not precisely the rope-and-tire combination the customer wanted, it is a quality design because it will allow the customer to swing. *Quality of conformance*—conformance to requirements—refers to the manufacture of the product or the provision of the service that meets the specific requirements set by the consumer. From this viewpoint, only a tire-and-rope swing will meet the customer's requirements. *Quality of performance* means that the product or service performs its intended function as identified by the consumer. If the customer performance requirements stated the desire to swing, with no specific type of swing mentioned, then either the simple tire-and-rope design or a flat swing performs the function desired by the user. Clearly communicating the needs, requirements, and expectations of the consumer requires a more complex definition of quality.

Armand Feigenbaum, author of *Total Quality Control*, states that *quality is a customer determination which is based on the customer's actual experience with the product or service, measured against his or her requirements—stated or unstated, conscious or merely sensed, technically operational or entirely subjective—and always representing a moving target in a competitive market*. Several key words stand out in this definition:

- Customer determination: Only a customer can decide if and how well a product or service meets his or her needs, requirements, and expectations.
- Actual experience. The customer will judge the quality of a product or service not only at the time of purchase but throughout usage of the product or service.
- Requirements: Necessary aspects of a product or service called for or demanded by the customer may be stated or unstated, conscious or merely sensed.
- Technically operational: Aspects of a product or service may be clearly identified in words by the consumer.
- Entirely subjective: Aspects of a product or service may only be conjured in a consumer's personal feelings.

Feigenbaum's definition shows how difficult it is to define quality for a particular product or service. Quality definitions are as different as people. In many cases, no two customers will have exactly the same expectations for the same product. Notice that Feigenbaum's definition also recognizes that a consumer's needs, requirements, and expectations change over time and with different situations. Under some circumstances customer expectations will not remain the same from purchase to purchase or encounter to encounter. To produce or supply a quality product or service, a company must be able to define and meet the customer's reasonable needs, requirements, and expectations, even as they change over time. This is true whether the product is tangible (automobiles, stereos, or computers) or intangible (airplane schedules, hospital care, or repair service). Because of Feigenbaum's broad emphasis on customer requirements, this text will use his description of quality as its guide.

INGREDIENTS FOR SUCCESS

Corporate Culture

Companies seeking to remain competitive in today's global markets must integrate quality into all aspects of their organization. Successful companies *focus on customers* and their needs, requirements, and expectations. The *voice of the customer* serves as a significant source of information for making improvements to a company's products and services. A successful enterprise has a **vision** of how it sees itself in the future. This vision serves as a guide, enabling company leaders to create strategic plans supporting the organization's objectives. A clear vision helps create an atmosphere within an organization that is cohesive, with its members sharing a common culture and value system focused on the customer. Teamwork and a results-oriented, problem-solving approach are often mainstays in this type of environment.

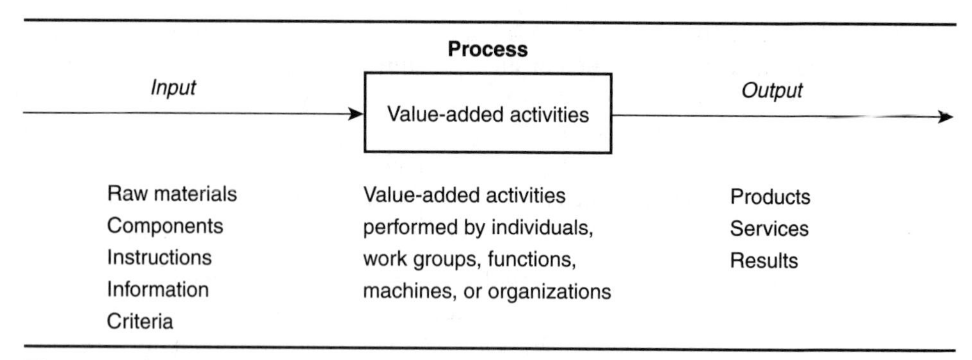

Figure 1.1 Processes

Processes and Process Improvement

Processes provide us with a way to get things done. In general, a ***process** takes inputs and performs value-added activities on those inputs to create an output* (Figure 1.1). Most of us do not realize how many processes we perform on a day-to-day basis. For instance, you go through a process when you select a movie to see. The input is the information about show times and places, whom you are going with, and what criteria you have for choosing a movie. The value-added activities are driving to the movie theater, buying a ticket, and watching the movie. And the output is the result, the entertainment value of the movie.

Industries have innumerable processes that enable them to provide products or services for customers. Think about the number of processes necessary to provide a shirt by mail-order. The company must have a catalog-preparation process, a catalog-distribution process, a process for obtaining the goods it plans to sell, an ordering process, a credit-check process, a packaging process, a mailing process, and a billing process, to name a few. Other processes typically found in organizations include financial management, involving planning, account servicing, and receivables; customer service; equipment maintenance and installation; production and inventory control; employee hiring, training, reviewing, firing, and payroll; software development; and product or service design, creation, inspection, packaging, delivery, and improvement.

Many processes develop over time, with little concern for whether or not it is the most effective manner in which to provide a product or service. To remain competitive in the world marketplace, companies must seek out wasteful processes and improve them. The processes providing the products and services will need to be quality-engineered, with the aim of preventing defects and increasing productivity by reducing process cycle times and eliminating waste. Many of the quality techniques discussed in this text support process improvement.

Variation

In any process that produces a product or provides a service, rarely are two products or service experiences exactly alike. Even identical twins have their differences. *Because **variation** is present in any natural process, no two products or occurrences are* *exactly alike*. In manufacturing, variation is often identified as the difference between the specified target dimension and the actual part dimension. In service industries, variation may be the difference between the type of service received and the type of service expected. Companies interested in providing a quality product or service use statistical process-control techniques to carefully study the variation present in their processes. Determining the reasons why differences exist between similar products or services and then removing the causes of these differences from the processes that produce them enable a company to more consistently provide a high-quality product or service. Think of it this way: If you are carpooling with an individual who is sometimes late, sometimes early, and sometimes on time, it is difficult to plan when you should be at the car. If, however, the person is always five minutes late, you may not like it, but you can plan around it. The first person exhibits a lot of variation; you never know when to expect them at the car. The second person, although late, has very little variation in his or her process; hence you know that if you need to leave at exactly 5 P.M., you had better tell that person to be ready at 4:55. The best situation would be to be on time every time. It is this best situation at which companies are aiming when they seek to eliminate or reduce the variation present in a process. Methods of improving processes by removing variation are the focus of this text.

Product and service designers translate customer needs, requirements, and expectations into tangible requirements called *specifications*. ***Specifications*** *state product or service characteristics in terms of a desired target value or dimension*. In service industries, specifications may take the form of descriptions of the types of services that are expected to be performed (Table 1.1). In manufacturing, specifications may be given as nominal target dimensions (Table 1.2), or they may take the form of tolerance limits (Table 1.3). ***Tolerance limits*** *show the permissible changes in the dimension of a quality characteristic*. Parts manufactured between the tolerance, or specification, limits are considered acceptable. Designers should seek input from the customer, from engineering and manufacturing professionals, and from any others who can assist in determining the appropriate specifications and tolerances for a given item.

To manufacture products within specifications, the processes producing the parts need to be stable and predictable. A process is considered to be *under control* when the variability (variation) from one part to another or from one service to another

Table 1.1 Specifications for Banking Transactions at Teller Windows Serving 10,000 Customers Monthly

Item	*Specification*
Customer Perception of Service/Quality	2 or Fewer Complaints per Month
Downtime of Teller Window Due to Teller Absence	Not to Exceed 5 min per Day
Deposits not Credited	1 or Fewer per Month
Accounts not Debited	1 or Fewer per Month
Errors on Cash In and Out Tickets	1 or Fewer per Month
Missing and Illegible Entries	2 or Fewer per Month
Inadequate Cash Reserves	2 or Fewer Occurrences per Month

Table 1.2 Specifications for Nominal Dimensions

Item	*Nominal Dimension*
Door Height	6 ft 6 inch
Theater Performance Start Time	8:00 P.M.
Wheelbase Length of a Car (in Catalog Shown to Customer)	110 inch
Frequency for a Radio Station	102.6 Hz
Calories in a Serving	100
Servings in a Package	8

is stable and predictable. Just as in the car pooling example, predictability enables those studying the process to make decisions concerning the product or service. When a process is predictable, very little variation is present. Statistical process-control practitioners use a variety of techniques to locate the sources of variation in a process. Once these sources are located, process improvements should be made to eliminate or reduce the amount of variation present.

EXAMPLE 1.1 Reducing Variation

Maps help us find the best route to get where we're going. At Transitplan Inc., mapmaking is serious business. Transitplan prints and distributes multicolor regional highway maps for locales all across the United States. Printing multicolored maps is tricky business. Rolls of white paper pass through printing presses containing plates etched with map designs. For multicolored maps, eight printing plates, each inked with a different color, are etched with only the items that will appear in that particular color. In many cases, the items are interrelated, as in the case of a yellow line inside two green lines to delineate an expressway. If the map printing plates are not aligned properly with each other and with the map paper, the colors may be offset, resulting in a blurry, unreadable map.

Transitplan's process engineers have worked diligently to reduce or eliminate variation from the map-printing process. Sources of variation include the printing plates, the inks, the paper, and the presses containing the plates. The engineers

Item	*Specifications*
Priority Overnight Delivery Time	Before 10:00 A.M.
Metal Hardness	R_c-44–48
Car Tire Pressure	30–35 psi
Heat-Treat Oven Temperature	1300–1400°C
Wheelbase Length of a Car	110 ± 0.10 in.
Household Water Pressure	550 ± 5 psi
Diameter of a Bolt	10.52–10.55 mm

Table 1.3 Specifications for Tolerance Limits

have improved the devices that hold the plates in place, eliminating plate movement during press cycles. They have developed an inventory-control system to monitor ink freshness in order to ensure a clean print. They have tested different papers to determine which ones hold the best impressions. These improvements have been instrumental in removing variation from the map-printing process.

With a stable printing process that exhibits consistent variation, process managers at Transitplan can predict future production rates and costs. They can respond knowledgeably to customer inquiries concerning map costs and delivery dates. If the map-production process were unstable, exhibiting unpredictable variation and producing both good and bad maps, then the process managers could only guess at what the future would bring (Figure 1.2).

Productivity

Some people believe that quality and productivity are the same or very similar. Actually, there is a difference between the two. To be productive, one must work efficiently and operate in a manner that best utilizes the available resources. Productivity's principal focus is on doing something more efficiently. Quality, on the other hand, focuses on being effective. Being effective means achieving the intended results or goals while meeting the customer's requirements. So quality concentrates not only on doing things right (being productive), but on doing the right things right (being effective). In manufacturing terms, if a company can produce 10,000 table lamps in 13 hours instead of in 23 hours, this is a dramatic increase in productivity.

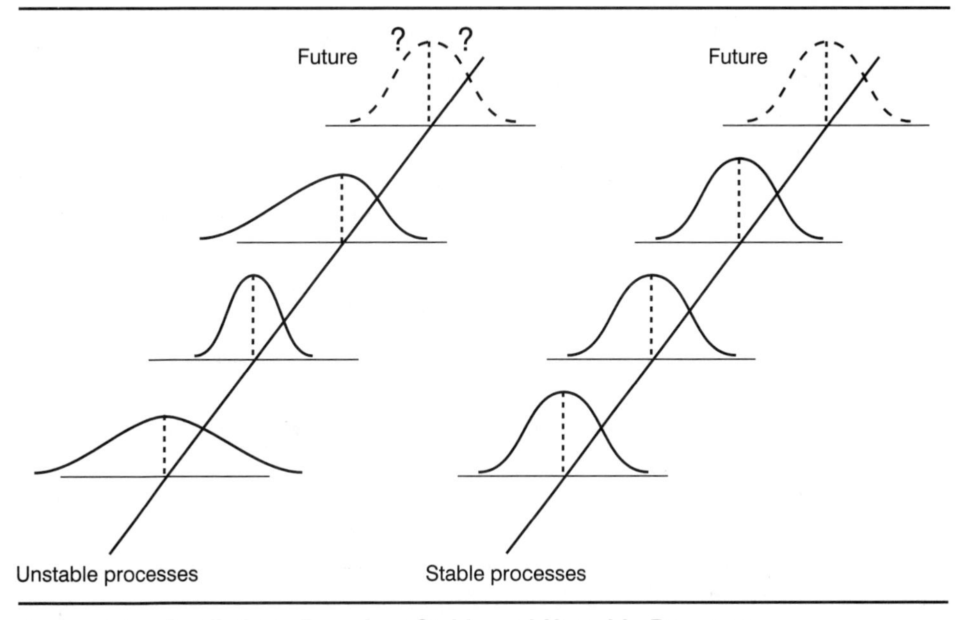

Figure 1.2 Predictions Based on Stable and Unstable Processes

However, if customers are not purchasing these table lamps because they are ugly, then the company is not effective, and the increased productivity is meaningless. To remain competitive, companies must focus on effectively meeting the reasonable needs and expectations of their customers. Productivity and quality improvements come from managing work activities as processes. As process performance is measured and sources of variation are removed, the effectiveness of the process increases. Process-improvement methods are discussed in greater detail in future chapters.

THE EVOLUTION OF QUALITY

Up until the advent of mass production, artisans completed individual products and inspected the quality of their own work or that of an apprentice before providing the product to the customer. If the customer experienced any dissatisfaction with the product, he or she dealt directly with the artisan.

Firearms were originally created individually. The stock, barrel, firing mechanism, and other parts were fabricated for a specific musket. If part of the musket broke, a new part was painstakingly prepared for that particular firearm or the piece was discarded. In 1798 Eli Whitney began designing and manufacturing muskets with interchangeable parts. Firing mechanisms, barrels, or other parts could be used on any musket of the same design. By making parts interchangeable, Eli Whitney created the need for quality control.

In a mass production setting, the steps necessary to create a finished product are divided among many individuals who each perform a single repetitive operation. Any mass production assembly area will have two needs. First, in order to be interchangeable, the parts must be nearly identical. This allows the assembler to randomly select a part from a group and assemble it with a second randomly selected part. For this to occur without problems, the machines must be capable of producing parts with minimal variation, within the specifications set by the designer. Producing with minimal variation and within specifications is the second need. If the parts are not made to specification, during assembly a randomly selected part may or may not fit together easily with its mating part. This situation defeats the idea of interchangeable parts. Quality principles evolved to meet these two needs (Figure 1.3).

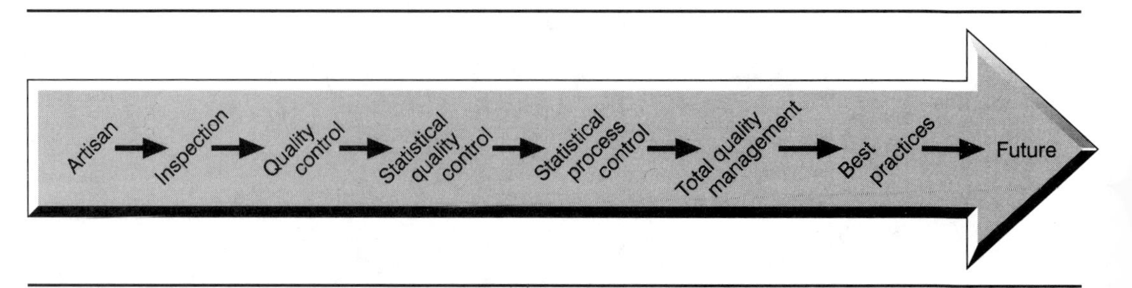

Figure 1.3 Evolution of Quality Principles

Inspection

As the variety of items being mass-produced grew, so did the need for monitoring the quality of the parts produced by those processes. The customer no longer dealt directly with the individuals responsible for creating the product, and industries saw a need to ensure that the customer received a quality product. At first, inspection was the primary method of ensuring that a quality product or service was provided. ***Inspection*** *refers to those activities designed to detect or find nonconformances existing in already completed products and services*. Inspection, the detection of defects, is a regulatory process.

Inspection involves the measuring, examining, testing, or gauging of one or more characteristics of a product or service. This result is compared with established standards to determine whether or not the product or service conforms. In a detection environment, inspection, sorting, counting, and grading of products compose the major aspects of a quality professional's position. This results in the general feeling that the responsibility for quality lies in the inspection department. Philosophically, this approach encourages the belief that good quality can be inspected into a product and bad quality can be inspected out of the product.

Inspection occurring only after the part or assembly has been completed can be costly. If a large number of defective products has been produced and the problem has gone unnoticed, then scrap or rework costs will be high. For instance, if a preliminary operation is making parts incorrectly, and inspection does not occur until the product has been through a number of other operations, a large number of defective items will have been fabricated before the defect is discovered by the inspectors. Any work done after the first operation will be wasted because the part used in the assembly is defective. This type of mistake is very costly to the producer because it involves not only the defective aspect of the part but also the cost of performing work on that part by later workstations (Figure 1.4).

The same is true in a service environment. If the service has been incorrectly provided, the customer receiving the service must spend additional time in the system having the problem corrected. Imagine a builder who inadvertently creates a house with a roof that leaks. If the leak is discovered after the customer moves in and rain has damaged the furniture, carpeting, drywall, and personal items, the costs of the

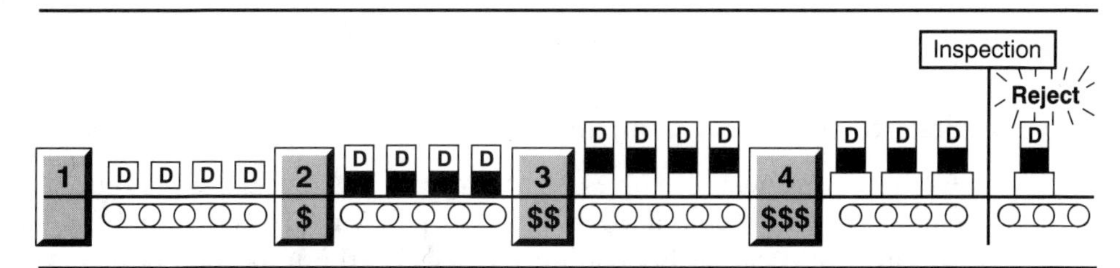

Figure 1.4 Consequences of Defects Compounded in Process

poorly done job involve significantly more than repairing the roof. Even if the problems have been fixed, there is always the threat that the customer receiving the incorrect service may choose not to return.

Quality Control

Quality control (QC) *refers to the use of specifications and inspection of completed parts, subassemblies, and products to design, produce, review, sustain, and improve the quality of a product or service.* Quality control goes beyond inspection by

1. Establishing standards for the product or service, based on the customer needs, requirements, and expectations.
2. Ensuring conformance to these standards. Poor quality is evaluated to determine the reasons why the parts or services provided are incorrect.
3. Taking action if there is a lack of conformance to the standards. These actions may include sorting the product to find the defectives. In service industries, actions may involve contacting the customer and correcting the situation.
4. Implementing plans to prevent future nonconformance. These plans may include design or manufacturing changes; in a service industry they may include procedural changes.

These four activities work together to improve the production of a product or provision of a service. Quality-control efforts can be enhanced by the use of statistics to help with decision making.

Statistical Quality Control

Building on the four tenets of quality control, statistics were added to map the results of parts inspection. In the 1920s, statistical charts used to monitor and control product variables were developed by Walter A. Shewhart of Bell Telephone Laboratories. At the same time H. F. Dodge and H. G. Romig, also of Bell Telephone Laboratories, developed the area of acceptance sampling as a substitute for 100 percent inspection. The use of statistical methods of production monitoring and parts inspection became known as ***statistical quality control (SQC),*** *wherein statistical data are collected, analyzed, and interpreted to solve quality problems.* The primary concern of individuals involved in quality is the monitoring and control of variation in the product being produced or service being provided.

Statistical Process Control

As the area of quality evolved, it became obvious that there was a need to become more proactive when dealing with quality problems. Thus the emphasis shifted from utilizing statistical quality control methods for the inspection or detection of poor quality to their use in the prevention of poor quality. *Prevention of defects by applying statistical methods to control the process is known as* ***statistical process control (SPC).***

Statistical process control emphasizes the prevention of defects. ***Prevention*** *refers to those activities designed to prevent defects, defectives, and nonconformance in products and services*. The most significant difference between prevention and inspection is that with prevention, the process—rather than solely the product—is monitored, controlled, and adjusted to ensure correct performance. By using key indicators of product performance and statistical methods, those monitoring the process are able to identify changes that affect the quality of the product and adjust the process accordingly. To do this, information gained about the process is fed back to those involved in the process. This information is then used to prevent defects from occurring. Emphasis shifts away from inspecting quality into a completed product or service toward making process improvements to design and manufacture quality into the product or service. The responsibility for quality moves from the inspectors to the design and manufacturing departments.

Statistical process control also seeks to limit the variation present in the item being produced or the service being provided. While it once was considered acceptable to produce parts that fell somewhere between the specification limits, statistical process control seeks to produce parts as close to the nominal dimension as possible and to provide services of consistent quality from customer to customer. To relate loss to only those costs incurred when a product or service fails to meet specifications is unrealistic. The losses may be due to reduced levels of performance, marginal customer service, a slightly shorter product life, lower product reliability, or a greater number of repairs. In short, losses occur when the customer has a less-than-optimal experience with the product or service. The larger the deviation from the desired value, the greater the loss. These losses occur regardless of whether or not the specifications have been met. Reducing process variation is important because any reduction in variation will lead to a corresponding reduction in loss.

Statistical process control can be used to help a company meet the following goals:

- To create products and services that will consistently meet customer expectations and product specifications.
- To reduce the variability between products or services so that the results match the desired design quality
- To achieve process stability that allows predictions to be made about future products or services
- To allow for experimentation to improve the process and to know the results of changes to the process quickly and reliably
- To further the long-term philosophy of continual improvement
- To minimize production costs by eliminating the costs associated with scrapping or reworking out-of-specification products
- To place the emphasis on problem solving and statistics
- To support decisions with statistical information concerning the process
- To give those closest to the process immediate feedback concerning current production
- To assist with the problem-solving process

Uniformity of Output
Reduced Rework
Fewer Defective Products
Increased Output
Increased Profit
Lower Average Cost
Fewer Errors
Predictable, Consistent Quality Output
Less Scrap
Less Machine Downtime
Less Waste in Production Labor Hours
Increased Job Satisfaction
Improved Competitive Position
More Jobs
Factual Information for Decision Making
Increased Customer Satisfaction
Increased Understanding of the Process
Future Design Improvements

Figure 1.5 Positive Results of Statistical Process Control

- To increase profits
- To increase productivity

The positive results provided by statistical process control can be seen in Figure 1.5. Many of the techniques associated with SPC are covered in this text.

Total Quality Management

As the use of statistical process control grew in the 1980s, industry saw the need to monitor and improve the entire system of providing a quality product or service. Sensing that meeting customer needs, requirements, and expectations involved more than providing a product or service, industry began to integrate quality into all areas of operations, from the receptionist to the sales and billing departments to the manufacturing, shipping, and service departments. Methods of quality management have been developed and utilized to encourage the design, production, marketing, sales, and service of quality products and services. The philosophy revolves around building in quality. This integrated approach, involving all departments in a company in providing a quality product or service, became known as "total quality management" (Figure 1.6).

Total quality management (TQM) *is a management approach that places emphasis on continuous process and system improvement as a means of achieving customer satisfaction to ensure long-term company success.* Total quality management utilizes the strengths and expertise of all the employees of a company as well as the statistical problem-solving and charting methods of statistical process control. TQM is based on and relies on the participation of all members of an organization to continuously improve the processes, products, and services their company provides as well as the culture they work in.

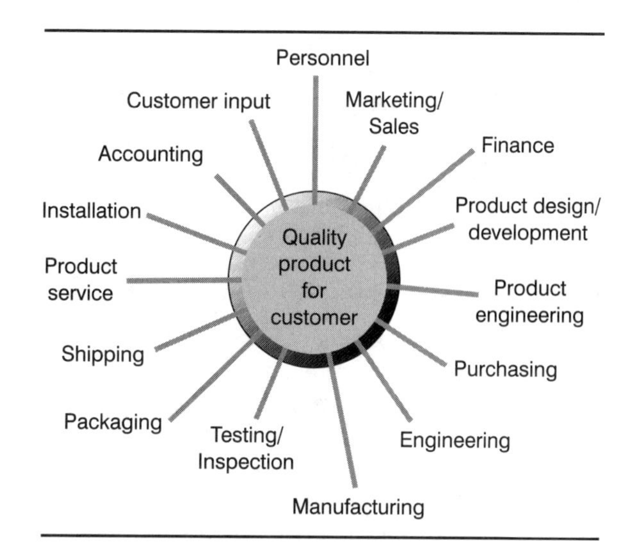

Figure 1.6 Department Involvement in Total Quality Management

Most importantly, quality management encourages a long-term, never-ending commitment to the improvement of the process, not a temporary program to be begun at one point in time and ended at another. The total quality process is a culture that top-level management develops within a company to replace the old management methods. Chapter 2 expands on the idea that the commitment to quality must come from upper management to guide corporate activities year after year toward specified goals. Given this long-term commitment to integrating quality into all aspects of the organization, companies pursuing quality management have an unwavering focus on meeting the customer's needs, requirements, and expectations. Since the customer's needs, requirements, and expectations are always changing, total quality management must be adaptable in order to pursue a moving target.

BEYOND TOTAL QUALITY MANAGEMENT

Over time, as consumers became more quality conscious, companies expanded their quality-management practices beyond the traditional manufacturing arena. Total quality management continues to evolve and embrace such concepts as optimization of processes, elimination of waste, and creation of a customer focus. Companies seeking to optimize business processes take a systems approach, emphasizing improving the systems and processes that enable a company to provide products or services for their customers. Examples of systems include ordering processes, billing processes, manufacturing processes, and shipping processes, among others. Improving processes means finding and eliminating sources of waste, such as idle time, rework time, excess variation, and underutilized resources. As competition becomes more and more global, companies must develop a more customer-oriented approach to quality, studying how their product or service is used from the moment a customer first comes in contact with the product or service until the moment that the product is disposed of or the service is complete. Global competition has also encouraged companies to seek out and emulate *best practices*. The term "best practices" refers to choosing a method of work that has been found to be the most effective and efficient, i.e., with no waste in the process. As long as there is competition, companies will continue to seek ways in which to improve their competitive position. The quality concepts presented in this text provide a firm foundation for any company seeking to continually improve the way it does business.

EXAMPLE 1.2 Tying It All Together*

This example, as well as the questions and case study at the end of the chapter, exist to try to encourage you to think about quality from the viewpoints of customers and providers of products and services.

Air travel is a common method of transportation today. Scheduled air carriers handle about two million pieces of luggage each day. That's a lot of luggage to keep track of, and unfortunately sometimes the passenger and the bag don't end up going to the same place at the same time. Each day, approximately one-half of 1 percent (10,000) of the two million bags are mishandled. When a bag is labeled mishandled, it either doesn't arrive with the passenger or it is damaged while en route. Each day, 100 unlucky travelers learn that their bags have been irretrievably lost or stolen. As late as 1998, on domestic flights airlines were liable for only $1250 of the cost of lost, damaged, or delayed luggage. International travelers received $9.07 per pound of checked luggage. The good news is that passenger reports of lost, damaged, or delayed baggage on domestic flights steadily decreased throughout the 1990s (Figure 1.7). What would be your definition of quality for this example? One hundred percent of the bags arriving with their passengers? What about the enormous amount of luggage being handled each day? Two million bags versus 10,000 bags mishandled? Seven hundred thirty million bags correctly handled annually versus 36,500 bags irretrievably lost or stolen? Is one-half of 1 percent (99.5 percent good) an acceptable level of quality? Do we hold other industries to such a high performance level? Do you turn in perfect term papers?

Can you identify the customers in this example? Are they the passengers or the airlines (as represented by the baggage handlers, the check-in clerks, and the lost-luggage finders). What are their needs, requirements and expectations? In order to do a quality job in the eyes of passengers—i.e., in order to get the luggage to the right place at the right time—the airlines must label bags correctly, bar code them to ensure efficient handling, employ trained baggage handlers and clerks, and determine ways to reduce the number of mishandled bags. Passengers also play a role in the overall quality of the system. At a minimum, passengers must label bags clearly and correctly. They must arrive in ample time before the flight and not overfill their suitcases so the suitcases do not unexpectedly open while being handled.

How will a customer recognize quality? Feigenbaum's definition of quality stresses that quality is a customer determination based on the customer's actual

*K. Choquetter, "Claim Increase for Lost Baggage Still Up in the Air," *USA Today,* March 17, 1998.

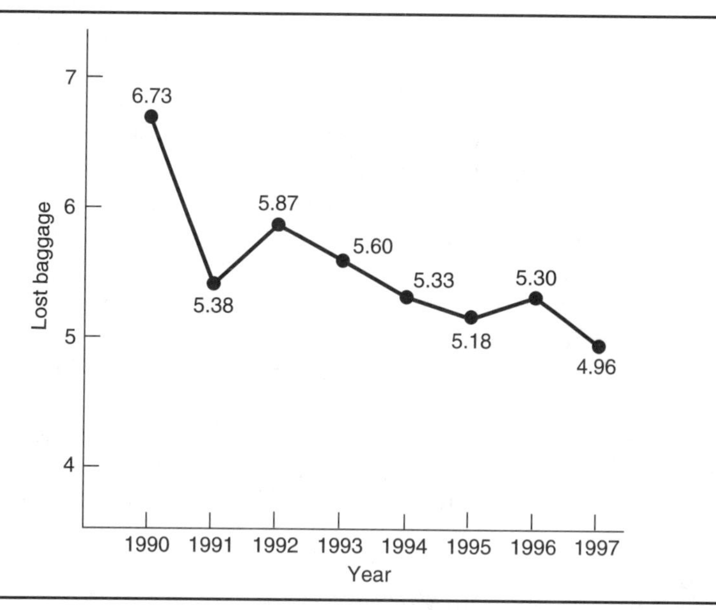

Figure 1.7 Mishandled Luggage (per 1000 passengers)
SOURCE: K. Choquetter, "Claim Increase for Lost Baggage Still Up in the Air," *USA Today,* March 17, 1998.

experience with the product or service and measured against his or her requirements. Different people will see this example differently. An individual about to make a very important presentation without needed clothing lost with a suitcase will be very angry. A traveler who receives his or her baggage on the next flight with only a small delay may be annoyed but minimally inconvenienced.

There are many aspects to quality. In this text, we explore techniques that enable us to improve the way we do business.

SUMMARY

In order to meet the challenges of a global economy, manufacturers and providers of services must balance the economic and profit aspects of their businesses with the goal of achieving total customer satisfaction. Quality must be designed into, built into, and maintained for each product or service provided by the company. Variation must be removed from the processes involved in providing products and services. Knowledge of consumer needs, requirements, and expectations will allow the company to succeed in the marketplace. Statistical process control, with its inherent emphasis on the creation of a quality product, is paramount in helping companies meet the challenges of global markets. Total quality management, with its pursuit of continuous improvement, is the management system that can help ensure the success of a company. Quality is a business strategy that is critical for success.

Lessons Learned

1. Defining quality is a complex endeavor. Quality is defined by a consumer's individual and reasonable needs, requirements, and expectations.
2. Processes perform value-added activities on inputs to create outputs.
3. Variation is present in any natural process. No two products or occurrences are exactly alike.
4. Specifications are used to help define a customer's needs, requirements, and expectations.
5. Productivity is doing something efficiently; quality is doing the right thing efficiently.
6. The monitoring and control of quality has evolved over time. Inspection, quality control, statistical quality control, statistical process control, and total quality management are all aspects of the evolution of quality. ■

Chapter Problems

1. What is your definition of quality?
2. Describe the differences between the definitions for quality given by the American Society for Quality Control, Dr. W. Edwards Deming, and Armand Feigenbaum.
3. Using Feigenbaum's definition, discuss how a customer may define quality for having a muffler put on her car. Be sure to discuss the key terms identified in the definition of quality.
4. Describe a product or service that you have purchased recently. In terms of Feigenbaum's and your own definitions, was it a quality product or service? How do you know?
5. Define variation. Describe three examples of variation in your own life.
6. Define process. Describe three examples of process in your own life.
7. In your own words, describe the difference between productivity and quality.
8. A professor has asked that students complete a term paper on the meaning of quality. The paper is to be four pages in length and typed. The paper may not contain any typographical, grammar, spelling, or punctuation errors. Treating the professor as a customer, what should the student do to ensure that quality is maintained?
9. Every day a dry cleaner receives a wide variety of clothes to clean. Some items may be silk, others may be wool, others rayon. Some fabrics may be delicate, other fabrics may be sturdier. Some clothing may contain stains. As a customer bringing your clothes in to be cleaned, what needs, requirements, and expectations can you identify? What must the dry cleaner do to do a quality job? How much would you be willing to pay for the service?

10. Answer the following questions about your own work environment:
    a. Who are your customers? These may be customers you directly serve or they may be customers of the overall company.
    b. How are the needs, requirements, and expectations of your customers identified?
    c. How does your customer recognize quality?
11. Describe the evolution of total quality management.
12. Define the following: specifications, tolerances, inspection, prevention.
13. Describe the philosophical differences between inspection, prevention, quality, quality control, statistical quality control, statistical process control, and total quality management.
14. Describe the differences between the actions necessary in inspection, quality control, statistical quality control, statistical process control, and total quality management.
15. Use your own work experience to describe a company you worked for and where they are philosophically (inspection, quality control, statistical quality control, statistical process control, or total quality management). Be sure to justify your decision by describing the company's actions that support its philosophy.
16. Choose and describe a quality (or nonquality) situation that you or someone close to you has experienced. What role did quality—and the customer's needs, requirements, and expectations—play in this situation? Describe how the creator of the product or the provider of the service could have dealt with the incident.
17. The following is a list of specifications for operating a hotel. Add four or five of your own customer specifications to this list.

Item	*Specification*
Customer Perception of Service/Quality:	2 or Fewer Complaints per Month
Downtime of Reservation/ Check-in Computer	Not to Exceed 15 minutes per Month
Room Reservations Incorrect/ Overbooked	1 or Fewer Occurrences per Month
Credit Card Billing/ Transaction Errors	1 or Fewer per Month

CASE STUDY 1.1
Specifications and Customers' Needs

THE CLEAN AIR ACT OF 1970 AND THE 1990 AMENDMENTS

In order to protect the environment and the air we breathe, in 1970 the U.S. government passed the National Environmental Policy Act (NEPA). At the end of the year, the Clean Air Act of 1970 was added as an amendment to the NEPA. For the next 20 years, air quality standards were established, monitored, and enforced by the Environmental Protection Agency (EPA). Air quality standards have been set to protect health, economic values (of crops and buildings), and the environment. Carbon monoxide, lead, nitrogen oxides, ozone, particulates, sulfur oxides, arsenic, mercury, benzene, beryllium, asbestos, radioactive substances, and vinyl chloride are all considered undesirable air pollutants.

In 1990 the Clean Air Act was updated and revised to reflect what the world had learned about air quality and the atmosphere. This Act continues the emphasis on controlling emissions from the smokestacks of utilities and other combustion facilities. It also lists 189 toxic compounds whose emissions must be controlled and reduced.

The combustion of fossil fuels generates four types of recognized pollutants: particulates, sulfur oxides, nitrogen oxide compounds, and volatile organic compounds and toxics. Particulates contribute to the visibly darkened, polluted skies because the tiny pieces of ash appear as black smoke leaving the stack. In 1971 the EPA set a performance standard limiting outlet of particulate emissions to 0.1 lb/10^6 Btu heat input. Since 1979 new generating facilities have been permitted a maximum of 0.03 lb/10^6 Btu heat input. From 1970 to 1990, particulate emissions have dropped from nearly 25 million tons per year to under 10 million tons.*

Sulfur oxides are formed when sulfur is released from coal burned for fuel. Sulfur dioxide (SO_2) is a suffocating gas, and sulfur trioxide (SO_3), when combined with a small amount of moisture, forms sulfuric acid. Known as acid rain, this deadly combination eats away at human lungs and defaces buildings. By the year 2000, sulfur oxide emissions must be reduced from 20 million tons per year discharged into the atmosphere (1980 levels) to 10 million tons per year as a nationwide total.† Based on the number of units emitting sulfur oxides, this represents an annual allowance of 1.2 lb of SO_2 per million Btus generated.

*M. Cooper. "Environmental Movement at 25: Will Congress Weaken Environmental Regulations?" *Congressional Quarterly Researcher*, March 31, 1995, p. 275.
†E. Edelson. *Clean Air*. New York: Chelsea House, 1992, p. 56.

1300-MW Generating Station

One megawatt of power will light ten thousand 100-W lightbulbs. So 1300 MW of power will light 13 million 100-W lightbulbs. The average house uses 960 per kWh/month. A 1300-MW power plant can power 975,000 houses per month around the clock.

The electric power required to run the flue gas desulfurization and the electrostatic precipitator systems as well as the pumps and motors for grinding and conveying material will require about 3 percent of the power generated by a 1300-MW generating station.

Cost per Ton of SO_2 Removed: $250 to $500

The high-sulfur midwestern plants tend to have the lowest cost per ton. Their locations allow for river delivery of bulk materials and a readily available water supply. The western low-sulfur plants, whose low-sulfur fuels yield a higher gas flow, do not experience the same economies of scale since the equipment is sized for gas flow and there is less SO_2 to remove with the same equipment.

Capital Costs

These costs, at $145 per kilowatt, include construction, foundations, and reagent storage and waste disposal systems. The primary process equipment (absorber towers, reagent prep systems, and spent reagent systems) is about $50 per kilowatt generated.

Figure 1 Flue Gas Desulfurization System Costs

Nitrogen oxide compounds, while odorless and colorless, are unstable and combine easily with other elements in the atmosphere to form smog. The 1990 Clean Air Act requires that nitrogen oxide emissions be diminished by 2 million tons each year.*

Volatile organic compounds and toxics are poisonous substances such as the heavy metals of mercury and arsenic. This relatively new area of the 1990 Clean Air Act requires that a portion of these elements be removed from flue gas emissions. At this point in time, while mercury and arsenic are a measurable part of coal, from an emissions point of view these elements occur in trace amounts and are difficult to measure. Precipitator ash contains arsenic, and fortunately modern precipitators are able to remove 99.99 percent of that ash.

INDUSTRY'S RESPONSE

A number of different types of environmental equipment can be used to remove undesirable pollutants from the flue gas of utility boilers that use fossil fuels to provide electricity. These different systems are often combined to remove the greatest variety of pollutants from the exhaust gases. Before any environmental protection devices are installed, a typical 1300-megawatt (MW) plant, described in Figure 1, that burns a 3.5 percent sulfur coal with 8.6 percent ash at 85 percent efficiency will produce 227,000 tons of SO_2 and 201,000 tons of precipitate (ash) per year.

*Ibid.

Electrostatic precipitators are designed to remove particulates from flue exhaust gases. Electrostatically charged wires lend a charge to the ash particles which then collect on grounded steel plates. Some 99.99 percent of the ash particles can be captured by the plates. Hammers then knock the ash loose to be caught in ash hoppers for disposal. Because electrostatic precipitators remove the "visible" pollutants from the exhaust gas, facilities installing these units note a big difference in air quality. These systems have already been retrofitted in power plants across the nation and are required as standard equipment on new plants. The estimated cost to remove a ton of ash is $61. The capital costs associated with purchasing and installing such a system vary between $30 and $50 per kilowatt (kW) generated.

A different type of equipment is needed to remove sulfur dioxide (SO_2). Flue gas desulfurization systems are often called scrubbers because the combustion gases are sprayed with water and then allowed to react with fine particles of limestone, or calcium carbonate (Figure 2). When the SO_2, water, and limestone react in the presence of oxygen, they form a calcium sulfite solid similar to calcium sulfate, which is commonly known as gypsum. Flue gas desulfurization systems can remove up to 95 percent of the SO_2 contained in the flue gas. General cost and emission information is given in Figure 1.

Nitrogen oxides (NO_X) discharges can be controlled by converting the NO_X in the flue gases to water (H_2O) and nitrogen (N_2), both components of normal air. Some systems vaporize ammonia in the presence of a catalyst and the flue gas to create N_2 and H_2O. The costs associated with removing a ton of NO_X from exhaust gas vary between $600 and $1000 a ton. For a gas-fired boiler, this translates to a capital cost of approximately $20 per kilowatt generated.

To remove trace amounts of heavy metals like mercury and arsenic as well as to recover heat from exhaust gas, some companies are offering condensing heat exchangers. These units recover heat from the flue gas which can be used in industrial processes. Lowering the flue gas temperature to condense the water vapor causes volatile organic compounds to condense at the same time. Much of the work in this area is still experimental because of the great expense of retrofitting the flues and stacks for lowered flue gas temperatures.

OTHER CONSIDERATIONS

As important as it is to have clean air, given the lifestyle enjoyed by most Americans, clean air doesn't come cheap. In 1990 the Environmental Protection Agency projected that the United States spends approximately 2 percent of its gross national product to control pollution. This translates to spending approximately $30 billion on air pollution a year. New requirements to be met by the year 2000 push costs even higher. The extra costs, running into the hundreds of dollars per consumer, will show up in the form of utility rate hikes. Even with the potential to sell the calcium sulfite for use in making drywall or as a soil conditioner, the costs of waste disposal are very high. For example, each ton of SO_2 removed from flue gas requires the production (digging and pulverizing) of 1.3 tons of limestone and large quantities of

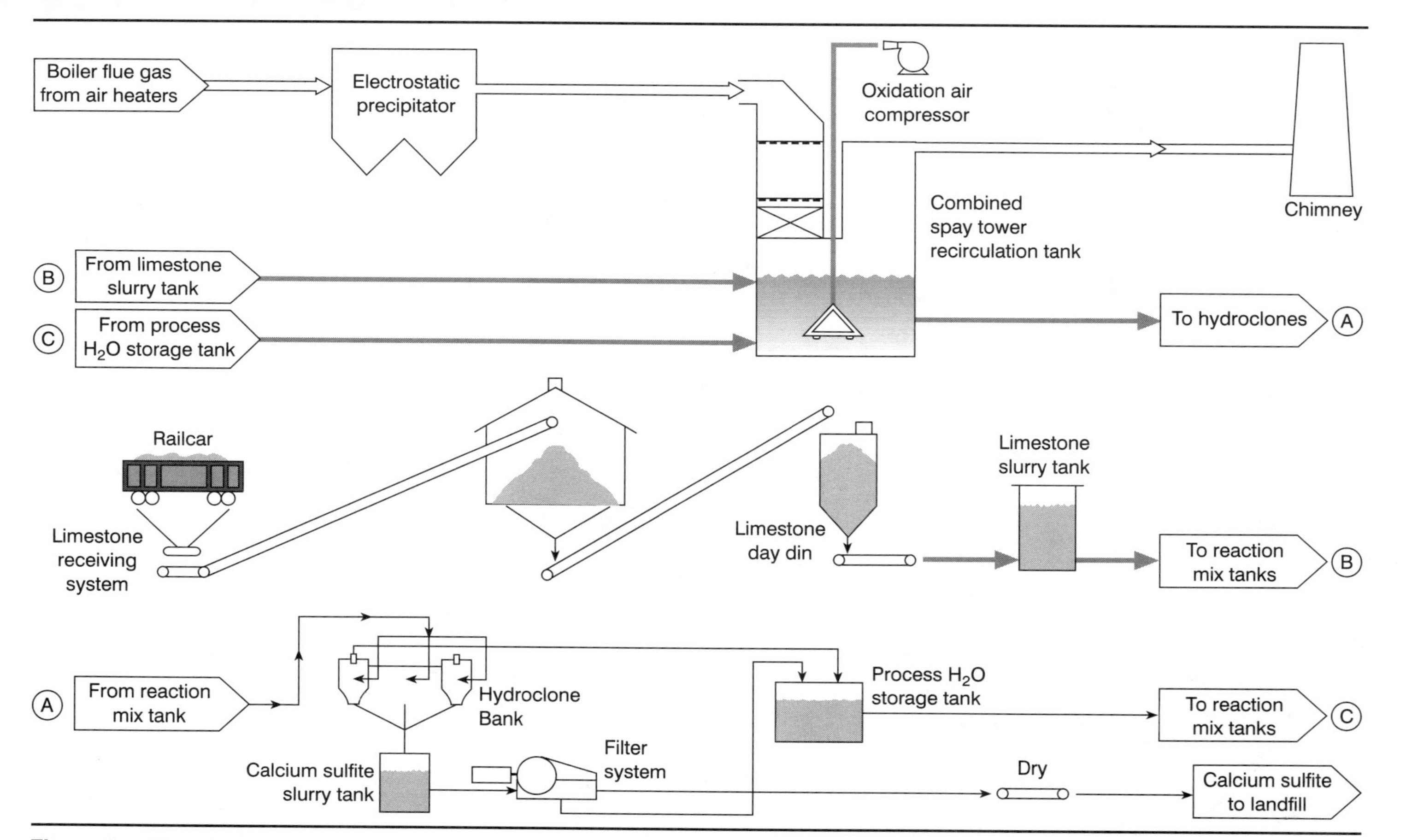

Figure 2 Flue Gas Desulfurization Process

water. The by-product, calcium sulfite, when mixed with lime and ash to be bound into a cement form is approximately equal in value to fill dirt. Very few utilities manage to sell their calcium sulfite to wallboard and cement companies. Selling calcium sulfite as a soil conditioner has come under scrutiny by the EPA, who may declare the material a hazardous waste because of the minute traces of metals carried from the coal. A decision to call the material hazardous waste will complicate the disposal of the remainder in a landfill.

Costs for future air quality improvements can only be estimated because much of the technology has yet to be invented and tested. This is particularly true for systems removing volatile compounds from flue gas emissions. It will be interesting to see if the same type of legislation that brought about innovations like the catalytic converter and oxygen-rich unleaded fuel—a combination that removes 96 percent of the hydrocarbons, 76 percent of the nitrogen oxides, and 25 percent of carbon monoxide from vehicle exhaust—will produce similar innovations for exhaust flue gas emissions.

Benefits from cleaner air are also hard to determine. The EPA estimates that between 1700 and 2700 cancer cases are caused by outdoor air pollution.* This number represents approximately 1 percent of all cancers nationwide. Asthma and allergy sufferers also feel the effects of pollution and incur associated health care costs. Economists estimate that Americans receive between $8 billion and $24 billion† in benefits from clean air in the form of increased crop yields, decreased building decay, lower health care costs, and improved industrial productivity. The Clean Air Act of 1970 and its 1990 amendments have resulted in significantly cleaner air for the United States.

Assignment

To enhance your answers to the following questions, you may wish to do your own library research, using the bibliography below as a guide.

1. What are the specifications set by the Clean Air Act as presented in this case? Investigate whether or not any new specifications have been set.
2. What are the capabilities of the equipment provided by the manufacturers to meet the specifications set by the act?
3. What other specifications or product capabilities are there in the case?
4. Given the information in the case, create a spreadsheet that contains the following information:

   - Who are the customers who would be interested in and affected by the Clean Air Act? Be sure to include all those who will be affected by the act.

*G. Cohen, K. Sheets, and B. Carpenter. "Costs and Benefits: Fresh Questions About Clean Air." *U.S. News & World Report*, July 30, 1990, p. 40.
†Ibid.

- Describe the needs, requirements, and expectations of each of the different customers.
- What are the potential types of losses (costs) to be incurred by each different customer?

CASE STUDY BIBLIOGRAPHY

Begley, S., and M. Hagar. "Keep Holding Your Breath." *Newsweek*, June 4, 1990, pp. 68, 70.

"Clean Air Amendments Estimated at $54 Billion." *Nation's Business*, April 1990, p. 8.

Cohen, G., K. Sheets, and B. Carpenter. "Costs and Benefits: Fresh Questions about Clean Air." *U.S. News & World Report*, July 30, 1990, p. 40.

Cook, J. "Rain from Heaven." *Forbes*, March 4, 1991, pp. 90–91.

Cooper, M. "Environmental Movement at 25: Will Congress Weaken Environmental Regulations?" *Congressional Quarterly Researcher*. March 31, 1995, pp. 275–92.

Edelson, E. *Clean Air*. New York: Chelsea House, 1992, pp. 50–60.

Gibbs, W., and G. Stix. "The Price of Clean Air." *Scientific American*, May 1994, pp. 113–14.

Graham, J., and J. Merline. "Air Toxics: How Serious a Threat?" *Consumers' Research Magazine*, March 1990, pp. 20–21.

Harris, J. "How to Sell Smoke." *Forbes*, June 11, 1990, pp. 204–206.

"How Do You Know the Clean Air Act Is Working?" Office of Air Quality Planning and Standards, U.S. Environmental Protection Agency, January 15, 1996.

Krupnick, A., and P. Portney. "Cleaning Up Smog: Costs versus Benefits." *Consumers' Research Magazine*, August 1991, pp. 23–27.

Licata, A., and L. Benson. "Controlling Acid Gases, Mercury and Dioxins from Municipal Waste Combustors Using the Sorbalit Technology." *Clean Air Technology News*, Winter 1995, pp. 3, 4.

McKee, B. "Clean-Air Rules Affect Small Firms." *Nation's Business*, July 1991, p. 28.

Merline, J. "How Deadly Is Air Pollution?" *Consumers' Research Magazine*, February 1, 1997, pp. 10–14.

Regan, M. B. "May Old Clean-Air Laws Be Forgot." *Business Week*, December 26, 1994, p. 64.

Romm, J., and C. Ervin. "How Energy Policies Affect Public Health." *Public Health Reports*, September 19, 1996, pp. 390–400.

Sayer, J. "Energy Efficient Air Pollution Controls for Fossil-Fueled Plants: Technology Assessment." New York: New York State Energy Research and Development Authority, May 1995.

Stultz, S., and J. Kitto. *Steam: Generation and Use*, 40th ed. Akron, Ohio: Babcock and Wilcox, 1992.

Sweeney, D. "Babcock and Wilcox's Environmental Equipment: Cleaning Up after the Fire." *B&W News*, Spring 1995, pp. 14–16.

2

Quality Advocates and Quality Management

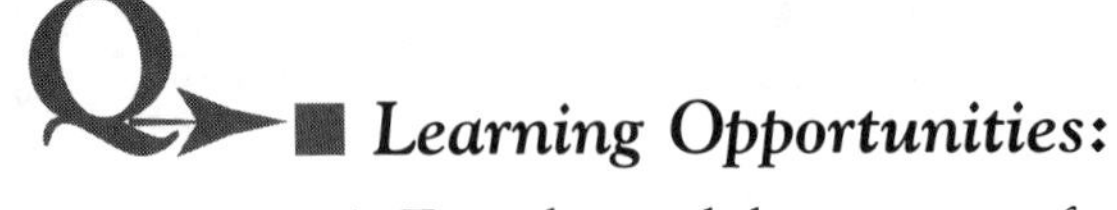

Learning Opportunities:

1. To understand the concepts of managing a total quality environment in order to improve quality, increase productivity, and reduce costs
2. To learn the basics involved in the creation of a system to support an overall quality environment
3. To become familiar with seven quality masters and their philosophies of quality management and continuous improvement ■

Awesome Odds

Most individuals are motivated to do the best they can. Sometimes though, the goal seems too far away and too hard to reach. The problem appears too big to tackle. How does a company deal with these issues? How can managers enable their employees to do their best? This chapter discusses total management and the quality advocates who have encouraged its use. Through effective management, companies can significantly improve their levels of quality, productivity, efficiency, and customer and employee satisfaction.

QUALITY ADVOCATES

Many individuals have proclaimed the importance of quality. The seven discussed in this chapter are among the most prominent advocates. You'll see that basic similarities exist among their ideas. Following a brief introduction to the seven men, we'll look at the concept of total quality management and explain some of the similarities found in the philosophies of these leading professionals.

Dr. Walter Shewhart

While working at Bell Laboratories in the 1920s and 1930s, Dr. Walter Shewhart (1891–1967) was the first to encourage the use of statistics to identify, monitor, and eventually remove the sources of variation found in repetitive processes. Dr. Shewhart identified two sources of variation in a process. ***Controlled variation,*** *also termed common causes, is variation present in a process due to the very nature of the process.* This type of variation can be removed from the process only by changing the process. For example, consider a person who has driven the same route to work dozens of times and determined that it takes about 20 minutes to get from home to work, regardless of minor changes in weather or traffic conditions. If this is the case, then the only way the person can improve upon this time is to change the process by finding a new route. ***Uncontrolled variation,*** *also known as special or assignable causes, comes from sources external to the process.* This type of variation is not normally part of the process. It can be identified and isolated as the cause of a change in the behavior of the process. The commuter described above would experience uncontrolled variation if a major traffic accident stopped traffic or a blizzard made traveling nearly impossible. Uncontrolled variation prevents the process from performing to the best of its ability.

It was Dr. Shewhart who put forth the fundamental principle that once a process is under control, exhibiting only controlled variation, future process performance can be predicted, within limits, on the basis of past performance. He wrote:

> A phenomenon will be said to be controlled when, through the use of past experience, we can predict, at least within limits, how the phenomenon may be expected to vary in the future. Here it is understood that prediction within limits means that we can state, at least approximately, the probability that the observed phenomenon will fall within the given limits.*

Based on this principle, Dr. Shewhart developed the formulas and table of constants used to create the most widely utilized statistical control charts in quality: the $\overline{X}$ and R charts (Chapter 5 and Appendix 2). These charts (Figure 2.1) first appeared in a May 16, 1924 memo of Dr. Shewhart's and later in his 1931 text, *Economic Control of Quality of Manufactured Product*. It is in this text that Shewhart presented the foundation principles upon which modern quality control is based.

*Walter Shewhart, *Economic Control of Quality of Manufactured Product*. New York: Van Nostrand Reinhold, 1931, p. 6.

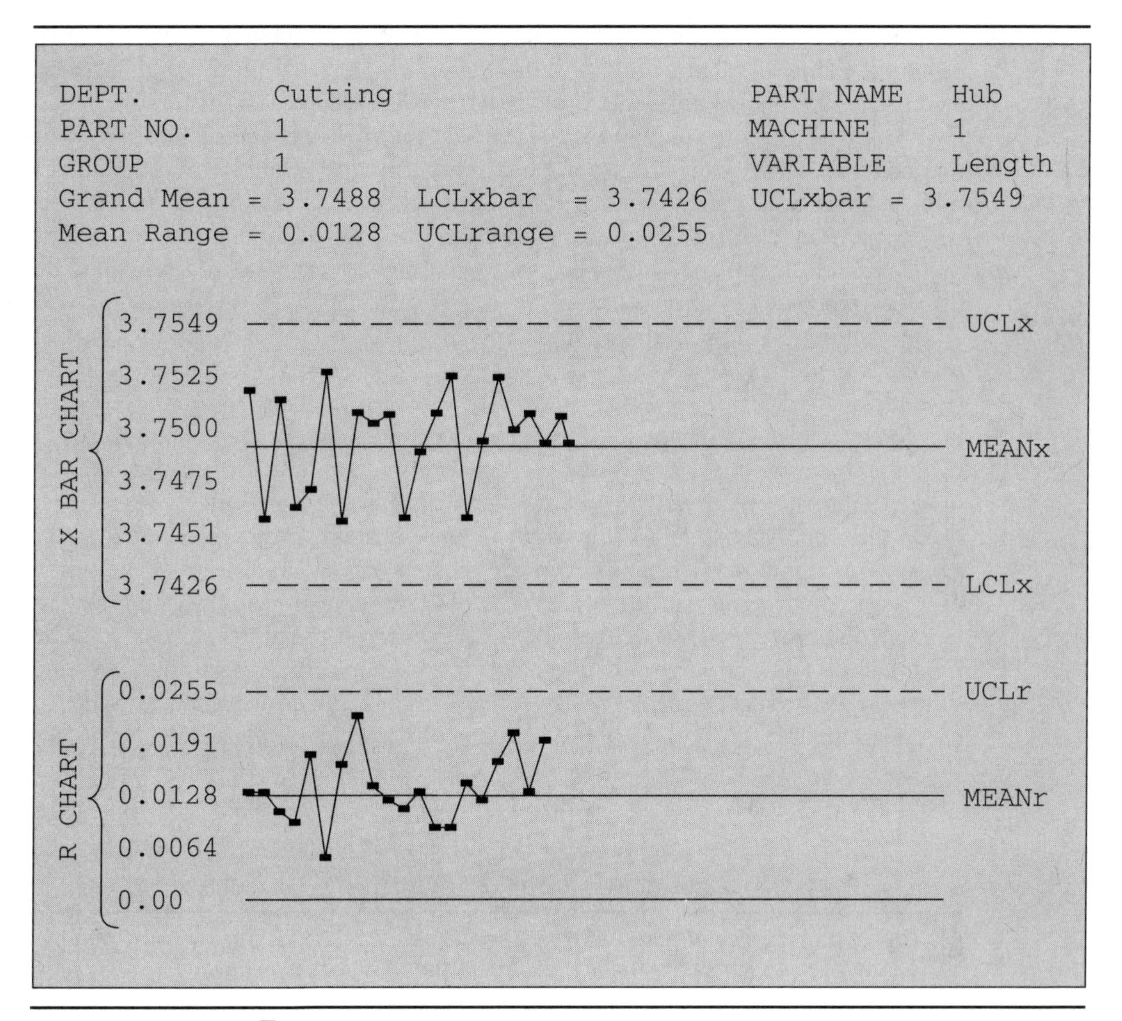

Figure 2.1 Typical $\overline{X}$ and R Charts

The control charts, as designed by Dr. Shewhart, have three purposes: to define standards for the process, to aid in problem-solving efforts to attain the standards, and to serve to judge if the standards have been met. These charts are covered in detail in Chapters 5, 6, 7, and 9. Although Dr. Shewhart concentrated his efforts on manufacturing processes, his ideas and charts are applicable to any process found in nonmanufacturing environments.

Dr. W. Edwards Deming

Dr. W. Edwards Deming (1900–1993) made it his mission to spread the gospel of quality management. Dr. Deming encouraged top-level management to get involved in the process of creating an environment that supports continuous improvement. A statistician by training, Dr. Deming graduated from Yale University in 1928. He first began spreading his quality message shortly after World War II. In the face of American prosperity following the war, his message was not accepted in the United States. His work with the Census Bureau and other government agencies led to his eventual contacts with Japan as that nation was beginning to rebuild. There he helped turn Japan into an industrial force to be reckoned with. It was only after his early 1980s appearance on the TV program "If Japan Can, Why Can't We?" that Deming found an audience in the United States. Over time he became one of the most influential experts on quality assurance.

Dr. Deming considered quality improvement activities as the catalyst to start an economic chain reaction. Improving quality leads to decreased costs, fewer mistakes, fewer delays and better use of resources, which in turn leads to improved productivity, which enables a company to capture more of the market, which enables the company to stay in business, which results in providing more jobs. He felt that without quality improvement efforts to light the fuse, this process would not begin.

Dr. Deming, who described his work as "management for quality," felt that the consumer is the most critical aspect in the production of a product or the provision of a service. Listening to the voice of the customer and utilizing the information learned to improve products and services is an integral part of his teachings. Dr. Deming's theories focus heavily on management involvement, continuous improvement, statistical analysis, goal setting, and communication. His message is aimed primarily at management. Using his 14 points as a guideline (Figure 2.2), Deming's philosophy encourages company leaders to dedicate themselves and their companies to the long-term improvement of the quality of their products or services.

1. Create a constancy of purpose toward improvement of product and service, with the aim to become competitive and to stay in business and to provide jobs.
2. Adopt the new philosophy.
3. Cease dependence on inspection to achieve quality.
4. End the practice of awarding business on the basis of price tag alone. Instead minimize total cost.
5. Constantly and forever improve the system of production and service.
6. Institute training on the job.
7. Institute leadership.
8. Drive out fear.
9. Break down barriers between departments.
10. Eliminate slogans, exhortations, and targets for the work force.
11. Eliminate arbitrary work standards and numerical quotas. Substitute leadership.
12. Remove barriers that rob people of their right to pride of workmanship.
13. Institute a vigorous program of education and self-improvement.
14. Put everybody in the company to work to accomplish the transformation.

Figure 2.2 Deming's 14 Points

SOURCE: Reprinted from *Out of the Crisis* by W. Edwards Deming by permission of MIT and The W. Edwards Deming Institute. Published by MIT, Center for Advanced Educational Services, Cambridge, MA 02139. Copyright © 1986 by The W. Edwards Deming Institute.

Dr. Deming's first point—create a constancy of purpose—encourages management to accept the obligation to constantly improve the product or service through innovation, research, education, and continual improvement in all facets of the company. Imagine your company as an Olympic athlete who must constantly improve in order to attain a gold medal. Lack of constancy of purpose is one of the deadly diseases Dr. Deming warns about. Without dedication, performance of any task is not at its best. Dr. Deming's second point: Adopt a new philosophy that rejects "acceptable" quality levels and poor service as a way of life; support continuous improvement in all that we do.

The 12 other points ask management to rethink past practices, such as awarding business on the basis of price tag alone, using mass inspection, setting arbitrary numerical goals and quotas, enforcing arbitrary work time standards, allowing incomplete training or education, and using outdated methods of supervision. Mass inspection has limited value because quality cannot be inspected into a product. Awarding business on the basis of price tag alone is shortsighted and fails to establish mutual confidence between the supplier and the purchaser. Low-cost items can sometimes lead to losses in productivity elsewhere. Dr. Deming's points about driving out fear and removing barriers stress the importance of communication. Barriers are any aspect of a job that prevent people from performing at their best levels. By removing any barriers preventing employees from doing their jobs well, management creates an environment supportive of continuous improvement. Throughout his life, Dr. Deming has encouraged management to create systems that enable people to find joy in their work.

The concepts of authority and responsibility play a role in all Deming's points. Improved management-employee interaction as well as increased communication between departments will lead to more effective solutions to the challenges of creating a product or providing a service. Education and training also play an integral part in Deming's plan. Continual education creates an atmosphere that encourages the discovery of new ideas and methods. This translates as innovative solutions to problems. Training ensures that products and services are provided that meet standards based on customer requirements.

Reducing the variation present in a process is one of the most critical messages Dr. Deming sent to management. To do this, he emphasized the use of the statistics and quality techniques espoused by Dr. Shewhart. According to Dr. Deming, process improvement is best carried out in three stages:

Stage 1: Get the process under control by identifying and eliminating the sources of uncontrolled variation. Remove the special causes responsible for this variation.

Stage 2: Once the special causes have been removed and the process is stable, improve the process. Tackle the common causes responsible for the controlled variation present in the process. Determine if process changes can remove them from the process. Investigate whether or not waste exists in the process.

Stage 3: Monitor the improved process to determine if the changes that were made are working.

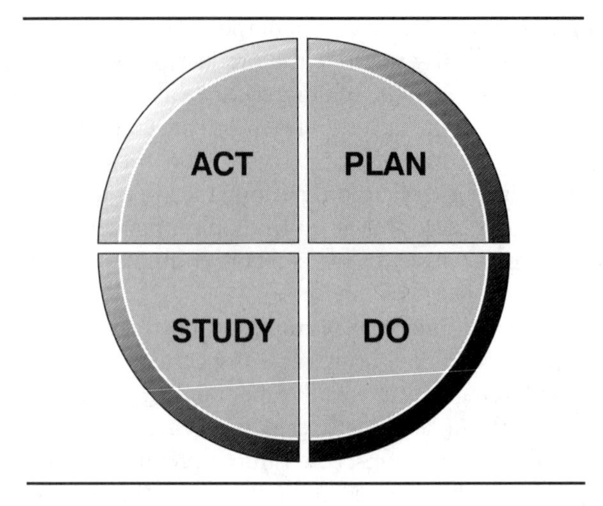

Figure 2.3 The Deming Cycle
SOURCE: Reprinted from *Out of the Crisis* by W. Edwards Deming by permission of MIT and The W. Edwards Deming Institute. Published by MIT, Center for Advanced Educational Services, Cambridge, MA 02139. Copyright © 1986 by The W. Edwards Deming Institute.

When tackling process improvement, it is important to find the root cause of the variation. Rather than apply a Band-Aid sort of fix, when seeking the causes of variation in a process, Dr. Deming encouraged the use of the Plan-Do-Study-Act (PDSA) cycle (Figure 2.3). Originally developed by Dr. Walter Shewhart, the PDSA cycle is a systematic approach to problem solving. Users of the cycle study a problem and plan a solution (Plan). This should be the portion of the cycle that receives the most attention since good plans lead to well-thought-out solutions. The solution is implemented during the Do phase of the cycle. The results of the change to the process are studied during the Study phase. Finally, when the results of the Study phase reveal that the root cause of the problem has been isolated and removed from the process permanently, the changes are made permanent (Act). If the problem has not been resolved, a return trip to the Plan portion of the cycle for further investigation is undertaken. In Chapter 3, we cover problem solving in more detail.

Deming used two experiments to help individuals visualize process improvement. The red bead experiment is aimed at improving management's understanding of how a process with problems can inhibit an individual's ability to perform at his or her best. The red beads represent problems in the process that can be changed only through the work of management. For Dr. Deming, it is the job of managers to work on the system to continuously improve it. The funnel experiment describes how tampering with a process can actually make the performance of that process worse. Tampering can be avoided by isolating and removing the root causes of process variation.

In his last book, *The New Economics*, Dr. Deming tied much of his work together when he introduced the concept of profound knowledge. There are four interrelated parts to a system of profound knowledge:

- An appreciation for a system
- Knowledge about variation
- Theory of knowledge
- Psychology

Those who have an appreciation for systems create alignment between the systems that produce products and services and the company's purpose. Knowledge of variation means being able to distinguish between controlled and uncontrolled variation. The theory of knowledge involves using data to understand situations. An understanding of psychology enables us to understand each other. Knowledge of all these areas enables companies to expand beyond small process-improvement efforts and to optimize their systems in their entirety rather than suboptimize only their parts.

Living the continuous improvement philosophy is not easy. The level of dedication required to become the best is phenomenal. Dr. Deming warned against the "hope for instant pudding." Improvement takes time and effort and does not happen instantly. The hope for instant pudding is one that afflicts us all. After all, how many of us wouldn't like all our problems taken care of by just wishing them away? As evidenced by his fourteenth point, Dr. Deming's quality system is really an ongoing process of improvement. To him, quality must be an integral part of how you do business. Companies must continuously strive to improve; after all, the competition isn't going to wait for us to catch up!

Dr. Joseph M. Juran

Born in 1904, Dr. Joseph M. Juran immigrated from Rumania to the United States in 1912. Dr. Juran's approach involves creating awareness of the need to improve, making quality improvement an integral part of each job, providing training in quality methods, establishing team problem solving, and recognizing results. Juran emphasizes the need to improve the entire system. To improve quality, individuals in a company need to develop techniques and skills and understand how to apply them.

The Juran trilogy makes use of three managerial processes: Quality Planning, Quality Control, and Quality Improvement (Figure 2.4 and Table 2.1). By following Dr. Juran's approach, companies can reduce the costs associated with poor quality and remove chronic waste from their organizations. *Quality Planning* encourages the development of methods to stay in tune with customers' needs and expectations. *Quality Control* involves comparing products produced with goals and specifications. *Quality Improvement* involves the ongoing process of improvement necessary for the company's continued success.

In his text *Juran on Leadership for Quality: An Executive Handbook*, Dr. Juran discusses the importance of achieving world-class quality by identifying the need for improvement, selecting appropriate projects, and creating an organizational structure that guides the diagnosis and analysis of the projects. Successful improvement efforts encourage breakthroughs in knowledge and attitudes. The commitment and personal leadership of top management must be assured in order to break through cultural resistance to change.

In the project-by-project implementation procedure (Table 2.2), project teams are set up to investigate and solve specific problems. To guide the project teams, the Juran program establishes a steering committee. The steering committee serves three purposes: to ensure emphasis on the company's goals, to grant authority to diagnose and investigate problems, and to protect departmental rights.

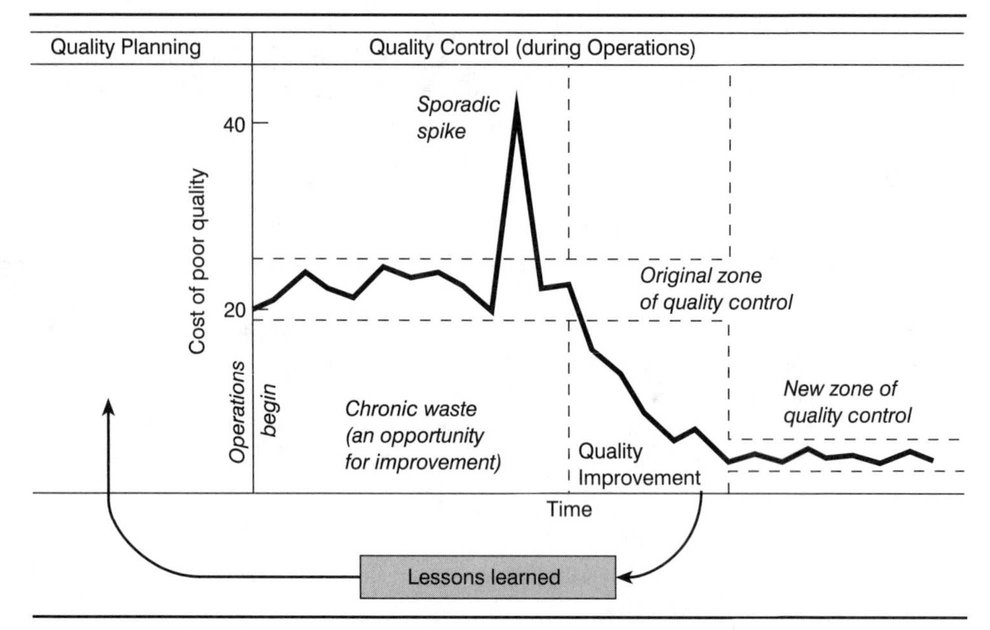

Figure 2.4 The Juran Trilogy Diagram
SOURCE: Reprinted with the permission of the Free Press, a division of Simon & Schuster, from *Juran on Leadership for Quality: An Executive Handbook* by J. M. Juran. Copyright © 1989 by Juran Institute Inc.

Table 2.1 The Three Universal Processes of Managing for Quality

Quality Planning	*Quality Control*	*Quality Improvement*
Determine Who the Customers Are	Evaluate Actual Product Performance	Establish the Infrastructure
Determine the Needs of the Customers	Compare Actual Performance to Product Goals	Identify the Improvement Projects
Develop Product Features that Respond to Customers' Needs	Act on the Difference	Establish Project Teams
Develop Processes Able to Produce the Product Features		Provide the Teams with Resources, Training and Motivation to Diagnose the Causes Stimulate Remedies Establish Controls to Hold the Gains
Transfer the Plans to the Operating Forces		

SOURCE: Reprinted with the permission of the Free Press, a division of Simon & Schuster, from *Juran on Leadership for Quality: An Executive Handbook* by J. M. Juran. Copyright © 1989 by Juran Institute Inc.

Table 2.2 Juran's Journey from Symptom to Cause: Quality Improvement in Action

Process	***Activity***	***Steering Arm***	***Diagnostic Arm***
	Assign Priority to Projects	X	
	Pareto Analysis of Symptoms		X
Journey from Symptom to Cause	Theorize on Causes of Symptoms	X	
	Test Theories: Collect, Analyze Data		X
	Narrow List of Theories	X	
	Design Experiment(s)		X
	Approve Design; Provide Authority	X	
	Conduct Experiment; Establish Proof of Cause		X
	Propose Remedies	X	
Journey from Cause to Remedy	Test Remedy		X
	Actions to Institute Remedy; Control at New Level	X	

SOURCE: Adapted with the permission of the Free Press, a division of Simon & Schuster, from *Juran on Leadership for Quality: An Executive Handbook* by J. M. Juran. Copyright © 1989 by Juran Institute Inc.

The project teams should be composed of individuals with diverse backgrounds. Diversity serves several purposes. It allows for a variety of viewpoints, thus avoiding preconceived answers to the problem. Having a diversified group also aids in implementing the solutions found. Group members are more willing to implement the solution because they have a stake in the project. The different backgrounds of the group members can also assist in breaking down the cultural resistance to change.

Juran's project teams are encouraged to use a systematic approach to problem solving. Group members use Pareto analysis, cause-and-effect diagrams, and experiments to clarify the symptoms and locate the true cause(s) of the problem. When the cause is determined, finding a solution becomes a process of proposing remedies, testing them, and instituting the remedy that most effectively solves the problem. Controlling the process once changes have been made is important to ensure that the efforts have not been wasted. Improvements continue as the groups study and resolve other problems.

Dr. Armand Feigenbaum

Armand Feigenbaum (1920–) is considered to be the originator of the total quality movement. His landmark text, *Total Quality Control,* published in 1951 while Feigenbaum was still in graduate school at the Massachusetts Institute of Technology, has significantly influenced industrial practices. In his text he predicted that quality would become a significant customer-satisfaction issue, even to the point of surpassing price in importance in the decision-making process. As he predicted, consumers have come to expect quality to be an essential dimension of the product or service they are purchasing.

To Feigenbaum, quality is more than a technical subject; it is a mind-set that makes an organization more effective. He has consistently encouraged treating quality as a fundamental element of a business strategy. In his article "Changing Concepts and Management of Quality Worldwide," from the December 1997 issue of *Quality Progress*, he asserts that quality is not a factor to be managed but a method of "managing, operating, and integrating the marketing, technology, production, information, and finance areas throughout a company's quality value chain with the subsequent favorable impact on manufacturing and service effectiveness." Quality systems are a method of managing an organization to achieve higher customer satisfaction, lower overall costs, higher profits, and greater employee effectiveness and satisfaction. Company leadership is responsible for creating an atmosphere that enables employees to provide the right product or service the first time, every time. Feigenbaum encourages companies to eliminate waste, which drains profitability, by determining the costs associated with failing to provide a quality product (see Chapter 11). Quality efforts should emphasize increasing the number of experiences that go well for a customer versus handling things when they go wrong. Statistical methods and problem-solving techniques should be utilized to effectively support business strategies aimed at achieving customer satisfaction. In its newest edition, his text serves as a how-to guide for establishing a quality system.

Philip Crosby

Philip Crosby's (1926–) message to management emphasizes four absolutes (Figure 2.5). The four absolutes of quality management set expectations for a continuous improvement process to meet. The first absolute defines quality as *conformance to requirements*. Crosby emphasizes the importance of determining customer requirements, defining those requirements as clearly as possible, and then producing products or providing services that conform to the requirements as established by the customer.

Prevention of defects, the second absolute, is the key to the system that needs to be in place in order to ensure that the products or services provided by a company

Figure 2.5 Crosby's Absolutes of Quality Management

Quality Definition: Conformance to Requirements
Quality System: Prevention of Defects
Quality Performance Standard: Zero Defects
Quality Measurement: Costs of Quality

meet the requirements of the customer. Prevention of quality problems in the first place is much more cost-effective in the long run. Determining the root causes of defects and preventing their recurrence are integral to the system.

According to Crosby, the performance standard against which any system must be judged is zero defects. This third absolute, ***zero defects,*** *refers to making products correctly the first time, with no imperfections.* Traditional quality control centered on final inspection and "acceptable" defect levels. Systems must be established or improved that allow the worker to do it right the first time.

His fourth absolute, ***costs of quality,*** *are the costs associated with providing customers with a product or service that conforms to their expectations.* Quality costs, to be discussed in more detail in Chapter 11, are found in prevention costs; detection costs; costs associated with dissatisfied customers; rework, scrap, downtime, and material costs; and costs involved anytime a resource has been wasted in the production of a quality product or the provision of a service. Once determined, costs of quality can be used to support investments in equipment and processes that reduce the likelihood of defects.

In some circumstances, quality may seem intangible. By discussing five erroneous assumptions about quality, Crosby attempts to make quality more understandable and tangible. The first erroneous assumption, quality means goodness, or luxury, or shininess, or weight, makes quality a relative term. Only when quality is defined in terms of customer requirements can quality be manageable. The second incorrect assumption about quality is that quality is intangible and therefore not measurable. If judged in terms of "goodness," then quality is intangible; however, quality is measurable by the cost of doing things wrong. More precisely, quality costs involve the cost of failures, rework, scrap, inspection, prevention, and loss of customer goodwill.

Closely related to the first two assumptions is the third, which states that there exists "an economics of quality." Here again, one errs in thinking that quality means building "luxuries" into a product or service; rather, quality means that it is more economical to do things right the first time.

Often workers are blamed for being the cause of quality problems. This is the fourth erroneous assumption about quality. Without the proper tools, equipment, and raw materials, workers cannot produce quality products or services. Management must ensure that the necessary items are available to allow workers to perform their jobs well.

The final erroneous assumption that Crosby discusses is that quality originates in the quality department. According to Crosby, the quality department's responsibilities revolve around educating and assisting *other* departments in monitoring and improving quality.

Dr. Kaoru Ishikawa

One of the first individuals to encourage total quality control was Dr. Kaoru Ishikawa (1915–1989). He played a prominent role in refining the application of different statistical tools to quality problems. Ishikawa felt that all individuals employed by a company should become involved in quality problem solving. He advocated the use of seven quality tools: histograms; check sheets; scatter diagrams; flowcharts; control charts; Pareto charts; and cause-and-effect, or fish-bone, diagrams.

These tools, shown in Figure 2.6, are covered in detail in Chapter 3. Dr. Ishikawa developed the *cause-and-effect diagram* in the early 1950s. This diagram, used to find the root cause of problems, is also called the *Ishikawa diagram*, after its creator, or the *fish-bone diagram*, because of its shape.

Dr. Ishikawa also promoted the use of ***quality circles,*** *teams that meet to solve quality problems related to their own work*. The quality circle concept has been adapted and modified over time to include problem-solving team activities. Membership in a quality circle is often voluntary. Participants receive training in the seven tools, determine appropriate problems to work on, develop solutions, and establish new procedures to lock in quality improvements.

Dr. Genichi Taguchi

Dr. Genichi Taguchi (1924–) developed methods that seek to improve quality and consistency, lower losses, and identify key product and process characteristics before production. Taguchi methods emphasize consistency of performance and significantly reduced variation. Dr. Taguchi introduced the concept that the total loss to society generated by a product is an important dimension of the quality of a product. In his "loss function" concept, Taguchi expressed the costs of performance variation (Figure 2.7). Any deviation from target specifications causes loss, he said, even if the variation is within specifications. When the variation is within specifications, the loss may be in the form of poor fit, poor finish, undersize, oversize, or alignment problems. Scrap, rework, warranties, and loss of goodwill are all examples of losses when the variation extends beyond the specifications. Knowing the loss function helps designers to set product and manufacturing tolerances. Capital expenditures are more easily justified by relating the cost of deviations from the target value to quality costs. Minimizing losses is done by improving the consistency of performance.

Using Taguchi methods, statistically planned experiments can identify the settings of product and process parameters that reduce performance variation. Taguchi's methods design the experiment to systematically weed out a product's or process's insignificant elements. The focus of experiment efforts is then placed on the significant elements. There are four basic steps:

1. Select the process/product to be studied.
2. Identify the important variables.
3. Reduce variation on the important variables through redesign, process improvement, and tolerancing.
4. Open up tolerances on unimportant variables.

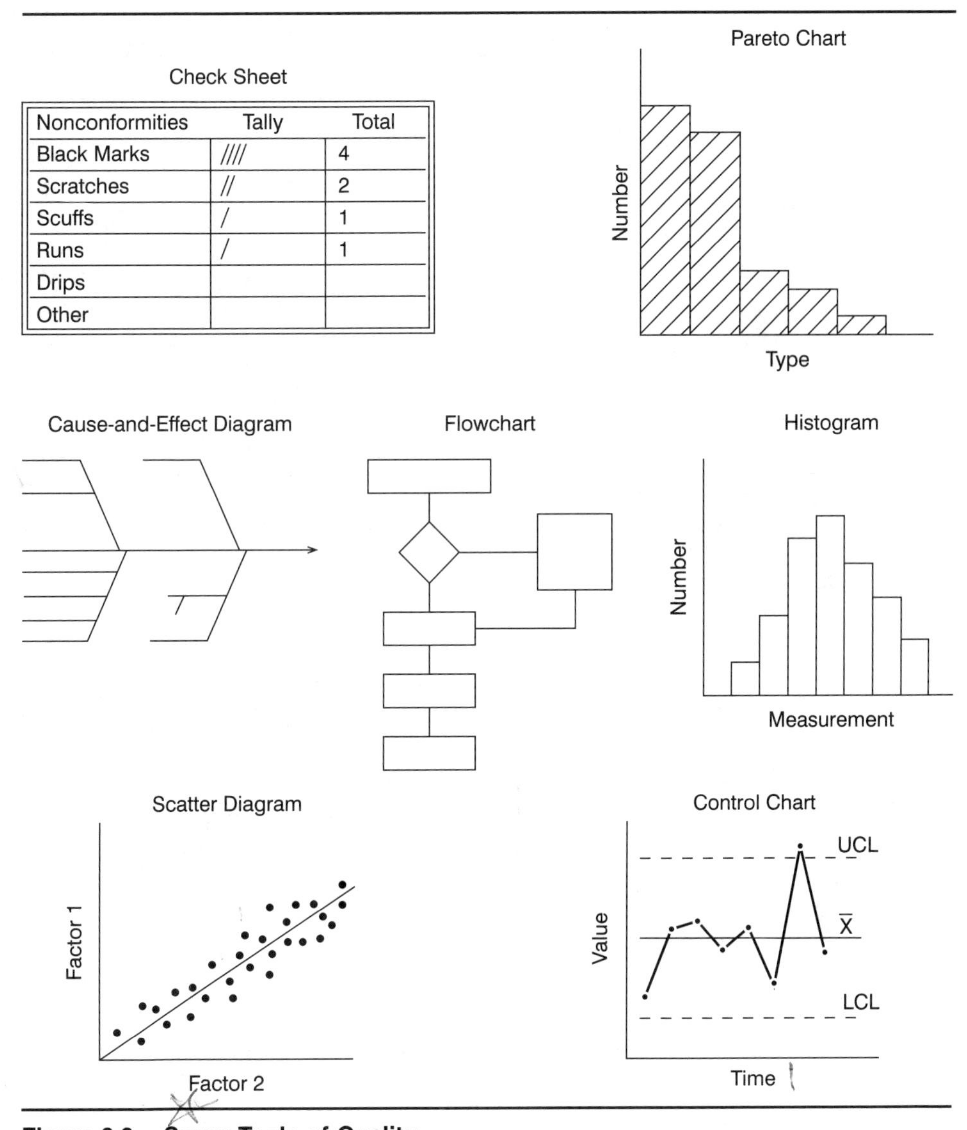

Figure 2.6 Seven Tools of Quality

The final quality and cost of a manufactured product are determined to a large extent by the engineering designs of the product and its manufacturing process.

CONTINUOUS IMPROVEMENT

The continuous improvement (CI) philosophy focuses on improving processes to enable companies to give the customer what they want the first time, every time. This customer-focused, process-oriented approach to doing business results in increased satisfaction and delight for both customers and employees. Sometimes known as "best practices," quality management, or continuous improvement efforts are characterized

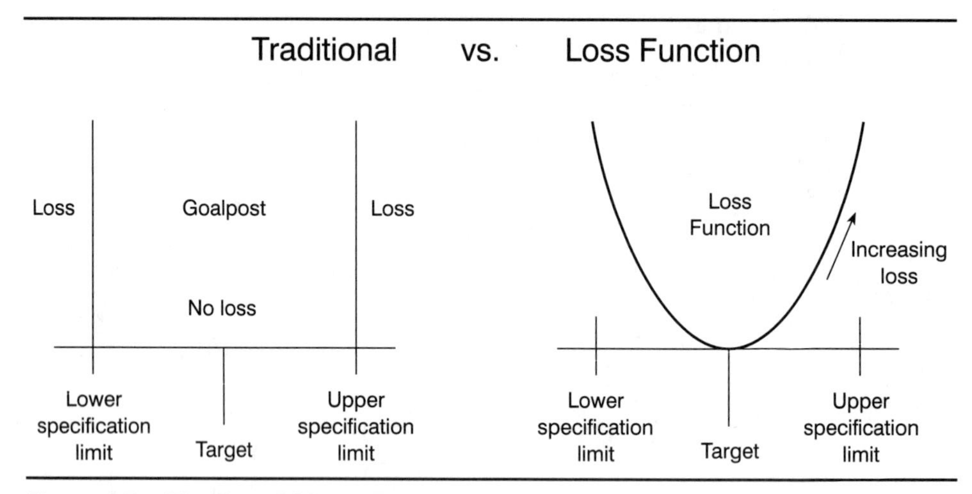

Figure 2.7 The Taguchi Loss Function
SOURCE: Adapted from *Taguchi Techniques for Quality Engineering* by Phillip J. Ross (New York: McGraw-Hill Book Co., Copyright © 1988).

by their emphasis on determining the best method of operation for a process or system. The key words to note are *continuous* and *process*. Continuous improvement represents an ongoing, *continuous* commitment to improvement. Because the quest for continuous improvement has no end, only new directions in which to head, continuous improvement is a *process*.

One of the strengths of the CI process is that a company practicing these methods develops flexibility. A company focusing on continuous improvement places greater emphasis on customer service, teamwork, attention to detail, and process improvement. Table 2.3 shows some other differences between a traditional company and one that practices continuous improvement.

Table 2.3 Continuous Improvement versus Traditional Orientation

Company Oriented Toward Continuous Improvement	***Traditional Company***
Customer Focus	Market-Share Focus
Cross-Functional Teams	Individuals
Focus on "What" and "How"	Focus on "Who" and "Why"
	Judgmental Attitudes
Attention to Detail	
Long-Term Focus	Short-Term Focus
Continuous Improvement Focus	Status Quo Focus
Process Improvement Focus	Product Focus
Incremental Improvements	Innovation
Problem Solving	Fire Fighting

Management Commitment and Involvement

The foundation of continuous improvement is a management philosophy that supports meeting customer requirements the first time, every time. Perhaps the most obvious similarity among the various quality advocates is the consistent insistence that management be actively involved with and committed to improving quality within the corporation. Merely stating that quality is important is not sufficient. Philosophies are easy to preach but difficult to implement. The strongest continuous improvement processes are the ones that begin with and have the genuine involvement of top-level management. Substitutions from lower levels of management will send employees the message that this new effort is not important enough to require the time and commitment of top management.

A variety of different approaches exist for integrating continuous improvement efforts into everyday business activities. The flowchart in Figure 2.8 illustrates a typical CI process. Note that many activities are ongoing and that several overlap. Most CI efforts begin with a vision. ***Visions,*** which are developed and supported by senior management, are statements describing how a company views itself now and in the future. A company's vision is the basis for all subsequent strategies, objectives, and decisions. If the company management envisions itself as a world-class leader, then continuous improvement will need to be integrated into the organization's vision and strategic objectives.

Top management often develops a mission statement to support the organization's vision. The mission statement sets the stage for improvement by making a strong statement about the corporation's goals. The mission statement should be short enough for its essence to be remembered by everyone, but it should also be complete. The mission statement should be timeless and adaptable to organizational changes. Examples of mission statements are shown in Figure 2.9.

The trouble with visions, missions, and philosophies, though, is that they are easier to dream up than to turn into reality. Creating a world-class corporation with both a customer focus and a quality focus takes commitment. A bacon-and-egg breakfast can be used to illustrate the difference between involvement and commitment. The chicken is merely involved, whereas the pig is committed. This sort of direct involvement is crucial to the success of any quality process. Consider the situation given in Example 2.1.

EXAMPLE 2.1 How Committed Is Management?

Sara's day began with a memo requiring her presence at the first of what promised to be many meetings about quality. Sara was pleased to see management had finally been awakened to the need for an organized quality improvement program. However, as the meeting progressed, she began to question management's sincerity. After the personnel officer delivered a presentation containing considerable hoopla about the importance of producing a quality product, the assistant to the assistant vice president spoke. The message that Sara got from him was vague

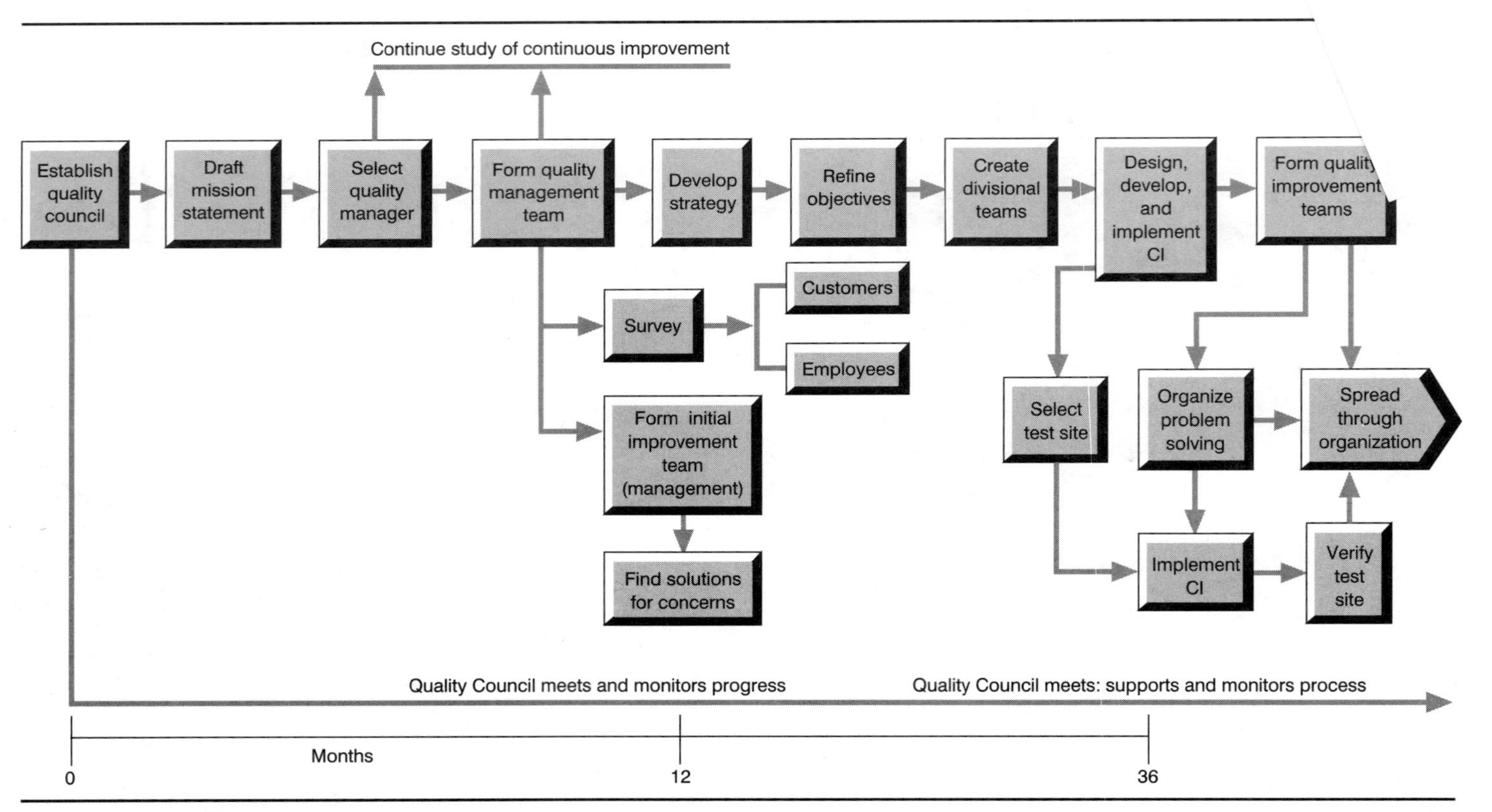

Figure 2.8 A Typical CI Process

Hospital

As a major teaching institution, we will continue to be a leader in providing a full range of health care services. Working together with our medical staff, we will meet and exceed our patients' needs for high-quality health care given in an efficient and effective manner.

Grocery Store

The mission of our store is to maintain the highest standards of honesty, trust, and integrity toward our customers, associates, community, and suppliers. We will strive to provide quality merchandise consistent with market values.

Student Project

Our mission is to provide an interesting and accurate depiction of our researched company. Our project will describe their quality processes as compared with the Malcolm Baldrige Award standards.

Pet Food Company

Our mission is to enhance the health and well-being of animals by providing quality pet foods.

Customer Service Center

The mission of our center is to gain and retain all customers who contact us.

Figure 2.9 Examples of Mission Statements

and spiritless: "We don't have time to show you any examples of how this stuff will work. . . . We don't see how it applies anywhere but the shop floor. . . . We're sure you can do it if you try. Good luck."

Later at lunch Sara and several of her peers gathered to discuss the morning's presentation. They decided it was all just another program exhorting the need to work harder but providing no way to improve. Lack of direct involvement, training, and encouragement from management resulted in failure of the process before it had even begun.

Motivation and Change

Change is a difficult thing to accomplish. Anyone who has attempted to break a habit, become proficient at a sport or a language, or taken on a large-scale project will understand that there are different levels of commitment, and only the most committed make a vision a reality. Many possible attitudes toward change exist. Some people respond to change with negativism or apathy. Others react with tolerance or verbal support. It is only when commitment takes the form of providing resources and active participation that change begins to occur. In essence, change takes place not by wishing for it but by getting your hands dirty. Management must visibly support the change by doing it themselves. They cannot turn to subordinates and say the magic words: Make it so! Instead they must participate in the change process themselves.

EXAMPLE 2.2 Make It So!

A sales manager at a make-to-order company investigated the causes of late quotes to customers. His investigations revealed that it was not unusual for customer information to be misplaced or buried beneath other paperwork on someone's desk. He was about to blame his staff for the problem when he happened to notice the edge of a missing quote peeking out from under some disorganized stacks on his own desk. It seems that a disorganized workplace was prevalent throughout the department.

Losing customer information and quotes could not be tolerated. Because disorganization throughout the department was the culprit, the sales manager decided to cancel his tennis match and spend Saturday organizing his office. On Monday, he showed his staff his newly organized office and decreed that everyone should take a Saturday to completely clean and organize their offices. They would be paid overtime for their efforts and the manager would personally purchase any needed supplies. With the manager setting the example, the new level of organization would continue from that moment on. Over the four weeks that it took to reorganize the office activities, many other opportunities to improve processes emerged. Now instead of spending their weekly staff meetings pointing fingers at each other for losing customer information, they spend their time implementing new methods to provide better customer service.

Motivation for change can come from a variety of sources. Deming suggested that there are three reasons to change: fear, financial incentive, and desire. Continuous improvement practitioners prefer that it is the latter reason that serves as a motivator for change. One powerful source involves listening to the employees and their needs and then acting on their concerns. Since actions speak louder than words, taking action will go a long way toward creating an environment that encourages continuous improvement. Management needs to create the feeling that each employee's job is important and has an impact on the company's bottom line. In a continuous improvement process, management must meet the obligations of educating the workforce; supporting innovation and research; and encouraging the improvement of product design, processes, and service. To create an environment that fosters the continuous improvement spirit, management should support the following ideas:

- Goal setting
- Planning to support accomplishment of goals
- Designated responsibilities
- Shared authority
- Challenging tasks

- Opportunities for growth
- Feelings of accomplishment
- Respect for everyone
- Sense of trust
- Creativity
- Teamwork

Goal setting establishes what must be accomplished. An equally important, but often overlooked, aspect of goal setting is creating a plan or method to reach the goals. Goals without a supporting plan on how to meet them can severely undermine employee motivation. Improvement driven by objectives without realistic plans to reach them results in fear and competitiveness that actually obstruct improvement efforts. Managers must be accessible, fair, and involved in the process. As leaders, they have the ability to remove the barriers that hinder success. Feedback concerning progress toward goals must be constantly and consistently provided. Timely, sincere, fair recognition and appreciation must also be given. Being immediately recognized in front of their peers is often more important to people than the actual form such recognition takes.

When any new method or process is introduced, it takes time for the old methods to be converted. During this time, members of the organization often feel as if they are doing two jobs at once. Since the old methods are comfortable and familiar and the new are not, people resist change and tend to give most of their attention to continuing with the old ways of doing things. Management must recognize the need to deal with the resistance that accompanies most changes. Finding the time to use the new methods is critical to the success of any new process. Creative improvements, supported by management, will be necessary to free people from other tasks to allow them to focus their efforts on quality. Previous work methods, forms, procedures, and activities should be investigated to determine if the work is really necessary. Questions should be asked to determine if versatility and efficiency can be improved. Quality should be treated as an integral part of the workday, not relegated to once-a-week meetings or brought up when a problem arises. Demonstrated support and involvement of top management will go a long way toward neutralizing the resistance to change.

Responsibility and Authority

Assigning responsibility and granting authority is an important aspect of creating a CI environment. Assigning responsibilities removes the "not my job" syndrome. Granting authority enables those with responsibilities to meet them. Responsibility and authority go hand in hand. ***Responsibility,*** or *the obligation to get something done*, is separate from ***authority,*** or *the power or influence to make something happen*. When employees receive an assignment, they are responsible for getting it done. Their supervisor will hold them accountable, even if the work is to be delegated. If employees have no authority, then essentially they must use persuasion to get the job done. Given responsibility without authority, the employee is placed between a rock and a hard place. Consider the following example.

EXAMPLE 2.3 Who Has the Authority?

A process engineer's supervisor has just informed her that the vice president of the company is coming to see their department the next day. Her supervisor has given her the responsibility of getting the department cleaned up. But when the engineer speaks with the supervisor of housekeeping, he responds that he is unable to provide cleaners for her until the next week. Nothing can persuade him to prioritize her request, for he is very understaffed and other departments have made their requests first. Since she has no crew of her own to clean the department and has no authority over the housekeeping department, she is unable to fulfill her responsibility. It's not that she doesn't *want* to get the job done—it's that she doesn't have the needed authority to carry it out. Even so, her supervisor will hold her accountable for the mess, without enabling or empowering her.

Often the granting of authority and responsibility is considered empowering and enabling. People who are *enabled* are able to perform at the best of their ability. When enabling someone, a company removes the obstacles that prevent that employee from making improvements. Managers *empower* employees when they let them make improvements and decisions on their own, without fear of retribution. Employees have a clear understanding of their positions and the decision-making responsibilities that accompany their position.

Consider another example.

EXAMPLE 2.4 Problems with the Printer *and* with Management

An employee is asked to prepare computer printouts for timely delivery. While the information is printing out, the paper keeps tearing and jamming in the printer. The employee knows that this particular brand of paper is incompatible with this printer. When he checked on this in the past, he was told that the paper was the least costly, and since the company was facing a budget crunch, purchase orders were being filled with the least expensive items possible. Unfortunately, no one thought to check on what kind of impact those choices would have on productivity in various departments.

Because of jams and tears, the document preparer now has to work longer on the job than expected. Though his boss complains about the quality of the printouts, he makes no request for better paper so the preparer has no chance of doing a better job in the future. Unless management steps in to make changes in the paper purchasing process, the preparer will face the same situation the next time documents are needed. Although his management is proclaiming CI with words, their actions are not supporting it, and employees are far more likely to focus on the actions of management than on their words.

In a stable process, the variability from one part to another or from one service to another is predictable. Such a process is viewed as "being under control." When a process is stable, it is producing to the best of its abilities. A process can be stable and still produce nonconforming products or services that do not meet specifications. When a process is stable and still producing nonconforming items, it cannot improve unless management commits the necessary time and resources to find and resolve the root causes of the problems. In Example 2.4 the printer and paper are a stable system performing to the best of its ability, but it is unable to meet the standards set. Improvement will not come by punishing the employee in the system, only through adjustment of the system. It is at this point that Deming, Juran, Crosby, and other quality advocates stress the need for management involvement to remedy the situation.

Training and Education

A difference exists between the words "education" and "training." ***Training*** *refers to being taught to perform a function so as to be qualified to perform that function or to be proficient at performing that function.* Essentially, it is teaching someone how to do a job or task correctly. To train someone for a position correctly, the trainer should reference specific guidelines, methods, or specifications. Training should not take the form of one operator showing another, unless the operator doing the training is considered a master at the craft. Failure to train specific methods will have the same result as the children's whispering game where the message is diluted with each successive whisper. Training must be transmitted correctly if customers are to receive what they ordered.

Education has a broader-based meaning. ***Education*** *is focused on developing a person's level of understanding about a topic, job, or task.* It is teaching a person to understand why that job or task is important and why it is important to do that job right. Consider just one example.

EXAMPLE 2.5 The Importance of a Training Program

John has just been hired as a color separator in a printing company. He is familiar with many types of separators but each has its own processes that must be learned. The person doing the hiring assured John that a training program was in place and that he would begin his work there.

John's first day at work was a disaster. He was assigned to separator 3 and told to "run it." When he explained to the supervisor that he needed some training to get up to speed on the machine, the supervisor grudgingly asked another employee to spend 30 minutes showing John the process. Wanting to get back to her own machine, the other employee spoke quickly, and John didn't understand all the steps. He asked questions but was still unsure of some things when the 30 minutes were up.

Later, when he was running the machine, everything seemed to go wrong, and he worked an extra four hours just correcting the problems. The company lost money, was late on delivery, and blamed John for the problems. But who was really to blame?

Both education and training are important to providing a quality product or service. A person who can perform the job or task correctly—and who understands why that job is being performed and the importance of performing it correctly—becomes an integral part of the quality improvement process. Ignoring education and training will result in errors. Errors are the source of poor quality.

EXAMPLE 2.6 CI in Action

Frustrated at her present job, Sara began interviewing at competing organizations. She was surprised and pleased when during one interview she learned that the company's executives had spent many hours developing a quality training program specifically for their corporation. This program was developed by the company's top 10 executives after months of research. Each executive had read several books on quality improvement and personnel management. Each executive had also attended Dr. Deming's five-day seminar and at least two other quality seminars. The executives had then formed their own quality team and tackled a list of 10 top problems as identified by their employees through a confidential survey. Solutions had been identified and plans were underway to implement the changes.

The executives were now ready to present the mission statement to all employees and begin the training programs. Using their own problem-solving efforts as teaching examples, all executives, including the president, would be leading a group of 25 employees through a quality training program. This training would continue until all employees were trained. Some employees would receive additional training as their job required. Volunteer quality circles would begin after the training. Each circle would become the responsibility of a different executive.

Sara was awestruck. Here was a company whose executives believed in quality. She couldn't wait to begin her new job.

Communication

Everyone in the process of making a finished product or providing a service fills three roles: the role of receiver of unfinished goods or information; the role of worker adding value to the goods or service; and the role of supplier to the next consumer in line. When those involved in the process fail to communicate, these roles become distorted. Without communication, ideas are lost, improvements are not implemented, and the continuous improvement process is stalled. As seen in the swing cartoon at the beginning of the first chapter, distortion and lack of understanding cause problems with the design, construction, and delivery of a product or service. Stress should be placed on ensuring communication between the departments involved in any aspect of the product: manufacturing, purchasing, sales, engineering, design, customer service, shipping, etc. Communication between departments is key to providing the customer with the desired product or service.

Quality in All Aspects of the Company

Often the emphasis on quality is evident only on the production floor or where the service is actually being performed. While many believe that a quality product is all that should be judged by the customer, customers often look at the total package provided by the company. This package will include sales, service during the sale, packaging, delivery, and service after the sale. A customer's impression of quality begins with the initial contact with the company and continues through the life of the product. Quality extends to how the receptionist answers the phone, how the product is serviced after the sale, how courteous sales and repair people are, how managers treat subordinates. . . . *All* departments of the company must strive to improve the quality of their operations.

Companies implementing a total quality management or continuous improvement process have discovered many benefits. Top-management leadership of the quality improvement process establishes quality as the first priority. Managers view business decisions for the long term, thus increasing the importance of investment in the future. High quality and productivity standards are met through the use of teams to solve problems. These teams also uncover opportunities to remove barriers to quality work, leading to decreased bureaucracy. Management and employee education supports the quest to improve quality. Managers serve as leaders rather than as supervisors. This change results in greater concern for the individual employee. Employee involvement increases, as does job satisfaction. Individual and team contributions to improved quality and productivity are recognized and rewarded, leading to greater job satisfaction. Often the CI process involves sharing prosperity with the employees. Close customer relationships result in the ability to meet and exceed customer needs, requirements, and expectations. In all, companies pursuing quality excellence throughout their firm will find many rewards.

SUMMARY

Why emphasize world-class quality? Most will agree that world-class quality is needed to remain competitive in any market, foreign or domestic. This means being able to give customers what they want at prices they can afford and allowing the quality of the product or service to speak for itself. The total quality management or continuous improvement process has many attributes, including the ones discussed in this text: management commitment, a positive organizational culture, a customer focus, problem solving, and teamwork.

■ *Lessons Learned*

1. Top management must be intimately involved. They must "get their hands dirty."
2. Companies must develop and pursue an understanding of a customer's reasonable needs, requirements, and expectations.
3. Management commitment and involvement, communication, education and training, and support of the employee are all vital to achieving quality in all aspects of a company. ■

Chapter Problems

1. Describe the three purposes of Shewhart's control charts.
2. How do Deming's 14 points interact with each other?
3. Which point from Deming's 14 points do you agree with the most strongly? Why?
4. Which of Deming's 14 points do you have a hard time understanding? Why do you think that is?
5. Which of Deming's points do you disagree with? Why?
6. Describe Juran's approach to quality improvement.
7. How do the steering/diagnostic arms of Juran's program work together?
8. a. What is Crosby's definition of quality?
   b. Explain Crosby's system of quality.
   c. What is Crosby's performance standard?
   d. Why do you believe this can or cannot be met?
9. People tend to make five erroneous assumptions about quality. What are two of these assumptions and how would you argue against them? Have you seen one of Crosby's erroneous assumptions at work in your own life? Describe the incident(s).
10. Deming, Juran, and Crosby all believe in striving toward world-class quality. Each has his own set of guidelines for achieving such quality. Choose and discuss one area where all three men agree. This choice must be backed up by points cited from each man's plan.
11. What follows is a short story about a worker who has requested additional education and training. Read the story and discuss which point or points of Deming's, Juran's, and Crosby's philosophies are not being followed. Cite at least one point from each man's plan. How did you reach your conclusions? Back up your answers with statements from the story. Support your argument.

    Inspector Simmons has been denied permission to attend an educational seminar. Although Simmons has attended only one training course for plumbing inspectors in his 15 years on the job, he will not be permitted to attend a two-week skills enhancement and retraining session scheduled for the coming month. The course devotes a significant amount of time to updating inspectors on the new plumbing regulations. While the regulations concerning plumbing have changed dramatically in the past five years, this is the third request for training in recent years that has been denied.

    City commissioners have voted not to send Simmons for the $1150 course, even though the plumbing guild has offered to pay $750 of the cost. The commissioners based their decision on a lack of funds and a backlog of work resulting from stricter plumbing standards enacted earlier this year. City commissioners do not believe that Simmons's two-week salary should be paid during the time that he is "off work." They also feel that the $400 cost to the city as well as travel expenses are too high. Although the city would benefit from Simmons's enhanced knowledge of the regulations, one city official was quoted as saying, "I don't think he really needs it anyway."

The one dissenting city commissioner argued that this is the first such course to be offered covering the new regulations. She has said, "Things change. Materials change. You can never stop learning, and you can't maintain a quality staff if you don't keep up on the latest information."

12. Describe the Taguchi loss function versus the traditional approach to quality.
13. You are in charge of setting up and working with a quality improvement team. Why would it be smart for you to begin with a small problem to tackle and solve?
14. Describe the concept of management leadership as it pertains to quality improvement. Specifically, what things would you be looking for if you were interviewing a company about their quality leadership?
15. Research Deming's "red bead" experiment. What does it show people?
16. Discuss the concepts at the base of continuous improvement.
17. Locate the vision and mission statements from several organizations. What do you think of them? What do you think the vision statements are trying to tell you about the companies? Do the mission statements support the vision statements? Do the companies' day-to-day activities support their vision and mission?
18. Research Deming's funnel experiment. What is the experiment trying to show people?
19. Research Deming's profound knowledge system. What are its components? How do they work together? Describe each component's critical concept.

CASE STUDY 2.1
The Total Quality Person

Companies today are turning to total quality management (TQM) to improve their capabilities. To adapt to TQM, management styles have had to change. Discussions in texts and articles focus on the new form of employee-employer relationships. But what about the individuals involved in this transformation? Who are they? Are they TQM people? Take the following survey and discover insights into your TQM self!

ARE YOU A TOTAL QUALITY PERSON?*

Since 1987, American companies have jumped on the bandwagon to meet the criteria for the Malcolm Baldrige National Quality Award. This award, which is named after the late Secretary of Commerce Malcolm Baldrige, was established by Congress in 1987. The award's intent was to encourage U.S. companies to adopt the total quality concept in their quest to become more globally competitive. The goal of the award is for companies to move toward customer satisfaction and continuous quality improvement.

Now it is your turn to be tested. Are you a total quality person? Wouldn't it be great if there were a total quality award for which American citizens could apply? I would like to see all Americans striving for continuous improvement and customer satisfaction with each other. And now the time has come. If you are curious as to where you stand, take this test to see whether you meet the criteria for this individual total quality award.

Good luck.

(Circle the appropriate number for each item.)

Personal Leadership

1. I treat other people fairly and with respect.

Rarely			Sometimes				Always		
1	2	3	4	5	6	7	8	9	10

2. I actively listen to other people and don't interrupt to give my point of view.

Rarely			Sometimes				Always		
1	2	3	4	5	6	7	8	9	10

*This article appeared in *Quality Progress*, the journal of the American Society for Quality, September 1993, and is reproduced here with the permission of the author, Craig Nathanson, a continuous improvement specialist at the Intel Corporation in Folsom, California, and a member of the American Society for Quality.

3. I take on responsibility for my actions and don't rely on others to plan my future.

<table><tr><th colspan="3">Rarely</th><th colspan="4">Sometimes</th><th colspan="3">Always</th></tr><tr><td>1</td><td>2</td><td>3</td><td>4</td><td>5</td><td>6</td><td>7</td><td>8</td><td>9</td><td>10</td></tr></table>

4. I volunteer my services to help others in need.

<table><tr><th colspan="3">Rarely</th><th colspan="4">Sometimes</th><th colspan="3">Always</th></tr><tr><td>1</td><td>2</td><td>3</td><td>4</td><td>5</td><td>6</td><td>7</td><td>8</td><td>9</td><td>10</td></tr></table>

5. I maintain a healthy, positive outlook on life.

<table><tr><th colspan="3">Rarely</th><th colspan="4">Sometimes</th><th colspan="3">Always</th></tr><tr><td>1</td><td>2</td><td>3</td><td>4</td><td>5</td><td>6</td><td>7</td><td>8</td><td>9</td><td>10</td></tr></table>

6. I understand my values and apply them in my daily living.

<table><tr><th colspan="3">Rarely</th><th colspan="4">Sometimes</th><th colspan="3">Always</th></tr><tr><td>1</td><td>2</td><td>3</td><td>4</td><td>5</td><td>6</td><td>7</td><td>8</td><td>9</td><td>10</td></tr></table>

7. My long- and short-term goals are tied to my values to ensure that what I am doing in my life is important to me.

<table><tr><th colspan="3">Rarely</th><th colspan="4">Sometimes</th><th colspan="3">Always</th></tr><tr><td>1</td><td>2</td><td>3</td><td>4</td><td>5</td><td>6</td><td>7</td><td>8</td><td>9</td><td>10</td></tr></table>

8. My daily activities are in harmony with my values.

<table><tr><th colspan="3">Rarely</th><th colspan="4">Sometimes</th><th colspan="3">Always</th></tr><tr><td>1</td><td>2</td><td>3</td><td>4</td><td>5</td><td>6</td><td>7</td><td>8</td><td>9</td><td>10</td></tr></table>

9. I enjoy the people and things in my environment.

<table><tr><th colspan="3">Rarely</th><th colspan="4">Sometimes</th><th colspan="3">Always</th></tr><tr><td>1</td><td>2</td><td>3</td><td>4</td><td>5</td><td>6</td><td>7</td><td>8</td><td>9</td><td>10</td></tr></table>

10. I practice good customer service with all the people with whom I come into contact.

<table><tr><th colspan="3">Rarely</th><th colspan="4">Sometimes</th><th colspan="3">Always</th></tr><tr><td>1</td><td>2</td><td>3</td><td>4</td><td>5</td><td>6</td><td>7</td><td>8</td><td>9</td><td>10</td></tr></table>

Planning

11. Every day I take time to plan my daily activities around that which is important to me.

Rarely			Sometimes				Always		
1	2	3	4	5	6	7	8	9	10

12. I try to align my long- and short-term goals with my values to ensure that my daily activities are in harmony with my goals.

Rarely			Sometimes				Always		
1	2	3	4	5	6	7	8	9	10

13. During my daily planning time, I prioritize both important and routine activities that I need to accomplish.

Rarely			Sometimes				Always		
1	2	3	4	5	6	7	8	9	10

14. Each day I plan to accomplish only those activities for which I have allocated enough time.

Rarely			Sometimes				Always		
1	2	3	4	5	6	7	8	9	10

15. I strive for continuous learning and have plans to further my education in areas that interest me.

Rarely			Sometimes				Always		
1	2	3	4	5	6	7	8	9	10

16. I strive to work up to the standards set by the most accomplished people in areas that interest me.

Rarely			Sometimes				Always		
1	2	3	4	5	6	7	8	9	10

17. I try to exceed the expectations of all the customers with whom I come into contact in my activities.

Rarely			Sometimes				Always		
1	2	3	4	5	6	7	8	9	10

18. When I plan my activities, I have knowledge of my environment and take any changing elements into consideration.

Rarely			Sometimes				Always		
1	2	3	4	5	6	7	8	9	10

19. I have a good sense of how my personal values, strengths, and weaknesses align with what I am doing.

Rarely			Sometimes				Always		
1	2	3	4	5	6	7	8	9	10

20. I have thought-out, realistic goals with achievable targets for my major activities.

Rarely			Sometimes				Always		
1	2	3	4	5	6	7	8	9	10

Improvement

21. I can document three major processes that I use in accomplishing my personal goals.

Rarely			Sometimes				Always		
1	2	3	4	5	6	7	8	9	10

22. I constantly strive to improve my skills, knowledge, and sense of purpose in my life's work.

Rarely			Sometimes				Always		
1	2	3	4	5	6	7	8	9	10

23. I constantly strive to measure whether I am meeting my personal goals.

Rarely			Sometimes				Always		
1	2	3	4	5	6	7	8	9	10

24. I constantly strive to eliminate activities that have no value in my life and focus only on activities that enrich my life.

Rarely			Sometimes				Always		
1	2	3	4	5	6	7	8	9	10

25. I admit my mistakes, acknowledge the reasons, and then move on with the goal to not make the same mistakes again.

Rarely			Sometimes				Always		
1	2	3	4	5	6	7	8	9	10

26. I celebrate my successes and improvements.

Rarely			Sometimes				Always		
1	2	3	4	5	6	7	8	9	10

27. I measure my successes by achieving my goals on time.

Rarely			Sometimes				Always		
1	2	3	4	5	6	7	8	9	10

28. I constantly strive to improve in areas that are important to me and learn to accept my weaknesses in areas that don't interest me.

Rarely			Sometimes				Always		
1	2	3	4	5	6	7	8	9	10

29. I am a role model for continuous improvement in everything I do.

Rarely			Sometimes				Always		
1	2	3	4	5	6	7	8	9	10

30. I am open to changes in my life that will enable me to learn new things.

Rarely			Sometimes				Always		
1	2	3	4	5	6	7	8	9	10

Scoring

Add up the numbers you have circled and write the total in the box below.

Your score: ☐
Maximum points: 300

How to Interpret Your Score

60–89 Points Grade F. You might want to adopt some of these individual total quality strategies to get your life back on track.

90–128 Points Grade D. You might want to analyze your daily living patterns and goals in life. You do not demonstrate an individual total quality philosophy.

129–158 Points Grade C. You demonstrate some patterns of a total quality person but need to be more consistent on a daily basis.

159–229 Points Grade B. You have a good individual foundation in total quality principles and could serve as a role model for others.

230–300 Points Grade A. You are a great total quality role model, with a solid set of principles in leadership, planning, and continuous improvement.

CASE STUDY 2.2
Quality and Ethics

In this chapter, the issues of commitment, involvement, motivation, responsibility, authority, training, education, and communication were discussed as they relate to quality in all aspects of a company. The following article presents a disastrous situation that could have been avoided had the principles of TQM been applied.

THE QUALITY-ETHICS CONNECTION*

There is a generalized and widespread perception that the United States is suffering from moral malaise—a breakdown in ethics that has pervaded every corner and stratum of society.

Signposts of this breakdown are everywhere. The current generation in power wrestles ineffectually with the problems it faces, such as hunger, poverty, environmental degradation, urban decay, the collapse of economic systems, and corruption in business and government.

This moral malaise is infecting U.S. institutions at the highest levels. Along with murderers, rapists, muggers, and thieves, there are religious leaders, political leaders, banking officials, and other business executives being carted off to jail.

A younger generation is questioning at a very early age whether it is realistic to expect morality in contemporary society. The answer to this question—and possibly a solution to this dilemma—can be found in the cross-application of quality management theory and the realm of ethics. There is a striking similarity between the issues Americans are facing in ethics and the issues that quality professionals are facing in U.S. businesses.

Ethical Base Not at Fault

In many ways, the apparent decline in individuals' and institutions' ethical behavior parallels the now well-understood decline in global competitiveness of the nation's industrial base. The United States became the world's role model in part because of a political system that recognized, for the first time in human history, the importance of individual rights and responsibilities in maintaining a free (and moral) society. Yet there is now a growing frustration among Americans because their ability to exercise those rights and responsibilities has been seriously impaired.

*This article by Marion W. Steeples, president of Resources for Quality, Denver, Colorado, appeared in the June 1994 issue of *Quality Progress*, the journal of the American Society for Quality, and is reproduced here with the permission of the Society.

Many people attribute the moral crisis problem to a breakdown in the ethical character of too many individuals within the society. They often cite the decline of the nuclear family, increased tolerance for alternative lifestyles, drug and alcohol abuse, disrespect for authority, or some other lapse of traditional values as the source of the malaise.

Yet, if the public were surveyed on their ethical beliefs, the results would likely show that, overall, Americans hold moral beliefs similar in most respects to those of their parents and grandparents. While some of the particulars of what constitutes moral behavior might have changed, Americans still hold to a personal ethic that emphasizes honesty, personal responsibility, tolerance, and good citizenship.

So the question becomes: If personal ethics have not substantially changed, what is the source of the ethical breakdown? The national rhetoric about ethics has overtones of despair in it; there is a belief that the individual has somehow lost the will to act ethically. Solutions tend to center on the need to indoctrinate students, from the earliest ages through college, in the finer points of their civic and ethical responsibilities.

This is a familiar tune to quality professionals. In U.S. factories, employees are repeatedly called to account for every sort of problem when, in fact, the source of the problem is not the employees or a department, but the system itself. A typical response to problems is exhorting employees to work harder, more diligently, and with greater care and attention to detail. The implication is that employees don't care about the outcome of their work.

To the contrary, employees typically come to work with the intention of doing the best job possible but are stymied and discouraged at every level in the system. Employees are consistently prevented from doing the right things by systems that discourage individual initiative, improved efficiency, and improved quality.

What if America's so-called "ethical crisis" were the result of similar structural deficiencies? What if the crisis were simply a matter of societal structures that do not support and sustain ethical behavior?

Societal and Corporate Structures

In my work as a quality practitioner and an examiner for the Malcolm Baldrige National Quality Award, I have seen a strong correlation between quality and ethics. Quality is the standard by which Americans measure the goods and services they value. Ethics is the standard by which Americans measure their own behavior and that of institutions.

In virtually every case, when a company improved quality, ethics also improved. This was evident not only in the employees' actions (e.g., decreased absenteeism, decreased internal thefts, and increased participation), but also in the company's actions (e.g., examining such conceptual problems as defining corporate purpose, introducing long-term thinking and integrated planning, and determining internal and external customers' needs and acting on those needs). These improvements occurred merely as a latent benefit of quality improvement. Improved ethics was rarely a stated goal of the quality improvement programs when they were initiated. Yet the benefit is real and universal.

The Great Chicago Flood

The Chicago flood of 1992 is a classic example of how a system breakdown resulted in what is typically attributed to individual moral lapses. In mid-April 1992, the Chicago River broke through a crack in a tunnel beneath the Chicago Loop's business district, and businesses in the nation's third largest city came to a halt. Water snaked through the 50-mile labyrinth of century-old freight tunnels, and 250 million gallons rushed into commercial-area basements. Electrical power was shut off to avert the possibility of explosions from transformers shorting out. More than 200,000 people were evacuated, and more than 120 buildings were dark for two days.

To stem the tide, crews labored around the clock drilling holes and plugging them with concrete. It took more than two weeks for the U.S. Army Corps of Engineers to drain the water.

This devastating underground flood took a heavy toll. The Chicago Board of Trade shut down, hampering worldwide trading, with crippling economic effects. City Hall, several office towers, and many retailers were closed. Fifteen buildings were unable to operate for at least a week. Estimates put the price tag of this snafu at $1.7 billion.

The irony is that, for an estimated $10,000, the disaster could have been prevented. On Jan. 14, 1992, two cable TV workers discovered a 20-foot-by-6-foot crack in the tunnel. Standing knee deep in water and mud, the crew videotaped the event, recording "This is a cave-in!"

The cable TV crew, however, had trouble finding the correct city government official to which they could report the incident—the local government was in the midst of a major reorganization to increase efficiency.

In late February, the cable TV workers were finally able to discuss the situation with the appropriate city official; they urged that the tunnel site be inspected. But when they talked to the city official again on March 2, they learned that little had been done. Finally, a city worker led an inspection, took photographs of the subterranean leak, and then waited a week for a drugstore to develop the prints.

On April 2, the city's bridge engineer sent a memo urging immediate action to his superior, the acting transportation commissioner. Two bids were obtained to repair the crack, but both were turned down in an attempt to get a lower price. It was business as usual.

On April 13, the tunnel burst. On April 14, the governor of Illinois declared Chicago a disaster area.

The city's response was predictable: Heads rolled in an effort to assign blame to individuals within the system. On April 15, Chicago's mayor fired the acting commissioner. Subsequently, an engineer was discharged and five others were disciplined. The city blamed individuals in the system without addressing the structural problems that made such a debacle possible in the first place. Yet, from a quality viewpoint, the system's inefficiencies are immediately apparent.

How Chicago Went Wrong

Like many traditional U.S. corporations, the city of Chicago suffered from structural problems generated by the specialization of functions and a horizontal management structure that made individual initiative to take positive action next to impossible. As quality slips through the functional cracks of these outmoded systems, ethics are not far behind. Any system that values efficiency over effectiveness also devalues ethical behavior.

Unfortunately, for the individual citizen or bureaucrat who still attempts to behave ethically, the fragmentation of the structures that make up an organization, such as a large city government, makes taking action difficult, if not impossible.

Before history has made its final judgment on the Chicago flood, it behooves those within the quality profession to point out that the emperor had no clothes on. Far from being assignable to the incompetence of one or more city employees, the disaster that overtook downtown Chicago was the obvious—and predictable—result of the fragmentation of the city's organizational structure.

Chicago was not acting as one organization but as many—each one with its own set of rules, agenda, and rewards for success. No individual within those separate departments has a stake in the final outcome, only a stake in his or her part.

The Chicago flood dramatizes what is the rule and not the exception in most U.S. organizations, public or private. Whenever there is a break in responsibilities and accountabilities, there is no structural way to ensure that responsible, ethical actions will result. To the contrary, the system in Chicago consistently frustrated every attempt by individuals to behave ethically.

What Could Have Made It Right?

For openers, the organization needed clearer direction and goals. The city government needed to build its structure based on these goals to ensure that connections within and between the departments and agencies were sound. These human connections work well only if an integrated organizational structure that aligns functions is in place.

When each department stands alone, the support needed to enable employees to do the right thing is non-existent. Employees cannot be expected to continually and repeatedly go to heroic measures simply to perform their jobs. Some people can't do it; they simply don't know how to maneuver around the system. Others won't do it; it's simply too hard and asks too much, and the rewards of bucking the system are dubious. A faulty system can make employees powerless to make positive contributions. There are simply too many barriers.

It is important to reiterate that, at every stage in the flood fiasco, attempts were made to act ethically. The cable TV crew recognized the problem and attempted to notify the appropriate authorities, but the news was not quickly relayed or responded to. The various levels of engineers attempted to solve the problem, but they couldn't break through the layers of bureaucracy in time. Like the crumbling concrete in the tunnel, the decades-old policies and procedures became deadly impediments to the exercise of good judgment. The deluge of water that stopped downtown Chicago served only to mark the beginning of the challenging road that now lies before it and every city government. The old system is inadequate and, as thousands of Chicago residents can attest, the results have been disastrous.

The dollar cost for correcting things gone wrong, as in the Chicago flood, takes a heavy toll on the taxpayers. Moreover, there is the toll of disillusionment and a loss of trust, as was witnessed in the savings-and-loan debacle, in which the old system of checks and balances collapsed completely.

How Can Americans Make a Change?

Chicago's century-old tunnel system was once used to carry freight; now it carries electronic and communications gear. A system that once helped Chicago be productive is now a problem. Unfortunately, Chicago's situation represents business as usual in much of America. The organizational structures that Americans created during the industrial age don't always translate well in the information age. An earlier age of specialization has led to the current age of frustration.

Out of seemingly separate movements, quality and ethics have emerged as top issues in the national agenda. The issue of quality has been forced by economic reality: In a global economy, quality-leveraged companies are simply more competitive. Quality, in other words, provides a system for living up to the expectations and addressing the needs of customers.

In a strikingly similar fashion, the issue of ethics has emerged out of a sense that the United States' long-standing reputation as the model of democracy and as a moral force in the world has eroded. Americans consider themselves a moral people, dedicated to the high ideas laid out in such documents as the Constitution and Bill of Rights.

But citizens from Los Angeles, CA, to Washington, DC, are grappling with the apparent inability of U.S. systems to live up to their ethical expectations or to reasonably address their legitimate needs. They feel helpless when dealing with institutions that seem out of touch with reality and that don't have an innate sense of right and wrong. Such frustrations have grown beyond "big government" and "big industry." People encounter unresponsive, apparently amoral systems daily in such places as local public schools, grocery stores, doctors' offices, and banks.

The old systems make it difficult at best and impossible at worst for institutions to live up to their ethical expectations and to address Americans' legitimate needs. Thus, new systems need to be created—and that is where total quality management (TQM) comes in.

TQM provides a model for systematically creating responsive societal structures. It can build individual responsibility and initiative back into impervious and rigid systems. Reconfigured systems, based on a solid foundation of vision and purpose, ensure that rational, long-term thinking, rather than expediency, guides organizational decisions.

TQM offers a systematic way for organizations to link values and value. Experience shows that total quality companies are successful because they translate what customers value into quality requirements and practices. In that context, ethics can be viewed simply as a primary set of customer values. Creating ethical systems is a matter of building ethical expectations into systems and providing support to employees so that they can live up to those expectations.

By integrating the functions of an organization and by connecting quality and ethics, institutions can provide what is of value to customers and provide what is valuable to society. They must ensure the integrity of systems (quality) and the integrity of people (ethics). The absence of one affects the other.

TQM has proven itself as a way to rebuild infrastructure by integrating and aligning operations to provide value. But the paradox is that, inasmuch as quality systems can assist Americans' search for continued ethical improvement, quality itself is not possible without ethics. Quality theory provides a means for integrating systems to provide value. But only an ongoing discussion of ethics will provide a notion of what is meant by "value."

Assignment

1. Discuss how the specific issues of communication, responsibility, authority, commitment, and motivation relate to quality in all aspects of a company's operation.
2. Discuss quality issues as they pertain to the Chicago incident. To organize your discussion, consider the following questions:
   - Who is management in this instance?
   - What was their level of commitment and involvement?
   - How were they at motivating employees?
   - Who had the responsibility?
   - Who had the authority?
   - Were the employees appropriately educated and trained?
   - Why did communication break down?

3

Quality Improvement: Problem Solving

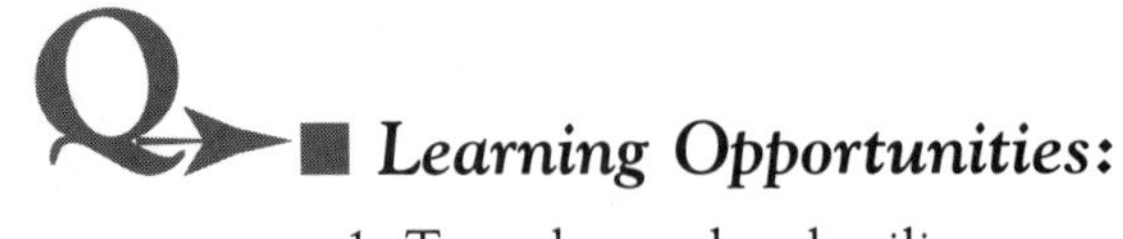

■ *Learning Opportunities:*

1. To understand and utilize a systematic problem-solving process; to learn to ask the right questions, present information clearly and unambiguously, and make judgments based on the information
2. To understand and utilize a variety of techniques for effective problem diagnosis and problem solving
3. To learn to diagnose and analyze problems that cause variation in the manufacturing, process, and service industries
4. To introduce and explain the seven tools of quality ■

Have you ever been lost? Being lost is different from not being able to find something. In one case, you're unable to locate an object or a place; in the other, you don't know where you are. When you find out where you are, you can figure out where you need to go.

Dealing with problems can be similar to being lost. Problem solvers need to know where they stand, what the problem really is, and what the cause of the problem is before any solutions can be proposed. This chapter seeks to teach problem-solving methods to help identify the problem and its causes.

PROBLEM SOLVING

Problem solving, *the isolation and analysis of a problem and the development of a permanent solution*, is an integral part of the quality-improvement process. Like drivers randomly turning on different streets in the hope of finding their destination, people often find solutions to problems by following a hit-or-miss approach. Sometimes these solutions attack the symptoms associated with the problem rather than the root cause of the problem. This leaves the real problem unsolved, the real destination unreached. A hit-or-miss type approach to tackling problems is not very effective. Problem solving is not an automatic process; people need to be trained in correct problem-solving procedures. Problem-solving efforts should be objective and focused on finding root causes. Proposed solutions should prevent a recurrence of the problem. Controls should be present to monitor the solution. Teamwork, motivation, coordinated and directed problem solving, problem-solving techniques, and statistical training are all part of ensuring that problems are isolated, analyzed, and corrected.

EXAMPLE 3.1 A Problem

One hot summer evening, a woman returned home from work and was surprised when a harried man dashed from the side of the house and gasped:

"Are you Mrs. G.?"
"Yes."

With this admission, the man launched into his tale of his truly horrible day. It seems that he was employed by an air-conditioning company and had been sent out to disconnect and remove an air conditioner at 10 Potter Lane. He had been informed that no one would be at home and that the cellar door would be left unlocked. He had just finished disconnecting the air conditioner when his office paged him. The office had given him the wrong address. What should he do? Reconnect it, the office replied. It wasn't as simple as that, because he damaged the air conditioner while removing it. After all, since the air conditioner was being replaced, he did not consider it necessary to be careful.

"How could this have happened?" asked Mrs. G.

We often read about or experience life's little errors. Sometimes they are minor disturbances to our routines and sometimes they are even funny, but occasionally mistakes cost significant time and money. How do companies avoid errors? How do they find the answer to the question: how could this have happened? The problem-solving techniques presented in this chapter can help.

STEPS IN PROCESS IMPROVEMENT

Problem solving should follow a logical, systematic method. This will place emphasis on locating and eliminating the root or real cause of a problem. Other, less systematic attempts at problem solving run the risk of attempting to eliminate the symptoms associated with the problem rather than eliminating the problem at its cause.

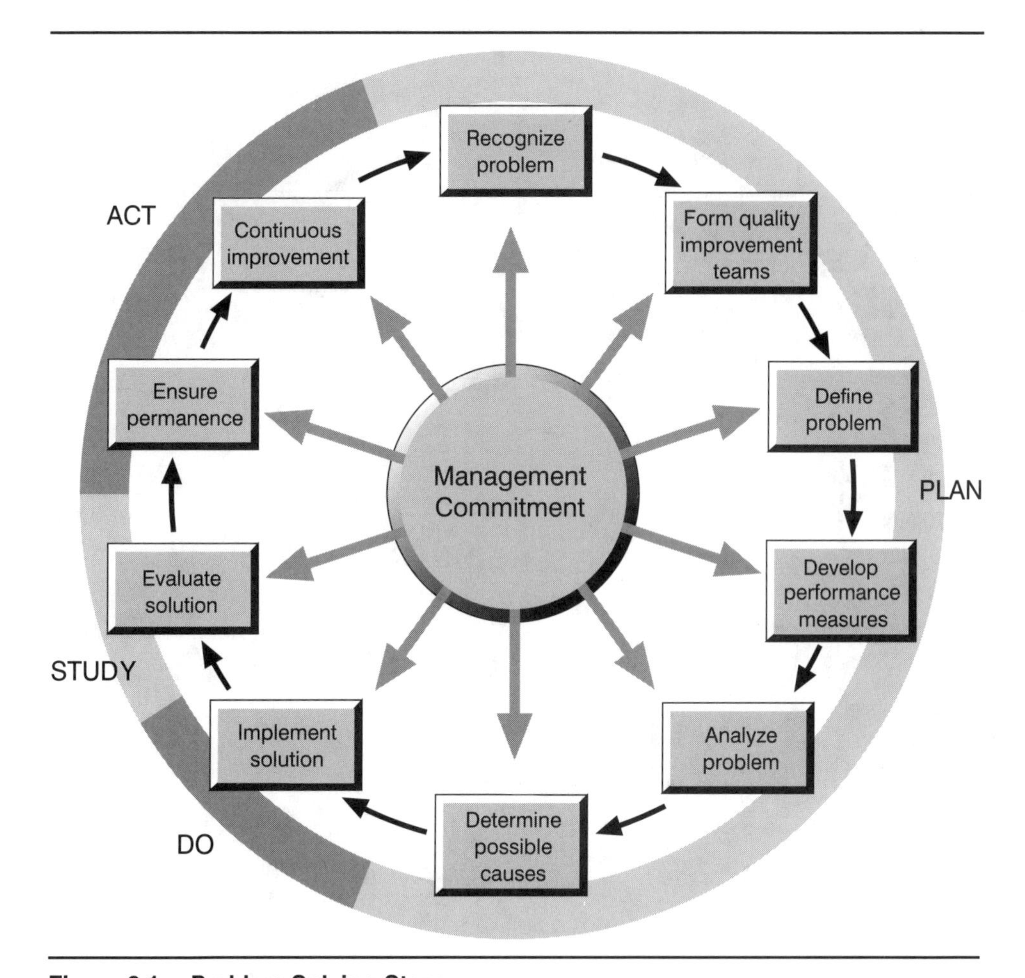

Figure 3.1 Problem-Solving Steps

Introduced in Chapter 2, Dr. Deming's Plan-Do-Study-Act (PDSA) cycle is the systematic approach to problem solving that will be followed in this chapter (Figure 3.1).

Organized problem-solving efforts utilize a variety of quality tools for problem analysis. As mentioned in Chapter 2, some of the tools are flowcharts, histograms, Pareto charts, cause-and-effect diagrams, check sheets, control charts, and scatter diagrams. Of these seven, five are discussed in this chapter. Histograms and control charts are covered in greater detail in later chapters. We will look here at other problem-solving techniques as well: brainstorming, WHY-WHY diagrams, force-field analysis, and run charts. The particular situation needing a solution will dictate which problem-solving techniques will be used. The techniques are presented in this chapter in the section of the PDSA cycle where they are most frequently used (Figure 3.2). This does not prevent their use in other areas of the problem-solving process.

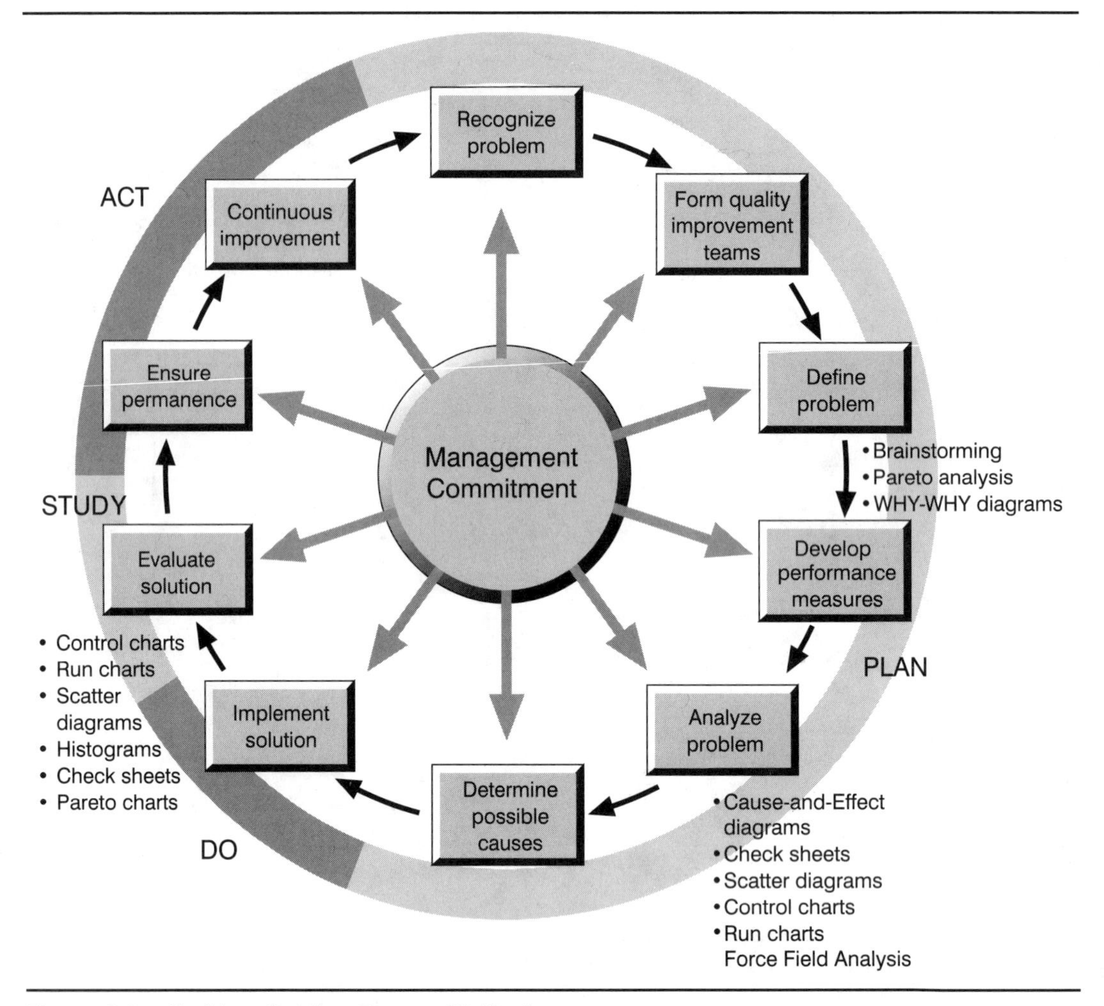

Figure 3.2 Problem-Solving Steps with Tools

PLAN

Step 1. Recognizing the Problem and Establishing Priorities

In problem solving, the PDSA cycle places a strong emphasis on determining the current conditions and planning how to approach a problem. In the Plan phase, problem investigators are looking at the processes and products involved to determine how they are presently performing. The activities engaged in at this stage are similar to those of drivers who must determine where they are so that they know which roads will take them to their destinations in the quickest, easiest way. Planning is the most time-consuming portion of the cycle. In this stage, the problem will be investigated and actions considered.

Management involvement and commitment is crucial to the success of any major problem-solving process. Management is ultimately responsible for seeing that problems are isolated and solved. To give emphasis to the importance of solving problems, management should participate in the recognition and identification of problems. Information concerning the problem(s) may have come from a number of different sources, including but not limited to manufacturing, assembly, shipping, or product design departments or employees, or customers.

During the problem recognition stage, the problems will be outlined in very general terms. At this point in the problem-solving process, management has recognized or identified that a problem or problems exist. As yet, the specifics of the problem(s) have not been clearly defined.

An excellent way for management to get directly involved in the problem-solving process is for management to select problem-solving teams and give direction as to the problems to be tackled. Management should determine if the problem is solvable and who should be involved in determining a solution. "Solvable" means that the organization has the personnel and financial resources as well as the knowledge to solve the problem. For example, a customer may want a cure for the common cold (in its many varieties) that a pharmaceutical company simply cannot develop. In all likelihood, the management group will have a list of potential opportunities or problems that outnumber the teams created to solve them.

After management selects a problem or group of problems to investigate and determines that the problem is solvable within a given time period and with the available resources, an interdisciplinary team needs to be established to find a solution to the problem.

Step 2. Forming Quality Improvement Teams

Once a problem situation has been recognized and before the problem is attacked, an ***interdisciplinary problem-solving or quality improvement team*** must be created. *This team will be given the task of investigating, analyzing, and finding a solution to the problem situation within a specified time frame.* Sometimes called a **quality circle,** *this problem-solving team consists of people who have knowledge of the process or problem under study.* The project teams are given a mandate to focus on a particular process, area, or problem. Generally, this team is composed of those closest to the problem as well as a few individuals from middle management with enough power to effect change. The team may consist of people from engineering, manufacturing, purchasing, sales, and/or design departments. It may even include an outside vendor or a representative from the customer base. During the problem-solving process, the team can be supplemented on an as-needed basis, with people who have expertise in the areas related directly to the problem solution. Upon the resolution of a project, the team will be disbanded or reorganized to deal with another problem.

Upper-management involvement in problem selection has a very positive benefit. Where management sets the direction, the team is much more focused and tends not to get bogged down in the problem-selection process. An additional benefit of management involvement is that it is a direct showing of management support and management buy-in on finding a workable solution. Under certain circumstances, upper management provides the team with an objective statement. Upper management's sincere interest and support in the resolution of the problem is evidenced by their willingness to commit money and time for training in problem solving and facilitation. In any case, upper management must monitor and encourage these teams to solve problems. The teams will quickly become ineffective if the solutions they propose are consistently turned down or ignored. Management support will be obvious in management's visibility, diagnostic support, recognition, and limited interference.

EXAMPLE 3.2 Steps 1 and 2. Recognizing the Problem and Forming a Quality Improvement Team

Plastics and Dashes Inc. supplies instrument panels and other plastic components for automobile manufacturers. Recently their largest customer informed them that there have been an excessive number of customer complaints and warranty claims concerning the P&D instrument panel. The warranty claims have amounted to over $200,000, including the cost of parts and labor. In response to this problem, Plastics and Dashes has formed an improvement team to investigate. The steps they will take to solve this problem are detailed in the examples throughout this chapter.

Step 3. Defining the Problem

Once established, the quality-improvement or problem-solving team sets out to clearly define the problem and its scope. A clear problem definition will help the team focus on the problem and avoid chasing causes that are contributing to but are not the true cause of the problem. As Example 3.3 shows, a Pareto analysis can often point to significant areas to investigate.

EXAMPLE 3.3 An Incomplete Picture

Tensions are high at PL Industries. Several departments have been reprimanded for low customer-satisfaction ratings. The catalog department, in particular, has been under a lot of pressure to improve customer service on their toll-free service lines. Customers are complaining loudly about the amount of time they are kept waiting on hold for the next available operator. The catalog department's defense that they are doing the best they can has failed to impress upper management. Recently the catalog department manager was replaced.

The new manager, who has training in statistics, offered no excuses for her new department's poor performance, asking only for time to study the situation. Within a few weeks time, she returned to upper management with information concerning toll-free-number users. The Pareto chart in Figure 3.3 shows that a significant number of the calls received on the toll-free line actually have nothing to do with the catalog department and must be transferred to other departments. An equally significant number of calls to the toll-free number are due to billing statement questions, which are not the responsibility of the catalog department.

By the conclusion of the meeting, statistical information has allowed PL Industries' upper management to develop a clearer understanding of the true nature of

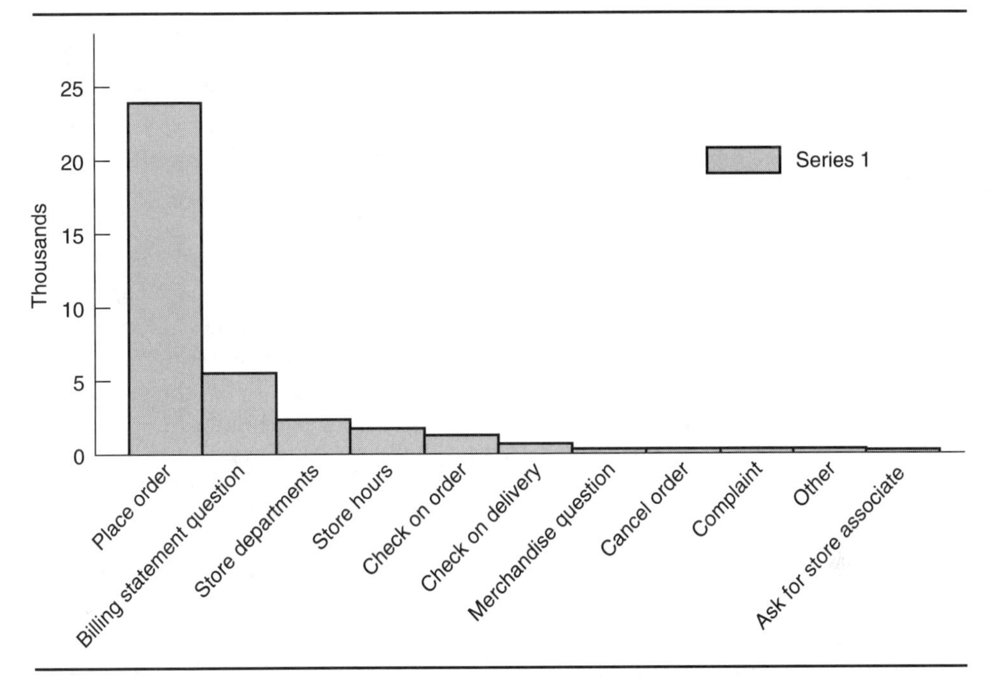

Figure 3.3 Toll-Free Number Calls to Catalog Department, June

the problem. In the future, rather than waste their efforts on berating the catalog department, their focus will be on determining how to create a better billing statement as well as determining which departments should also have a toll-free number.

How will they know their efforts have been successful? Look at Figure 3.4, which provides information about the toll-free number six months after the service has been expanded to include numbers for customer service, the stores, and the billing department. Notice the dramatic reduction in customer wait times with no additional catalog department staff. It is also important to notice that due to the significantly reduced wait time, the number of calls handled nearly doubled; however, the cost of operating the toll-free numbers increased by only a little more than $1500.

Technique: Pareto Analysis

The ***Pareto chart*** *is a graphical tool for ranking causes of problems from the most significant to the least significant.* Named after the Italian economist Vilfredo Pareto, Pareto

	June	November
Total No. of Calls	36,633	58,801
Cost	$24,269	$25,823
Average Cost/Call	$0.66	$0.32
Average Wait (min)	3.86	0.32
Staff	15	15

Figure 3.4 Toll-Free Number: June versus November

charts are a graphical display of the 80–20 rule. Pareto, during his study of the Italian economy, found that 80 percent of the wealth in Italy was held by 20 percent of the people, thus the name "80–20 rule." In 1950 Dr. Joseph Juran applied this principle to quality control when he noticed that 80 percent of the dollar loss due to quality problems was found in 20 percent of the quality problems. Since then, the 80–20 rule, through Pareto charts, has been applied to a number of areas, including scrap rates, sales, and billing errors.

Pareto charts are a helpful tool for problem analysis. Problems and their associated costs are arranged according to their relative importance in bar-chart form. While the split is not always 80–20, the chart is a visual method of identifying which problems are most significant. Pareto charts allow users to separate the vital few problems from the trivial many. The use of Pareto charts also limits the tendency of people to focus on the most recent problems rather than on the most important problems.

A Pareto chart is constructed using the following steps:

1. Select the subject for the chart. This can be a particular product line exhibiting problems, or a department, or a process.
2. Determine what data need to be gathered. Determine if numbers, percentages, or costs are going to be tracked. Determine which nonconformities or defects will be tracked.
3. Gather data related to the quality problem. Be sure that the time period referenced is the same. Use the number of defects or nonconformities per hour, per shift, or per week, but keep it standard.
4. Make a table of the gathered data and record the total numbers in each category. Categories will be the types of defects or nonconformities.
5. Determine the total number of nonconformities and calculate the percent of the total in each category.
6. Determine the costs associated with the nonconformities or defects.
7. Select the scales for the chart. Y-axis scales are typically the number of occurrences, number of defects, dollar loss per category, or percent. The x axis usually displays the categories of nonconformities or defects.
8. Draw a Pareto chart by organizing the data from the largest category to the smallest. Include all pertinent information on the chart.
9. Analyze the chart or charts. The largest bars represent the vital few problems. If there does not appear to be one or two major problems, recheck the categories to determine if another analysis is necessary.

EXAMPLE 3.4 Constructing a Pareto Chart

The team members working on the instrument panel warranty issue first discussed in Example 3.2 have decided to begin their investigation by creating a Pareto chart.

Step 1. Select the Subject for the Chart. The subject of the chart is instrument panel warranty claims.

Step 2. Determine What Data Need to Be Gathered. The data to be used to create the chart are the different reasons customers have returned their instrument panels for warranty work. Cost information on instrument panel warranty work is also available.

Step 3. Gather the Data Related to the Quality Problem. The team has determined it is appropriate to use the warranty information for the preceding six months. Copies of warranty information have been distributed to the team.

Step 4. Make a Table of the Gathered Data and Record the Total Numbers in Each Category. Based on the warranty information, the team has chosen the following categories for the chart: loose instrument panel components, noisy instrument panel components, electrical problems, improper installation of the instrument panel or its components, inoperative instrument panel components, and warped instrument panels.

Step 5. Determine the Total Number of Nonconformities and Calculate the Percent of the Total in Each Category. From the warranty information, they also have the number of occurrences for each category:

1. Loose instrument panel components	355	41.5%
2. Noisy instrument panel components	200	23.4%
3. Electrical problems	110	12.9%
4. Improper installation of the instrument panel or its components	80	9.4%
5. Inoperative instrument panel components	65	7.6%
6. Warped instrument panel	45	5.2%

Warranty claims for instrument panels total 855.

Step 6. Determine the Costs Associated with the Nonconformities or Defects. The warranty claims also provided cost information associated with each category.

1. Loose instrument panel components	$115,000
2. Noisy instrument panel components	$25,000
3. Electrical problems	$55,000
4. Improper installation of the instrument panel or its components	$10,000
5. Inoperative instrument panel components	$5,000
6. Warped instrument panel	$1,000

Step 7. Select the Scales for the Chart. The team members have decided to create two Pareto charts, one for number of occurrences and the other for costs. On each chart, the x axis will display the warranty claim categories. The y axis will be scaled appropriately to show all the data.

Step 8. Draw a Pareto Chart by Organizing the Data from the Largest Category to the Smallest. The Pareto charts are shown in Figures 3.5 and 3.6. A Pareto chart for percentages could also be created.

Step 9. Analyze the Charts. When analyzing the charts, it is easy to see that the most prevalent warranty claim is loose instrument panel components. It makes sense that loose components might also be noisy and the Pareto chart (Figure 3.5) reflects this, noisy instrument panel components being the second most frequently occurring warranty claim. The second chart, in Figure 3.6, tells a slightly different

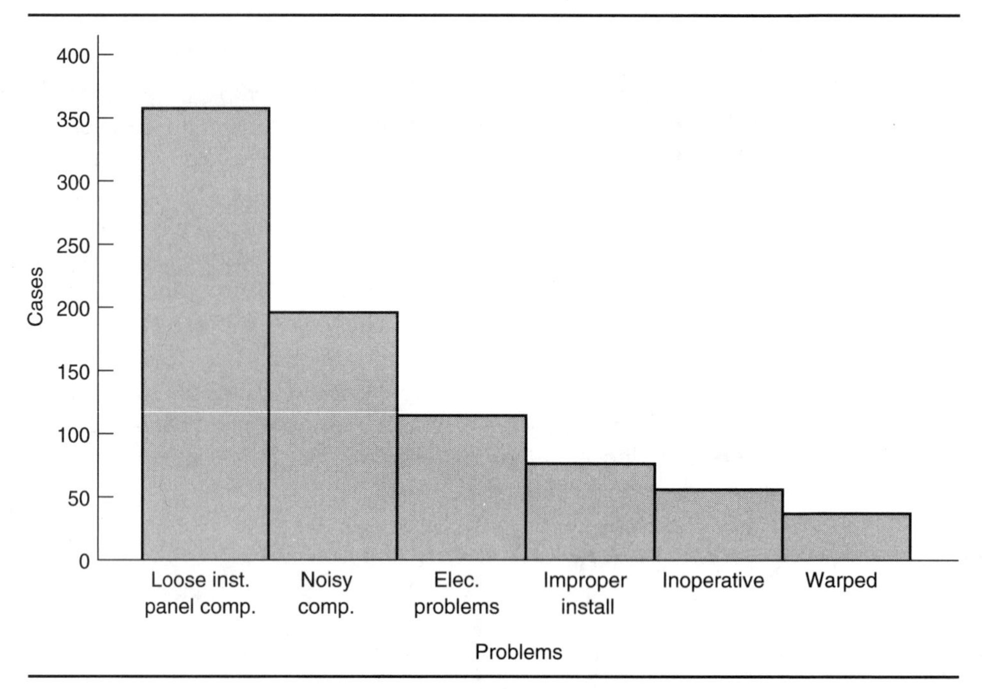

Figure 3.5 Pareto Chart of Problems Related to the Instrument Panel

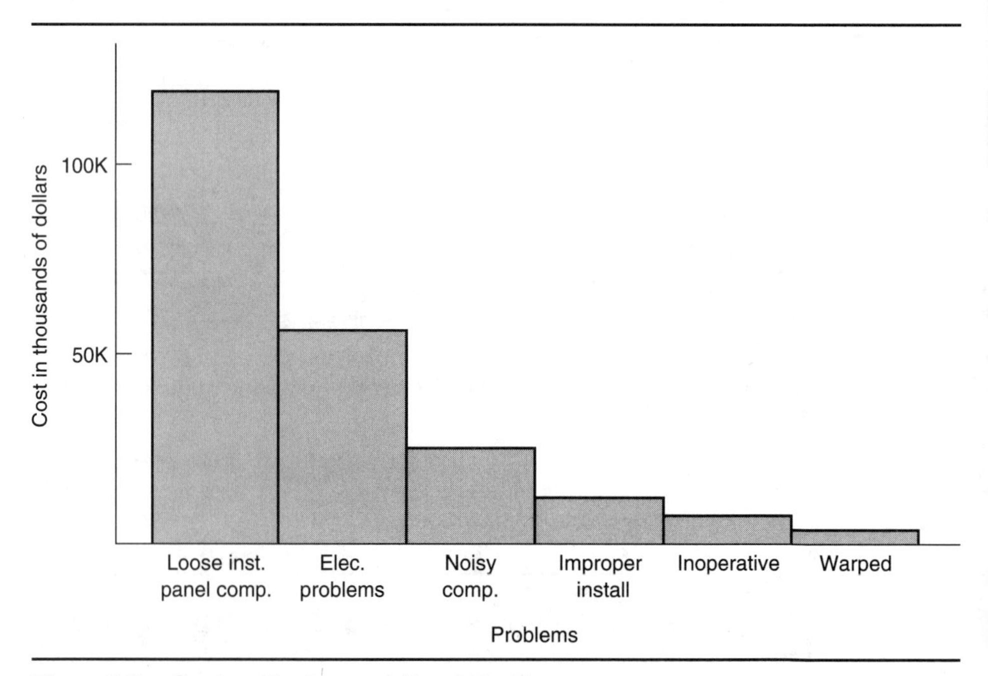

Figure 3.6 Costs of Instrument Panel Problems

story. The category loose instrument panel components has the highest costs, as might be expected; however, electrical problems has the second-highest costs.

At this point, although all the warranty claims are important, the Pareto chart has shown that efforts should be concentrated on investigating the causes of loose instrument panel components. Solving this warranty claim would significantly affect warranty numbers and costs.

After a Pareto analysis reveals an area to study, team members must begin the search for the root cause behind the problem. Brainstorming is one technique helpful in this search.

Technique: Brainstorming

The purpose of ***brainstorming*** *is to generate a list of problems, opportunities, or ideas from a group of people.* Everyone present at the session should participate. The discussion leader must ensure that everyone is given an opportunity to comment and add ideas. Critical to brainstorming is that no arguing, no criticism, no negativism, and no evaluation of the ideas, problems, or opportunities take place during the session. It is a session devoted purely to the generation of ideas and opportunities.

The length of time allotted to brainstorming varies; sessions may last from 10 to 45 minutes. Some team leaders deliberately keep the meetings short to limit opportunities to begin problem solving. A session ends when no more items are brought up. The result of the session will be a list of ideas, problems, or opportunities to be tackled. After being listed, the items are sorted and ranked by category, importance, priority, benefit, cost, impact, time, or other consideration.

EXAMPLE 3.5 Brainstorming

The team at Plastics and Dashes Inc. conducted a further study of the causes of loose instrument panel components. Their investigation revealed that the glove box in the instrument panel was the main problem area (Figure 3.7). They were led to this conclusion when further study of the warranty data allowed them to create the Pareto chart shown in Figure 3.8. This figure displays problems related specifically to the glove box.

In order to better understand why the glove box might be loose, the team assembled to brainstorm the variables associated with the glove box.

JERRY: I think you all know why we are here today. Did you all get the opportunity to review the glove box information? Good. Well, let's get started by concentrating on the relationship between the glove box and the instrument panel. I'll list the ideas on the board here, while you folks call them out. Remember, we are not here to evaluate ideas. We'll do that next.
SAM: How about the tightness of the latch?
FRANK: Of course the tightness of the latch will affect the fit between the glove box and the instrument panel! Tell us something we don't know.

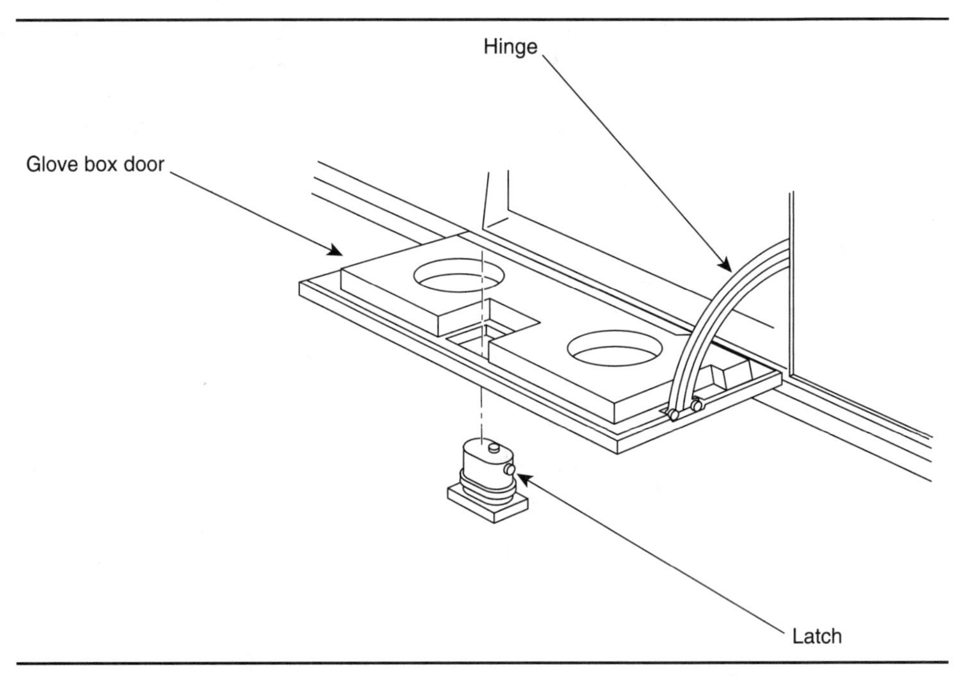

Figure 3.7 Glove Box

JERRY: Frank, have you forgotten the rules of a brainstorming session? No criticizing. Sam, can you expand on your concept?
SAM: I was thinking that the positioning of the latch as well as the positioning of the hinge would affect the tightness of the latch.
JERRY: Okay. (Writes on board.) Tightness of Latch, Positioning of Latch, Positioning of Hinge. Any other ideas?
SUE: What about the strength of the hinge?
JERRY: (Writes on board.) Strength of Hinge.
SHARON: What about the glove box handle strength?
FRANK: And the glove box handle positioning?
JERRY: (Writes on board.) Glove Box Handle Strength. Glove Box Handle Positioning.

The session continues until a variety of ideas have been generated (Figure 3.9). After no more ideas surface or at subsequent meetings, discussion and clarification of the ideas can commence.

An excellent technique for finding the root cause(s) of a problem is to ask "Why" five times. This is also an excellent method for determining what factors have to be in place in order to respond to an opportunity. ***WHY-WHY diagrams*** *organize the thinking of a problem-solving group and illustrate a chain of symptoms leading to the true cause of a problem.* By asking "why" five times, the problem solvers are stripping away the symptoms surrounding the problem and getting to the true cause of the problem.

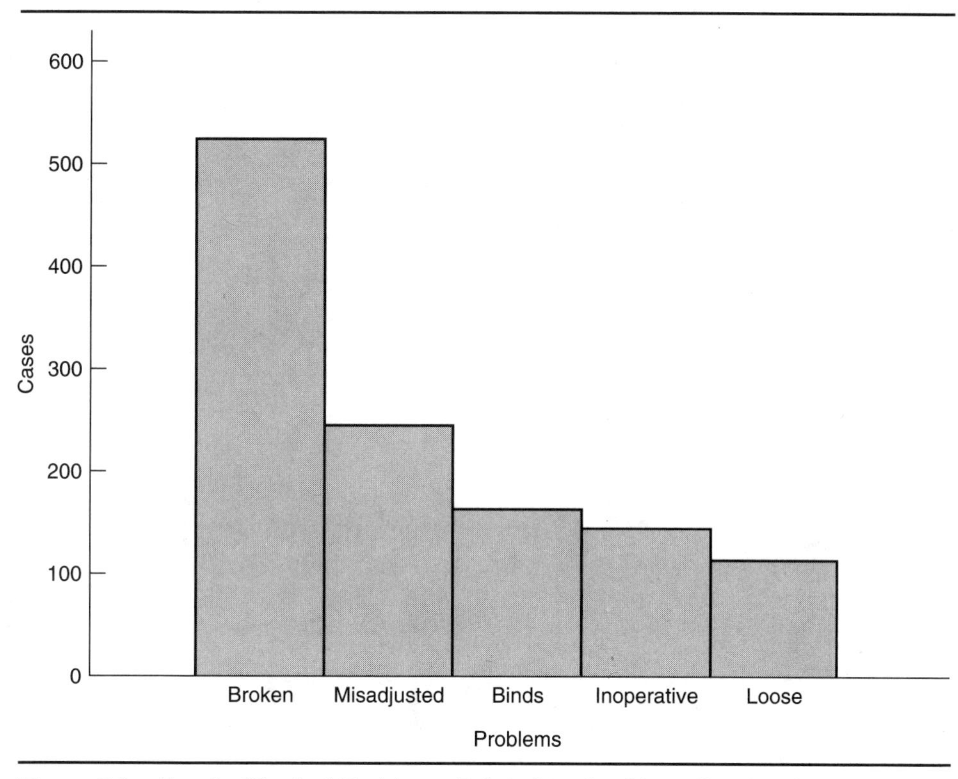

Figure 3.8 Pareto Chart of Problems Related to the Glove Box Latch

At the end of a session it should be possible to make a positively worded, straightforward statement defining the true problem to be investigated.

Technique: WHY-WHY Diagrams

Developed by group consensus, the WHY-WHY diagram flows from left to right (Figure 3.10). The diagram starts on the left with a statement of the problem to be resolved. Then the group is asked why this problem might exist. The responses will be statements of causes that the group believes contribute to the problem under discussion. There may be only one cause or there may be several. Causes can be separate

Positioning of the Glove Box
Strength of the Glove Box
Tightness of the Latch
Positioning of the Latch
Strength of the Latch
Positioning of the Hinge
Strength of the Hinge
Glove Box Handle Strength
Glove Box Handle Positioning
Glove Box Construction Materials

Figure 3.9 Variables Associated with the Glove Box

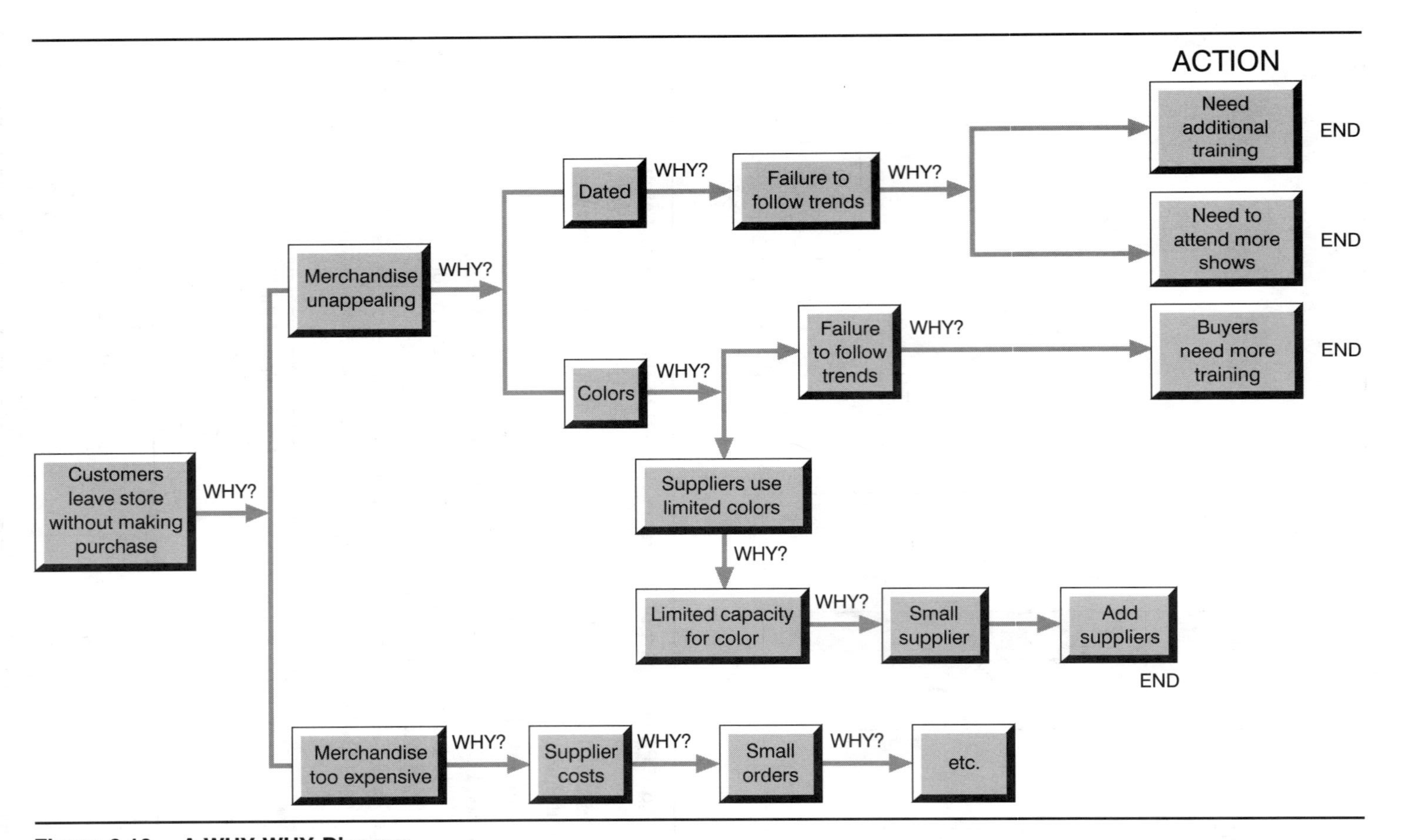

Figure 3.10 A WHY-WHY Diagram

or interrelated. Regardless of the number of causes or their relationships, the causes should be written on the diagram in a single, clear statement. "Why" statements should be supported by facts as much as possible and not by hearsay or unfounded opinions.

This investigation is continued through as many levels as needed until a root cause is found for each of the problem statements, original or developed during the discussions. Frequently five levels of "why" are needed to determine the root cause. In the end, this process leads to a network of reasons the original problems occurred. The ending points indicate areas that need to be addressed to resolve the original problem. These become the actions the company must take to address the situation. WHY-WHY diagrams can be expanded to include notations concerning who will be responsible for action items and when the actions will be completed.

The WHY-WHY process is not meant to locate solutions; it is used primarily to identify the root causes of problems. Care should be taken to not propose solutions to the problem yet. Not enough is known about the problem to create a truly workable solution. Solutions enacted at this point may be shortsighted and might not include answers for the complexities involved in the problem. The problem analysis stage needs to come next, before any solutions are proposed.

Step 4. Developing Performance Measures

Measures of performance enable problem solvers to answer the question, How do we know the right changes have been made? Measures may be financial in nature, customer-oriented, or pertinent to the internal workings of the organization. Examples of financial measures are costs, return on investment, value added, and asset utilization. Financial measures usually focus on determining whether or not the changes that have been made enhance an organization's financial performance. Companies use customer-oriented measures to determine whether or not their plans and strategies keep the existing customers satisfied, bring in new customers, and encourage customers to return. These measures may include response times, delivery times, product or service functionality, price, quality, or other intangible factors. Measures pertinent to the internal workings of an organization concentrate on the internal business processes that are critical for achieving customer satisfaction. These measures focus on process improvement and productivity; employee and information system capabilities; and employee satisfaction, retention, and productivity. They seek to provide the answers to questions such as these: Do our employees have the right equipment or information to do their jobs well? Do they receive recognition and support? What are their skills and competencies? What will be needed in the future?

When creating measures it is important not to create measures for measurement sake. Measures require gathering and analyzing data. Most organizations already have measures in place. Those measures need to be refined by asking questions: What does the organization need to know? What is important to our customers? What data are currently being gathered? How are the data being used? What measures currently exist? Are they useful? How does the organization use the information and measures already in existence? The following example provides some insight into process measures.

EXAMPLE 3.6 Measures of Performance

A copy center is considering replacing a copy machine. Here is their proposed action and the measures they intend to use to answer the following question: How do we know we have made the correct change?

Proposed Action: Replace copy machine 815 with copy machine 1215.

Why? A. Copy machine 1215 can run a larger number of impressions faster than copy machine 815.

B. The copy center has or could quickly develop a profitable customer base with the larger machine.

Measures of Performance: (How will we know A and B are true?)

1. Measure number of impressions made by existing 815 machine.
2. Measure average time it takes to complete a copying job with 815 machine.
3. Measure customer copy job delays caused by 815 (lead time).
4. Measure number of customers who must take job elsewhere to meet deadlines.

Compare these measures with the projected number of impressions and the average job time of the 1215 machine. If the comparison is favorable, install the 1215 machine.

After the installation of the 1215, how do we know whether or not it was a good idea?

Apply the four measures of performance to the 1215 and compare. What is the result of this comparison?

EXAMPLE 3.7 Measures of Performance for Instrument Panel Warranty Issues

The instrument panel warranty team has decided that warranty claim information will provide the best measure of performance. They intend to track the number of warranties, the reasons behind the warranty claims, warranty costs, and improvement implementation time. The number of warranties and their reasons will provide customer information. Financial information will come from tracking warranty costs. The company's ability to make process changes will be measured by the improvement in implementation time.

Step 5. Analyzing the Problem/Process

Now that the problem is defined, the problem and its processes are investigated to identify the potential constraints and determine the sources of difficulties. Investigators are seeking a deeper understanding of the problem. Information gathered at this stage will help determine potential solutions. The analysis must be thorough to uncover any intricacies involved in or hidden by the problem. To understand an involved process, problem solvers often utilize flowcharts.

Technique: Flowcharts

A ***flowchart*** *is a graphical representation of all of the steps involved in an entire process or a particular segment of a process*. Diagramming the flow of a process or system aids in understanding it. Flowcharting is effectively used in the first stages of problem solving because the charts enable those studying the process to quickly understand what is involved in a process from start to finish. Problem-solving team members can clearly see what is being done to a product at the various steps in the process. Flowcharts clarify the routines used to serve customers. Problems nested within a process or within a process segment are easily identified by using a flowchart.

Flowcharts are fairly straightforward to construct. The steps to creating such a chart are the following:

1. Define the process steps. Use brainstorming to identify the steps for new processes. For existing processes actually observe the process in action.
2. Sort the steps into the order of their occurrence in the process.
3. Place steps in appropriate flowchart symbols and create the chart.
4. Evaluate the steps for completeness, efficiency, and possible problems.

Flowcharts can be constructed with or without the symbols shown in Figure 3.11. This is especially true when first creating a flowchart. Because processes and systems are often complex, in the early stages of flowchart construction, removable 3 by 5 sticky notes placed on a large piece of paper or board allow creators greater flexibility when creating and refining a flowchart. When the flowchart is complete, a final

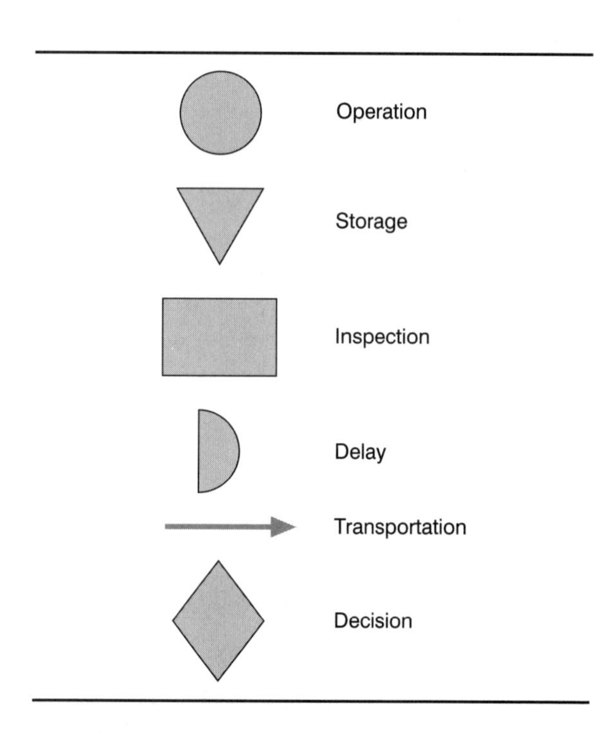

Figure 3.11 Flowchart Symbols

copy can be made utilizing the correct symbols. The symbols can either be placed next to the description of the material, processes, and controls or they can surround the information (Figure 3.12).

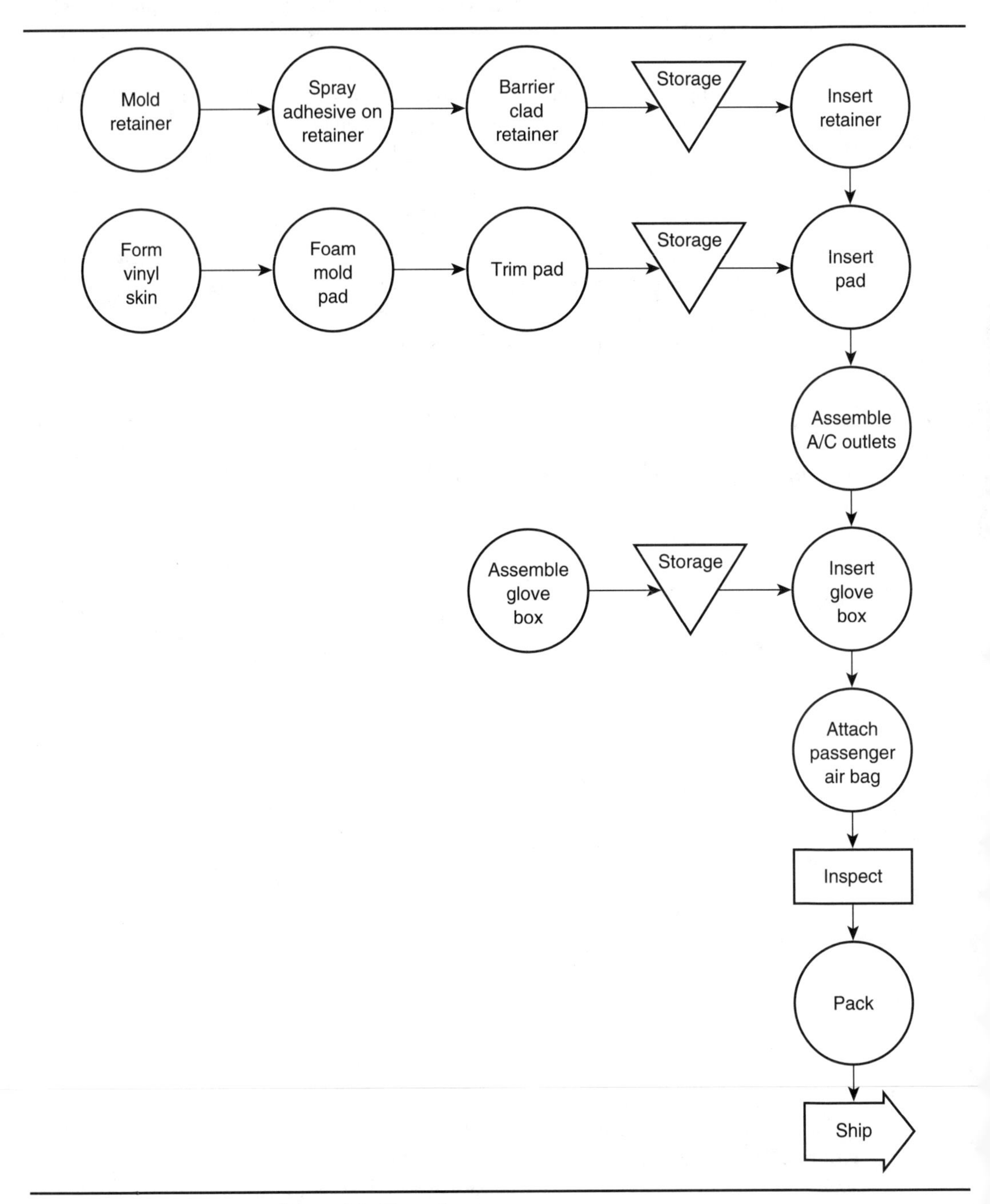

Figure 3.12 Glove Box Assembly Flowchart

EXAMPLE 3.8 Creating a Flowchart

The instrument panel warranty team has decided to create a flowchart of the instrument panel assembly process.

Step 1. Define the Process Steps. First, the team members brainstormed the steps in the assembly process. They doublechecked their steps by observing the actual process. They wrote down each step on 3 × 5 paper.

Step 2. Sort the Steps into the Order of Their Occurrence in the Process. After reconciling their observations with their brainstorming efforts, the team sorted the steps into the order of their occurrence.

Step 3. Place the Steps in Appropriate Flowchart Symbols. With the steps in the correct order, it was a simple task to add the appropriate flowchart symbols and create the chart (Figure 3.12).

Step 4. Evaluate the Steps for Completeness, Efficiency, and Possible Problems. The team reviewed the finished chart for completeness. Several team members were unaware of the complete process. Because it creates a greater understanding of the process, this diagram will be helpful during later problem-solving efforts.

During the flowcharting process, members of the problem-solving team gain a greater understanding of their process. They will also begin to identify possible causes of problems within the process. It is now time to more clearly identify those possible causes and measure the process.

Step 6. Determining Possible Causes

Determining the possible causes of a problem requires that the problem be clearly defined. In step 3, problem solvers used the WHY-WHY diagram to clarify the problem statement. The flowchart in step 5 gave the problem solvers a greater understanding of the processes involved. Now the problem statement can be combined with knowledge of the process to isolate potential causes of the problem. An excellent technique for determining causes is the cause-and-effect diagram.

Technique: Cause-and-Effect Diagrams

The ***cause-and-effect diagram*** is also called the Ishikawa diagram after Kaoru Ishikawa, who developed it, and the fish-bone diagram because the completed diagram resembles a fish skeleton (Figure 3.13). A chart of this type will help *identify causes for nonconforming or defective products or services*. Cause-and-effect diagrams can be used after flowcharts and Pareto charts to identify the cause(s) of the problem.

This chart is useful in a brainstorming session because it organizes the ideas that are presented. Problem solvers benefit from using the chart by being able to separate a large problem into manageable parts. It serves as a visual display to aid the understanding of problems and their causes. The problem or effect is clearly identified on the right-hand side of the chart, and the potential causes of the problem are organized on the left-hand side. The cause-and-effect diagram also allows the session leader to logically organize the possible causes of the problem and to focus on one

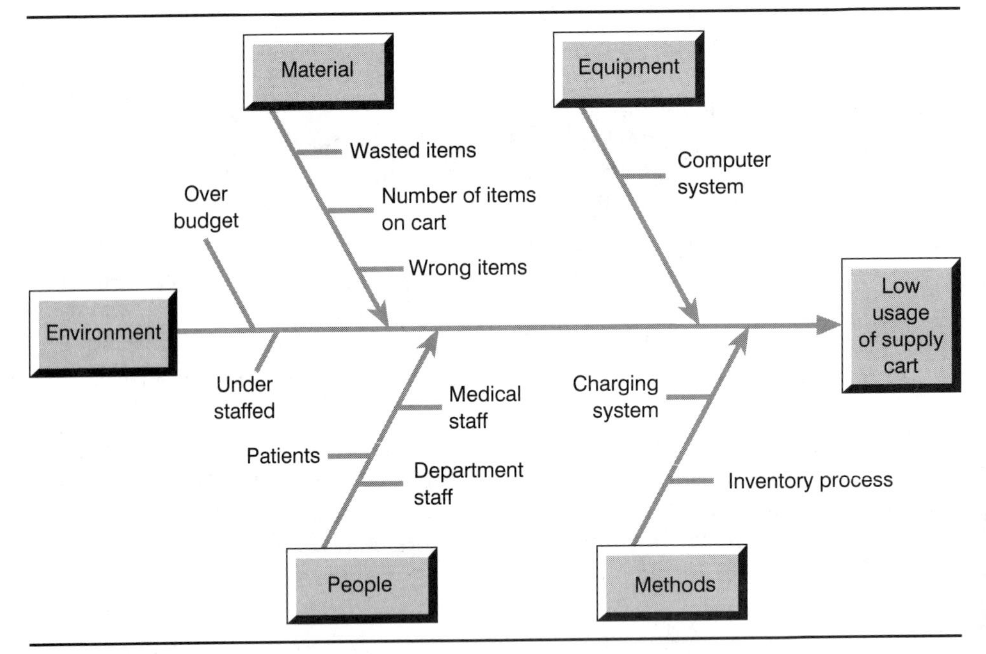

Figure 3.13 A Cause-and-Effect Diagram

area at a time. Not only does the chart permit the display of causes of the problem, it also shows subcategories related to those causes.

To construct a cause-and-effect diagram:

1. Clearly identify the effect or the problem. The succinctly stated effect or problem statement is placed in a box at the end of a line.
2. Identify the causes. Discussion ensues concerning the potential causes of the problem. To guide the discussion, attack just one possible cause area at a time. General topic areas are usually methods, materials, machines, people, environment, and information, although other areas can be added as needed. Under each major area, subcauses related to the major cause should be identified. Brainstorming is the usual method for identifying these causes.
3. Build the diagram. Organize the causes and subcauses in diagram format.
4. Analyze the diagram. At this point, solutions will need to be identified. Decisions will also need to be made concerning the cost-effectiveness of the solution as well as its feasibility.

EXAMPLE 3.9 Constructing a Cause-and-Effect Diagram

At a furniture-manufacturing facility, the upholstery department is having trouble with pattern alignment. They have decided to use a cause-and-effect diagram to help them determine the root causes of pattern misalignment.

Step 1. Clearly Identify the Effect or Problem. The upholstery department team identified the problem as an incorrectly aligned fabric pattern.

Step 2. Identify the Causes. The team, representatives from all of the areas affected by pattern misalignment, brainstormed to identify the causes of pattern misalignment. To guide the discussion, they attacked just one possible cause at a time, starting with methods, materials, machines, people, and environment and ending with information. Under each major area, subclauses related to the major cause were identified.

Step 3. Build the Diagram. During the brainstorming process, the diagram emerged (Figure 3.14).

Step 4. Analyze the Diagram. At this point, solutions need to be identified to eliminate the causes of misalignment. Decisions also need to be made concerning the cost-effectiveness of the solutions as well as their feasibility.

EXAMPLE 3.10 Constructing a Cause-and-Effect Diagram

As the Plastics and Dashes Inc. instrument panel warranty team continued its investigation further, it was determined that defective latches were causing most of the warranty claims associated with the categories of loose instrument panel components and noise.

Step 1. Identify the Effect or Problem. The team identified the problem as defective latches.

Step 2. Identify the Causes. Rather than use the traditional methods, materials, machines, people, environment, and information, this team felt that the potential areas to search for causes related directly to the latch. For that reason, they chose these potential causes: broken, misadjusted, binds, inoperative, loose.

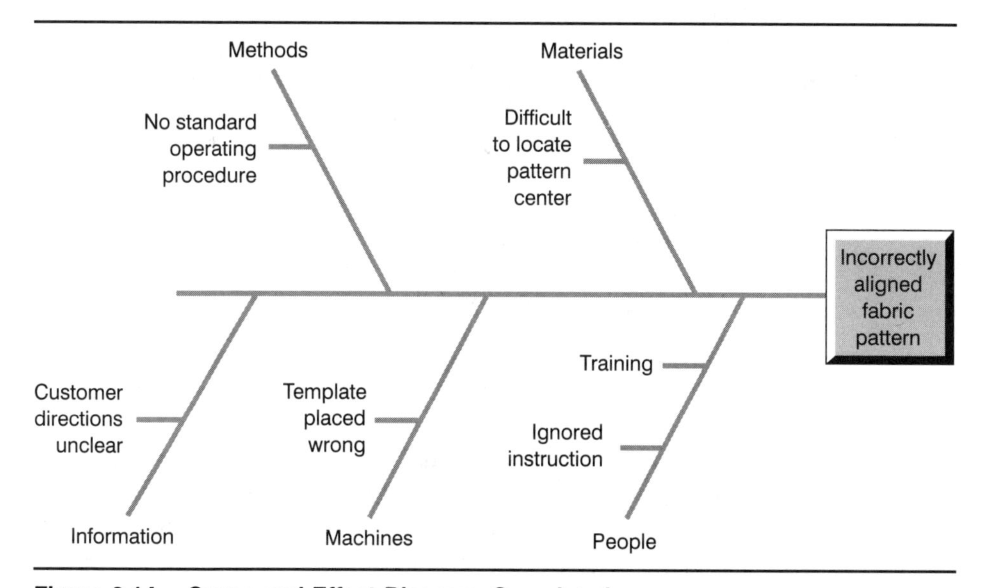

Figure 3.14 Cause-and-Effect Diagram: Completed

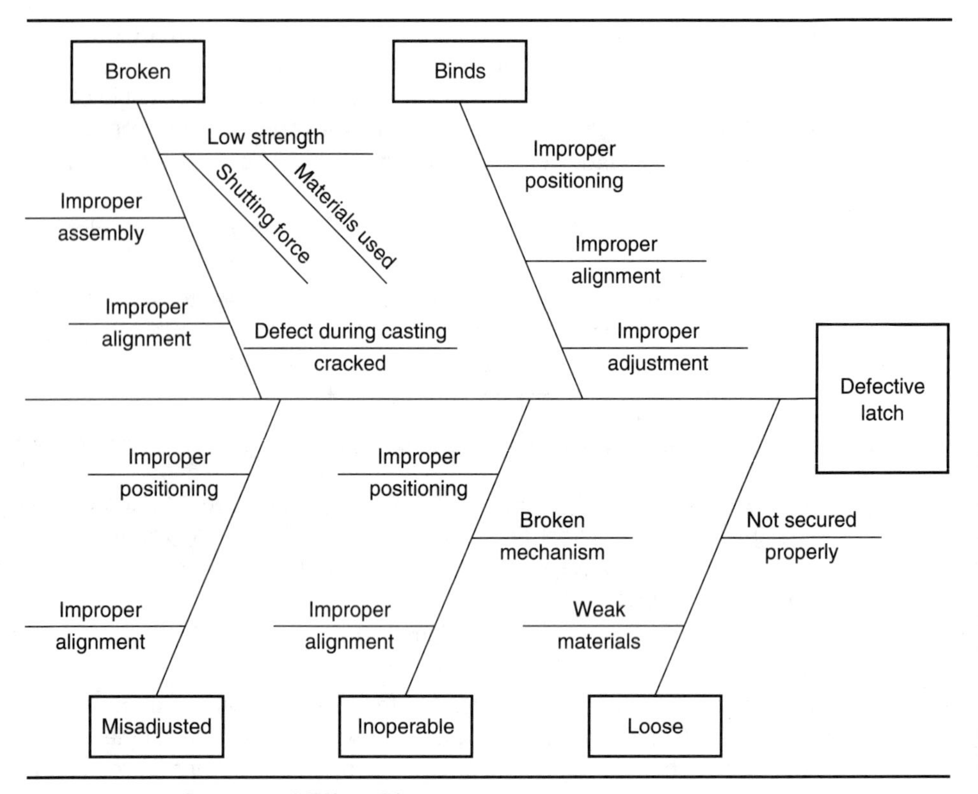

Figure 3.15 Cause-and-Effect Diagram

Step 3. Build the Diagram. The team brainstormed root causes for each category (Figure 3.15).

Step 4. Analyze the Diagram. The team discussed and analyzed the diagram. After much discussion, they came to the following conclusions. Latches that were either broken, misadjusted, or inoperable or those that bind have two root causes in common: improper alignment and improper positioning. Latches that were loose or broken had a root cause of low material strength (those materials supporting the latch were low in strength). From their findings, the team determined that there were three root causes associated with defective latches: improper alignment; improper positioning; and low material strength.

Cause-and-effect diagrams allow us to isolate potential causes of problems. Once identified, these causes need to be investigated by measuring and organizing the data associated with the process. Measuring the process will help refine the investigators' understanding of the problem and help sort out relevant from nonrelevant information. It will also help individuals involved with the problem maintain objectivity. Measuring can be done through the use of check sheets, histograms, scatter diagrams, control charts, and run charts.

Technique: Check Sheets

A ***check sheet*** *is a data-recording device*. It is not to be confused with a checklist, which lists all of the important steps, actions that need to take place, or things that need to be remembered. Given a list of items or events, the user of a check sheet marks down the number of times a particular item or event occurs. In essence, the user checks off occurrences. A check sheet has many applications and can be custom-designed by the user to fit the needs of the situation.

EXAMPLE 3.11 Making a Check Sheet

To gather information about errors and nonconformities that are occurring in their shop, the upholsterers at the furniture-manufacturing facility from Example 3.9 have decided to utilize check sheets. They brainstormed a list in a meeting of representatives from the various areas in the plant. This check sheet will be used by employees to record errors and nonconformities. When they encounter a problem, the employees are to check the appropriate item on the check sheet (Table 3.1). Once a week, these sheets will be collected and turned over to the problem-solving teams to be tallied. The information from these sheets will help direct the problem-solving efforts of the teams.

Table 3.1 Check Sheet

Nonconformity	*Number*
Item Dirtied in Shipping	III
Item Shipped to Incorrect Address	𝍸 II
Customer Billed Incorrectly	
Labor-Hours Calculated Incorrectly	
Material Needs Miscalculated	
Sewing Machine Down	I
Incorrect Setup	
Material Error	
Pattern Cut Incorrectly	II
Loose Threads	𝍸
Incorrect Hemming	
Material Flaws	
Stitching Flaws	IIII
Framing Loose	
Webbing Overstretched	
Item Parts Missing or Lost	
Customer Changes Mind	
Pattern Alignment Errors	
Color Mismatch	I
Trim Errors	
Miscellaneous	

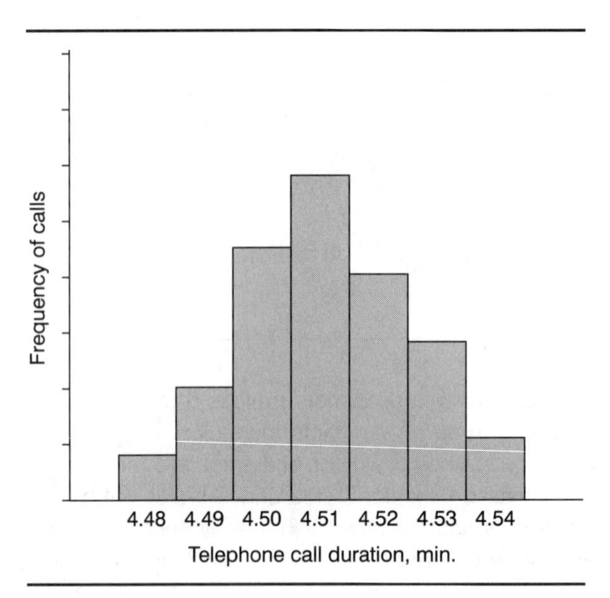

Figure 3.16 A Histogram

Technique: Histograms

*A **histogram** is a graphical summary of the frequency distribution of the data.* When measurements are taken from a process, they can be summarized by using a histogram. Data are organized in a histogram to allow those investigating the process to see any patterns in the data that would be difficult to see in a simple table of numbers. The data are separated into classes in the histogram. Each interval on a histogram shows the total number of observations made in each separate class. Histograms display the variation present in a set of data taken from a process (Figure 3.16). Histograms are covered in more detail in Chapter 4.

Technique: Scatter Diagrams

*The **scatter diagram** is a graphical technique that is used to analyze the relationship between two different variables.* Two sets of data are plotted on a graph. The independent variable—i.e., the variable that can be manipulated—is recorded on the x axis. The dependent variable, the one being predicted, is displayed on the y axis. From this diagram, the user can determine if a connection or relationship exists between the two variables being compared. If a relationship exists, then steps can be taken to identify process changes that affect the relationship. Figure 3.17 shows different interpretations of scatter diagrams.

EXAMPLE 3.12 Creating a Scatter Diagram

Shirley is the setup operator in the shrink-wrap area. In this area, 5-ft-by-5-ft cartons of parts are sealed with several layers of plastic wrap before being loaded on the trucks. Shirley's job is to load the plastic used in the shrink-wrapping operation onto the shrink-wrap machine, set the speed at which the unit will rotate, and set

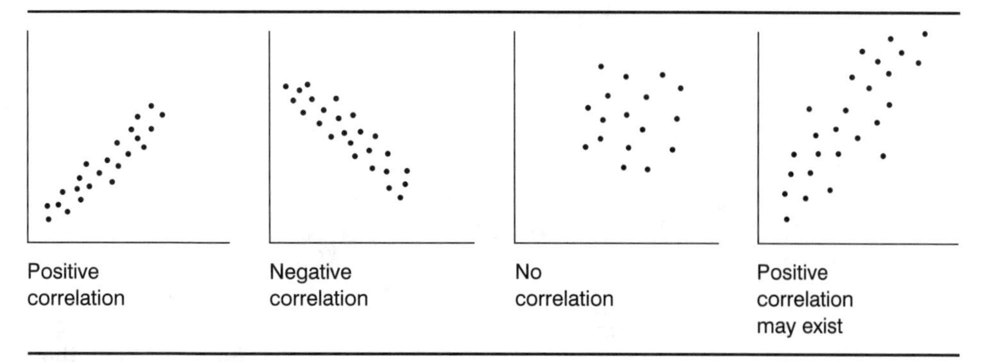

Figure 3.17 Scatter Diagram Interpretations

the tension level on the shrink-wrap feeder. To understand the relationship between the tension level on the feeder and the speed of the rotating mechanism, Shirley has created a scatter diagram.

The rotator speed is most easily controlled, so she has placed the most typically used speed settings (in rpm) on the x axis. On the y axis, she places the number of tears (Figure 3.18).

The diagram reveals a positive correlation: As the speed increases, the tension has to be reduced in order to prevent the wrap from tearing. Using the diagram, Shirley is able to determine the optimal speed for the rotor and the best tension setting.

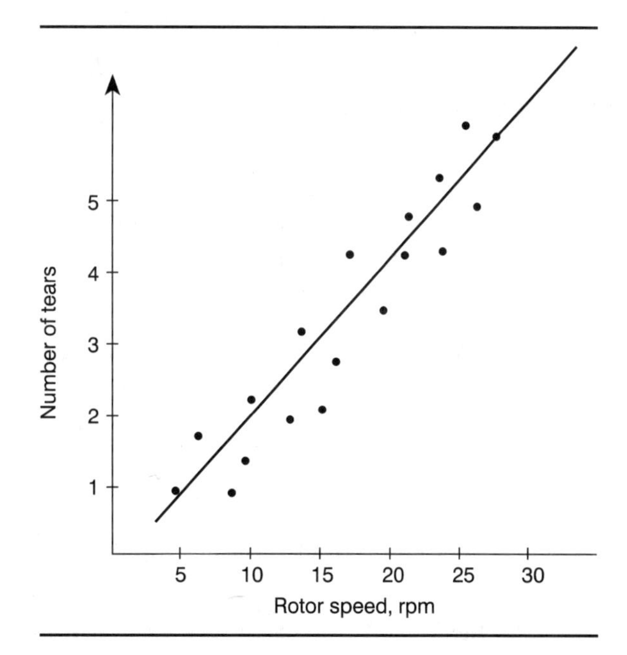

Figure 3.18 The Scatter Diagram for Example 3.12

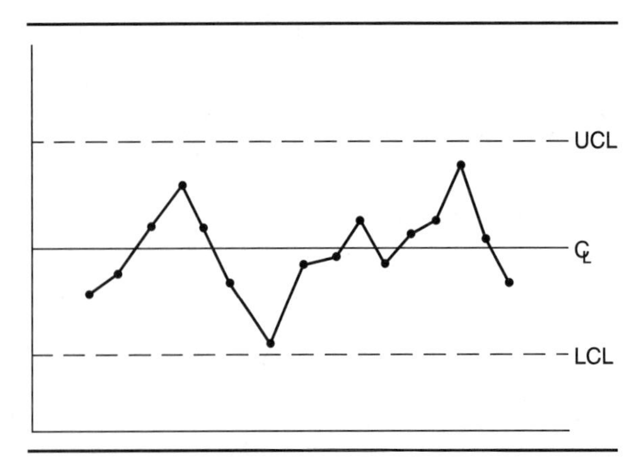

Figure 3.19 A Control Chart, Showing Centerline, Upper Control Limits, and Lower Control Limits

Technique: Control Charts

A ***control chart*** *is a chart with a centerline showing the average of the data produced.* It has upper and lower control limits that are based on statistical calculations (Figure 3.19). It is used to determine process centering and process variation and to locate any unusual patterns or trends in the data. Control charts are covered in more detail in Chapters 5, 6, 7, and 9.

Technique: Run Charts

Run charts are similar to control charts. *They follow a process over time, reflected on the* x *axis.* The run chart is very good at reflecting trends in the measurements over time. Normally, one measurement is taken at a time and graphed on the chart. The y axis shows the magnitude of that measurement.

EXAMPLE 3.13 Making a Run Chart

Joe and Sue want to use a run chart to track the gasoline mileage that their truck gets. On their chart, the y axis shows the miles per gallon that they calculate at each fill-up on Monday. The x axis reflects the dates that they have filled up the tank with gas (Figure 3.20). On their chart they have also noted any unusual uses for their truck that might affect gas mileage. Examples of these uses include pulling a horse trailer, using the truck to pull out stumps, and a long freeway trip to move some lightweight furniture. By using this chart, they hope to establish a feel for what the usual gas mileage is for the truck. Knowing this information, they will be able to make intelligent decisions on when to have the truck serviced.

DO

Step 7. Selecting and Implementing the Solution

We have been applying problem-solving techniques to find the root cause of a problem. Once the cause has been identified, it is time to begin proposing potential

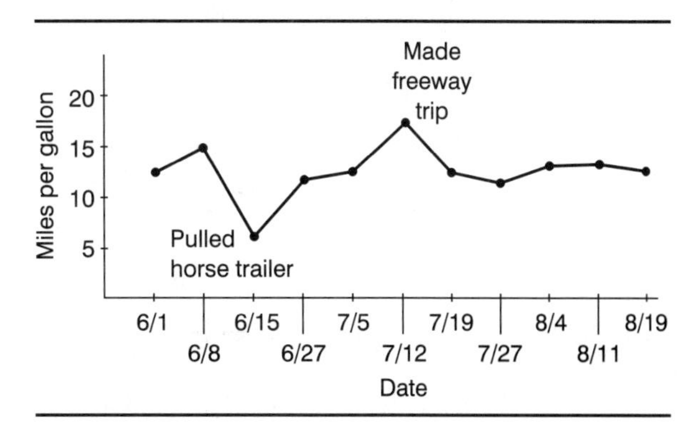

Figure 3.20 A Run Chart

solutions. This begins the Do section of the PDSA cycle. This is the portion of the cycle that attracts everyone's attention. So great is the desire to act that many problem solvers are tempted to reduce the amount of time spent on planning to virtually nothing. The temptation to immediately propose solutions must be ignored. The best solutions are those that solve the true problem, and they are found only after the root cause of that problem has been identified. The most significant portion of the problem-solving effort must be concentrated in the Plan phase.

It is important to recognize that applying these techniques does not mean that taking care of the immediate problem should be ignored. Immediate action should be taken to rectify any situation that does not meet the customer's reasonable needs, requirements, and expectations. These quick fixes are just that, a quick fix of a problem for the short term; they simply allow time for a long-term solution to be found. In no situation should a quick fix be considered the end of a problem. Problems are solved only when recurrences do not happen.

Selecting and implementing the solution is a matter of the project team's choosing the best solution for the problem under examination. The solution should be judged against four general criteria:

1. The solution should be chosen on the basis of its potential to prevent a recurrence of the problem. A quick or short-term fix to a problem will only mean that time will be wasted in solving this problem again when it recurs in the future.
2. The solution should address the root cause of the problem. A quick or short-term fix that focuses on correcting the symptoms of a problem will waste time because the problem will recur in the future.
3. The solution should be cost-effective. The most expensive solution is not necessarily the best solution for the company's interests. Solutions may necessitate determining the company's future plans for a particular process or product. Major changes to the process, system, or equipment may not be an appropriate solution for a process or product that will be discontinued shortly.

Technological advances will need to be investigated to determine if they are the most cost-effective solutions.

4. The solution should be capable of being implemented within a reasonable amount of time. A timely solution to the problem is necessary to relieve the company of the burden of monitoring the current problem and its associated quick fixes.

EXAMPLE 3.14 Solutions

Because the team was able to identify the root causes of the glove box latch problem as improper alignment, improper positioning, and low material strength, they decided to make the following changes part of their solution:

1. Redesign the glove box latch. This solution was chosen to counteract low material strength.
2. Reposition the glove box door, striker, and hinge. This solution was chosen to counteract improper positioning and alignment. They also hoped this change would eliminate potential squeaks and rattles.
3. Reinforce the glove box latch. This solution was chosen to counteract breakage. By increasing the material at the latch position on the glove box door, they hoped to eliminate breakage. They also decided to use a stronger adhesive to reinforce the rivets securing the latch to the door.

Implementing the solution is often done by members of the problem-solving team. Critical to ensuring the success of the solution implementation is assigning responsibilities to specific individuals and holding them accountable for accomplishing the task. Knowing who will be doing what and when will help ensure that the project stays on track.

Technique: Force-Field Analysis

A ***force-field analysis*** *is a chart that helps teams separate the driving forces and the restraining forces associated with a complex situation.* These easy-to-develop charts help a team determine the positive or driving forces that are encouraging improvement of the process as well as the forces that restrain improvement. When the improvements possibly outweigh the restraining forces, then a force-field analysis can be used to encourage action on a problem. Teams may also choose to use force-field analysis as a source of discussion issues surrounding a particular problem or opportunity. Once the driving and restraining forces have been identified, the team discusses how to enhance the driving forces and remove the restraining forces. This sort of brainstorming leads to potential solutions to be investigated.

EXAMPLE 3.15 Using Force-Field Analysis

At a local bank, a team has gathered to discuss meeting the customers' demands for improved teller window service. One of the major customer desires, as revealed by a survey, is to have more windows open at 9 A.M. So far the meeting has been a less-than-organized discussion of why more tellers are not at their windows at 9. To remove some of the finger-pointing and blame-laying, the leader has decided to use force-field analysis to focus the discussion.

The team members identify the following as driving forces:

1. Improving customer service by having more windows open
2. Creating shorter lines for customers by serving more customers at one time
3. Relieving the stress of morning teller activities

Notice that the team has identified both internal (less stress) and external (happier customers) factors as the driving forces behind wanting to solve this problem.

The discussion of restraining forces is much more heated. Management has been unaware of the number of activities that need to be completed before opening the teller windows at 9 A.M. Slowly, over time, the amount of paperwork and number of extra duties have increased. Since this change has been gradual, it has gone unnoticed by all but the tellers. They accept it as part of their job but are under considerable stress to complete the necessary work by 9 A.M., given that their workday starts at 8:30 A.M. The restraining forces identified by the tellers are the following:

1. Money must be ready for Brinks pickup.
2. Audit department needs balance sheet by 9 A.M.
3. Balance sheet is complicated, cumbersome, and inadequate.
4. Comptroller needs petty cash sheets, payroll, and other general ledger information.
5. Day begins at 8:30.

A variety of suggestions are given to remove the restraining forces, including changing the starting and ending times of the day, adding more people, and telling the audit and comptroller departments to wait until later in the day. Table 3.2 presents the completed force-field diagram.

Since it isn't economically feasible to add more people or increase the work hours, emphasis is now placed on asking the audit and comptroller departments what their needs are. Interestingly enough, while the bank has surveyed their external customers, they have not investigated the needs of the people working within the bank. In the discussions that follow with the audit's and comptroller's offices, it is determined that only the balance sheet has to be completed by 9 A.M. The other paperwork can be received as late as 11 and not affect the performance of the other departments. This two-hour time space will allow the tellers to open their windows at 9 for the morning rush and still complete the paperwork for the comptroller's office before 11. Having achieved a successful conclusion to the project, the

Table 3.2 Force-Field Analysis

Driving Forces	*Restraining Forces*
1. Improving Customer Service by Having More Windows Open	1. Money Must Be Ready for Brinks Pickup.
2. Creating Shorter Lines for Customers by Serving More Customers at One Time	2. Audit Department Needs Balance Sheet by 9 A.M.
3. Relieving the Stress of Morning Teller Activities	3. Balance Sheet Is Complicated, Cumbersome, and Inadequate.
	4. Comptroller Needs Petty Cash Sheets, Payroll, and Other General Ledger Information.
	5. Day Begins at 8:30 A.M.

Actions
1. Change the Starting and Ending Times of the Day.
2. Add More Tellers.
3. Tell the Audit and Comptroller Departments to Wait until Later in the Day.

team decides to work on reducing the complexity of the balance sheet as its next problem-solving activity. Perhaps a simpler balance sheet could reduce stress even more!

STUDY

Step 8. Evaluating the Solution: The Follow-up

Once implemented and given time to operate, problem-solving actions are checked to see if the problem has truly been solved. During the Study stage we study the results and ask, What did we learn? Is the solution we've chosen working? To determine if the solution has worked, prior data collected during the analysis phase of the project should be compared with present data taken from the process. Control charts, histograms, and run charts can be used to monitor the process. If these formats were used in the original problem analysis, a direct comparison can be made to determine how well the solution is performing. If the solution is not correcting the problem, then the process should begin again to determine a better solution.

EXAMPLE 3.16 Evaluating the Solution

The instrument panel warranty team implemented their solutions and used the measures of performance developed in Example 3.7 to study the solutions in order to determine whether or not the changes were working. The Pareto chart in

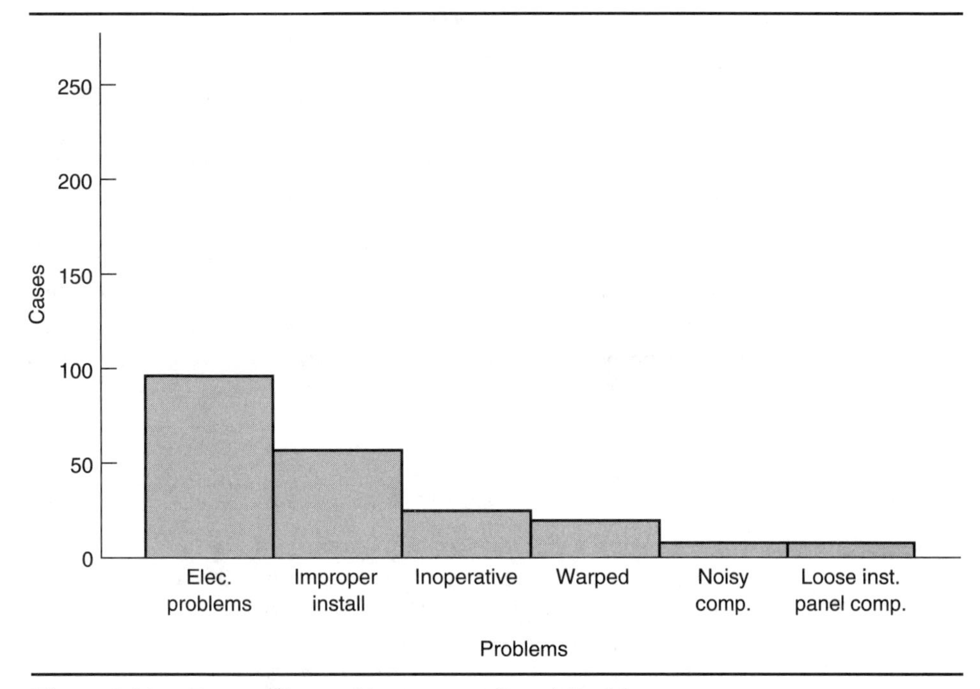

Figure 3.21 Pareto Chart of Instrument Panel Problems

Figure 3.21 provides information about warranty claims made following the changes. When this figure is compared to Figure 3.5, showing warranty claims before the problem-solving team went into action, the improvement is obvious. Warranty costs declined in proportion to the decreased number of claims, to just under $25,000. The process measure also tracked the length of time to implement the changes, a very speedy five days.

ACT

Step 9. Ensuring Permanence

The final stage, Action, involves making the decision to adopt the change, abandon it, or repeat the problem-solving cycle. If the change is adopted, then efforts must be made to ensure that the new methods have been established so that the new level of quality performance can be maintained. Now that a follow-up investigation has revealed that the problem has been solved, it is important that improved performance continue. From Figure 3.1 it can be seen that "ensuring permanence" is part of the action phase. This phase of the quality-improvement process exists to ensure that the new controls and procedures stay in place. It is easy to believe that the "new and better" method should be utilized without fail; however, in any situation where a change has taken place there is a tendency to return to old methods, controls, and procedures when stress is increased. It is a bit like switching to an automatic shift after having driven a manual shift car for a number of years. Under normal driving conditions, the driver will not utilize her left leg to operate a clutch pedal that does not exist in an automatic gear shift car. But place that same individual in an emergency stop situation, and chances are she will attempt to activate a nonexistent clutch pedal at the same time as the brake. Under stress, people have a tendency to revert to their original training.

To avoid a lapse back into old routines and methods, controls must be in place to remind people of the new method. Extensive training and short follow-up training are very helpful in ingraining the new method. Methods must be instituted and follow-up checks must be put in place to prevent problem recurrences from lapses to old routines and methods.

EXAMPLE 3.17 Standardize Improvements

Plastics and Dashes asks each team to write up a brief but formal discussion of the problem-solving steps their team took to eliminate the root causes of problems. These discussions are shared with others involved in problem solving to serve as a guide for future problem-solving efforts.

Step 10. Continuous Improvement

A project team's tasks don't end with the solution of one particular problem. The quality- and productivity-improvement process never ends. Once a problem is solved, teams are reformed to "do it all over again," this time with a new problem, opportunity, or project. Only through continual improvement can a company hope to move toward the future, improve its customer base, and ensure future profits.

SUMMARY

Teaching the tools of quality improvement and problem solving is actually the easiest part of the quality-improvement process. Helping individuals and groups apply those techniques in a problem-solving format is critical and difficult. Upper-management involvement in selecting issues to be investigated is important to the success of a quality-improvement program. Brainstorming and Pareto analyses also help identify where problem-solving efforts should be concentrated. Teach people the techniques and then use brainstorming to encourage them to uncover problems in their own area and begin to solve them. Figure 3.22 lists the problem-solving tools by type: problem/opportunity development tools, quantitative tools, control or tracking tools, and appraisal tools. Upper management should be involved in the entire process, from education to implementation. They should be the ones providing their people with a push in the right direction.

Problem/Opportunity Development Tools
- Brainstorming
- WHY-WHY Diagrams
- Flowcharts
- Force-Field Analysis

Quantitative Tools
- Histograms and Statistics
- Cause-and-Effect Diagrams
- Pareto Analysis

Control or Tracking Tools
- Control Charts
- Run Charts
- Scatter Diagrams

Appraisal Tools
- Check Sheets

Figure 3.22 Problem-Solving Tools

Lessons Learned

1. Problem solving is the isolation and analysis of a problem and the development of a permanent solution. Problem solving should be logical and systematic.
2. The following steps should be taken during the problem-solving process:

   Step 1. Recognize the problem; establish the priorities.
   Step 2. Form quality-improvement teams.
   Step 3. Define the problem.
   Step 4. Develop performance measures.
   Step 5. Analyze the problem/process.
   Step 6. Determine possible causes.
   Step 7. Select and implement the solution.
   Step 8. Evaluate the solution and institute the follow-up.
   Step 9. Ensure permanence.
   Step 10. Continuous improvement.

3. The following are techniques used in problem-solving: brainstorming, Pareto analysis, WHY-WHY diagrams, flowcharts, force-field analysis, cause-and-effect diagrams, check sheets, histograms, scatter diagrams, control charts, and run charts.
4. Problem solvers are tempted to propose solutions before identifying the problem and performing an in-depth study of the situation. Adhering to a problem-solving method avoids this tendency.
5. Brainstorming is designed for idea generation. Ideas should not be discussed or criticized during a brainstorming session.

6. Flowcharts are powerful tools that allow problem solvers to gain in-depth knowledge of the process.
7. Cause-and-effect diagrams enable problem solvers to identify the root causes of succinctly stated problems.
8. Steps must be taken to ensure that the new methods or changes to the process are permanent. ■

Chapter Problems

Problem Solving

1. Describe the 10 steps of problem solving.
2. An orange juice producer has found that the fill weights (weight of product per container) of several of its orange juice products do not meet specifications. If the problem continues, unhappy customers will stop buying their product. Outline the steps that they should take to solve this problem. Provide as much detail as you can.
3. Bicycles are being stolen at a local campus. Campus security is considering changes in bike rack design, bike parking restrictions, and bike registration to try to reduce thefts. Thieves have been using hacksaws and bolt cutters to remove locks from the bikes. Create a problem statement for this situation. How will an improvement team use the problem statement?

Pareto Charts

4. During the past month, a customer-satisfaction survey was given to 200 customers at a local fast-food restaurant. The following complaints were lodged:

Complaint	*Number of Complaints*
Cold Food	105
Flimsy Utensils	20
Food Tastes Bad	10
Salad Not Fresh	94
Poor Service	15
Food Greasy	9
Lack of Courtesy	5
Lack of Cleanliness	25

   Create a Pareto chart with this information.

5. A local bank is keeping track of the different reasons people phone the bank. Those answering the phones place a mark on their check sheet in the rows

most representative of the customers' questions. Given the following check sheet, make a Pareto diagram:

Credit Card Payment Questions	245
Transfer Call to Another Department	145
Balance Questions	377
Payment Receipt Questions	57
Finance Charges Questions	30
Other	341

Comment on what you would do about the high number of calls in the "Other" column.

6. Once a Pareto chart has been created, what steps would you take to deal with the situation given in Problem 5 in your quality-improvement team?

7. Two partners in an upholstery business are interested in decreasing the number of complaints from customers who have had furniture reupholstered by their staff. For the past six months, they have been keeping detailed records of the complaints and what had to be done to correct the situations. To help their analysis of which problems to attack first, they decide to create several Pareto charts. Use the following table to create Pareto charts for the number of complaints, the percentage of complaints, and the dollar loss associated with the complaints. Discuss the charts.

Category	*Number of Complaints*	*Percent of Complaints*	*Dollar Loss*
Loose Threads	14	28	294.00
Incorrect Hemming	8	16	216.00
Material Flaws	2	4	120.00
Stitching Flaws	6	12	126.00
Pattern Alignment Errors	4	8	240.00
Color Mismatch	2	4	180.00
Trim Errors	6	12	144.00
Button Problems	3	6	36.00
Miscellaneous	5	10	60.00
	50	100	

8. Create a Pareto chart for the following safety statistics from the security office of a major apartment building. What does the chart tell you about their safety record and their efforts to combat crime? Where should they concentrate their efforts? In this example, does a low number of occurrences necessarily mean those areas should be ignored? Does the magnitude of the injury done mean anything in your interpretation?

	1989	1990	1991	1992	1993	1994	1995	1996	1997	1998	1999
Homicide	0	0	0	0	0	0	0	0	0	1	0
Aggravated Assault	2	1	0	1	3	4	5	5	6	4	3
Burglary	80	19	18	25	40	26	23	29	30	21	20
Grand Theft Auto	2	2	6	15	3	0	2	10	5	2	3
Theft	67	110	86	125	142	91	120	79	83	78	63
Petty Theft	37	42	115	140	136	112	110	98	76	52	80
Grand Theft	31	19	16	5	4	8	4	6	2	4	3
Alcohol Violations	20	17	16	16	11	8	27	15	10	12	9
Drug Violations	3	4	2	0	0	0	0	2	3	2	1
Firearms Violations	0	0	0	0	0	1	0	0	1	0	0

9. Max's Barbeque Tools manufactures top-of-the-line barbeque tools. These tools are sold in sets that include knives, long-handled forks, and spatulas. During the past year, nearly 240,000 tools have passed through final inspection. Create a Pareto chart with the information provided in Figure 1. Comment on whether or not any of the categories may be combined. What problems should Max's tackle first?

Brainstorming

10. Apply the 10 problem-solving steps described in this chapter to a problem you face(d) at work or in school.
11. Brainstorm 10 reasons why the university computer might malfunction.
12. Brainstorm 10 reasons why a customer may not feel the service was adequate at a department store.

WHY-WHY Diagrams

13. Create a WHY-WHY diagram for how you ended up taking this particular class.
14. A mail-order company has a goal of reducing the amount of time a customer has to wait in order to place an order. Create a WHY-WHY diagram about waiting on the telephone. Now that you have created the diagram, how would you use it?
15. Apply a WHY-WHY diagram to a project you face at work or in school.

Flowcharts

16. Create a flowchart for registering for a class at your school.
17. Create a flowchart for solving a financial aid problem at your school.
18. WP Uniforms provides a selection of lab coats, shirts, trousers, uniforms, and outfits for area businesses. For a fee, WP Uniforms will collect soiled garments once a week, wash and repair these garments, and return them the following week while picking up a new batch of soiled garments.

    At WP Uniforms, shirts are laundered in large batches. From the laundry, these shirts are inspected, repaired, and sorted. To determine if the process

Final Inspection
Department 8
February

Defect	Totals	Defect	Totals
Bad handle rivets	1753	Buff concave	33
Grind in blade	995	Bent	32
Etch	807	Hit handle	27
Hit blades	477	Burn	16
Bad tines	346	Wrap	15
Bad steel	328	Burned handles	11
Rebend	295	Re-color concave	10
Cracked handle	264	Rivet	9
Rehone	237	Handle color	6
Cracked steel	220	Hafting marks handles	5
Nicked and scratched	207	Raw fronts	4
Haft at rivets	194	Pitted blades	4
High handle rivets	170	Scratched blades	4
Dented	158	Raw backs	3
Pitted bolsters	130	Water lines	3
Vendor rejects	79	Open handles	3
Holder marks	78	Seconds	3
Heat induct	70	Reruns	2
Open at rivet	68	DD edge	2
Finish	56	Bad/bent points	1
Edge	54	Stained rivets	1
Honing	53	Seams/holes	1
Cloud on blades	46	Open steel	1
Hafting marks steel	43	Narrow blades	1
Scratched rivets	41		
Burrs	41	Total inspected	238,385
High-speed buff	41	Total accepted	230,868
Stained blades	35	Total rejected	7,517
Crooked blades	34		

Figure 1 Problem 9

can be done more effectively, the employees want to create a flowchart of the process. They have brainstormed the following steps and placed them in order. Create a flowchart with their information. Remember to use symbols appropriately.

Shirts arrive from laundry.
Pull shirts from racks.
Remove shirts from hangers.
Inspect.
Ask: Does shirt have holes or other damage?
Make note of repair needs.
Ask: Is shirt beyond cost-effective repair?
Discard shirt if badly damaged.
Sort according to size.
Fold shirt.
Place in proper storage area.
Make hourly count.

19. Create a flowchart using symbols for the information provided in the table below.

Materials	Process	Controls
	Buy Filter	
	Buy Oil	
Wrench to Unscrew Oil Plug		
Oil Filter Wrench		
Oil Drain Pan		
	Start Car and Warm It Up (5 min)	
	Shut Off Car When Warm	
	Utilize Pan for Oil	
	Remove Plug	
	Drain Oil into Pan	
	Remove Old Filter with Wrench When Oil Is Completely Drained	
	Take Small Amount of Oil and Rub It around Ring on New Filter	
	Screw New Filter On by Hand	
	Tighten New Filter by Hand	
	Add Recommended Number of New Quarts of Oil	
	Start Engine and Run for 5 min to Circulate Oil	
		Look for Leaks While Engine Is Running
	Wipe Disptick Off	
		Check Oil Level
		Add More if Low
	Place Used Oil in Recycling Container	
	Take Used Oil to Recycling Center	
	Dispose of Filter Properly	

Force-field Diagrams

20. Create a force-field diagram for Problem 3, concerning bike thefts.

Cause-and-Effect Diagrams

21. Create cause-and-effect diagrams for (*a*) a car that won't start, (*b*) an upset stomach, and (*c*) a long line at the supermarket.
22. A customer placed a call to a mail-order catalog firm. Several times the customer dialed the phone and received a busy signal. Finally, the phone was answered electronically, and the customer was told to wait for the next available operator. Although it was a 1-800 number, he found it annoying to wait on the phone until his ear hurt. Yet he did not want to hang up for fear he would not be able to get through to the firm again. Using the problem statement "What makes a customer wait?" as your base, brainstorm to create a cause-and-effect diagram. Once you have created the diagram, how would you use it?

CASE STUDY 3.1
Problem Solving

PART 1

WP Inc. is a manufacturer of small metal parts. Using a customer's designs, WP creates the tools, stamps, bends, forms the metal parts, deburrs, washes, and ships the parts to the customer. WP has been having a recurring problem with the automatic parts washer, used to wash small particles of dirt and oil from the parts. The parts washer (Figure 1) resembles a dishwasher. The recurrent problem involves the spray nozzles, which frequently clog with particles, causing the parts washer to be shut down.

Although the parts washer is not the most time-consuming or the most important operation at WP, it has been one of the most troublesome. The nozzle-clogging problem causes serious time delays, especially since almost every part manufactured goes through the parts washer. The parts-washing operation is a critical aspect of WP's

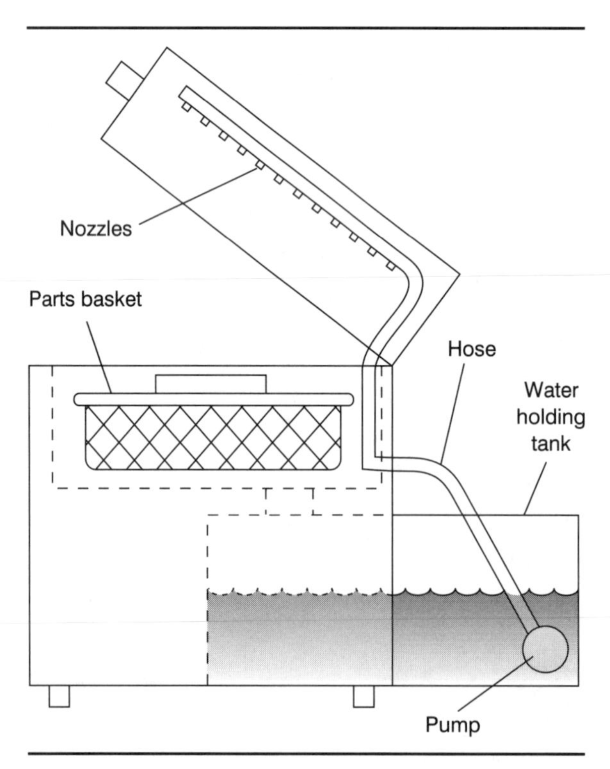

Figure 1 Automatic Parts Washer

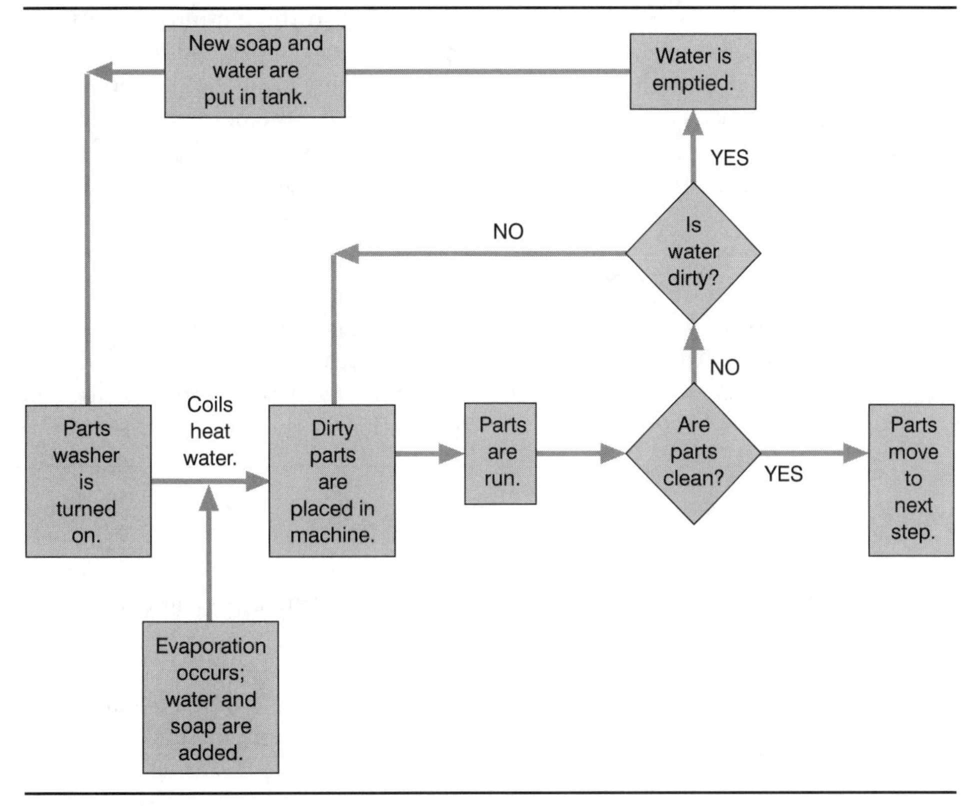

Figure 2 Process Flowchart

quality process. It is also the only alternative for cleaning parts since EPA regulations no longer allow the use of a vapor degreaser using tricloroethane 1,1,1. To better understand the process, study the process flowchart seen in Figure 2.

Assignment

Create a problem-solving group and design a problem statement. Follow the problem-solving steps from Figure 3.2 throughout this case.

PART 2

Intensive questioning of the operator monitoring the process yields the following information:

For about a month, the operator has been observing the parts washer and recording the behavior of the nozzles. A significant portion of the time, the nozzles do not spray as freely as they should. On seven occasions, the nozzles clogged completely. Two of these clogs happened midday: the remainder occurred at the end of the day.

During the month, the nozzles did not clog in the morning. Each time the nozzles clogged, the operator took the opportunity to remove and inspect them. It appears that small particles become clogged in the tiny orifices of the nozzles. This restricts the flow of the soap-and-water solution through the nozzles, causing in turn the nozzles to clog further and eventually shutting the parts washer down. When dry, the small particles are white and flakelike in appearance.

Assignment

Return to your problem-solving group and brainstorm possible identities for the particles. Revise your problem-solving statement on the basis of what you learned above. Detail the problem-solving steps that you would follow to discover the cause of clogged nozzles. Given the information in the case, be as specific as possible.

PART 3

It is possible that the particles could be one or several of the following: (1) hard-water buildup, (2) calcium, (3) chips from the parts, (4) paint chips, (5) soap flakes, (6) something caused by a chemical reaction, (7) some type of gravel or dirt. Add your group's ideas to this list.

Further questioning of the operator discloses that the problem occurs most often after the old cleansing solution has been drained from the tank and new solution put in. This doesn't seem to make sense because, if the cleansing solution has just been recently changed, the liquid in the tank should be free of particles.

Although this doesn't seem to make sense, tests are conducted in which new cleansing solution is put in the tank. After running the parts washer only two minutes, the nozzles are removed and inspected. To everyone's surprise, many particles have been collected by the nozzles. A study of the particles establishes that these particles are soap! Now it must be determined where these soap particles are coming from.

At the suggestion of one of the problem-solving team members, the tank is drained. The bottom of the tank is found to be coated with a layer of hard soap. Your team has decided to return to the conference room and use a cause-and-effect diagram to guide a brainstorming session to establish why there is a buildup of soap on the bottom of the tank.

While constructing a cause-and-effect diagram, note the following:

1. The water in the tank is heated, causing evaporation to occur. This in turn increases the concentration of the soap content in the solution. Following the written procedures, when the operator notes a decrease in the fluid level in the tank, he or she adds more of the soap-and-water solution. This activity increases the soap concentration even more. Once the soap reaches a certain concentration level, it can no longer be held in solution. The soap particles then precipitate to the bottom of the tank. In the cleaning process, the

tank is drained and new solution is put in. This causes a disturbance to the film of soap particles at the bottom of the tank, and the particles become free-floating in the tank.

2. The soap dissolves in the water best at an elevated temperature. A chart of the daily temperatures in the parts washer shows that the tank has not been held at the appropriate temperature. It has been too low.
3. No filtering system exists between the tank and the nozzles.

Assignment

Using the above information, create a cause-and-effect diagram for the problem of soap buildup on the bottom of the tank.

Assignment

Draw up a force-field analysis listing potential driving and restraining forces involved in correcting this problem. Brainstorm potential corrective actions.

Assignment

Combine the information from the case and your solutions to the assignments to create a management summary detailing the problem-solving steps taken in this case. Relate your summary to Figure 3.2.

CASE STUDY 3.2
Process Improvement

This case is the beginning of a four-part series of cases involving process improvement. The other cases are found at the end of Chapters 4, 5, and 6. Data and calculations for this case establish the foundation for the future cases; however, it is not necessary to complete this case in order to complete and understand the cases in Chapters 4, 5, and 6. Completing this case will provide insight into the use of problem-solving techniques in process improvement. The case can be worked by hand or with the software provided.

PART 1

Figure 1 provides the details of a simplified version of a bracket used to hold a strut in place on an automobile. Welded to the auto-body frame, the bracket cups the strut and secures it to the frame with a single bolt and a lock washer. Proper alignment is necessary for both smooth installation during assembly and future performance. For mounting purposes the left-side hole, A, must be aligned on center with the right-side hole, B. If the holes are centered directly opposite each other, in perfect alignment, then the angle between hole centers will measure 0°. The bracket is created by passing coils of flat steel through a series of progressive dies. As the steel moves through the press, the bracket is stamped, pierced, and finally bent into appropriate shape.

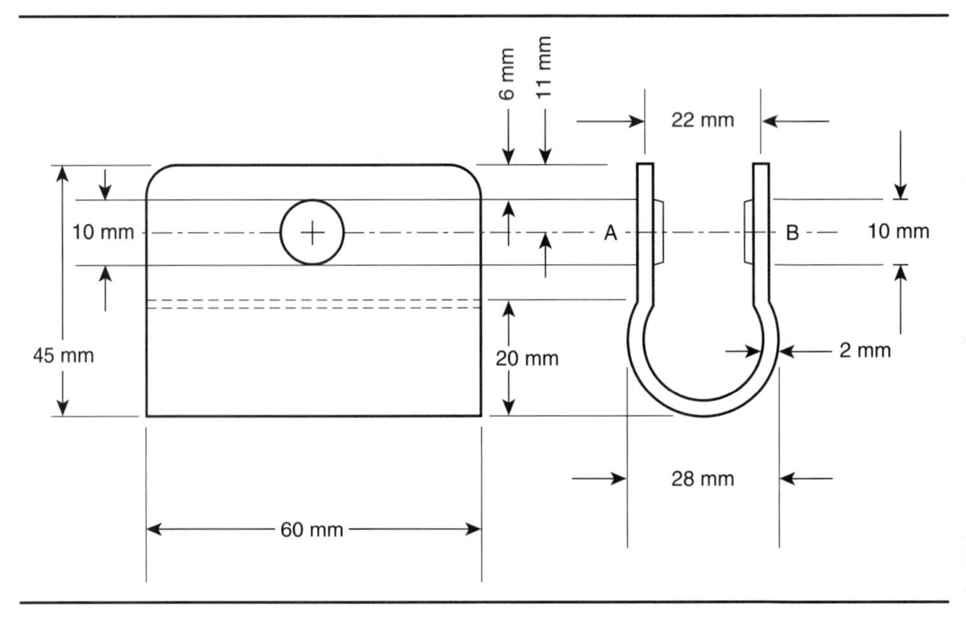

Figure 1 Bracket

Recently customers have been complaining about having difficulty securing the bracket closed with the bolt and lock washer. The bolts have been difficult to slide through the holes and then tighten. Assemblers complain of stripped bolts and snug fittings. The problem-solving cycle begins with "Recognizing the Problem and Establishing Priorities." Bracket customers have made management at WP Inc. well aware of the problem. Because management is unsure of the root cause of the problem, they proceed with the second step of problem-solving and assemble a team. The team consists of representatives from process engineering, materials engineering, product design, and manufacturing. Beginning with Step 3: Defining the Problem, the team has decided to brainstorm the reasons why this problem has occurred.

Assignment

Form a team and use the WHY-WHY diagram technique to determine why the bracket may be hard to assemble.

PART 2

In conjunction with the WHY-WHY diagram for why the bracket is not easily assembled, the team has decided to develop performance measures to answer the following question: How do we know that the changes we made to the process actually improved the process?

Assignment

What performance measures does your team feel are necessary to answer this question: How do we know that the changes we made to the process actually improved the process?

PART 3

Continuing in the problem-solving cycle, the team proceeds with Step 5. In order to analyze the current process, the team visited the customer's assembly plant to determine where the brackets were used in the process and how the assembly was actually performed. There they watched as the operator randomly selected a strut, a bracket, a bolt, and a locknut from different bins. The operator positioned the strut in place, wrapped the bracket around it, and secured it to the frame by finger-tightening the bolt and locknut. The operator then used a torque wrench to secure the assembly. While they watched, the operator had difficulty securing the assembly several times. Back at their plant, the team also created a flowchart for WP's process of fabricating the bracket (Figure 2).

After completing the flow diagram and verifying that it was correct, the team moved on to Step 6: Determining Possible Causes. They decided to use a cause-and-

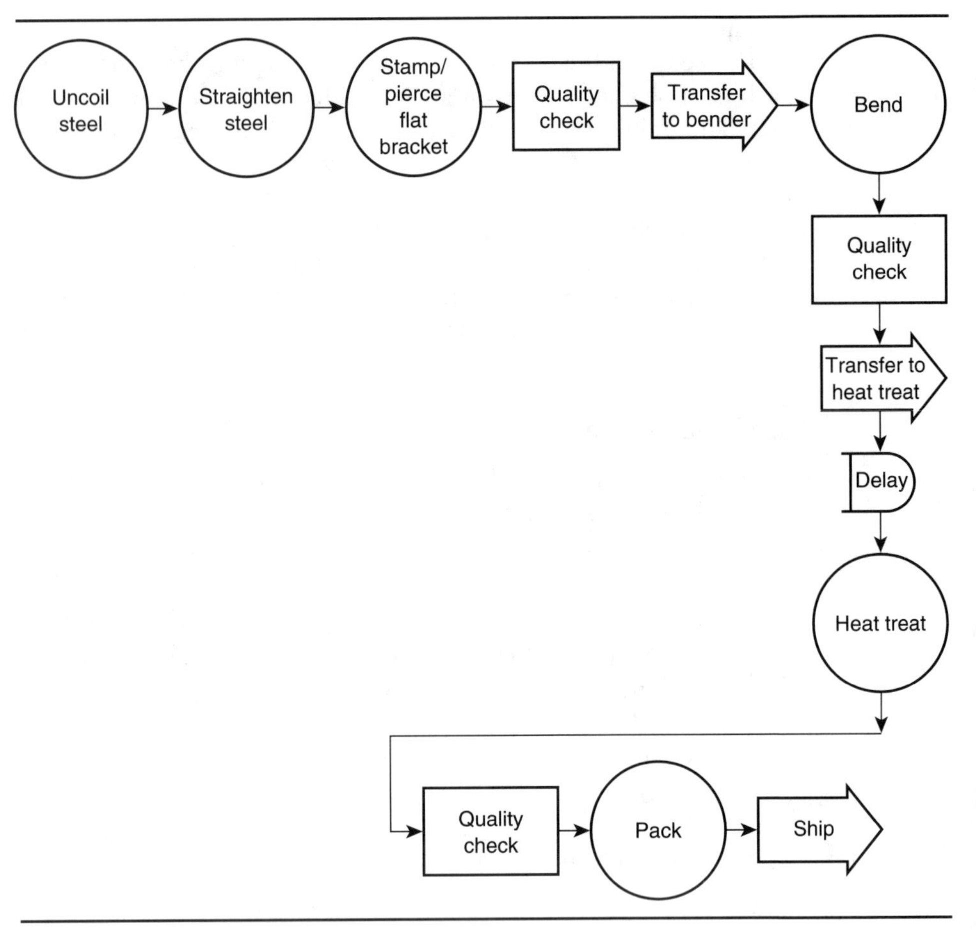

Figure 2 Flowchart of Bracket-Fabrication Process

effect diagram to guide their efforts in brainstorming potential root causes for the problem: difficulty securing brackets closed with bolt and locknut.

Assignment

Create a cause-and-effect diagram and brainstorm potential root causes for the problem: difficulty securing brackets closed with bolt and locknut.

PART 4

Through the use of a cause-and-effect diagram, the engineers determine that the most likely cause of the problems experienced by the customer is the alignment of the holes. At some stage in the formation process, the holes end up off center. Combining this information with the WHY-WHY diagram conclusion that hole alignment was critical for smooth installation during assembly narrows the search for a root cause. Unfortunately, the team still doesn't know why the holes are not properly aligned. They decide to create another cause-and-effect diagram that focuses on causes of improper hole alignment.

Assignment

Create a second cause-and-effect diagram that focuses on the root causes of improper hole alignment.

PART 5

At this point in the problem-solving process, it would be appropriate to use statistical information to determine whether the holes are truly not properly aligned. The team would confirm their suspicions during the next production run by having the press operator take samples and measure the angle between the centers of the holes for each sample. This data would then be utilized to create a histogram and compare the process performance with the specification for the angle between insert hole A and insert hole B of 0.00 mm with a tolerance of ±0.30. Hole alignment problems are confirmed through the use of histograms in Case 4.2, should you choose to use it. Histograms are one of the problem-solving techniques discussed in Step 6.

Determining Possible Causes

Assuming that hole alignment problems exist and are measurable, the team continues with their investigation. By studying the process, they determined that the fixture that holds the flat bracket in place during the bending operation does not securely hold the bracket in place. Changing the bracket fixture will be a relatively expensive undertaking. Although the engineers feel this change would eliminate the root cause of a problem as part of Step 7: Selecting and Implementing the Solution, the team has decided to create a force-field analysis before going to management to request funding to make the change.

Assignment

Create a force-field diagram that describes the forces driving the change to the fixture as well as the forces preventing the change from happening. Use your imagination; problem-solving is never as simple as "spend money."

Assignment

Describe the remaining steps that the team would take to finish the problem-solving process and ensure that the problem does not return.

II

Control Charts for Variables

4

Statistics

Learning Opportunities:

1. To review basic statistical concepts
2. To understand how to graphically and analytically study a process by using statistics
3. To know how to create and intercept a frequency diagram and a histogram
4. To know how to calculate the mean, median, mode, range, and standard deviation for a given set of numbers
5. To understand the importance of the normal curve and the central limit theorem in quality assurance
6. To know how to find the area under a curve using the standard normal probability distribution (Z tables)
7. To understand how to interpret the information analyzed ■

If things were done right just 99.9 percent of the time, then we'd have to accept

- One hour of unsafe drinking water per month
- Two unsafe plane landings per day at O'Hare International Airport in Chicago
- 16,000 pieces of mail lost by the U.S. Postal Service every hour
- 20,000 incorrect drug prescriptions per year
- 500 incorrect surgical operations each week
- 22,000 checks deducted from the wrong bank accounts per hour
- 32,000 missed heartbeats per person per year

Original source unknown.

Statistics

Each of the above statistics deals with the quality of life as we know it. We use statistics every day to define our expectations of life around us. Statistics, when used in quality assurance, define the expectations that the consumer and the designer have for the process. Processes and products are studied using statistics. This chapter provides a basic review of the statistical values most applicable to quality assurance.

STATISTICS

Every day, on the radio and TV, in newspapers and magazines, over the Internet, at work, and in personal conversations, we use statistics. ***Statistics,*** *the collection, tabulation, analysis, interpretation, and presentation of numerical data,* provide a viable method of supporting or clarifying a topic under discussion. Sometimes, without thinking, we accept statistics on their face value without really determining whether or not they are describing the whole truth. Misused statistics have led some people to distrust them completely, as Mark Twain once eloquently expressed:

> Figures often beguile me, particularly when I have the arranging of them myself; in which case the remark attributed to Disraeli would often apply with justice and force: "There are three kinds of lies: lies, damned lies, and statistics."*

Statistics provide beneficial information only when they have been properly collected, tabulated, analyzed, interpreted, and presented. In this chapter, we learn how to use statistical information to create a complete picture about a situation, process, or product. Sometimes it's obvious that the statistical information is incomplete. For instance, if a newscaster states that balanced trade has been achieved because the economy has experienced an 11% growth in both imports and exports and then mentions that imports totaled $90 billion, whereas exports were $30 billion, we could quickly discern the error. The $3.3 billion growth in exports does not balance the $9.9 billion growth in imports. The significantly different magnitudes of the original values reveal that trade is far from balanced. A majority of the time though, misunderstood, misused, or incomplete statistics are more subtle, as the following example shows.

EXAMPLE 4.1 Incomplete Statistical Data

Automotive manufacturers continue to seek ways to protect vehicle passengers in the event of an accident. Some of those changes, such as air bags, can be controversial. Air bags, designed to inflate automatically in accidents involving speeds of greater than 15 mph, have saved many lives. An air bag is a pretty impressive piece of equipment when you consider the protection it supplies. Even a collision at speeds as slow as 30 mph produces an impact on the body equivalent to falling off a three-story building. Essentially, an air bag inflates and stops your fall. Air bags first began to be offered on a wide variety of automobiles in the late 1980s. By 1998, 40 percent of the vehicles (70 million vehicles) on the road had air bags. Since their inception, records have been kept on accident statistics involving the deployment of air bags. As more and more air bags were installed in automobiles, it became obvious, unfortunately, that in some situations, air bags may kill. By 1998, 61 people, 38 of them children, had died from deployment of air bags. Alone, this

*Ayres, A., editor. *The Wit and Wisdom of Mark Twain*. New York: Harper and Row, 1987, p. 221.

statistic tells only part of the story. Missing are the statistics concerning how many lives have been saved by air bags (2,700 by 1998), how many of the children were unrestrained or improperly restrained in the front seat (34 of the 38 by 1998, nine infants in rear-facing restraints, 25 older children unrestrained), how many of the adults were not wearing seat belts at the time the air bag deployed (nearly all of the 23 by 1998). Although any death is a sad occurrence, statistical evidence points to a need to use seat belts in conjunction with air bags and to properly restrain children in the rear seat. Statistics also show that air-bag deployment in an accident is more likely to save lives than shorten them (2,700 lives saved versus 61 lost). Have these statistics changed in the past several years?*

*Allen, K., and Brehm, M. "Air Bag Switches Aren't the Answer; Seating Kids in the Back Protects Them." *USA Today,* March 7, 1997.
†Ledford, J. "The Lane Ranger: Measure Risk Before Deciding to Switch Air Bag." *The Atlanta Journal and Constitution,* November 22, 1997.
‡Nauss, D. "Less Powerful Air Bags Ok'd for Cars, Trucks." *Los Angeles Times,* March 15, 1997.

This first example encourages us to look beyond the face value of the statistics and ask questions. Asking questions to fully understand information is critical to correctly interpreting it. In the following example, statistical data are used to verify that the course of action taken in the example resulted in benefits that outweighed the costs. Here, statistics have been correctly collected, tabulated, analyzed, interpreted, presented, and used to clarify the benefits of using seat belts.

EXAMPLE 4.2 How Do We Know We're Doing the Right Thing?

Seat belts, like air bags, are known to save lives in accidents. According to A. Nomani's *Wall Street Journal* article entitled "How Aggressive Tactics Are Getting Drivers to Buckle Up" (December 11, 1997), highway safety research estimates that if 85 percent of Americans utilized seat belts, each year 4,200 lives would be saved and 100,000 injuries would be prevented. The resulting annual savings in medical costs would be about $6.7 billion. Given that nationally, only 68 percent of the population uses seat belts, some states have put efforts in place to educate drivers and encourage seat-belt use. These efforts cost money. How do the states implementing such programs know whether or not the programs are effective? One way to determine this is to monitor changes in statistical data before and after the program has begun. North Carolina, beginning in 1994, implemented its "Click It or Ticket" program. The results are impressive. Seat-belt use, just 64 percent in 1993, jumped to 83 percent. There has been a 14 percent reduction in serious injuries and deaths, resulting in an estimated $114 million in health-care–related costs. Even with program costs exceeding $4.5 million, these statistics verify that investment in seat-belt educational and enforcement programs have had a positive effect.

USE OF STATISTICS IN QUALITY ASSURANCE

In industry, quality improvement efforts have statistics as their foundation. Statistics are used to understand the behavior of processes. Correctly collected and analyzed, statistical information can be used to predict future process behavior. The five aspects of statistics—collection, tabulation, analysis, interpretation, and presentation—are equally important when analyzing a process. Once gathered and analyzed, statistical data can be used to aid in decisions about making process changes or pursuing a particular course of action. Assumptions based on incomplete information can lead to incorrect decisions, unwise investments, and uncomfortable working environments.

Being able to utilize statistics to make decisions, change processes, and improve company performance requires an understanding of basic statistical tools, beginning with an understanding of how to collect statistical information.

POPULATIONS VERSUS SAMPLES

Statistics can be gathered by studying either the entire collection of values associated with a process or only a portion of the values. A ***population*** *is a collection of all possible elements, values, or items associated with a situation.* A population can contain a finite number of things or it may be nearly infinite. The insurance forms a doctor's office must process in a day or the number of cars a person owns in a lifetime are examples of finite populations. All the tubes of popular toothpaste ever made by a manufacturer over the product's lifetime represent a nearly infinite population. Limitations may be placed on a collection of items to define the population.

As the size of a population increases, studying that population becomes unwieldy unless sampling can be used. A ***sample*** *is a subset of elements or measurements taken from a population.* The doctor's office may wish to sample 10 insurance claim forms per week to check the forms for completeness. The manufacturer of toothpaste may check the weight of a dozen tubes per hour to ensure that the tubes are filled correctly. This smaller group of data is easier to collect, analyze, and interpret.

EXAMPLE 4.3 Taking a Sample

An outlet store has just received a shipment of 1,000 shirts sealed in cardboard boxes. The store had ordered 800 white shirts and 200 blue. The store manager wishes to check that there actually are 20 percent blue shirts and 80 percent white shirts. She doesn't want to open all of the boxes and count all of the shirts, so she has decided to sample the population. Table 4.1 shows the results of 10 random samples of 10 shirts each.

A greater number of blue shirts is found in some samples than in others. However, when the results are compiled, the blue shirts comprise 19 percent, very close to the desired value of 20 percent. The manager of the outlet store is pleased to learn that the samples have shown that there are approximately 20 percent blue shirts and 80 percent white.

Table 4.1 A Sampling of Shirts

Sample Number	*Sample Size*	*Number of White Shirts*	*Number of Blue Shirts*	*Percentage of Blue Shirts*
1	10	8	2	20
2	10	7	3	30
3	10	8	2	20
4	10	9	1	10
5	10	10	0	0
6	10	7	3	30
7	10	8	2	20
8	10	9	1	10
9	10	8	2	20
10	10	7	3	20
Total	100	81	19	19

A sample will represent the population as long as the sample is random and unbiased. In a *random sample*, each item in the population has the same opportunity to be selected. A classic example of a *non*random sample was a newspaper poll conducted in the 1940s. The poll contacted only a limited number of people in a nonrandom format, causing the paper to incorrectly predict the results of a national presidential election (Figure 4.1). Statistical data quoted on TV and radio, in magazines and newspapers, and on the Internet, may present an incomplete picture of the situation. When someone tells you that 9 out of 10 experts agree with their findings, in order to interpret and use this information it is critical to know how many were sampled, the size of the whole group, and the conditions under which the survey was

Figure 4.1 The *Chicago Tribune,* Eager to Get the Scoop, Ran a Headline about Harry Truman That Proved False.

made. As previously discussed, in statistics, knowledge of the source of the data, the manner in which the data were collected, the amount sampled, and how the analysis was performed is necessary to determine the validity of the information presented. To put statistical values in perspective, you must find answers to such questions as

How was the situation defined?
Who was surveyed?
How many people were contacted?
How was the sample taken?
How were the questions worded?
Is there any ambiguity?

EXAMPLE 4.4 Fright Statistics

At least 60 million Americans have high blood pressure. One in every five will die of cancer. One in every three will die of heart disease. The chance of dying in an airplane accident is one in 4,000. One in every three Americans is obese. One in every 45 dies in an auto accident. Alarming statistics. When we look around at a room full of people, can we apply these statistics? When we start to examine these numbers using the questions just presented—How was the situation defined? Who was surveyed? How many people were contacted? How was the sample taken? How were the questions worded? Is there any ambiguity?—we realize that at best, these numbers are ambiguous. For example, consider the statistic one in five will develop cancer. How was the situation for this statistic defined? People under 40? People between 40 and 60? People over 60? People at any age? How were the questions worded? Is it developing cancer eventually? Developing cancer this year? Developing cancer in the next 10 years?

Because counting exactly who suffers from what is nearly impossible, health statistics are based on the results of samples and then projected over the entire population. The sensitive nature of health and lifestyle questions make the information difficult to get at best and nonrandom and untrue at worst. When viewing statistical information it is important to keep in mind the context in which the numbers were derived.

A second example of biased sampling could occur on the manufacturing floor. If, when an inspector receives a skid of goods, that inspector always samples from the top layer and takes a part from each of the four corners of the skid, the sample is biased. The parts in the rest of the skid have not been considered. Furthermore, operators observing this behavior may choose to place only the best-quality product in those corners. The inspector has biased the sample because it is not random and does not represent all of the parts of the skid. The inspector is receiving an incorrect impression about the quality in the entire skid.

Unbiased samples depend on other features besides randomness. Conditions surrounding the population should not be altered in any way from sample to sample.

The sampling method can also undermine the validity of a sample. Ensure the validity of a sample by asking such questions as

How was the problem defined?
What was studied?
How many items were sampled?
How was the sample taken?
How often?
Have conditions changed?

Analyzed correctly, a sample can tell the investigator a great deal about the population.

DATA COLLECTION

Two types of statistics exist: deductive and inductive. Also known as *descriptive statistics*, ***deductive statistics*** *describe a population or complete group of data*. When describing a population using deductive statistics, the investigator must study each entity within the population. This provides a great deal of information about the population, product, or process, but gathering the information is time-consuming. Imagine contacting each man, woman, and child in the United States, all 250 million of them, to conduct the national census!

When the quantity of the information to be studied is too great, inductive statistics are used. ***Inductive statistics*** *deal with a limited amount of data or a representative sample of the population*. Once samples are analyzed and interpreted, predictions can be made concerning the larger population of data. Quality assurance (and the U.S. census) relies primarily on inductive statistics. Properly gathered and analyzed, sample data provides a wealth of information.

In quality control, two types of numerical data can be collected. ***Variables data,*** *those quality characteristics that can be measured*, are treated differently from ***attribute data,*** *those quality characteristics that are observed to be either present or absent, conforming or nonconforming*. While both variables and attribute data can be described by numbers, attribute data are countable, not measurable.

Variables data tend to be continuous when measured. When data are ***continuous,*** *the measured value can take on any value within a range*. The range of values that the measurements can take on will be set by the expectations of the users or the circumstances surrounding the situation. For example, a manufacturer might wish to monitor the depth of a keyway as parts are stamped. During the course of the day, the samples may have values of 0.399, 0.402, 0.401, 0.400, 0.401, 0.403, and 0.398 inch.

Discrete data consist of distinct parts. In other words, when measured, ***discrete data*** *will be countable using whole numbers*. For example, the number of frozen vegetable packages found on the shelf during an inventory count—ten packages of frozen peas, eight packages of frozen corn, 22 packages of frozen Brussels sprouts—is discrete, countable data. Since vegetable packages can only be sold to customers when the packages are whole and unopened, only whole-numbered measurements will exist; continuous measurements would not be applicable in this case. Attribute data, since the data are seen as being either conforming or not conforming to specifications, are primarily discrete data.

A statistical analysis begins with the gathering of data about a process or product. Usually raw data gathered from a process take the form of ungrouped data. ***Ungrouped data*** *are easily recognized because when viewed, it appears that the data are without any order.* Unorganized data are difficult, if not impossible, to comprehend. ***Grouped data,*** on the other hand, *are grouped together on the basis of when the values were taken or observed.* Consider the following example.

EXAMPLE 4.5 Grouping Data

A company manufactures a flat round plate with four keyways stamped into it (Figure 4.2). Recently, a customer brought to the manufacturer's attention the fact that not all of the keyways are being cut out to the correct depth. The manufacturer asked the operator to measure each keyway in five parts every 15 minutes and record the measurements. Table 4.2 shows the results. When management started to analyze and interpret the data they were unable to do so. Why?

An investigation of this raw, ungrouped data reveals that there is no way to determine which measurements belong with which keyway. Which keyway is too deep? Too shallow? It is not possible to determine the answer.

To rectify this situation, during the stamping process, the manufacturer placed a small mark below one of the keyways (Figure 4.3). The mark labels that keyway as number 1. Clockwise around the part, the other keyways are designated 2, 3, and 4. The mark does not affect the use of the part. The operator was asked to measure the keyway depths again, five parts every 15 minutes (Table 4.3).

By organizing the data according to keyway, it could then be determined that keyway number 2 is too deep, and keyway number 4 is too shallow.

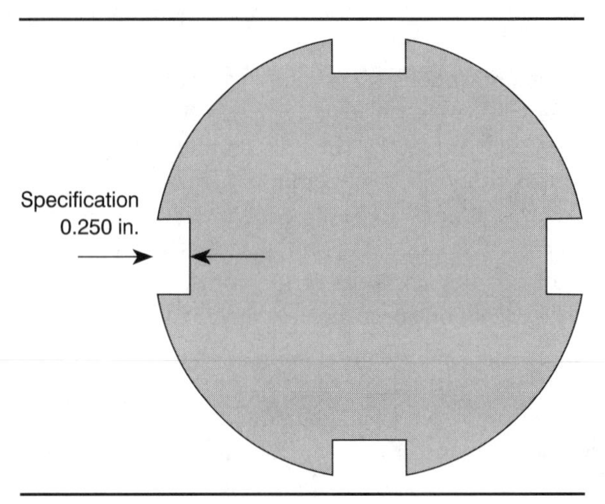

Figure 4.2 Flat Round Plate

Table 4.2 Ungrouped Data (in inches)

0.247	0.245	0.271
0.254	0.260	0.276
0.268	0.278	0.268
0.261	0.260	0.230
0.231	0.224	0.243
0.241	0.224	0.225
0.252	0.222	0.232
0.258	0.242	0.254
0.266	0.244	0.242
0.226	0.277	0.248
0.263	0.222	0.236
0.242	0.260	0.262
0.242	0.249	0.223
0.264	0.250	0.240
0.218	0.251	0.222
0.216	0.255	0.261
0.266	0.247	0.244
0.266	0.250	0.249
0.218	0.235	0.226
0.269	0.258	0.232
0.260	0.251	0.250
0.241	0.245	0.248
0.250	0.239	0.252
0.246	0.248	0.251

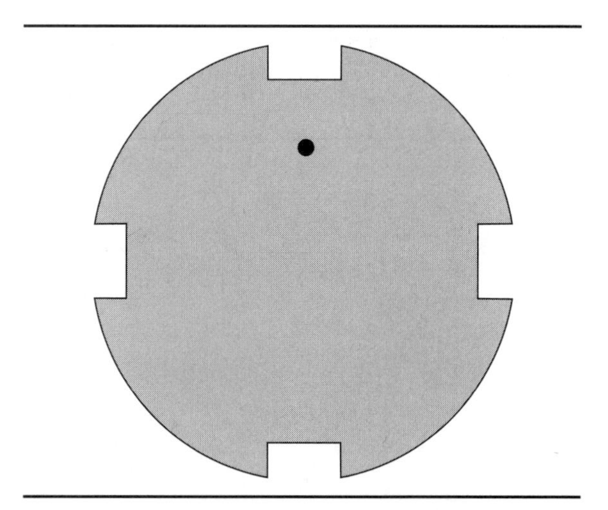

Figure 4.3 Flat Round Plate with Mark

Table 4.3 Grouped Data (in inches)

	Keyway 1	*Keyway 2*	*Keyway 3*	*Keyway 4*
Subgroup 1	0.250	0.261	0.250	0.240
Subgroup 2	0.251	0.259	0.249	0.242
Subgroup 3	0.250	0.258	0.251	0.245
Subgroup 4	0.249	0.257	0.250	0.243
Subgroup 5	0.250	0.257	0.250	0.244
Subgroup 6	0.251	0.260	0.249	0.245
Subgroup 7	0.251	0.258	0.250	0.243
Subgroup 8	0.250	0.259	0.249	0.247
Subgroup 9	0.250	0.257	0.250	0.245
Subgroup 10	0.248	0.256	0.251	0.244
Subgroup 11	0.250	0.260	0.250	0.243
Subgroup 12	0.251	0.259	0.251	0.244
Subgroup 13	0.250	0.257	0.250	0.245
Subgroup 14	0.250	0.256	0.249	0.245
Subgroup 15	0.250	0.257	0.250	0.246

MEASUREMENTS: ACCURACY, PRECISION, AND MEASUREMENT ERROR

The validity of a measurement comes not only from the selection of a sample size and an understanding of the group of data being measured, it also depends on the measurements themselves and how they were taken. Measurement error occurs while the measurements are being taken and recorded. ***Measurement error** is considered to be the difference between a value measured and the true value*. The error that occurs is one either of accuracy or of precision. ***Accuracy** refers to how far from the actual or real value the measurement is*. ***Precision** is the ability to repeat a series of measurements and get the same value each time*. Precision is sometimes referred to as ***repeatability.***

Figure 4.4*a* pictures the concept of accuracy. The marks average to the center target. Figure 4.4*b*, with all of the marks clustered together, shows precision. Figure 4.4*c*

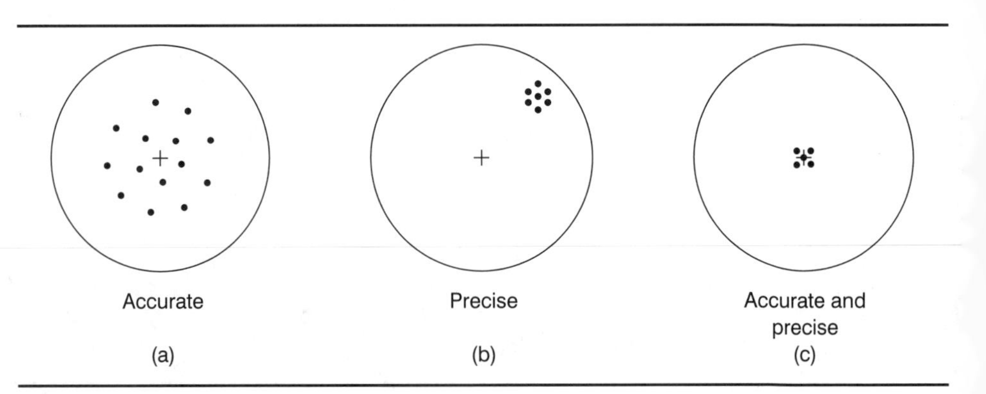

Figure 4.4 Accuracy and Precision

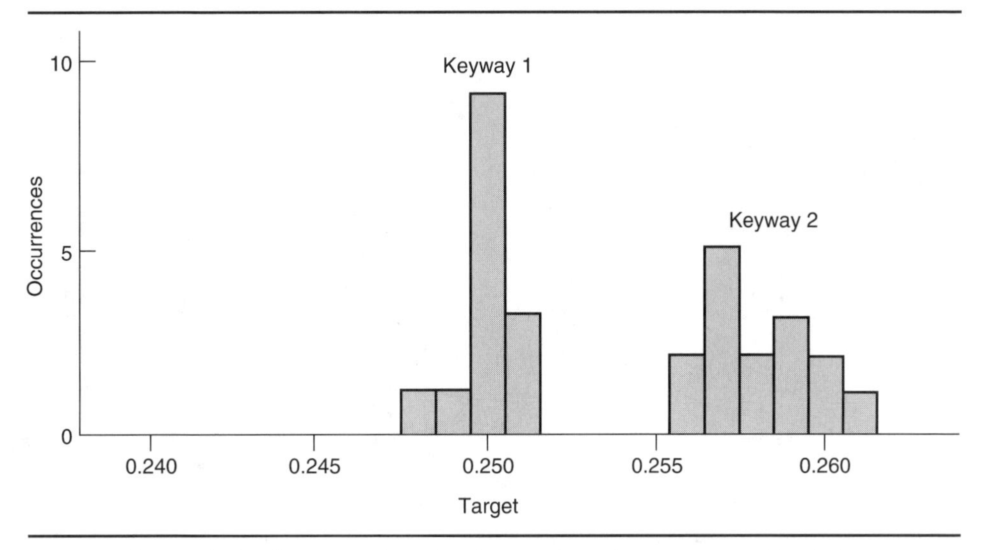

Figure 4.5 Data for Keyways 1 and 2 Comparing Accuracy and Precision

describes a situation in which both accuracy and precision exist. Example 4.6 and Figures 4.5, 4.6, and 4.7 illustrate the concepts of accuracy and precision.

EXAMPLE 4.6 Accuracy and Precision

Accuracy and precision describe the location and the spread of the data. Look at Figures 4.5, 4.6, and 4.7 showing the data from the Flat Round Plate example. When compared, the difference in the keyways' accuracy and precision becomes

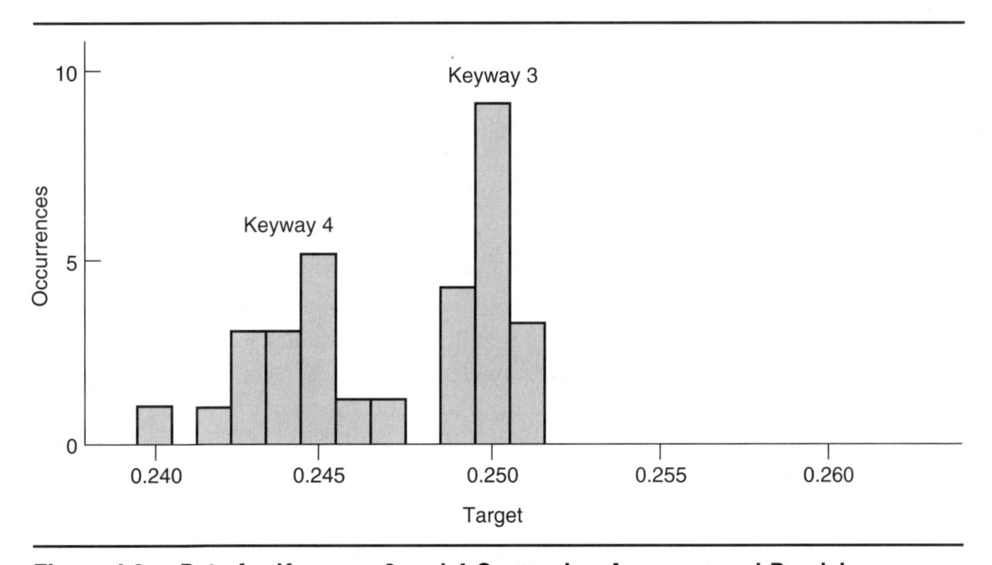

Figure 4.6 Data for Keyways 3 and 4 Comparing Accuracy and Precision

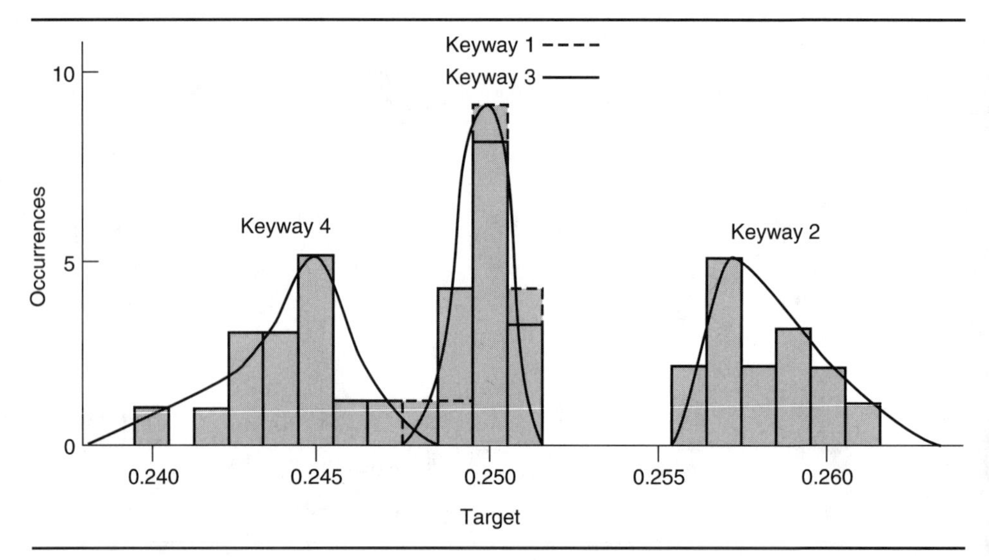

Figure 4.7 Data for Keyways 1 and 3 Are Both Accurate and Precise while Data for Keyways 2 and 4 Are Not.

apparent. The data for keyways 1 and 3 exhibit greater accuracy than that for keyways 2 and 4. Note how the data for keyways 1 and 3 are concentrated around the target specification of 0.250 inch. The data values for keyway 2 are greater than the desired target specification and those for keyway 4 are smaller. Notice, too, the difference in precision. Data for keyways 1 and 3 are precise, tightly grouped around the target. Data values for keyways 2 and 4 are not only far away from the target, they are also more spread out, less precise. Changes to this stamping process must be twofold, improving both the accuracy and the precision.

If accuracy and precision are missing during data collection, the validity of the information and conclusions will be questioned. Accuracy and precision play an important role in ensuring that the data collected are valuable to those interpreting and presenting it.

Measurement errors are not always the fault of the individual performing the measuring. In any situation, several sources of error exist, including environment and machine error. Environmental problems, such as with dust, dirt, temperature, and water, cause measurement errors by disturbing either the products or the measuring tools. A soiled measuring tool will produce a faulty reading as will a layer of dust or oil on a part. Temperature may change the dimensions of products.

Significant figures and associated rounding errors affect the viability of a measurement. ***Significant figures*** *are the numerals or digits in a number, excluding any leading zeros used to place the decimal point*. Zeros following a digit—for example, 9.700—are significant if they have truly been measured. When working a statistical problem, you should use only the number of digits that the measuring devices are able to provide. If a micrometer reads to three decimal places, then all mathematical calculations should be worked to no more than three decimal places. With today's computers and calculators, there is a temptation to use all the numbers that appear on the screen. Since the measuring device may not have originally measured to that many decimal places, the number should be rounded to the original number of significant figures. Calculations are performed and then rounded to the number of significant figures present in the value with the smallest number of significant figures. When rounding, round to the next highest number if the figure is a 5 or greater. If the figure is 4 or below, round down. Consider the following examples:

$23.6 \div 3.8 = 6.2$ (3.8 has fewest significant figures)

$3{,}456 \div 12.3 = 281$ (12.3 has three significant figures)

$3.2 \times 10^2 + 6{,}930 = 7.3 \times 10^3$ (3.2×10^2 has two significant figures)

$6{,}983 \div 16.4 = 425.79268 = 426$ when rounded

Human errors associated with measurement errors can be either unintentional or intentional. Unintentional errors result from poor training, inadequate procedures, or incomplete or ambiguous instructions. These types of errors can be minimized by good planning, training, and supervision. Intentional errors are rare and are usually related to poor attitudes; they will require improving employee relations and individual guidance to solve.

The choice and care of measuring devices is critical. The measuring instruments utilized must be able to measure at an accuracy level needed to monitor the process. For example, a meat thermometer would not be appropriate for monitoring the temperature in a hospital sterilizer. Proper care of measuring instruments includes correct handling and storage as well as calibration checks. A systematic calibration program verifies that the device is taking accurate readings of the items being measured. Calibration involves checking the gauge measurement against a known dimension. Recalibration should be done periodically to ensure that no measuring error exists. Recalibration should also be done after any improper handling occurs, such as dropping.

Proper measuring is critical to monitoring and controlling any process. A measurement is only as good as the person's reading of the measuring device, combined with the accuracy and precision of the measuring device itself. Those utilizing the measuring equipment must have the appropriate training and have been taught proper handling procedures. Proper care of measuring instruments and correct training of those using them will ensure that the raw data obtained with measuring instruments will be reliable. If this is an incorrect assumption, then any conclusions based on the data and subsequent analysis are useless.

DATA ANALYSIS: GRAPHICAL

A thorough statistical analysis of the data that has been gathered usually involves three aspects: graphical, analytical, and interpretive. The analytical or numerical work supports the more easily understood graphical presentation. By interpreting the graphical and analytical information, the data and the relevant calculations can be put to use solving quality problems. A variety of different graphical methods exist, including the frequency diagram and the histogram.

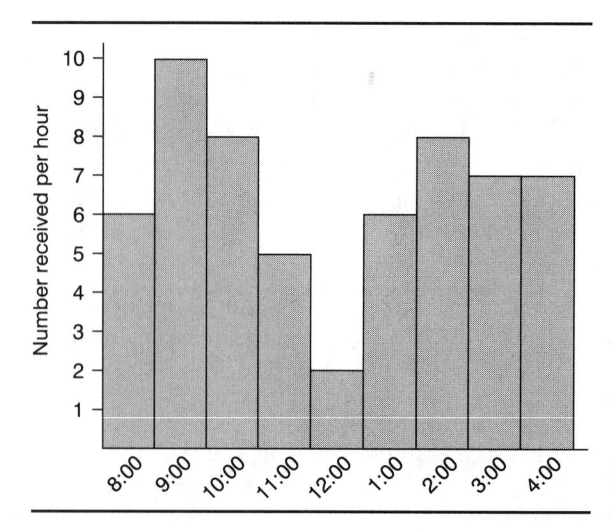

Figure 4.8 Frequency Diagram: Phone Calls Received Each Hour

Frequency Diagrams

A ***frequency diagram*** *shows the number of times each of the measured values occurred when the data were collected.* This diagram can be created either from measurements taken from a process or from data taken from the occurrences of events. When compared with the raw, ungrouped data, this diagram shows at a glance which values occur the most frequently as well as the spread of the data (Figure 4.8). To create a frequency diagram, the following steps are necessary:

1. Collect the data. Record the measurements or counts of the characteristics of interest.
2. Count the number of times each measurement or count occurs.
3. Construct the diagram by placing the counts or measured values on the x axis and the frequency or number of occurrences on the y axis. The x axis must contain each possible measurement value from the lowest to the highest, even if a particular value does not have any corresponding measurements. A bar is drawn on the diagram to depict each of the values and the number of times the value occurred in the data collected.
4. Interpret the frequency diagram. We learn more about interpreting frequency diagrams later in the chapter. Study the diagrams you create and think about the diagram's shape, size, and location in terms of the desired target specification.

EXAMPLE 4.7 Constructing a Frequency Diagram: Flat Round Plate

The engineers involved in the flat round plate problem have decided to study the thickness of the part. They plan to create a frequency diagram for the grouped data shown in Table 4.4. By doing this they hope to gain a clearer understanding of incoming material thickness.

Table 4.4 Grouped Data for Thickness (in inches)

Subgroup 1	0.0625	0.0626	0.0624	0.0625	0.0627
Subgroup 2	0.0624	0.0623	0.0624	0.0626	0.0625
Subgroup 3	0.0622	0.0625	0.0623	0.0625	0.0626
Subgroup 4	0.0624	0.0623	0.0620	0.0623	0.0624
Subgroup 5	0.0621	0.0621	0.0622	0.0625	0.0624
Subgroup 6	0.0628	0.0626	0.0625	0.0626	0.0627
Subgroup 7	0.0624	0.0627	0.0625	0.0624	0.0626
Subgroup 8	0.0624	0.0625	0.0625	0.0626	0.0626
Subgroup 9	0.0627	0.0628	0.0626	0.0625	0.0627
Subgroup 10	0.0625	0.0626	0.0628	0.0626	0.0627
Subgroup 11	0.0625	0.0624	0.0626	0.0626	0.0626
Subgroup 12	0.0630	0.0628	0.0627	0.0625	0.0627
Subgroup 13	0.0627	0.0626	0.0628	0.0627	0.0626
Subgroup 14	0.0626	0.0626	0.0625	0.0626	0.0627
Subgroup 15	0.0628	0.0627	0.0626	0.0625	0.0626
Subgroup 16	0.0625	0.0626	0.0625	0.0628	0.0627
Subgroup 17	0.0624	0.0626	0.0624	0.0625	0.0627
Subgroup 18	0.0628	0.0627	0.0628	0.0626	0.0630
Subgroup 19	0.0627	0.0626	0.0628	0.0625	0.0627
Subgroup 20	0.0626	0.0625	0.0626	0.0625	0.0627
Subgroup 21	0.0627	0.0626	0.0628	0.0625	0.0627
Subgroup 22	0.0625	0.0626	0.0628	0.0625	0.0627
Subgroup 23	0.0628	0.0626	0.0627	0.0630	0.0627
Subgroup 24	0.0625	0.0631	0.0630	0.0628	0.0627
Subgroup 25	0.0627	0.0630	0.0631	0.0628	0.0627
Subgroup 26	0.0632	0.0628	0.0631	0.0628	0.0627
Subgroup 27	0.0630	0.0628	0.0631	0.0628	0.0627
Subgroup 28	0.0632	0.0632	0.0628	0.0631	0.0630
Subgroup 29	0.0630	0.0628	0.0631	0.0632	0.0631
Subgroup 30	0.0632	0.0631	0.0630	0.0628	0.0628

Step 1. Collect the Data. The first step is performed by the operator, who randomly selects five parts each hour, measures the thickness, and records the values (Table 4.4).

Step 2. Count the Number of Times Each Measurement Occurs. A check sheet, or tally sheet (as described in Chapter 3) is used to make this step easier (Figure 4.9).

Step 3. Construct the Diagram. The count of the number of times each measurement occurred is placed on the y axis. The values, between 0.0620 and 0.0632, are each marked on the x axis. Even though no values exist for 0.0629, this number is on the diagram. The completed frequency diagram is shown in Figure 4.10.

0.0620	/
0.0621	/ /
0.0622	/ /
0.0623	/ / / /
0.0624	𝍸 𝍸 / /
0.0625	𝍸 𝍸 𝍸 𝍸 𝍸 /
0.0626	𝍸 𝍸 𝍸 𝍸 𝍸 𝍸
0.0627	𝍸 𝍸 𝍸 𝍸 𝍸 / /
0.0628	𝍸 𝍸 𝍸 𝍸 / / /
0.0629	
0.0630	𝍸 𝍸
0.0631	𝍸 / / /
0.0632	𝍸

Figure 4.9 Tally Sheet for Thickness of Flat Round Plate

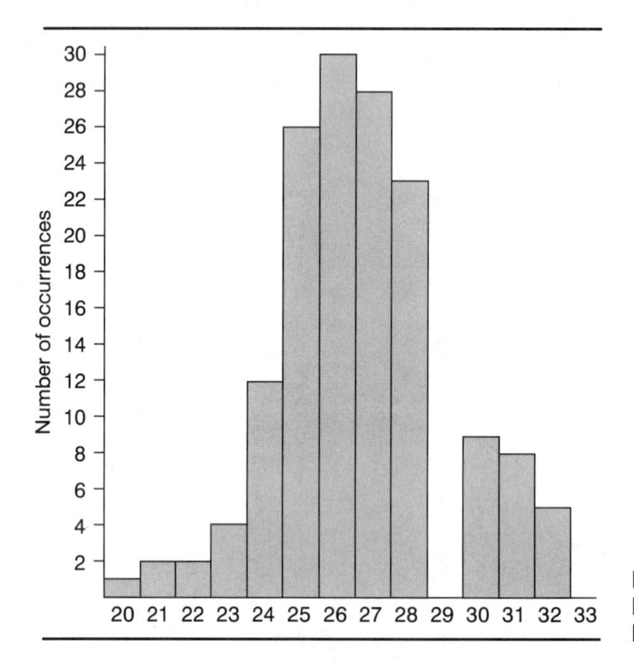

Figure 4.10 Frequency Distribution for Flat Round Plate Thickness (Coded 0.06)

Step 4. Interpret the Frequency Diagram. This frequency distribution is nearly symmetrical, but it is missing data for value 0.0629. The engineers should definitely investigate why this is so.

EXAMPLE 4.8 Monogramming and Embroidery Machine Arm

R & M Industries manufactures automated embroidery and monogramming machines. These machines are used to embroider or monogram designs on specialty items such as baseball caps, shirts, jackets, and other products. In order to create the embroidery work, the pantograph arm moves in the x and y directions through the use of servo motors. The arm has numerous through-holes for mounting, depending on the type of embroidery being created. R & M's engineers are testing the performance capabilities of a machine that produces a newly designed embroidery arm. This involves producing a limited number of parts, known as a runoff. The dimension being analyzed involves the mounting holes that hold a bearing and a mating part. The engineers are interested in the distance between the centers of the two holes (Figure 4.11). The specified distance between the holes is 11.25 cm. The measurements taken are shown in Figure 4.12. Create a frequency diagram with the data.

Step 1. Collect the Data. The engineers collected 25 arms machined during the runoff.

Step 2. Use a Tally Sheet. Count the number of times each measurement appeared (Figure 4.13).

Step 3. Construct the Diagram. Place the measured values (11.16, 11.17, etc.) on the x axis and the number of occurrences of those values on the y axis (Figure 4.14).

Step 4. Interpret the Frequency Diagram. The data are spread between 11.16 and 11.36, with a majority of them grouped between 11.21 and 11.25.

Histograms

Histograms and frequency diagrams are very similar. The most notable difference between the two is that on a histogram the data are grouped into cells. Each cell contains a range of values. This grouping of data results in fewer cells on the graph than with a frequency diagram. The x-axis scale on a histogram will indicate the cell ranges rather than individual values.

Construction of Histograms

Histograms are begun in the same manner as are frequency diagrams. The data are collected and grouped, then a check sheet is used to tally the number of times each measurement occurs. At this point, the process becomes more complex because of the need to create the cells. The following example details the construction of a histogram.

EXAMPLE 4.9 Constructing a Histogram: Flat Round Plate

The engineers working with the thickness of the flat round plate have decided to create a histogram to aid in their analysis of the process. They are following these steps:

Step 1. Collect the Data and Construct a Tally Sheet. The engineers will use the data previously collected (Table 4.4) as well as the tally sheet created during the construction of the frequency diagram (Figure 4.9).

Step 2. Calculate the Range. The ***range,*** *represented by the letter R, is calculated by subtracting the lowest observed value from the highest observed value.* In this case, 0.0620 is the lowest value and 0.0632 is the highest:

$$\text{Range} = R = X_h - X_l$$

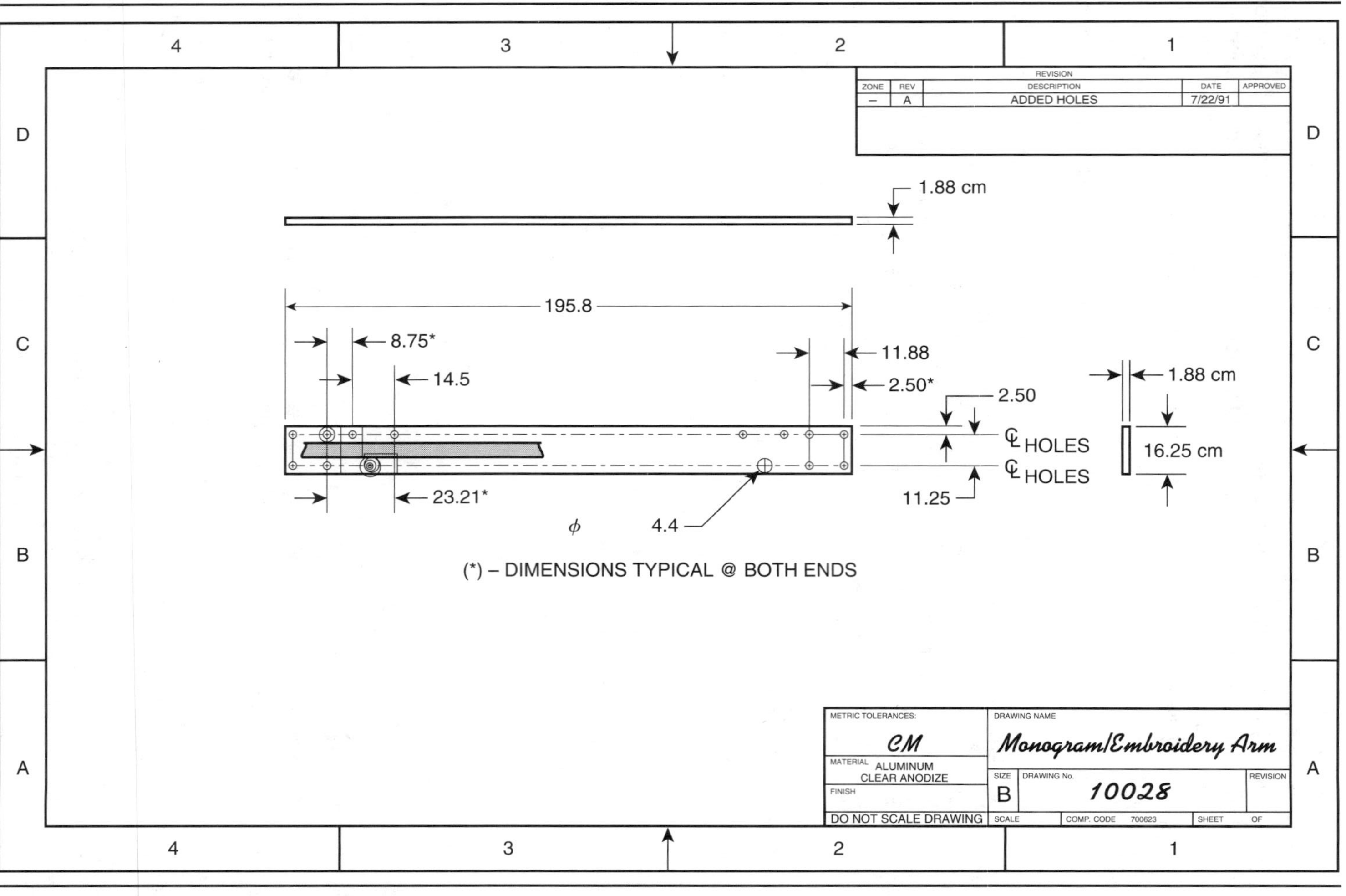

Figure 4.11 Monogram and Embroidery Arm

Figure 4.12 Data for Distance Between Hole Centers for the Monogramming and Embroidery Arm (in cm)

11.16	11.19	11.22	11.25	11.18
11.23	11.29	11.25	11.24	11.23
11.33	11.21	11.21	11.22	11.22
11.17	11.28	11.36	11.25	11.22
11.24	11.22	11.27	11.23	11.19

Figure 4.13 Tally Sheet of Data for Distance Between Hole Centers for the Monogramming and Embroidery Arm (in cm)

11.16	/
11.17	/
11.18	/
11.19	//
11.20	
11.21	//
11.22	~~////~~
11.23	///
11.24	//
11.25	///
11.26	
11.27	/
11.28	/
11.29	/
11.30	
11.31	
11.32	
11.33	/
11.34	
11.35	
11.36	/

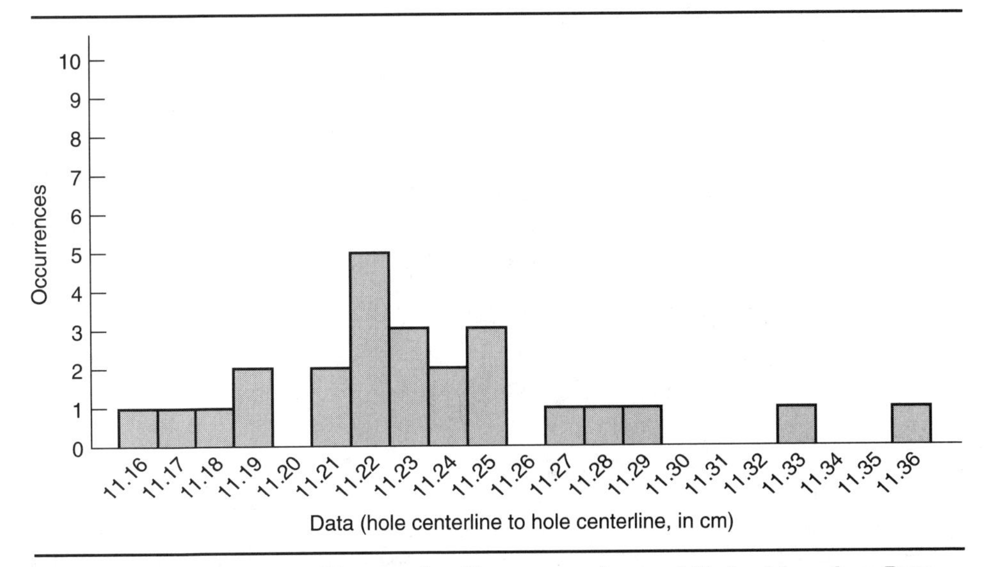

Figure 4.14 Frequency Diagram for Monogramming and Embroidery Arm Data

where

$$R = \text{range}$$
$$X_h = \text{highest number}$$
$$X_l = \text{lowest number}$$

$$R = 0.0632 - 0.0620 = 0.0012$$

Step 3. Create the Cells. In a histogram, data are combined into cells. Cells are composed of three components: cell intervals, cell midpoints, and cell boundaries (Figure 4.15). ***Cell midpoints*** *identify the centers of cells. A* ***cell interval*** *is the distance between the cell midpoints. The* ***cell boundary*** *defines the limits of the cell.*

Cell Intervals Odd-numbered cell intervals are often chosen for ease of calculation. For example, if the data were measured to one decimal place, then the cell intervals could be 0.1, 0.3, 0.5, 0.7, or 0.9. If the gathered data were measured to three decimal places, then the cell intervals to choose from would be 0.001, 0.003, 0.005, 0.007, 0.009. For this example, because the data were measured to four decimal places, the cell interval could be 0.0001, 0.0003, 0.0005, 0.0007, or 0.0009.

Cell interval choice plays a large part in the size of the histogram created. To determine the number of cells, the following formula is used:

$$h = \frac{R}{i} + 1$$

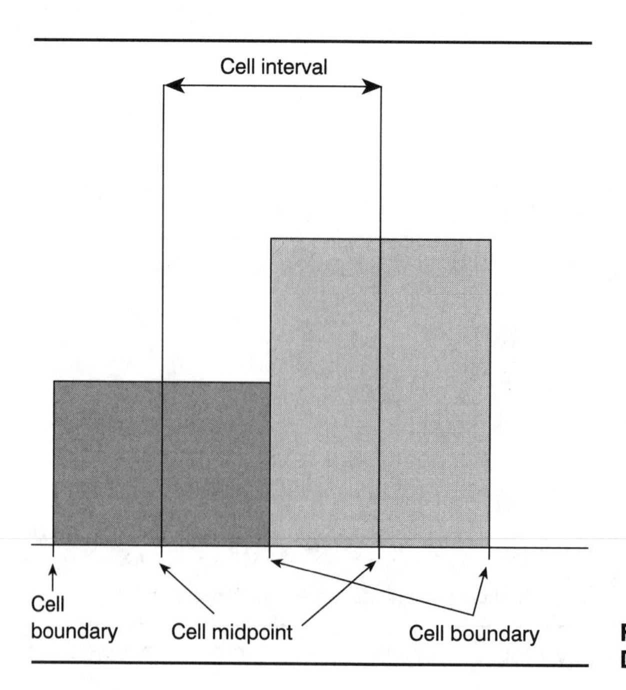

Figure 4.15 Histogram Cell Description

where

$$h = \text{number of cells}$$
$$i = \text{cell interval}$$
$$R = \text{range}$$

Since both i, the cell interval, and h, the number of cells, are unknown, creators of histograms must choose values for one of them and then solve for the other. For our example, if we choose a cell interval of 0.0001,

$$h = \frac{0.0012}{0.0001} + 1$$
$$h = 13$$

Using a cell interval of 0.0001 will result in 13 cells, the same as in the frequency diagram we created in Example 4.7 (Figure 4.10). For this reason, we will use either 0.0003, 0.0005, 0.0007, or 0.0009 as cell interval values.

For a cell interval value of 0.0003:

$$h = \frac{0.0012}{0.0003} + 1$$
$$h = 5$$

For a cell interval value of 0.0005:

$$h = \frac{0.0012}{0.0005} + 1$$
$$h = 3$$

For a cell interval value of 0.0007:

$$h = \frac{0.0012}{0.0007} + 1$$
$$h = 3$$

As the cell interval gets larger, the number of cells necessary to make a histogram decreases. For this example, we will use a cell interval of 0.0003. This will create a histogram that provides enough spread to analyze the data. When deciding the number of cells to use, it is sometimes helpful to follow this rule of thumb:

For fewer than 100 pieces of data, use 4 to 9 cells.
For 100 to 500 pieces of data, use 8 to 17 cells.
For 500 or more, use 15 to 20 cells.

Another helpful rule of thumb exists for determining the number of cells in a histogram. Use the square root of n ($\sqrt{n}$), where n is the number of data points, as an approximation of the number of cells needed.

Cell Midpoints When constructing a histogram, it is important to remember two things: (1) Histograms must contain all of the data; (2) one particular value cannot fit into two different cells. Cell midpoints are selected to ensure that these problems are avoided. To determine the midpoint values that anchor the histogram, use either one of two techniques.

The simplest technique is to choose the lowest value measured. In this example, the lowest measured value is 0.0620. Other midpoint values are determined by adding the cell interval of 0.0003 to 0.0620 first and then adding it to each successive new midpoint. If we begin at 0.0620, we find the other midpoints at 0.0623, 0.0626, 0.0629, and 0.0632.

A second method of determining the midpoint of the lowest cell in the histogram is the following formula:

$$MP_l = X_l + \frac{i}{2}$$

In this formula, MP_l is the midpoint of the lowest cell. The lowest value in the distribution (X_l) becomes the first value in the lowest cell.

Using this formula for this example, the first midpoint would be

$$MP_l = 0.0620 + \frac{0.0003}{2} = 0.06215$$

The remaining midpoint values are determined in the same manner as before. Beginning with the lowest midpoint, 0.06215, add the cell interval of 0.0003, and then add 0.0003 to each successive new midpoint. Starting with a midpoint of 0.06215, we find the other midpoints at 0.06245, 0.06275, 0.06305, and 0.06335.

If the number of values in the cell is high and the distance between the cell boundaries is not large, the midpoint is the most representative value in the cell.

Cell Boundaries The cell size, set by the boundaries of the cell, is determined by the cell midpoints and the cell interval. Locating the cell boundaries, or the limits of the cell, allows the user to place values in a particular cell. To determine the lower cell boundary, divide the cell interval by 2 and subtract that value from the cell midpoint. To calculate the lower cell boundary for a cell with a midpoint of 0.0620, the cell interval is divided by 2:

$$0.0003 \div 2 = 0.00015$$

Then, subtract 0.00015 from the cell midpoint,

$$0.0620 - 0.00015 = 0.06185, \text{ the first lower boundary}$$

To determine the upper cell boundary for a midpoint of 0.0620, add the cell interval to the lower cell boundary:

$$0.06185 + 0.0003 = 0.06215$$

The lower cell boundary of one cell is the upper cell boundary of another. Continue adding the cell interval to each new lower cell boundary calculated until all the lower cell boundaries have been determined.

Take a close look at the values for the cell boundaries. The first cell will have boundaries of 0.06185 and 0.06215. The second cell will have boundaries of 0.06215 and 0.06245. Where would a data value of 0.0621 be placed? In the first cell or the second? As stated earlier, values can be placed in only one cell of a histogram. In order to remove any doubt concerning which value belongs in which cell, the cell boundaries are made to be a half a decimal value greater in accuracy

0.06185
0.0620
0.06215
0.0623
0.06245
0.0626
0.06275
0.0629
0.06305
0.0632
0.06335

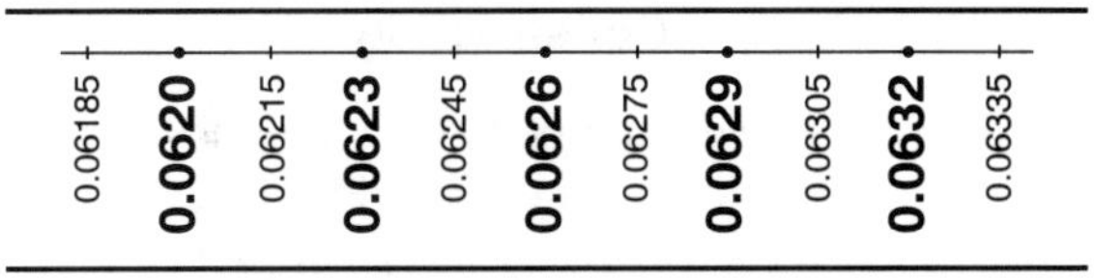

Figure 4.16 Cell Boundaries and Midpoints

than the measured values. This will make the boundaries have one more decimal place than the original values measured. Cell intervals, with their midpoint values starting at 0.0620, are shown in Figure 4.16.

Step 4. Label the Axes. Scale and label the horizontal axis according to the cell boundaries determined in step 3. Label the vertical axis to reflect the amount of data collected, in counting numbers.

Step 5. Post the Values. The final step in the creation of a histogram is to post the values from the check sheet to the histogram. The x axis is marked with the cell midpoints and, if space permits, the cell boundaries. The cell boundaries are used to guide the creator when posting the values to the histogram. On the y axis, the frequency of those values within a particular cell is shown (Figure 4.17). All the data must be included in the cells.

Step 6. Interpret the Histogram. As we can see in Figure 4.17, the data are grouped around 0.0626 and are somewhat symmetrical. In the following sections, we will study histogram shapes, sizes, and locations when compared to a desired target specification. We will also utilize measures such as means, modes, and medians to create a clear picture of where the data are grouped (the central tendency of the data). Standard deviations and ranges will be used to measure how the data are dispersed around the mean. These statistical values will be used to fully describe the data comprising a histogram.

Frequency of occurrence
80
70
60
50
40
30
20
10
0.06185
0.0620
0.06215
0.0623
0.06245
0.0626
0.06275
0.0629
0.06305
0.0632
0.06335

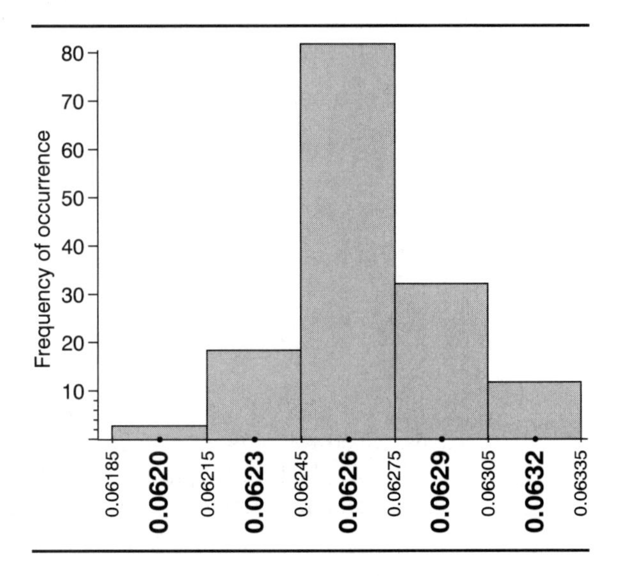

Figure 4.17 Flat Round Plate Thickness Histogram

EXAMPLE 4.10 Constructing a Histogram: Embroidery Arm

Having completed a frequency diagram, R & M's engineers have decided to create a histogram with the data.

Step 1. Collect the Data and Construct a Tally Sheet. This step was performed in Example 4.8 (Figures 4.12 and 4.13).

Step 2. Calculate the Range.

$$\text{Range} = R = X_h - X_l$$
$$R = 11.36 - 11.16 = 0.2$$

Step 3. Create the Cells. Cell intervals are found using the formula:

$$h = \frac{R}{i} + 1$$

where

$$h = \text{number of cells}$$
$$i = \text{cell interval}$$
$$R = \text{range}$$

As we know, a cell interval of 0.01 would result in a frequency diagram. Our choices for cell interval size are 0.03, 0.05, 0.07, and 0.09.

For a cell interval of 0.03:

$$h = \frac{0.2}{0.03} + 1$$
$$h = 8$$

For a cell interval of 0.05:

$$h = \frac{0.2}{0.05} + 1$$
$$h = 5$$

For a cell interval of 0.07:

$$h = \frac{0.2}{0.07} + 1$$
$$h = 4$$

For a cell interval of 0.09:

$$h = \frac{0.2}{0.09} + 1$$
$$h = 3$$

Though several choices exist, we will choose a cell interval of 0.05, resulting in five cells, which will present a clear picture of the data.

The simplest technique for determining cell midpoints is to choose the lowest value measured. Remember, histograms must contain all the data, and no single value can fit into two different cells. For the data, the lowest midpoint would be 11.16, increasing at a cell interval of 0.05 from there. Choosing the lowest value as the first midpoint ensures that all data will fit into the cells of the histogram.

Midpoints: 11.16 11.21 11.26 11.31 11.36

To determine the cell boundaries, divide the cell interval by 2 and subtract that value from the cell midpoint.

$$0.05 \div 2 = 0.025$$

For midpoint 11.16:

$11.16 - 0.025 = 11.135$, the first lower boundary.

For midpoint 11.21:

$11.21 - 0.025 = 11.185$, the second boundary.

Note that the boundaries have one more decimal place of accuracy. This ensures that a measured value fits into only one cell.

Step 4. Label the Axes. Scale and label the horizontal axis according to the cell boundaries and cell midpoints found in step 3. Scale the vertical axis to reflect the frequency of occurrence of the data.

Step 5. Post the Values. Post the values to the histogram, placing each value in the appropriate cell based on the cell boundaries (Figure 4.18).

Step 6. Interpret the Histogram. At first glance, this histogram has one peak and is not symmetrical. In the following sections we will discuss how to interpret and use the histogram's shape, size, and location as clues to process performance. Later, we will learn to use mathematical information to analyze the histogram in greater detail. The mean, mode, and median will tell us how the process is centered. The standard deviation and range will tell us about the amount of variation present in the process. We'll return to this topic.

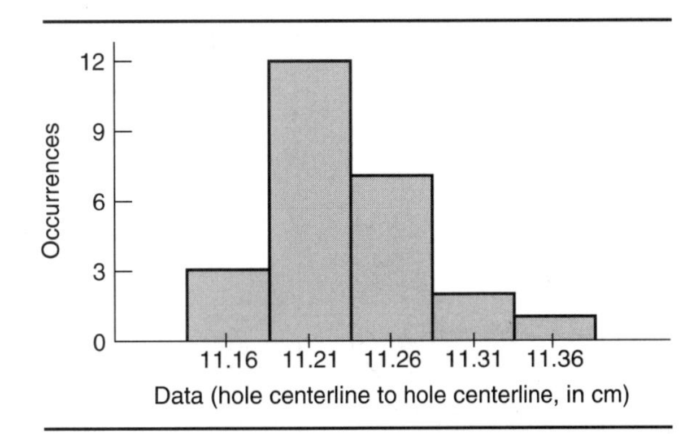

Figure 4.18 Histogram for Monogramming and Embroidery Arm Data

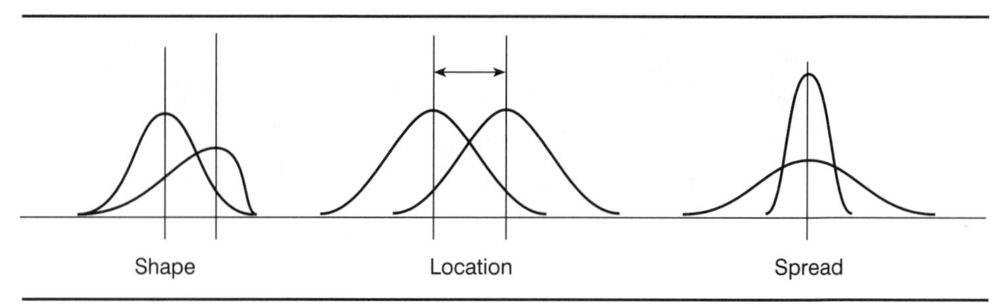

Figure 4.19 Shape, Location, and Spread

Analysis of Histograms

When analyzing a distribution, it is important to remember that it has the following characteristics: shape, location, and spread (Figure 4.19). These three characteristics combine to give us the ability to describe a distribution.

Shape: Symmetry, Skewness, Kurtosis ***Shape*** *refers to the form that the values of the measurable characteristics take on when plotted or graphed.* Tracing a smooth curve over the tops of the rectangular areas used when graphing a histogram clarifies the shape of a histogram for the viewer (Figure 4.20). Identifiable characteristics include ***symmetry,*** or, in the case of lack of symmetry, ***skewness*** of the data; ***kurtosis,*** or *peakedness of the data;* and **modes,** *the number of peaks in the data.*

When a distribution is ***symmetrical,*** *the two halves are mirror images of each other.* The two halves correspond in size, shape, and arrangement (Figure 4.21). When a distribution is not symmetrical, it is considered to be skewed (Figure 4.22). With a ***skewed distribution,*** *the majority of the data are grouped either to the left or the right of a center value, and on the opposite side a few values trail away from the center. When a distribution is* ***skewed to the right,*** *the majority of the data are found on the left side of the figure, with the tail of the distribution going to the right.* The opposite is true for a distribution that is *skewed to the left.*

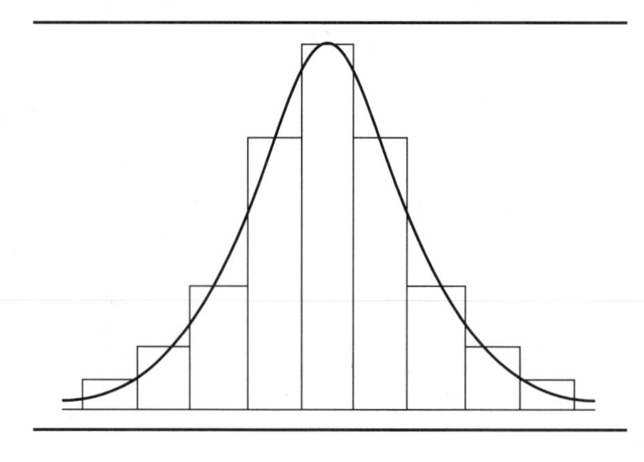

Figure 4.20 Histogram with Smooth Curve Overlay

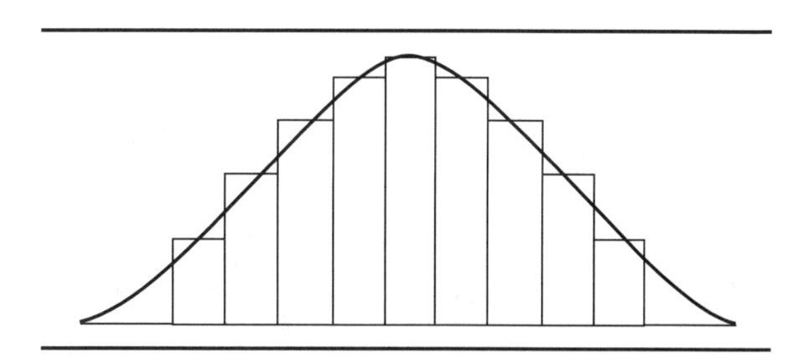

Figure 4.21 Symmetry

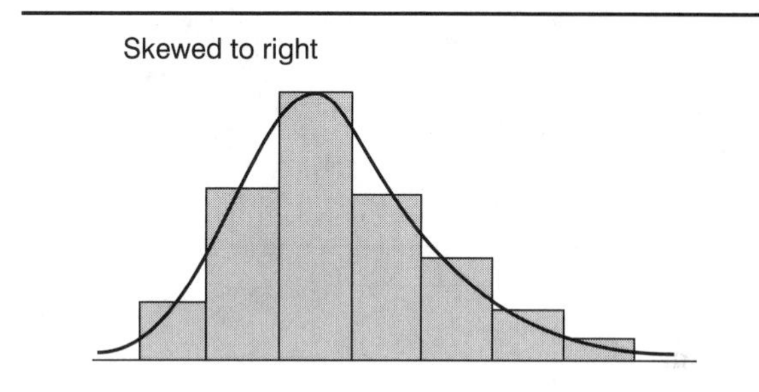

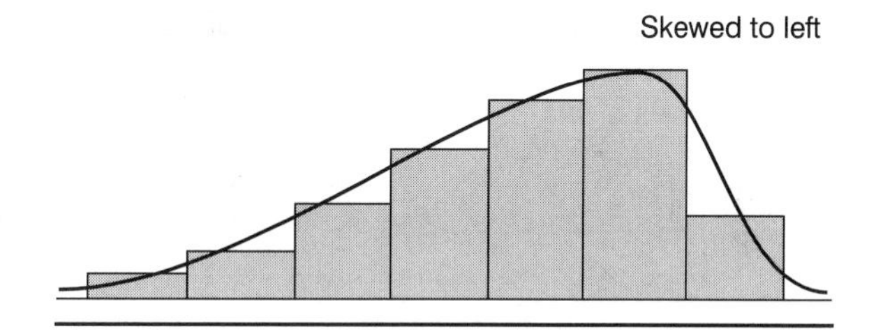

Figure 4.22 Skewness

"Leptokurtic" and "platykurtic" are words used to describe the peakedness of the distribution. A *distribution with a high peak is referred to as* ***leptokurtic;*** *a flatter curve is called* ***platykurtic*** (Figure 4.23). Typically, the kurtosis of a histogram is discussed by comparing it with another distribution. As we will see later in the chapter, skewness and kurtosis can be calculated numerically. Occasionally distributions will display unusual patterns. *If the distribution displays more than one peak,* it is considered ***multimodal.*** *Distributions with two distinct peaks are called* ***bimodal*** (Figure 4.23).

EXAMPLE 4.11 Analyzing the Distribution

Sue's cereal manufacturing department has been monitoring the sugar content in a production run of cereal. Measurements have been taken every 15 minutes and plotted in a histogram. At the end of each day, Sue and Elizabeth, the company's nutritionist, discuss the distribution over the phone. Figure 4.24 shows the distribution

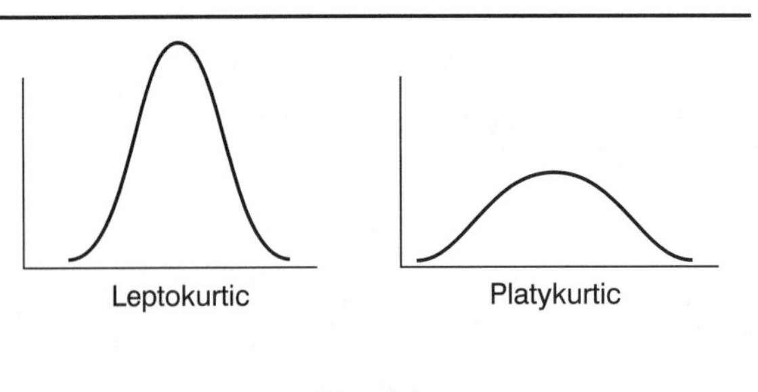

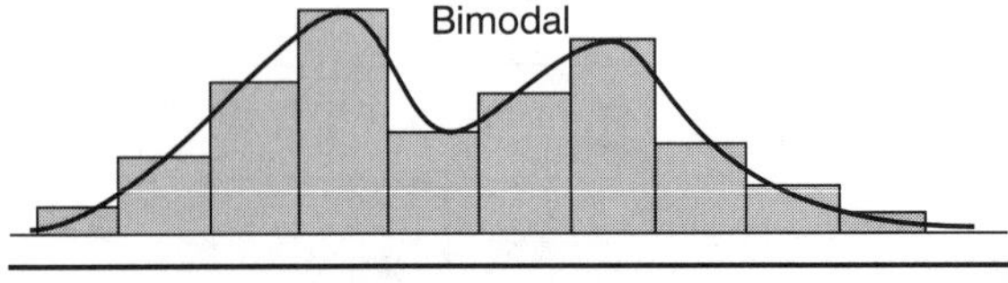

Figure 4.23 Leptokurtic, Platykurtic, and Bimodal Distributions

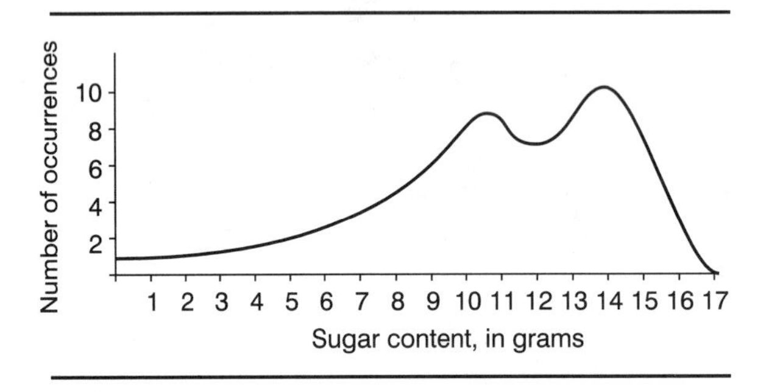

Figure 4.24 Cereal Sugar Content

that Sue now wants to describe to Elizabeth, who does not have the graph. Before she phones, what words should Sue jot down to clearly describe the distribution to Elizabeth?

In this case, the graph clearly shows two peaks in the data, one at 10.5 grams and the second at 14 grams. This bimodal distribution tells Sue and Elizabeth that the sugar concentration in the cereal produced this day peaked at two different points. The graph also shows a wide spread to the data, from 1 to 17 grams, as well as being skewed to the left. Sugar concentration in this particular cereal has not been consistent during the day's production run.

EXAMPLE 4.12 Analyzing the Histogram for the Flat Round Plate Data

Analyzing Figure 4.17 based on the three characteristics of shape, location, and spread reveals that the flat round plate thickness data are fairly consistent. The **shape** of the distribution is relatively symmetrical, skewed slightly to the right. The data are unimodal, centering on 0.0626 inch. Since we have no other distributions of the same type of product, we can not make any comparisons or comments on the kurtosis of the data. **Location,** or where the data are located or gathered, is around 0.0626. If the engineers have specifications of 0.0625 $\pm$ 0.0003, then the center of the distribution is higher than the desired value. Given the specifications, the **spread** of the data is broader than the desired 0.0622 to 0.0628 at 0.0620 to 0.0632. Further mathematical analysis with techniques covered later in this chapter will give us an even clearer picture of the data.

EXAMPLE 4.13 Analyzing the Histogram for the Monogramming and Embroidery Arm Data

The designers of the monogramming and embroidery arm have established the target specification at 11.25 cm. For this distribution, Figure 4.18, the shape is very nearly symmetrical, although the data are located around 11.23 rather than the preferred location of 11.25. The spread of the data is rather large, from 11.16 to 11.36.

Histograms provide a visual description of the information under study. Discrepancies in the data, such as gaps or unusual occurrences, can be seen at a glance (Figure 4.25). By comparing the spread of the data with the specifications, you can study

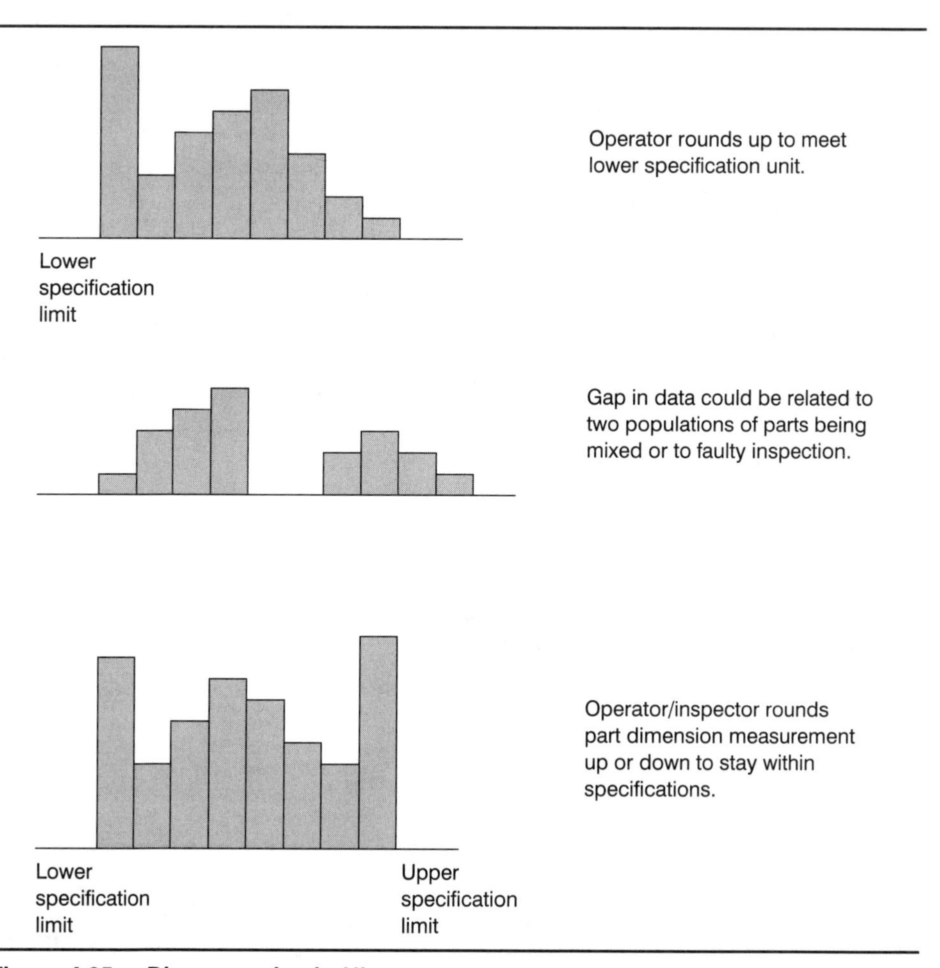

Figure 4.25 Discrepancies in Histograms

the capability of the process. Histograms made from sample data can also suggest what the actual population looks like. As we shall see in Chapters 5 and 6, histograms are very powerful statistical process control tools when combined with other statistical values.

EXAMPLE 4.14 Reading Histograms

Automobile manufacturers and their suppliers are constantly looking for new plastic and composite materials that will be resistant to cracking, chipping, and discoloration. At W.T. Plastics, Inc., a new dashboard is currently being tested. Scientists in W.T.'s labs are measuring the size of cracks in a group of dashboards that has completed a rigorous series of tests designed to mimic stressful environmental conditions. The dashboards have come from two different lots of plastic. Figure 4.26 shows a histogram of the crack sizes.

Upon learning that two different lots of dashboards have been combined, the head of the lab asks that the data be separated according to lot and two histograms be created (Figure 4.27). Data from two different populations should not be mixed into one statistical analysis. Now it is clear to all who view the chart that type A dashboards have a histogram that is slightly skewed to the left, while type B dashboards have a histogram that is skewed to the right. Type B's histogram is more leptokurtic than type A's. This means that type A's cracks vary more widely in size compared with type B's cracks.

Cumulative Frequency Distribution

A ***cumulative frequency distribution*** *shows the cumulative occurrences of all the values; the values from each preceding cell are added to the next cell until the uppermost cell boundary is reached.* Unlike a histogram, the cumulative frequency distribution adds the previous cell's number of occurrences to the next cell. Figure 4.28 shows a cumulative frequency distribution made from the information from Figure 4.27, type B dashboards.

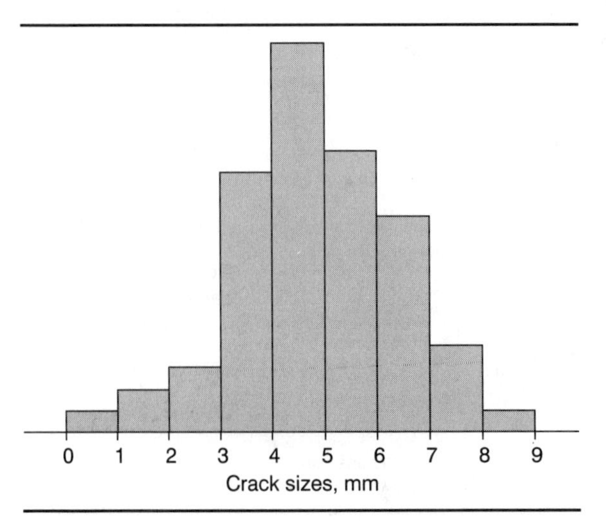

Figure 4.26 Histogram for Crack Sizes in Dashboards

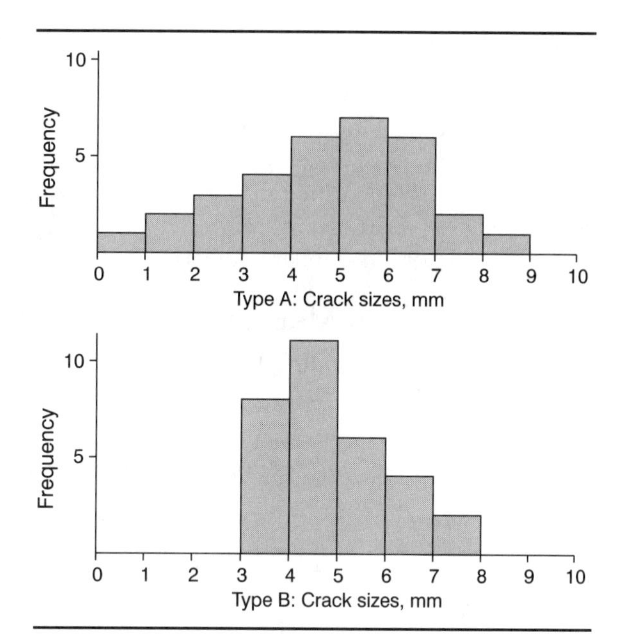

Figure 4.27 Histogram for Crack Sizes by Dashboard Type

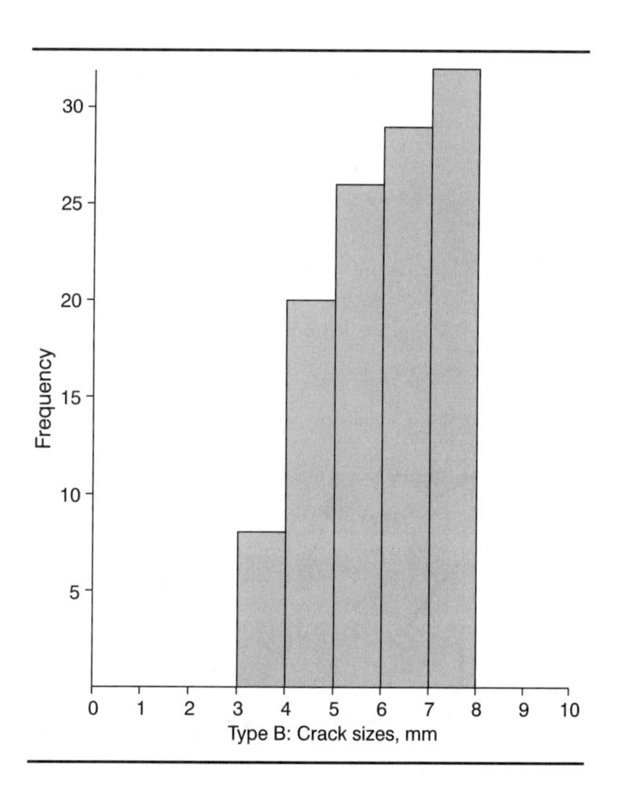

Figure 4.28 Cumulative Frequency Distribution

DATA ANALYSIS: ANALYTICAL

So far, distributions have been described by looking at a picture of the actual histogram. There are other, more analytical methods of describing histograms and distributions. Shape was easily seen from a picture: the location and spread can be more clearly identified mathematically. Location is described by measures of central tendency: the mean, mode, and median. Spread is defined by measures of dispersion: the range and standard deviation.

Location: Measures of Central Tendency

To fully describe a histogram, you must use calculated values as well as diagrams. Averages, medians, and modes are the statistical values that define the center of a distribution. *Since they reveal the place where the data tend to be gathered, these values are commonly called the* ***measures of central tendency.***

Mean

The **mean** *of a series of measurements is determined by adding the values together and then dividing this sum by the total number of values*. When this value is calculated for a population, it is referred to as the mean and is signified by μ. When this value is calculated for a sample, it is called the *average* and is signified by $\overline{X}$ (X bar). To calculate the mean of a population use the following formula:

$$\mu = \frac{X_1 + X_2 + X_3 + \cdots + X_n}{n} = \frac{\sum_{i=1}^{n} X_i}{n}$$

where

$$\begin{aligned} \mu &= \text{mean value of the series of measurements} \\ X_1, X_2, \cdots, X_n &= \text{values of successive measurements} \\ n &= \text{number of readings} \end{aligned}$$

The same formula can be used to calculate the average associated with a sample. Means and averages can be used to judge whether or not a group of values is accurate or on target.

EXAMPLE 4.15 Determining the Mean for the Flat Round Plate Data

Averages for each of the subgroups for the thicknesses of the flat round plate can be calculated.

1. Calculate the sum of each set of subgroup values:

   Subgroup 1:

   $$\begin{aligned} \Sigma X_i &= 0.0625 + 0.0626 + 0.0624 + 0.0625 + 0.0627 \\ &= 0.3127 \end{aligned}$$

2. Calculate the subgroup average by dividing the sum by the number of samples in the subgroup ($n = 5$):

Subgroup 1:

$$\overline{X} = \frac{0.0625 + 0.0626 + 0.0624 + 0.0625 + 0.0627}{5}$$
$$= \frac{0.3127}{5}$$
$$= 0.0625$$

Table 4.5 gives a list of the sums and averages calculated for this example. Once the averages for each subgroup have been calculated, a grand average for all of the subgroups can be found by dividing the sum of the subgroup sums by the total number of items taken in all of the subgroups (150). A grand average is designated as $\overline{\overline{X}}$ (X double bar):

$$\overline{\overline{X}} = \frac{0.3127 + 0.3122 + 0.3121 + 0.3114 + \cdots + 0.3149}{150}$$
$$= \frac{9.3997}{150}$$
$$= 0.0627$$

Notice that an average of the averages is not taken. Taking an average of the averages will work only when the sample sizes are constant. Use the sums of each of the subgroups to perform the calculation.

EXAMPLE 4.16 Determining the Mean for the Monogramming and Embroidery Arm Data

The data for the monogramming and embroidery arm represent the total population of parts produced during a runoff that judged machine performance. To calculate the average value,

$$\Sigma X_i = 11.16 + 11.17 + 11.18 + 2(11.19) + 2(11.21) + 5(11.22) + 3(11.23)$$
$$+ 2(11.24) + 3(11.25) + 11.27 + 11.28 + 11.29 + 11.33 + 11.36$$
$$= 280.86$$

Then:

$$\overline{X} = \frac{280.86}{25}$$
$$\overline{X} = 11.23$$

Notice that this value confirms our conclusion in Example 4.13 that the distance between the centers of the holes is smaller than the desired specification of 11.25.

Median

The ***median*** *is the value that divides an ordered series of numbers so that there is an equal number of values on either side of the center, or median, value.* An ordered series of data

Table 4.5 Flat Round Plate Thickness: Sums and Averages

						ΣX_i	$\overline{X}$
Subgroup 1	0.0625	0.0626	0.0624	0.0625	0.0627	0.3127	0.0625
Subgroup 2	0.0624	0.0623	0.0624	0.0626	0.0625	0.3122	0.0624
Subgroup 3	0.0622	0.0625	0.0623	0.0625	0.0626	0.3121	0.0624
Subgroup 4	0.0624	0.0623	0.0620	0.0623	0.0624	0.3114	0.0623
Subgroup 5	0.0621	0.0621	0.0622	0.0625	0.0624	0.3113	0.0623
Subgroup 6	0.0628	0.0626	0.0625	0.0626	0.0627	0.3132	0.0626
Subgroup 7	0.0624	0.0627	0.0625	0.0624	0.0626	0.3126	0.0625
Subgroup 8	0.0624	0.0625	0.0625	0.0626	0.0626	0.3126	0.0625
Subgroup 9	0.0627	0.0628	0.0626	0.0625	0.0627	0.3133	0.0627
Subgroup 10	0.0625	0.0626	0.0628	0.0626	0.0627	0.3132	0.0626
Subgroup 11	0.0625	0.0624	0.0626	0.0626	0.0626	0.3127	0.0625
Subgroup 12	0.0630	0.0628	0.0627	0.0625	0.0627	0.3134	0.0627
Subgroup 13	0.0627	0.0626	0.0628	0.0627	0.0626	0.3137	0.0627
Subgroup 14	0.0626	0.0626	0.0625	0.0626	0.0627	0.3130	0.0626
Subgroup 15	0.0628	0.0627	0.0626	0.0625	0.0626	0.3132	0.0626
Subgroup 16	0.0625	0.0626	0.0625	0.0628	0.0627	0.3131	0.0626
Subgroup 17	0.0624	0.0626	0.0624	0.0625	0.0627	0.3126	0.0625
Subgroup 18	0.0628	0.0627	0.0628	0.0626	0.0630	0.3139	0.0627
Subgroup 19	0.0627	0.0626	0.0628	0.0625	0.0627	0.3133	0.0627
Subgroup 20	0.0626	0.0625	0.0626	0.0625	0.0627	0.3129	0.0626
Subgroup 21	0.0627	0.0626	0.0628	0.0625	0.0627	0.3133	0.0627
Subgroup 22	0.0625	0.0626	0.0628	0.0625	0.0627	0.3131	0.0626
Subgroup 23	0.0628	0.0626	0.0627	0.0630	0.0627	0.3138	0.0628
Subgroup 24	0.0625	0.0631	0.0630	0.0628	0.0627	0.3141	0.0628
Subgroup 25	0.0627	0.0630	0.0631	0.0628	0.0627	0.3143	0.0629
Subgroup 26	0.0632	0.0628	0.0631	0.0628	0.0627	0.3149	0.0630
Subgroup 27	0.0630	0.0628	0.0631	0.0628	0.0627	0.3144	0.0629
Subgroup 28	0.0632	0.0632	0.0628	0.0631	0.0630	0.3153	0.0631
Subgroup 29	0.0630	0.0628	0.0631	0.0632	0.0631	0.3152	0.0630
Subgroup 30	0.0632	0.0631	0.0630	0.0628	0.0628	0.3149	0.0630
						9.3997	

has been arranged according to their magnitude. Once the values are placed in order, the median is the value of the number that has an equal number of values to its left and right. In the case of finding a median for an even number of values, the two center values of the ordered set of numbers are added together and the result is divided by 2. Figure 4.29 shows the calculation of several medians.

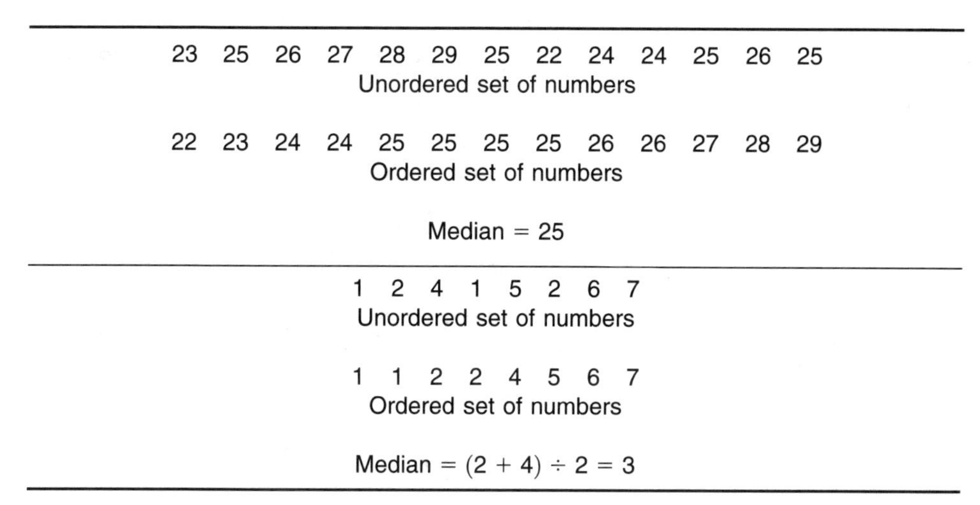

Figure 4.29 Calculating Medians

EXAMPLE 4.17 Determining the Median of the Flat Round Plate Data

From the check sheet (Figure 4.9) the median of the flat round plate thickness data can be found. When the data are placed in an ordered series, the center or median number is found to be 0.0626.

EXAMPLE 4.18 Determining the Median of the Monogramming and Embroidery Arm

Figure 4.13 can be used to determine the median of the monogramming and embroidery arm data. When the data are placed in an ordered series, the center, or median number, is 11.23.

Mode

The ***mode*** *is the most frequently occurring number in a group of values*. In a set of numbers, a mode may or may not occur (Figure 4.30). A set of numbers may also have two or more modes. If a set of numbers or measurements has one mode, it is said to be unimodal. If it has two numbers appearing with the same frequency, it is called bimodal. Distributions with more than two modes are referred to as multimodal. In a frequency distribution or a histogram, the cell with the highest frequency is the mode.

EXAMPLE 4.19 Determining the Mode for the Flat Round Plate Data

The mode can be found for the flat round plate thickness data. The check sheet (Figure 4.9) clearly shows that 0.0626 is the most frequently occurring number. It is tallied 30 times.

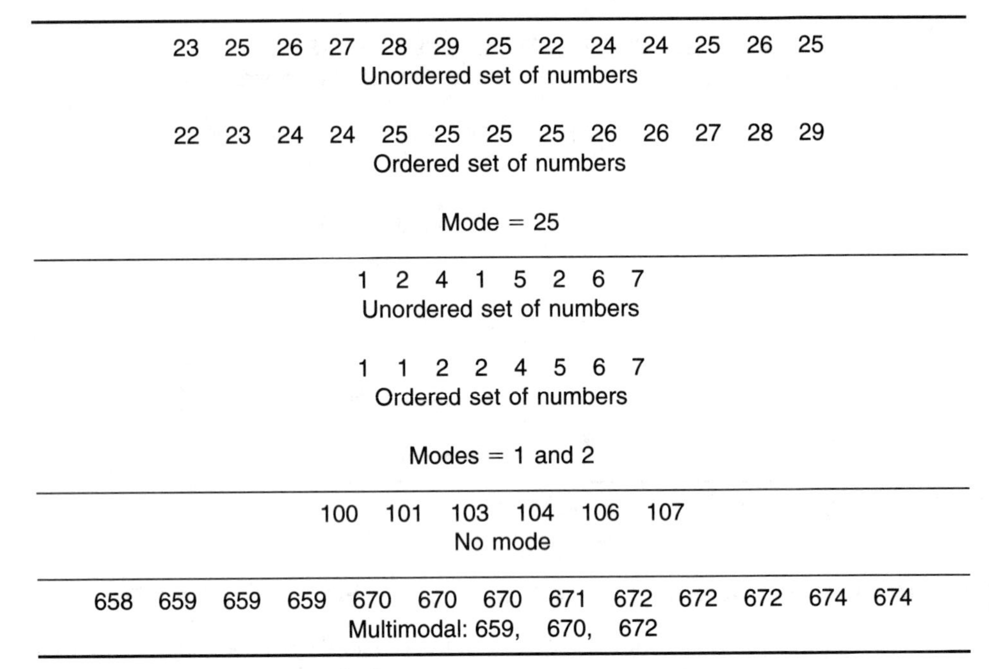

Figure 4.30 Calculating Modes

EXAMPLE 4.20 Determining the Mode for the Monogramming and Embroidery Arm Data

The mode for the monogramming and embroidery arm data can be found by looking at the frequency diagram created in Figure 4.14. This figure clearly shows that 11.22 cm is the most frequently occurring number, appearing five times.

The Relationship between the Mean, Median, and Mode

As measures of central tendency, the mean, median, and mode can be compared with each other to determine where the data are located. Measures of central tendency describe the center position of the data. They show how the data tend to build up around a center. When a distribution is symmetrical, the mean, mode, and median values are equal. For a skewed distribution, the values will be different (Figure 4.31). Comparing the mean (average), mode, and median determines whether or not a distribution is skewed and, if it is, in which direction.

EXAMPLE 4.21 Seeing the Relationship: Flat Round Plate Data

Knowing the average, median, and mode of the flat round plate data provides information about the symmetry of the data. If the distribution is symmetrical, the average, mode, and median values will be equal. A skewed distribution will have different values. From previous examples, the values for the round flat plate are

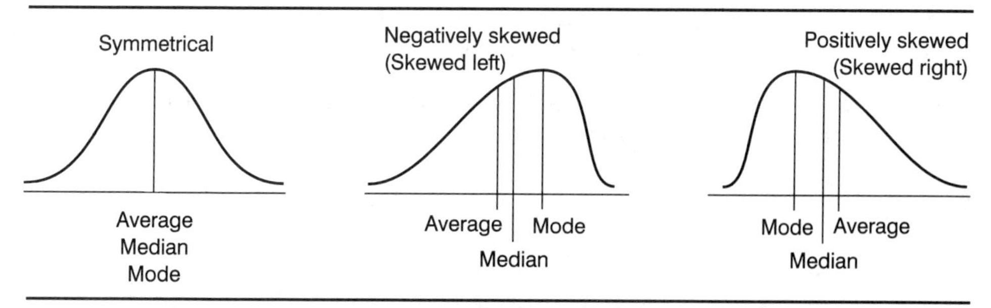

Figure 4.31 Comparison of Mean, Mode, and Median

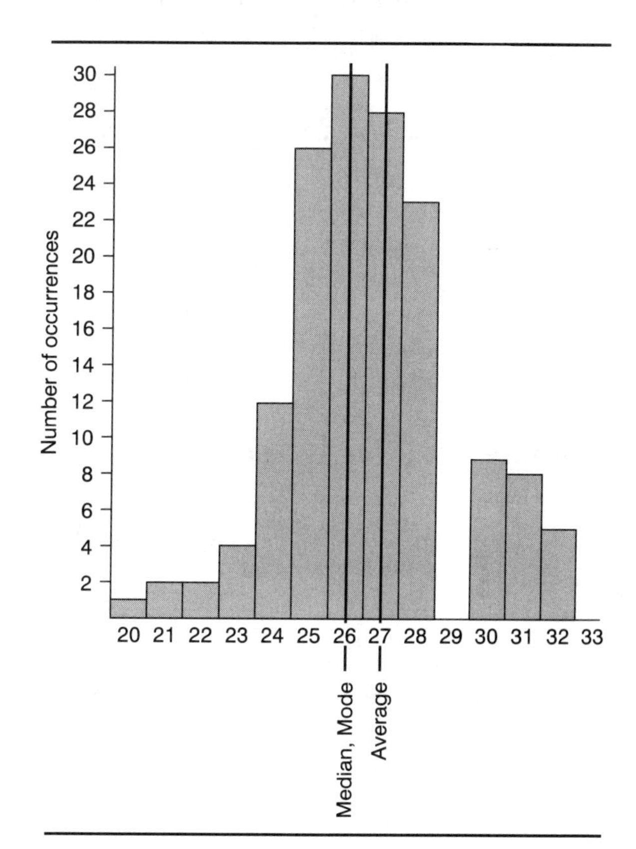

Figure 4.32 Comparison of Mean, Mode, and Median for the Flat Round Plate

Average = 0.0627 inch
Median = 0.0626 inch
Mode = 0.0626 inch

As seen in the frequency diagram (Figure 4.32), the mode marks the peak of the distribution. The average, slightly to the right of the mode and median, pulls the distribution to the right. This slight positive skew is due to the high values for plate thickness that occur in later samples.

EXAMPLE 4.22 Seeing the Relationship: Monogramming and Embroidery Arm Data

We can use the average, median, and mode of the monogramming and embroidery arm data to discuss the symmetry of the data.

Average = 11.23 cm
Median = 11.23 cm
Mode = 11.22 cm

The values for the average, median, and mode are nearly equal; thus the distribution is almost symmetrical. The mode skews the distribution slightly to the right. The central tendency of the data is lower than the target specification of 11.25 cm.

Spread: Measures of Dispersion

Spread refers to the span of values of a measurable characteristic, where they fall in relation to each other and the mean. The range and standard deviation are two measurements that enable the investigator to determine the spread of the data. Because these two describe where the data are dispersed on either side of a central value, they are often referred to as measures of dispersion. Used in conjunction with the mean, mode, and median, these values create a more complete picture of a distribution. The importance of this information becomes apparent in the next example.

EXAMPLE 4.23 The Significance of the Range

Sally and Harry, at the end of their vacation in New York City, are planning to take a taxi to the airport. At the front desk of their hotel, they ask the desk clerk how long it will take a taxi to drive from the hotel to the airport. The clerk answers that on average it takes 30 minutes to get to the airport by taxi. Just to be on the safe side, Sally and Harry decide to allow 45 minutes for the taxi ride, planning to arrive 45 minutes before their flight. But because of heavy Friday night traffic, the taxi ride to the airport takes over an hour. Angry that they have missed their flight, Harry and Sally call the hotel to complain. The desk clerk reminds them that he had cited 30 minutes as "an average." The time to the airport, he now explains, can range anywhere from 20 to 90 minutes. Had he given them the spread of the data, Harry and Sally could have made their flight.

Range

As was pointed out in the discussion of the histogram earlier in this chapter, the ***range** is the difference between the highest value in a series of values or sample and the lowest value in that same series*. A range value describes how far the data spread. All of the other values in a population or sample will fall between the highest and lowest values:

$$R = X_h - X_l$$

where

$$R = \text{range}$$
$$X_h = \text{highest value in the series}$$
$$X_l = \text{lowest value in the series}$$

EXAMPLE 4.24 Calculating Range Values for the Flat Round Plate Data

The flat round plate data comprises subgroups of sample size five (Table 4.5). For each sample, a range value can be calculated. For example:

Subgroup 1 0.0625 0.0626 0.0624 0.0625 0.0627

$$\text{Range} = X_h - X_l = 0.0627 - 0.0624 = 0.0003$$

Subgroup 2 0.0624 0.0623 0.0624 0.0626 0.0625

$$\text{Range} = X_h - X_l = 0.0626 - 0.0623 = 0.0003$$

The other ranges are calculated in the same manner. These range values are used in the next chapter to study the variation present in the process over time.

EXAMPLE 4.25 Determining the Range for the Monogramming and Embroidery Arm Data

The monogramming and embroidery arm data, taken from the machine runoff, represent a complete population. To calculate the range, simply subtract the highest value from the lowest:

$$\text{Range} = X_h - X_l = 11.36 - 11.16 = 0.2 \text{ cm}$$

With a target specification of 11.25, this range is a bit broad—some embroidery arms will have mounting holes that are farther apart than others. The range, which gives us a picture of how much variation is present in the process, is letting us know that the product being produced is inconsistent. The variation present in this process will make assembly difficult.

Standard Deviation

The range shows where each end of the distribution is located, but it doesn't tell how the data are grouped within the distribution. In Figure 4.33, the three distributions have the same average and range, but all three are different. *The **standard deviation** shows the dispersion of the data within the distribution*. It describes how the values fall in relation to the mean. The standard deviation, because it uses all of the measurements taken, provides more reliable information about the dispersion of the data. The range considers only the two extreme values in its calculation, giving no information concerning where the values may be grouped. Since it only considers the highest and lowest values, the range has the disadvantage of becoming a less accurate description of the data as the number of readings or sample values increases. The range is best used with small populations or small sample sizes of less than 10 values. However, since the range is easy to calculate, it is the most frequently used measure of dispersion.

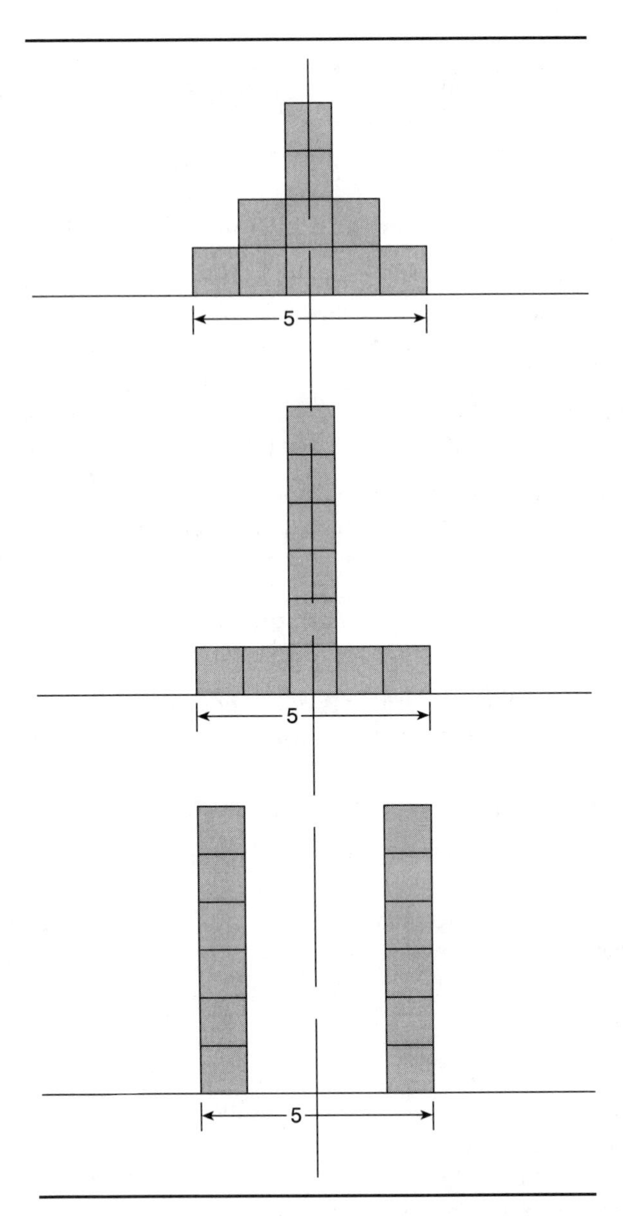

Figure 4.33 Different Distributions with Same Averages and Ranges

When the measurements have been taken from each and every item in the total population, the standard deviation is calculated through the use of the following formula:

$$\sigma = \sqrt{\frac{\sum_{i=1}^{n}(X_i - \mu)^2}{n}}$$

where

σ = standard deviation of the population
μ = mean value of the series of measurements
$X_i = X_1, X_2, \ldots, X_n$ = values of each reading
n = number of readings

The standard deviation of the population is sometimes known as the *root mean square deviation*. When populations increase in size it becomes difficult to calculate without help from a computer.

EXAMPLE 4.26 Determining the Standard Deviation of a Population: Monogramming and Embroidery Arm Data

Calculating the standard deviation of a population requires using the following formula:

$$\sigma = \sqrt{\frac{\Sigma(X_i - \mu)^2}{n}}$$

$$= \sqrt{\frac{\begin{array}{c}(11.16 - 11.23)^2 + (11.17 - 11.23)^2 + (11.18 - 11.23)^2 \\ + 2(11.19 - 11.23)^2 + 2(11.21 - 11.23)^2 + 5(11.22 - 11.23)^2 \\ + 3(11.23 - 11.23)^2 + 2(11.24 - 11.23)^2 + 3(11.25 - 11.23)^2 \\ + (11.27 - 11.23)^2 + (11.28 - 11.23)^2 + (11.29 - 11.23)^2 \\ + (11.33 - 11.23)^2 + (11.36 - 11.23)^2\end{array}}{25}}$$

$$= 0.05$$

Like the range, the standard deviation tells us how the data are spread around the mean. The magnitude of this standard deviation indicates that the data are not tightly grouped; that is, the distance between the centers of the holes varies from one embroidery arm to another. The variation present in the process will make assembly and use of the embroidery arm more difficult.

A smaller standard deviation is desirable because it indicates greater similarity between data values—that is, the data are more precisely grouped. In the case of products, a small standard deviation indicates that the products are nearly alike. As discussed with the Taguchi loss function, creating products or providing services that are similar to each other is optimal.

When the measurements are taken from items sampled from the total population, the previous formula is modified to reflect the fact that not every item in the population has been measured. This change is reflected in the denominator. The standard deviation of a sample is represented by the letter s:

$$s = \sqrt{\frac{\sum_{i=1}^{n}(X_i - \bar{X})^2}{n - 1}}$$

where

$$s = \text{standard deviation of the sample}$$
$$\overline{X} = \text{average value of the series of measurements}$$
$$X_i = X_1, X_2, \ldots, X_n = \text{values of each reading}$$
$$n = \text{number of readings}$$

EXAMPLE 4.27 Determining the Standard Deviation of a Sample: Flat Round Plate Data

In the case of subgroups comprising the flat round plate data, it is possible to calculate the sample standard deviation for each of the subgroups. For subgroup 1:

$$s_1 = \sqrt{\frac{\Sigma(X_i - X)^2}{n}}$$

$$= \sqrt{\frac{(0.0624 - 0.0625)^2 + 2(0.0625 - 0.0625)^2 + (0.0626 - 0.0625)^2 + (0.0627 - 0.0625)^2}{5}}$$

$$= 0.0001$$

Standard deviations for the remaining subgroups can be calculated in the same manner. These standard deviations are used in the next chapter to show the differences between the plates over time.

EXAMPLE 4.28 Determining the Standard Deviation of a Sample

At an automobile-testing ground, a new type of automobile was tested for gas mileage. Seven cars, a sample of a much larger production run, were driven under typical conditions to determine the number of miles per gallon the cars got. The following miles-per-gallon readings were obtained:

36 35 39 40 35 38 41

Calculate the sample standard deviation.
First calculate the average:

$$\overline{X} = \frac{36 + 35 + 39 + 40 + 35 + 38 + 41}{7}$$

$$= 37.7$$

which is rounded to 38. Then

$$s = \sqrt{\frac{(36 - 38)^2 + (35 - 38)^2 + (39 - 38)^2 + (40 - 38)^2 + (35 - 38)^2 + (38 - 38)^2 + (41 - 38)^2}{7 - 1}}$$

$$= 2.43$$

which is rounded to 2.4.

Using the Mean, Mode, Median, Standard Deviation, and Range Together

Measures of central tendency and measures of dispersion are critical when describing statistical data. As the following example shows, one without the other creates an incomplete picture of the values measured.

EXAMPLE 4.29 Seeing the Whole Picture

Two professors were keeping track of the ages of students taking part in their summer workshop. Over the past three summers ages were recorded and the frequency diagrams shown in Figure 4.34 were created. When they studied the data originally, the two professors had calculated only the mean, mode, and median for each workshop.

$$\text{Mean}_1 = 20 \quad \text{Median}_1 = 20 \quad \text{Mode}_1 = 20$$
$$\text{Mean}_2 = 20 \quad \text{Median}_2 = 20 \quad \text{Mode}_2 = 20$$
$$\text{Mean}_3 = 20 \quad \text{Median}_3 = 20 \quad \text{Mode}_3 = 20$$

On the surface, these distributions appear the same. It was not until the range and standard deviation for each of the three workshops were calculated that the differences became apparent:

$$\text{Range}_1 = 4$$
$$\text{Standard deviation}_1 = 1.03\text{, rounded to }1.0$$

$$\sigma = \sqrt{\frac{3(18-20)^2 + 7(19-20)^2 + 16(20-20)^2 + 7(21-20)^2 + 3(22-20)^2}{36}}$$

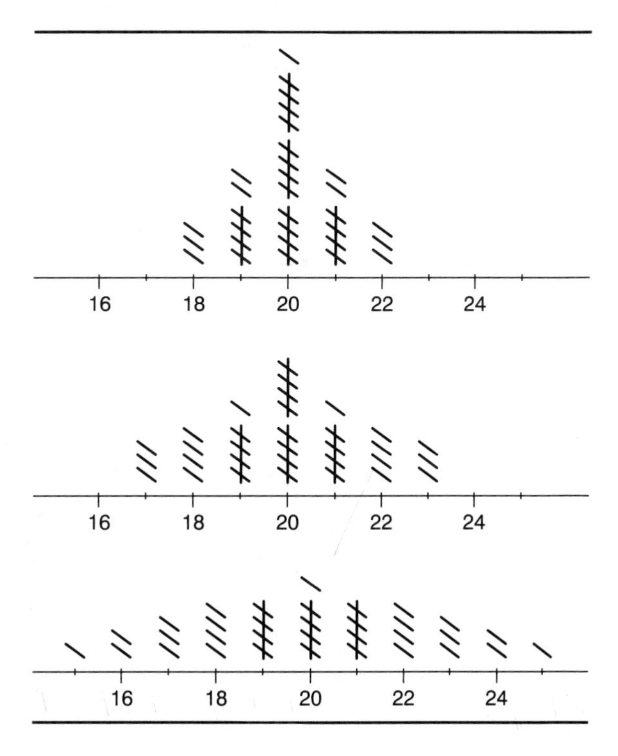

Figure 4.34 Frequency Diagrams of Student Ages

$\text{Range}_2 = 6$

$\text{Standard deviation}_2 = 1.65$, rounded to 1.7

$$\sigma = \sqrt{\frac{3(17-20)^2 + 4(18-20)^2 + 6(19-20)^2 + 10(20-20)^2 + 6(21-20)^2 + 4(22-20)^2 + 3(23-20)^2}{36}}$$

$\text{Range}_3 = 10$

$\text{Standard deviation}_3 = 2.42$, rounded to 2.4

$$\sigma = \sqrt{\frac{1(15-20)^2 + 2(16-20)^2 + 3(17-20)^2 + 4(18-20)^2 + 5(19-20)^2 + 6(20-20)^2 + 5(21-20)^2 + 4(22-20)^2 + 3(23-20)^2 + 2(24-20)^2 + 1(25-20)^2}{36}}$$

However, when the professors calculated the ranges and standard deviations it became apparent that as the program continued it appealed to a wider variety of ages.

EXAMPLE 4.30 Seeing the Whole Picture: Flat Round Plate Data

When we combine the analytical calculations with the graphical information from the previous examples, we see a more complete picture of the flat round plate data we are studying. The grand average, $\overline{\overline{X}} = 0.0627$ inches, median (0.0626), and mode (0.0626) confirm that the histogram is skewed slightly to the right. Because we know the grand average of the data, we also know that the distribution is not centered on the desired target value of 0.0625 inches. The frequency diagram gives us the critical information that there are no plates with a thickness of 0.0629 inches. The range of our data is fairly broad; the frequency diagram shows an overall spread of the distribution of 0.0012 inches. In the next two chapters we will learn how to use the ranges and standard deviation values for the individual subgroups. In general, through their calculations and diagrams, the engineers have learned that they are making the plates too thick, on average. They have also learned that the machining process is not producing plates of consistent thickness.

EXAMPLE 4.31 Seeing the Whole Picture: Monogramming and Embroidery Arm Data

The engineers involved in the monogramming and embroidery arm runoff are interested in whether or not the machine that they are considering for purchase is consistently able to produce arms with mounting holes that meet their target specification of 11.25 cm between centers. From the diagrams and calculations, they are able to determine that the machining process produces arms that have a distance between centers of 11.23 cm on average, narrower than their desired setting. The machining process also has a very slight skew to the right, as shown by the mode. The process is not consistent, as shown by a range of 0.2 cm and a standard deviation of 0.05 cm. Without collecting, tabulating, analyzing, and interpreting the data from the runoff, the engineers might have made an incorrect decision to purchase a piece of equipment that is not able to produce parts to their specifications.

Other Measures of Dispersion

Skewness

When a distribution lacks symmetry, it is considered ***skewed.*** A picture of the distribution is not necessary to determine skewness. Skewness can be measured by calculating the following value:

$$a_3 = \frac{\sum_{i=1}^{h} f_i(X_i - \overline{X})^3/n}{s^3}$$

where

a_3 = skewness
X_i = individual data values under study
$\overline{X}$ = average of individual values
n = sample size
s = standard deviation of sample
f_i = frequency of occurrence

Once determined, the skewness figure is compared with zero. A skewness value of zero means that the distribution is symmetrical. A value greater than zero means that the data are skewed to the right; the tail of the distribution goes to the right. If the value is negative, less than zero, then the distribution is skewed to the left, with a tail of the distribution going to the left (Figure 4.35). The higher the value, the stronger the skewness.

EXAMPLE 4.32 Determining the Skew

The engineers in Example 4.28 want to determine if the data gathered in their sample of 36, 35, 39, 40, 35, 38, 41 are skewed. They apply the following formulas:

$$\overline{X} = 38$$
$$s = 2$$
$$a_3 = \frac{[(36 - 38)^3 + 2(35 - 38)^3 + (39 - 38)^3 + (40 - 38)^3 + (38 - 38)^3 + (41 - 38)^3]/7}{8}$$
$$= -0.46$$

The negative value for a_3 means that the data are skewed to the left.

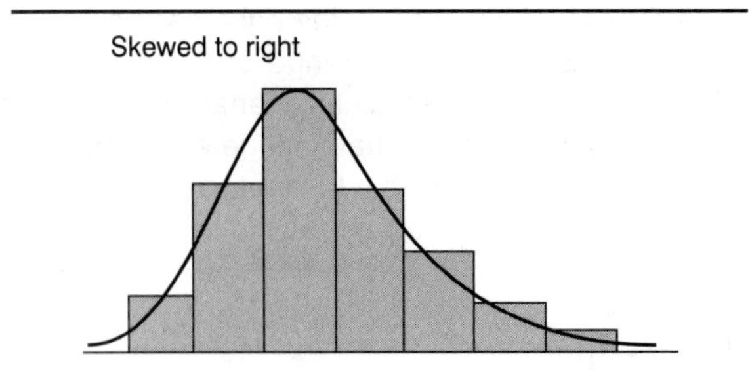

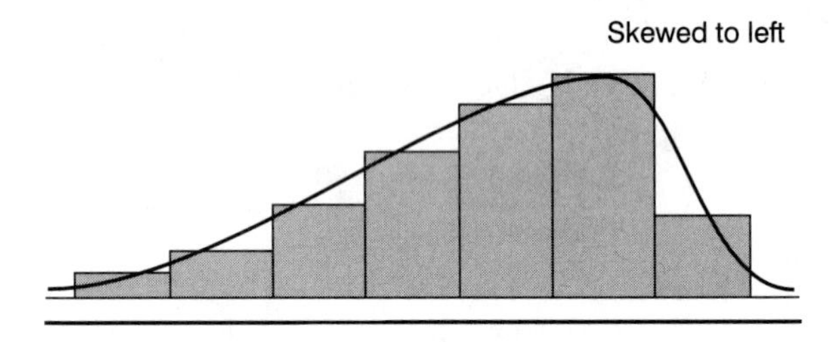

Figure 4.35 Skewness

Kurtosis

Kurtosis, the peakedness of the data, is another value that can be calculated:

$$a_4 = \frac{\sum_{i=1}^{h} f_i(X_i - \overline{X})^4/n}{s^4}$$

where

a_4 = kurtosis
X_i = individual data values under study
$\overline{X}$ = average of individual values
n = sample size
s = standard deviation of sample

Once calculated, the kurtosis value must be compared with another distribution or with a standard in order to be interpreted. In Figure 4.36, the distribution on the left side is more peaked than that on the right. The concept is similar to comparing heights of students; one is considered shorter (less peaked) only when there is a taller (more peaked) individual in the class. Kurtosis values are used to compare one distribution with another.

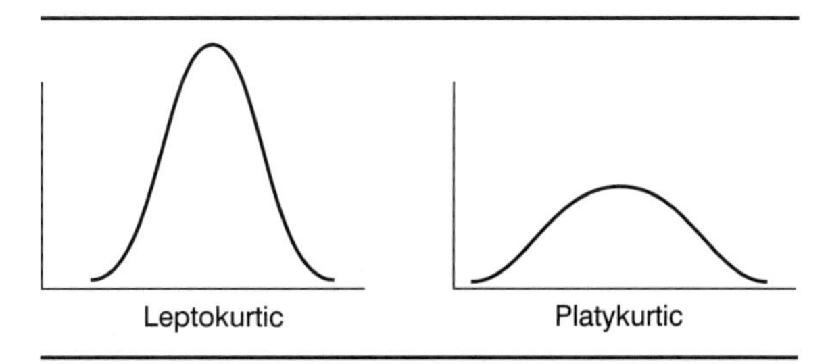

Figure 4.36 Kurtosis

EXAMPLE 4.33 Determining the Kurtosis

The engineers of Example 4.28 wanted to calculate the kurtosis:

$$a_4 = \frac{[(36-38)^4 + 2(35-38)^4 + (39-38)^4 + (40-38)^4 + (38-38)^4 + (41-38)^4]/7}{16}$$
$$= 2.46$$

The engineers compared these data with the previous day's test, where the kurtosis was 1.42. Those data spread more broadly (were less peaked) than these. The distribution is getting narrower, a sign that the miles-per-gallon data are becoming more uniform.

NORMAL FREQUENCY DISTRIBUTION

The normal frequency distribution, the familiar bell-shaped curve (Figure 4.37), is commonly called a *normal curve*. A ***normal frequency distribution*** *is described by the normal density function:*

$$f(x) = \frac{1}{\sqrt{2\pi}\sigma} e^{-(x-\mu)^2/2\sigma^2} \qquad -\infty < x < \infty$$

where

$$\pi = 3.14159$$
$$e = 2.71828$$

The normal frequency distribution has six distinct features:

1. A normal curve is symmetrical about μ, the central value.
2. The mean, mode, and median are all equal.
3. The curve is unimodal and bell-shaped.
4. Data values concentrate around the mean value of the distribution and decrease in frequency as the values get further away from the mean.
5. The area under the normal curve equals 1. One hundred percent of the data are found under the normal curve, 50 percent on the left-hand side, 50 percent on the right.

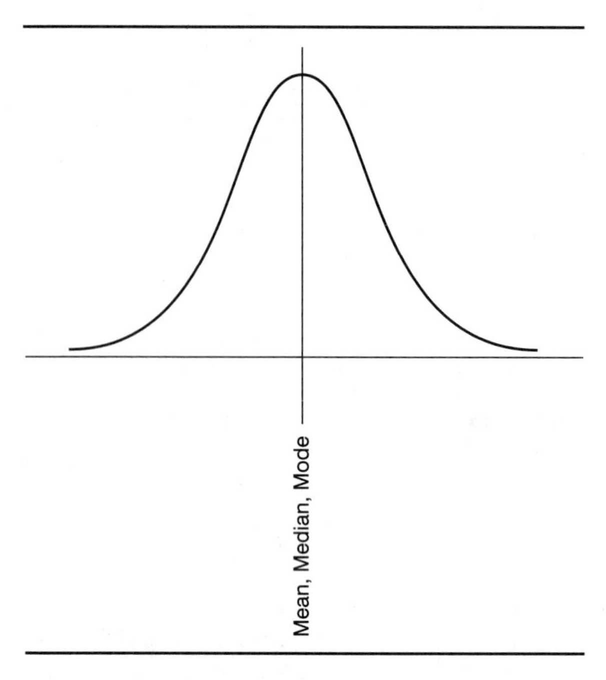

Figure 4.37 The Normal Curve

6. The normal distribution can be described in terms of its mean and standard deviation by observing that 99.73 percent of the measured values fall within ± 3 standard deviations of the mean ($\mu \pm 3\sigma$), that 95.5 percent of the data fall within ± 2 standard deviations of the mean ($\mu \pm 2\sigma$), and that 68.3 percent of the data fall within ± 1 standard deviation ($\mu \pm 1\sigma$). Figure 4.38 demonstrates the percentage of measurements falling within each standard deviation.

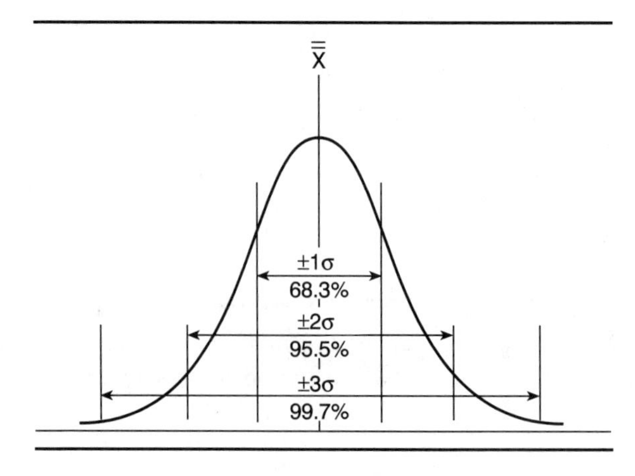

Figure 4.38 Percentage of Measurements Falling within Each Standard Deviation

These six features combine to create a peak in the center of the distribution, with the number of values decreasing as the measurements get farther away from the mean. As the data fall away toward the horizontal axis, the curve flattens. The tails of the normal distribution approach the horizontal axis and extend indefinitely, never reaching or crossing it (Figure 4.38).

While not all symmetrical distributions are normal distributions, these six features are general indicators of a normal distribution. (There is a chi square test for normality. Refer to a statistics text for a complete description of the chi square test.)

Standard Normal Probability Distribution: Z Tables

The area under the normal curve can be determined if the mean and the standard deviation are known. The mean, or in the case of samples, the average, locates the center of the normal distribution. The standard deviation defines the spread of the data about the center of the distribution.

The relationships discussed in features 5 and 6 of the normal frequency distribution make it possible to calculate the percentage of values that fall between any two readings. If 100 percent of the data are under the normal curve, then the amount of product above or below a particular value can be determined. These values may be dimensions like the upper and lower specification limits set by the designer or they can be any value of interest. The formula for finding the area under the normal curve is

$$f(x) = \frac{1}{\sigma\sqrt{2\pi}} e^{-(x-\mu)^2/2\sigma^2} \qquad -\infty < x < \infty$$

where

$$\pi = 3.14159$$
$$e = 2.71828$$

This formula can be simplified through the use of the standard normal probability distribution table (Appendix 1). This table uses the formula

$$f(Z) = \frac{1}{\sqrt{2\pi}} e^{-Z^2/2}$$

where

$$Z = \frac{X_i - \overline{X}}{s} = \text{standard normal value}$$
$$X_i = \text{individual X value of interest}$$
$$\overline{X} = \text{average}$$
$$s = \text{standard deviation}$$

This formula also works with population means and population standard deviations:

$$Z = \frac{X_i - \mu}{\sigma}$$

where

X_i = individual X value of interest
μ = population mean
σ = population standard deviation
Z = standard normal value

Z is used with the table in Appendix 1 to find the value of the area under the curve, which represents a percentage or proportion of the product or measurements produced. If Z has a positive value, then it is to the right of center of the distribution and X_i is larger than $\overline{X}$. If the Z value is negative, then it is on the left side of the center and X_i is smaller than $\overline{X}$.

To find the area under the normal curve associated with a particular X_i, use the following procedure:

1. Use the information on normal curves to verify that the measurements are normally distributed.
2. Use the mean, standard deviation, and value of interest in the formula to calculate Z.
3. Find the Z value in the table in Appendix 1.
4. Use the table to convert the Z values to the area of interest.
5. Convert the area of interest to a percentage by multiplying by 100.

The table in Appendix 1 is a left-reading table, meaning that it will provide the area under the curve from negative infinity up to the value of interest (Figure 4.39). These values will have to be manipulated to find the area greater than the value of interest or between two values. Drawing a picture of the situation in question and shading the area of interest often helps clarify the Z calculations. Values of Z should be rounded to two decimal places for use in the table. Interpolation between values can also be performed.

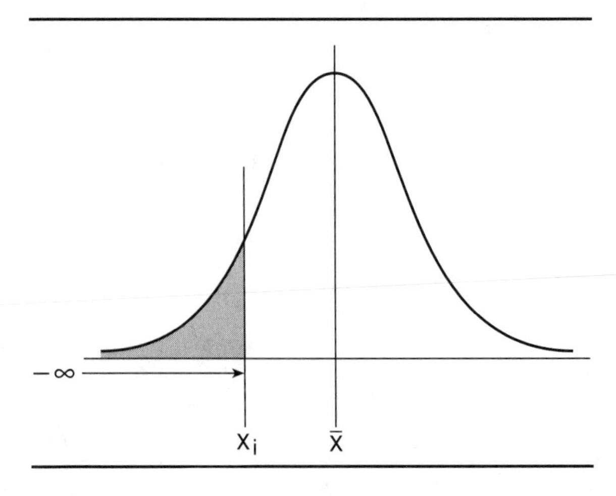

Figure 4.39 Normal Curve for Left-Reading Z Table

EXAMPLE 4.34 Using Standard Normal Probability Distribution: Flat Round Plate

The engineers working with the flat round plate thickness data would like to determine what percentage of parts from the samples taken are below 0.0624 inch.

1. Use the data in Table 4.5, an average of 0.0627, a standard deviation of 0.00023, and the Z tables to determine the percentage of parts under 0.0624 inch thick. In Figure 4.40, the area of interest is shaded:

$$Z = \frac{0.0624 - 0.0627}{0.00023} = -1.30$$

$$\text{Area} = 0.0968$$

   or 9.68 percent of the parts are thinner than 0.0624 inch.

2. To determine the percentage of the parts that are 0.0629 inch thick or thicker, it is important to note that the table in Appendix 1 is a left-reading table. Since the engineers want to determine the percentage of parts thicker than 0.0629 (the area shaded in Figure 4.41), they will have to subtract the area up to 0.0629 from 1.00.

$$Z = \frac{0.0629 - 0.0627}{0.00023} = 0.87$$

$$\text{Area} = 0.8079$$

   or 80.79 percent of the parts are thinner than 0.0629 inch. However, they want the area that is *thicker* than 0.0629 inch. To get this area they must subtract the area from 1.0 (remember 100 percent of the parts fall under the normal curve):

$$1.00 - 0.8079 - 0.1921$$

   or 19.21 percent of the parts are thicker than 0.0629.

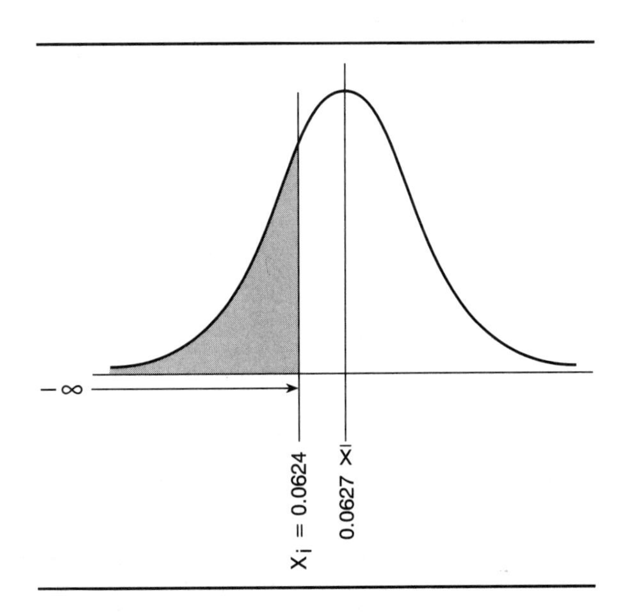

Figure 4.40 Example 4.34: Area Under the Curve, $X_i = 0.0624$

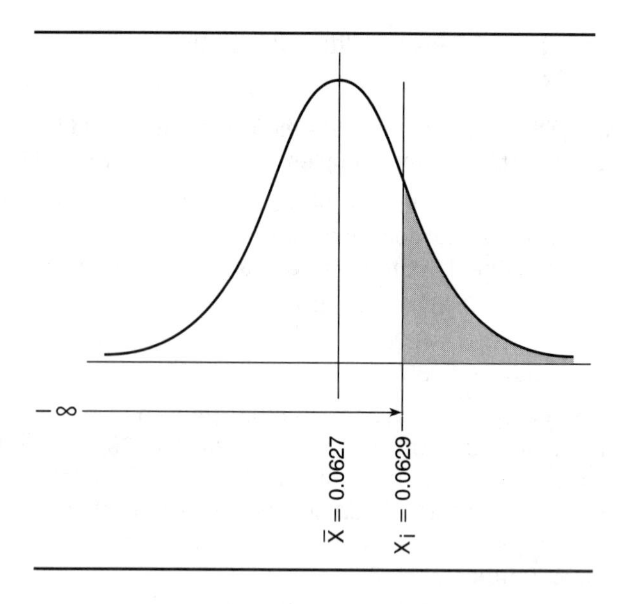

Figure 4.41 Example 4.35: Area Under the Curve, $X_i = 0.0629$

3. Now to find the percentage of the parts between 0.0623 and 0.0626 inch thick: The area of interest is shaded in Figure 4.42. In this problem the engineers must calculate two areas of interest, one for those parts 0.0623 inches thick or thinner and the other for those parts 0.0626 inch thick or thinner. The area of interest for those parts 0.0623 inch and below will be subtracted from the area of interest for 0.0626 and below:

$$Z_1 = \frac{0.0623 - 0.0627}{0.00023} = -1.74$$

$$\text{Area}_1 = 0.0409$$

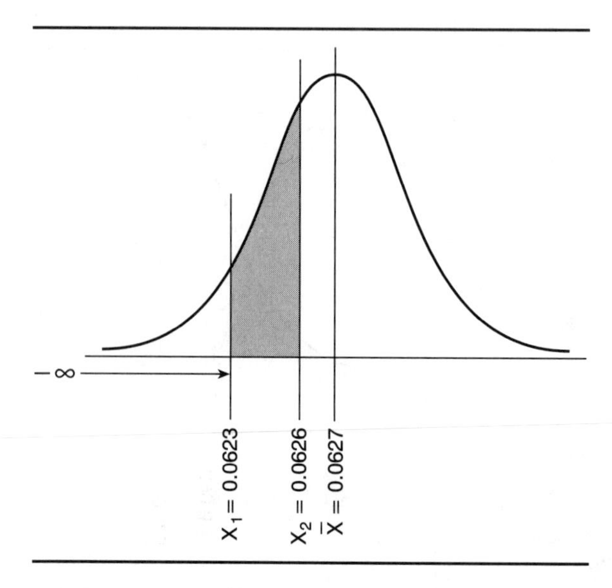

Figure 4.42 Example 4.35: Area Under the Curve Between 0.0623 and 0.0626

or 4.09 percent of the parts are thinner than 0.0623 inch.

$$Z_2 = \frac{0.0626 - 0.0627}{0.00023} = -0.44$$

$$\text{Area}_2 = 0.3300$$

or 33 percent of the parts are thinner than 0.0626 inch. Subtracting these two areas will determine the area in between:

$$0.3300 - 0.0409 = 0.2891$$

or 28.91 percent of the parts fall between 0.0623 and 0.0626 inch.

Central Limit Theorem

Much of statistical process control is based on the use of samples taken from a population of items. The central limit theorem enables conclusions to be drawn from the sample data and applied to a population. The ***central limit theorem*** *states that a group of sample averages tends to be normally distributed; as the sample size n increases, this tendency toward normality improves*. The population from which the samples are taken does not need to be normally distributed for the sample averages to tend to be normally distributed (Figure 4.43). In the field of quality, the central limit theorem

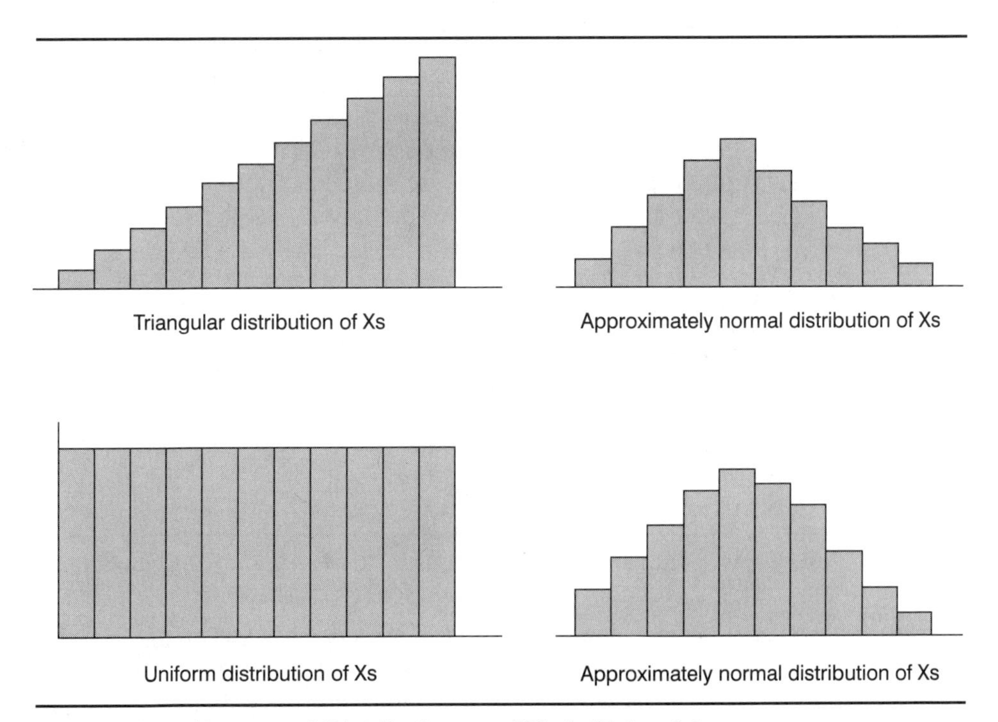

Figure 4.43 Nonnormal Distributions and Their Plots of Averages

1	11	13	10	12	7	10.6
2	1	7	12	9	3	6.4
3	9	5	12	1	11	7.6
4	11	5	7	9	12	8.8
5	7	12	13	7	4	8.6
6	11	9	5	1	13	7.8
7	1	4	13	12	13	8.6
8	13	3	2	6	12	7.2
9	2	4	1	10	13	6.0
10	4	5	12	1	9	6.2
11	2	5	7	7	11	6.4
12	6	9	8	2	12	7.4
13	2	3	6	11	11	6.6
14	2	6	9	11	13	8.2
15	6	8	8	9	1	6.4
16	3	4	12	1	6	5.2
17	8	1	8	6	10	6.6
18	5	7	6	8	8	6.8
19	2	5	4	10	1	4.4
20	5	7	12	7	8	7.8
21	9	1	3	6	12	6.2
22	1	13	9	3	6	6.4
23	4	5	13	5	7	6.8
24	3	7	9	8	10	7.4
25	1	7	6	6	1	4.2
26	8	10	1	9	10	7.6
27	9	13	2	2	2	5.6
28	12	4	12	3	13	8.8
29	12	4	7	6	9	7.6
30	1	12	3	12	11	7.8
31	12	3	10	11	6	8.4
32	3	5	10	2	7	5.4
33	9	1	2	3	11	5.2
34	6	8	6	13	9	8.4
35	2	12	5	10	4	6.6
36	6	4	8	9	12	7.8
37	9	13	3	10	1	7.2
38	2	1	13	7	5	5.6
39	10	11	5	12	13	10.2
40	13	2	8	2	11	7.2
41	2	10	5	4	11	6.4
42	10	4	12	7	11	8.8
43	13	13	7	1	10	8.8
44	9	10	7	11	11	9.6
45	6	7	8	7	4	6.4
46	1	4	12	11	13	8.2
47	9	11	8	1	11	8.0
48	8	13	10	13	4	9.6
49	12	11	11	2	3	7.8
50	2	12	5	11	9	7.8

1 ~~||||~~ ~~||||~~ ~~||||~~ ~~||||~~ ||
2 ~~||||~~ ~~||||~~ ~~||||~~ ||||
3 ~~||||~~ ~~||||~~ ||||
4 ~~||||~~ ~~||||~~ ~~||||~~ |
5 ~~||||~~ ~~||||~~ ~~||||~~ |
6 ~~||||~~ ~~||||~~ ~~||||~~ |||
7 ~~||||~~ ~~||||~~ ~~||||~~ ~~||||~~ |
8 ~~||||~~ ~~||||~~ ~~||||~~ |
9 ~~||||~~ ~~||||~~ ~~||||~~ ~~||||~~ |
10 ~~||||~~ ~~||||~~ ~~||||~~ ||
11 ~~||||~~ ~~||||~~ ~~||||~~ ~~||||~~ |||
12 ~~||||~~ ~~||||~~ ~~||||~~ ~~||||~~ ~~||||~~
13 ~~||||~~ ~~||||~~ ~~||||~~ ~~||||~~ ||

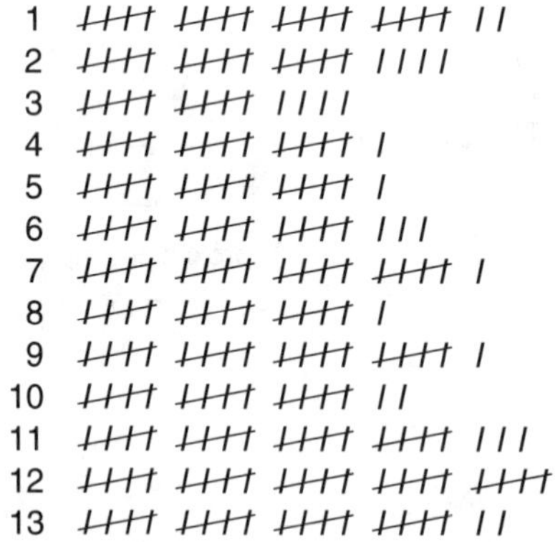

Figure 4.44 Numerical Values of Cards and Frequency Distribution

supports the use of sampling to analyze the population. The mean of the sample averages will approximate the mean of the population. The variation associated with the sample averages will be less than that of the population. It is important to remember that it is the sample *averages* that tend toward normality, as the following example shows.

EXAMPLE 4.35 Using the Central Limit Theorem

Roger and Bill are trying to settle an argument. Roger says that averages from a deck of cards will form a normal curve when plotted. Bill says they won't. They decide to try the following exercise involving averages. They are going to follow these rules:

1. They will randomly select five cards from a well-shuffled deck and write down the values (Figure 4.44). (An ace is worth 1 point, a jack 11, a queen 12, and a king 13.)
2. They will record the numerical values on a graph (Figure 4.45).
3. They will calculate the average for the five cards.
4. They will graph the results of step 3 on a graph separate from that used in step 2. (Figure 4.45).
5. They will then replace the five cards in the deck.
6. They will shuffle the deck.
7. They will repeat this process 50 times.

Figure 4.44 displays the results of steps 1 and 2. Since the deck was well shuffled and the selection of cards from the deck was random, each card had the same chance of being selected—1/52. The fairly uniform distribution of values in the frequency diagram in Figure 4.44 shows that each type of card was selected approximately the same number of times. The distribution would be even more uniform if a greater number of cards had been drawn.

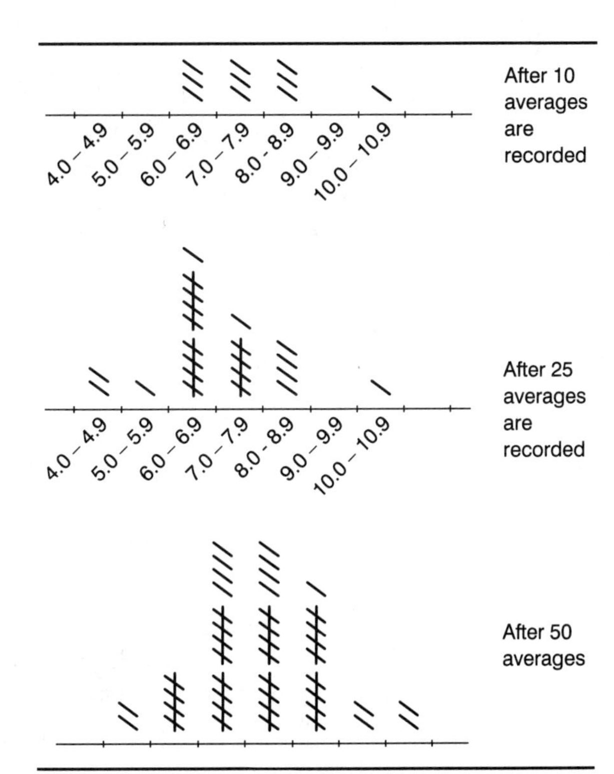

Figure 4.45 Distribution of Sample Averages

Figure 4.45 graphs the results of step 4. Notice that as the number of averages recorded increases, the results look more and more like a normal curve. As predicted by the central limit theorem, the distribution of the sample averages in the final diagram in Figure 4.45 is approximately normal. This has occurred even though the original distribution was not normal.

SUMMARY

Frequency diagrams and histograms graphically depict the processes or occurrences under study. Means, modes, medians, standard deviations, and ranges are powerful tools used to statistically describe processes and occurrences. Because of the central limit theorem, users of statistical information can form conclusions about populations of items based on the sample statistics. Since the behavior of the $\overline{X}$ values is predictable, when the values are different from an expected normal curve, there must be a reason why the process is not behaving normally. Knowing this, those working with a process can determine whether or not there is something wrong with the process that must be dealt with.

Lessons Learned

1. Quality assurance relies primarily on inductive statistics in analyzing samples.
2. Accuracy and precision are of paramount importance in quality assurance. Measurements should be both precise and accurate.
3. Histograms and frequency diagrams are similar. Unlike a frequency diagram, a histogram will group the data into cells.
4. Histograms are constructed using cell intervals, cell midpoints, and cell boundaries.
5. The analysis of histograms is based on shape, location, and spread.
6. Shape refers to symmetry, skewness, and kurtosis.
7. The location or central tendency refers to the relationship between the mean (average), mode, and median.
8. The spread or dispersion of data is described by the range and standard deviation.
9. Skewness describes the tendency of data to be gathered either to the left or right side of a distribution. When a distribution is symmetrical, skewness equals zero.
10. Kurtosis describes the peakedness of data. Leptokurtic distributions are more peaked than platykurtic ones.
11. A normal curve can be identified by the following five features: It is symmetrical about a central value. The mean, mode, and median are all equal. It is unimodal and bell-shaped. Data cluster around the mean value of the distribution and then fall away toward the horizontal axis. The area under the normal curve equals 1; 100 percent of the data is found under the normal curve.

12. In a normal distribution 99.73 percent of the measured values fall within ± 3 standard deviations of the mean ($\mu \pm 3\sigma$); 95.5 percent of the data fall within ± 2 standard deviations of the mean ($\mu \pm 2\sigma$); and 68.3 percent of the data fall within ± 1 standard deviation ($\mu \pm 1\sigma$).
13. The area under a normal curve can be calculated using the Z table and its associated formula.
14. The central limit theorem proves that the averages of nonnormal distributions have a normal distribution. ■

■ *Formulas*

$$R = X_h - X_l$$

$$\mu \text{ or } \overline{X} = \frac{X_1 + X_2 + X_3 + \cdots + X_n}{n} = \frac{\Sigma X_i}{n}$$

$$\sigma = \sqrt{\frac{\Sigma(X_i - \mu)^2}{n}}$$

$$s = \sqrt{\frac{\Sigma(X_i - \overline{X})^2}{n - 1}}$$

$$a_3 = \frac{\Sigma f_i(X_i - \overline{X})^3/n}{s^3}$$

$$a_4 = \frac{\Sigma f_i(X_i - \overline{X})^4/n}{s^4}$$

$$f(x) = \frac{1}{\sigma\sqrt{2\pi}} e^{-\frac{(x-\mu)^2}{2\sigma^2}} \qquad -\infty < x < \infty$$

where

$$\pi = 3.14159$$

$$e = 2.71828$$

$$f(Z) = \frac{1}{\sqrt{2\pi}} e^{-\frac{Z^2}{2}}$$

$$Z = \frac{X_i - \overline{X}}{s}$$

$$Z = \frac{X_i - \mu}{\sigma}$$

Chapter Problems

1. Describe the concepts of a sample and a population.
2. Give three examples each for continuous data and discrete data.
3. Describe a situation that is accurate, one that is precise, and one that is both. A picture may help you with your description.

Frequency Diagrams and Histograms

4. Make a frequency distribution of the following data. Is this distribution bimodal? Multimodal? Skewed to the left? Skewed to the right? Normal?

   225, 226, 227, 226, 227, 228, 228, 229, 222, 223, 224, 226, 227, 228, 225, 221, 227, 229, 230

5. NB Plastics uses injection molds to produce plastic parts that range in size from a marble to a book. Parts are pulled off the press by one operator and passed on to another member of the team to be cleaned up. This often involves trimming loose material, drilling holes, and painting. After a batch of parts has completed its cycle through the finishing process, a sample of five parts is chosen at random and certain dimensions are measured to ensure that each part is within certain tolerances. This information (in mm) is recorded for each of the five pieces and evaluated. Create a frequency diagram.

Part Name: Mount
Critical Dimension: 0.657 ± 0.001
Tolerance: ±0.001
Method of Checking: Caliper

Date	*Time*	*Press*	*Oper*	*Samp 1*	*Samp 2*	*Samp 3*	*Samp 4*	*Samp 5*	*Insp*
9/20/92	0100	#1	Jack	0.6550	0.6550	0.6540	0.6540	0.6545	Sam
9/20/92	0300	#1	Jack	0.6540	0.6540	0.6543	0.6543	0.6543	Sam
9/20/92	0500	#1	Jack	0.6540	0.6540	0.6540	0.6540	0.6535	Sam
9/20/92	0700	#1	Jack	0.6540	0.6540	0.6540	0.6540	0.6540	Sam
9/21/92	1100	#1	Mary	0.6595	0.6600	0.6600	0.6598	0.6595	Sam
9/21/92	1300	#1	Mary	0.6580	0.6580	0.6585	0.6590	0.6575	Sam
9/21/92	1500	#1	Mary	0.6580	0.6580	0.6580	0.6585	0.6590	Sam
9/22/92	0900	#1	Mary	0.6563	0.6570	0.6550	0.6543	0.6550	Sam

6. Create a frequency diagram using the following data:

1.116	1.122	1.125
1.123	1.122	1.123
1.133	1.125	1.118
1.117	1.121	1.123
1.124	1.136	1.122
1.119	1.127	1.122
1.129	1.125	1.119
1.121	1.124	
1.128	1.122	

7. PL Industries machines shafts for rocker-arm assemblies. One particularly complex shaft is shown in Figure 1. The existing machining cell is currently unable to meet the specified tolerances consistently. PL Industries has

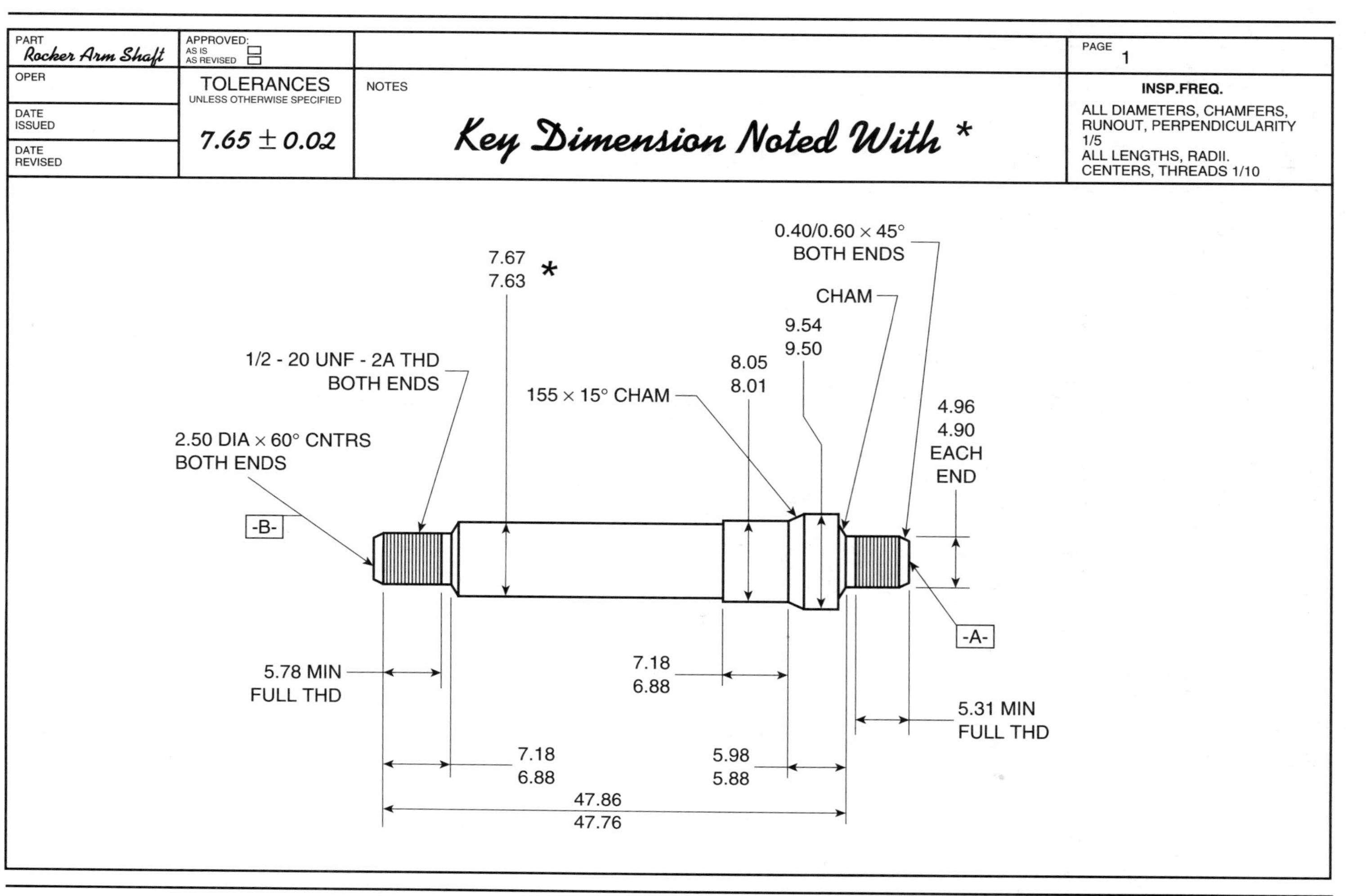

PART
Rocker Arm Shaft
APPROVED:
AS IS
AS REVISED
PAGE
1
OPER
TOLERANCES
UNLESS OTHERWISE SPECIFIED
7.65 ± 0.02
DATE ISSUED
DATE REVISED
NOTES
Key Dimension Noted With *
INSP.FREQ.
ALL DIAMETERS, CHAMFERS, RUNOUT, PERPENDICULARITY 1/5
ALL LENGTHS, RADII. CENTERS, THREADS 1/10
7.67
7.63
*
0.40/0.60 × 45°
BOTH ENDS
CHAM
9.54
9.50
1/2 - 20 UNF - 2A THD
BOTH ENDS
8.05
8.01
155 × 15° CHAM
4.96
4.90
EACH
END
2.50 DIA × 60° CNTRS
BOTH ENDS
-B-
-A-
5.78 MIN
FULL THD
7.18
6.88
5.31 MIN
FULL THD
7.18
6.88
5.98
5.88
47.86
47.76

Problem 7

decided to replace the existing machines with CNC turning centers. The following data are from a runoff held at the CNC vendor's facility. The indicated portion of the shaft was chosen for inspection. The diameter specifications for the round shaft are 7.650 $\pm$ 0.02 mm. The data are scaled from 7.650, meaning that a value of 0.021 is actually 7.671 mm. Create a frequency diagram with the data.

0.021	0.013	0.018	0.007	0.002	0.030	0.024	0.006	0.002
0.006	0.004	0.003	0.010	0.015	0.011	0.023	0.025	0.022
0.025	0.004	0.009	0.020	0.012	0.021	0.022	0.004	0.027
0.020	0.011	0.018	0.025	0.010				

8. Create a histogram with the shaft data in Problem 7. Describe the distribution's shape, spread, and location.
9. Make a histogram of the following sample data:
   225, 226, 227, 226, 227, 228, 228, 229, 222, 223, 224, 226, 227, 228, 225, 221, 227, 229, 230, 225, 226, 227, 229, 228, 224, 223, 222, 225, 226, 227, 224, 223, 222, 228, 229, 225, 226
10. Gold, measured in grams, is used to create circuit boards at MPL Industries. The following measurements of gold usage per batch of circuit boards have been recorded in a tally sheet. Create a histogram with the following information. Describe the shape, location, and spread of the histogram.

125	/
126	/
127	
128	///
129	////
130	
131	
132	~~////~~
133	
134	////
135	~~////~~ /
136	////
137	////
138	~~////~~ /
139	
140	~~////~~
141	~~////~~ /
142	~~////~~ /
143	~~////~~ /
144	~~////~~
145	~~////~~ ///
146	~~////~~ ~~////~~
147	~~////~~ //
148	~~////~~ /
149	////
150	~~////~~

11. At a local bank, the time a customer spends waiting to transact business varies between 1 and 10 minutes. Create and interpret a histogram with the following data:

Wait Time	*Frequency*
1 min	5
2 min	10
3 min	16
4 min	25
5 min	18
6 min	9
7 min	6
8 min	4
9 min	2
10 min	1

12. A manufacturer of CDs has a design specification for the width of the CD of 120 $\pm$ 0.3 mm. Create a histogram using the following data. Describe the distribution's shape, spread, and location.

Measurement	*Tally*
119.4	///
119.5	////
119.6	𝍸 /
119.7	𝍸 //
119.8	𝍸 𝍸
119.9	𝍸 ///
120.0	𝍸 //
120.1	𝍸
120.2	///

13. Use the concepts of symmetry and skewness to describe the histogram in Problem 9.

14. Why is it important to use both statistical measures and descriptive concepts when describing a histogram?

15. Create a histogram with the information from Problem 5.

Measures of Central Tendency and Dispersion

16. What is meant by the following expression: the central tendency of the data?

17. What is meant by the following expression: measures of dispersion?

18. Why does the salary information for the engineer provide more information than the salary information shown for the inspector?

	Min	*Max*	*Std. Dev.*	*Sample Size*	*Average*	*Median*
Engineer	$25,000	$80,000	$11,054	28	$43,298	$40,750
Inspector				5	$28,136	$27,000

19. Find the mean, mode, and median of the following numbers: 34, 35, 36, 34, 32, 34, 45, 46, 45, 43, 44, 43, 34, 30, 48, 38, 38, 40, 34
20. Using the following sample data, calculate the mean, mode, and median:

1.116	1.122	1.125
1.123	1.122	1.123
1.133	1.125	1.118
1.117	1.121	1.123
1.124	1.136	1.122
1.119	1.127	1.122
1.129	1.125	1.119
1.121	1.124	
1.128	1.122	

21. Determine the mean, mode, and median of the information in Problem 5.
22. Determine the mean, mode, and median of the following numbers. What is the standard deviation? Mark 1, 2, and 3 standard deviations (plus and minus) on the diagram.

    225, 226, 227, 226, 227, 228, 228, 229, 222, 223, 224, 226, 227, 228, 225, 221, 227, 229, 230

    (These are all the data; that is, the set comprises a population.)
23. For the shaft data from Problem 7, determine the mean, mode, median, standard deviation, and range. Use these values to describe the distribution. Compare this mathematical description with the description you created for the histogram problem.
24. For Problem 10, concerning the amount of gold used to create circuit boards, determine the mean, mode, median, standard deviation, and range. Use these values to describe the distribution. Compare this mathematical description with the description you created in Problem 10.
25. For the CD data of Problem 12, determine the mean, mode, median, standard deviation, and range. Use these values to describe the distribution. Compare this mathematical description with the description you created in Problem 12.
26. Determine the standard deviation and the range of the numbers in Problem 19.

27. Determine the standard deviation and the range of the numbers in Problem 5.
28. Describe the histogram created in Problem 15 using the standard deviation, range, mean, median, and mode. Discuss symmetry and skewness.
29. Four readings of the thickness of the paper in this textbook are 0.076, 0.082, 0.073, and 0.077 mm. Determine the mean and the standard deviation.
30. WT Corporation noticed a quality problem with their stainless steel product. Corrosion pits were forming after the product was already in the hands of the consumer. After several brainstorming sessions by the engineering and quality departments, the problem was traced to improper heat-treating. Work progressed to determine specifically what the problem was and give a recommendation for what could be done to eliminate it. It was determined that the small fluctuations in heating temperature and quenching time could be depriving the product of its full corrosion resistance. A simple and quick test was developed that allowed a heat-treat operator to roughly determine the product's ability to resist corrosion. The amount of retained austenite in a heat-treated stainless steel product is related to the product's ability to resist corrosion. The smaller the amount of retained austenite, the better the product will resist corrosion. What follows is one day's worth of test readings. Determine the range, standard deviation, mean, mode, and median for these data.

Time	*Reading*
7:00	12
	17
8:00	25
	23
9:00	21
	19
10:00	28
	25
11:00	31
	27
12:00	20
	28
1:00	23
	25
2:00	26
	29

31. How do the range, mean, standard deviation, median, and mode work together to describe a distribution?

32. Given the following histogram, describe it over the phone to someone who has never seen it before. Refer to kurtosis, skewness, mean, standard deviation, median, mode, and range. Be descriptive.

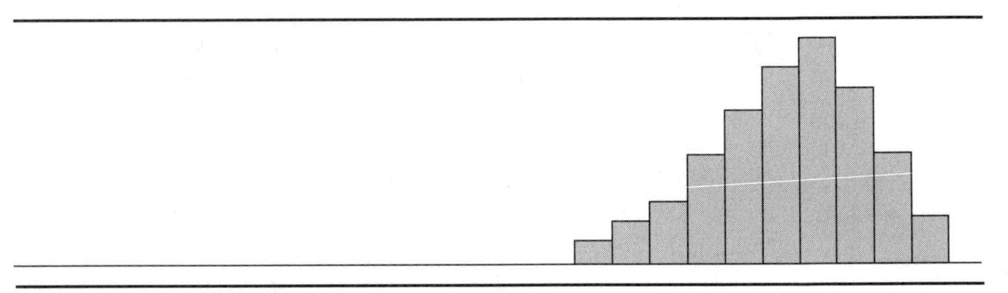

Problem 32

33. Pictured below are three normal curves. Assume the one on the left is the expected or acceptable curve for the process. How did the numerical magnitude of the standard deviation change for the middle curve? For the curve on the right?

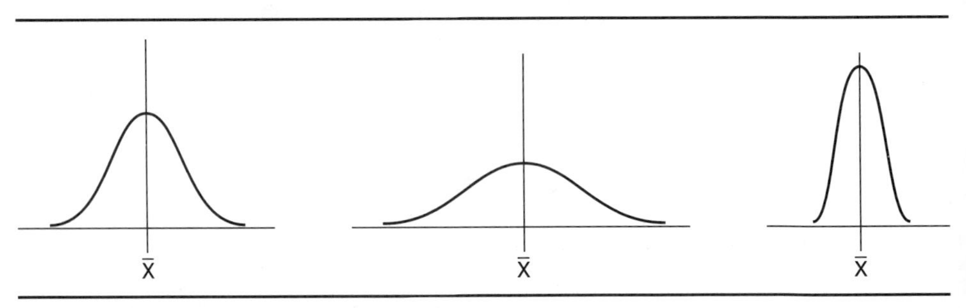

Problem 33

Normal Curve

34. If the average wait time is 12 minutes with a standard deviation of 3 minutes, determine the percentage of patrons who wait less than 15 minutes for their main course to be brought to their tables.
35. The thickness of a part is to have an upper specification of 0.925 and a lower specification of 0.87 mm. The average of the process is currently 0.917 with a standard deviation of 0.005.

    a. Determine the percentage of product above 0.93 mm.

b. Describe where the product is being produced in relation to the specifications set by the designer. Drawing a picture of the situation will help.

36. The Rockwell hardness of specimens of an alloy shipped by your supplier varies according to a normal distribution with mean 70 and standard deviation 3. Specimens are acceptable for machining only if their hardness is greater than 65. What percentage of specimens will be acceptable? Draw the normal curve diagram associated with this problem.

37. Given the specifications of 0.765 ± 0.002 inch diameter for the shaft, use the information from Problem 7 to determine the percentage of shafts able to meet the specification limits.

38. If the mean value of the weight of a particular brand of dog food is 20.6 lb and the standard deviation is 1.3, assume a normal distribution and calculate the amount of product produced that falls below the lower specification value of 19.7 lb.

39. To create the rocker arm for a car-seat recliner, the steel must meet a minimum hardness standard. The following histogram of data provides the Rockwell hardness test results from a year's worth of steel coil production. Given an average of 44.795 and a standard deviation of 0.402, calculate the percentage of coils whose hardness is less than the minimum requirement of 44.000.

40. Global warming is a concern for this and future generations. Measurements of the earth's atmospheric temperature have been taken over the past 100 years. If the average temperature has been 50°F with a standard deviation of 18°F, what percentage of the temperatures have been between 0 and 90°F?

41. For the CD data from Problem 12, determine what percentage of the CDs produced are above and below the specifications of 120 ± 0.3 mm.

42. Joe and Sally run a custom car wash. All the washes and waxes are done by hand and the interiors are vacuumed. Joe and Sally need to do some planning in order to schedule cars for appointments. If the mean time for a cleanup is 45 minutes and the standard deviation is 10 minutes, what percentage of the cleanups will take less than 65 and more than 35 minutes to complete? The data the couple has taken are normally distributed.

43. A lightbulb has a normally distributed light output with a mean of 3,000 foot-candles and standard deviation of 50 foot-candles. Find a lower specification limit such that only 0.5 percent of the bulbs will not exceed this limit.

44. The life of an automotive battery is normally distributed with a mean of 900 days and a standard deviation of 50 days. What fraction of these batteries would be expected to survive beyond 1,000 days?

45. Clearly describe the central limit theorem to someone who is unfamiliar with it.

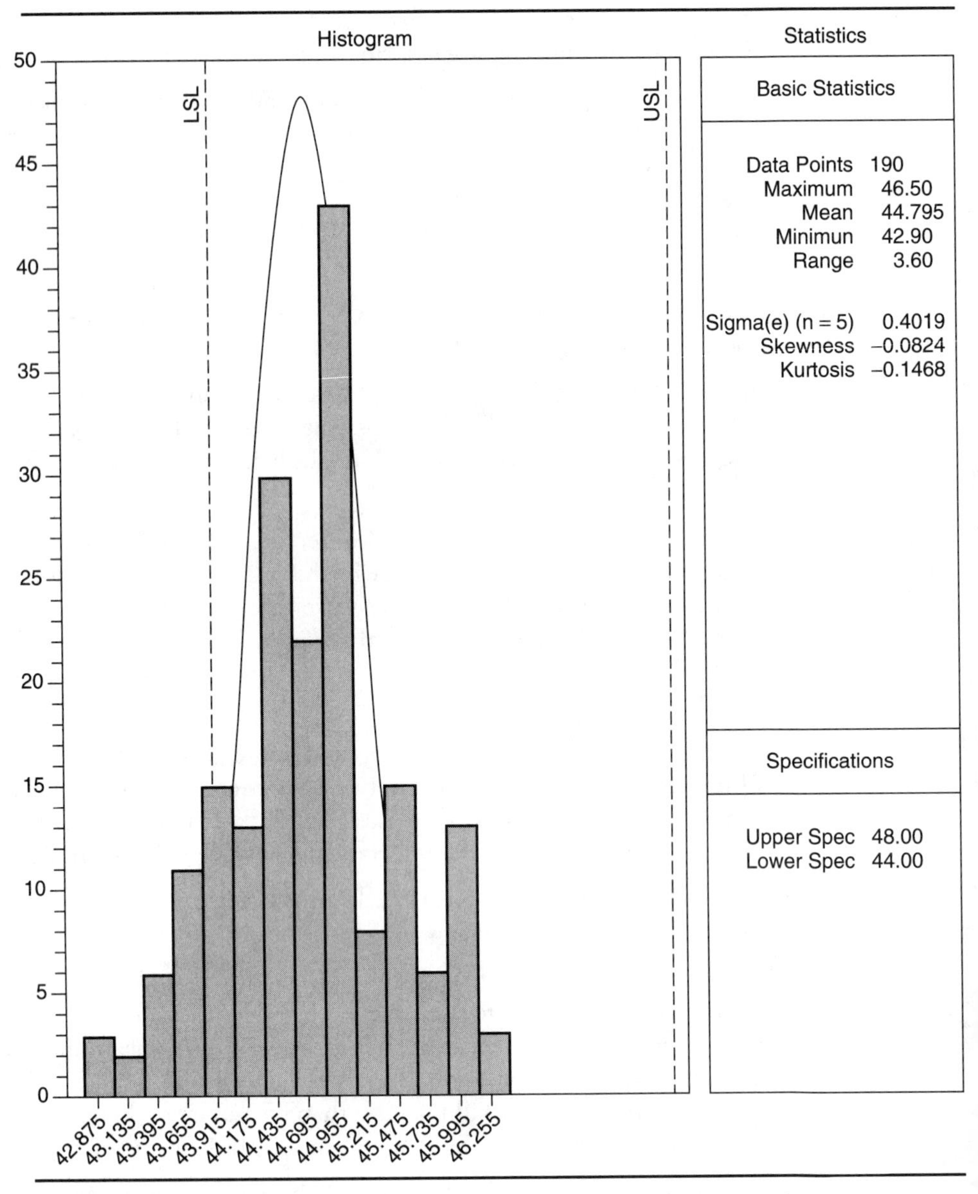

Problem 39

46. A sample distribution (closely approximating a normal distribution) of a critical part dimension has an average of 11.85 inches and a sample standard deviation of 1.33 inches. The lower specification limit is 11.41 inches. The upper specification limit is 12.83 inches. What percentage of parts will be produced below the lower specification limit of 11.41 inches?

47. NB Manufacturing has ordered the construction of a new machine to replace an older machine in a machining cell. Now that the machine has been built, a runoff is to be performed. The diameter on this piece of equipment was checked for runout. The upper tolerance for the runout is 0.002. What are the mean, mode, median, standard deviation, and range of the data? With an upper specification limit of 0.002, what percentage of the parts meet the specification?

Runout			
0.0021	0.0004	0.0009	0.0025
0.0013	0.0003	0.0020	0.0010
0.0018	0.0010	0.0012	
0.0007	0.0015	0.0021	
0.0002	0.0011	0.0022	
0.0030	0.0023	0.0004	
0.0024	0.0025	0.0027	
0.0006	0.0022	0.0020	
0.0002	0.0025	0.0011	
0.0006	0.0004	0.0018	

CASE STUDY 4.1
Statistics

PART 1

Angie's company has recently told her that her pension money can now be invested in several different funds. They have informed all of their employees that it is the employee's decision which of the six risk categories to invest in. The six risk categories include conservative, secure, moderate, balanced, ambitious, and aggressive. Within each risk category, there are two fund choices. This gives each employee twelve funds in which to invest capital.

As the names suggest, each fund category attempts to invest the person's capital in a particular style. A conservative fund will find investments that will not expose the investor's money to risk. This category is the "safest." Money invested here is not exposed to great potential losses but it is not exposed to great gains either. The aggressive fund is for investors who believe that when nothing is ventured, nothing is gained. It takes the greatest risks. With this fund, investors have a potential to make great gains with their capital. Of course, this is also the fund that exposes the investor to the potential for great losses. In between, the other four categories mix the amount of risk an investor is subject to.

Angie is rather at a loss. In the packet from the investment company she has received over 150 pages of information about each of the accounts. For the past three nights, she has been trying to read it all. Now all of the information is running together in her head, and she has to make up her mind soon.

Just as she is about to give up and throw a dart at the dartboard, she discovers an investment performance update (Table 1) among the many pages of data. Remembering her statistical training, she decides to study the performance of the different funds over the past 11 years. To fully understand the values she calculates, she decides to compare those values with the market performance of three market indicators: treasury bills, the Capital Markets Index, and the Standard & Poor's 500 Index. These indicators are used by investors to compare the performance of their investments against the market.

Assignment

From Table 1 calculate the average of each of the funds and market performance indicators.

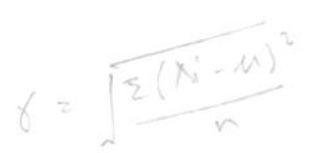

Table 1 Funds and Market Indices, 1983–1993

	'83	'84	'85	'86	'87	'88	'89	'90	'91	'92	'93
Conservative											
C1	10.4	12.8	18.4	15.0	3.0	8.2	11.3	8.0	13.2	7.3	10.7
C2	10.9	14.0	24.2	15.9	2.5	8.7	13.7	10.2	17.9	8.0	11.6
Secure											
S1	18.2	10.2	23.7	18.9	7.8	12.0	23.0	4.4	21.2	8.0	8.2
S2	16.3	9.3	27.8	19.1	0.5	12.8	15.3	1.4	27.9	2.7	13.1
Moderate											
M1	14.5	10.0	29.7	23.2	11.6	16.0	19.7	0.8	29.5	7.6	7.5
M2	21.9	14.4	27.2	17.6	6.4	15.7	28.2	4.5	27.6	6.6	4.5
Balanced											
B1	29.2	9.2	24.7	17.3	2.3	21.8	22.8	1.4	20.8	7.7	7.5
B2	21.5	8.9	23.2	13.2	7.8	12.7	24.8	2.3	25.6	9.2	11.9
Ambitious											
A1	36.2	6.3	32.4	11.8	2.9	14.7	22.9	3.5	38.6	10.8	15.8
A2	27.3	2.6	33.5	14.7	3.4	15.1	32.8	−6.1	24.1	13.8	25.2
Aggressive											
G1	23.7	7.6	33.4	25.8	5.5	13.3	28.6	6.9	51.5	4.0	−4.1
G2	22.3	0.3	38.3	19.7	6.0	14.1	37.0	−2.1	57.6	6.4	−1.2
Treasury bills											
	9.2	10.3	8.0	6.3	6.1	7.1	8.7	8.0	5.7	3.6	6.7
Capital Markets Index											
	17.7	8.3	28.6	16.5	3.4	13.2	21.6	1.0	25.0	8.4	13.4
Standard & Poor's 500											
	22.6	6.3	31.7	18.6	5.3	16.6	31.6	−3.1	30.4	7.6	10.1

PART 2

As her next step, Angie is going to interpret the results of her calculations. To make sure that she has a comfortable retirement, she has decided that she would like to have her money grow at an average of 15 percent or greater each year.

Assignment

Use the average values you calculated to determine which funds will average greater than 15 percent return on investment per year.

PART 3

Angie has narrowed her list to those funds averaging a 15 percent or greater return on investment per year. That still leaves her with several to choose from. Remembering that some of the funds are riskier than others, Angie is quick to realize that in some years she may not make 15 percent or more. She may add 25 percent or higher to her total amount of pension, or she may lose 5 percent of her pension. An average value doesn't guarantee that she will make exactly 15 percent a year. An average reflects how well she will do when all of the good years (gains) and bad years (losses) are combined.

Angie has 20 years to work before she retires. While this gives her some time to increase her wealth, she is concerned that a few bad years could significantly lower the amount of money that she will retire with. To find the funds that have high gains or high losses, she turns her attention to the study of the range and standard deviation.

Assignment

Calculate the range and standard deviation associated with each of the funds and market indicators with an average of 15 percent or greater return on investment.

PART 4

For all of the funds that she is interested in, Angie has noticed that the spread of the data differs dramatically. One of the ranges is almost 60 points! She has also noticed that the riskier the fund, the greater the range of values.

Since the ranges may have been caused by an isolated year that was particularly good or particularly bad, Angie turns her attention to the standard deviations, which also increase as the riskiness of the fund increases.

Decision time! Comparing the average return on investment yielded by the stock market (Standard & Poor's 500, Capital Markets Index) during this 11-year period, Angie notices that several of the investment funds have brought in returns that were less than the Standard & Poor's 500 market indicator. She decides to take them off her potential funds list. This removes M2 and B1 from consideration. She also removes from consideration any fund that has a large range and standard deviation. Because she is willing to take some risk, but not a dramatic one, she decides not to invest in the aggressive fund. This removes G1 and G2 from consideration. Now she can concentrate on a smaller group of funds: M1, A1, A2. So now, instead of having to devour hundreds of pages of information, Angie can look at the pages detailing the performance of these three funds and make her decision about which one to invest in.

To help make the choice between these funds, Angie has decided to use the Z tables to calculate the percentage of time that each fund's return on investment fell below the 12 percent that she feels that she absolutely has to make each year.

Assignment

Assume a normal distribution and use the Z tables to calculate the percentage of each of the remaining funds' return on investments that fell below 12 percent in the past 11 years. Do not perform this calculation for the fund indicators.

CASE STUDY 4.2
Process Improvement

This case is the second in a four-part series of cases involving process improvement. The other cases are found at the end of Chapters 3, 5, and 6. Data and calculations for this case establish the foundation for the future cases; however, it is not necessary to complete this case or the case in Chapter 3 in order to complete and understand the cases in Chapters 5 and 6. Completing this case will provide insight into the use of statistical calculations and histograms in process improvement. The case can be worked by hand or with the software provided.

PART 1

Figure 1 provides the details of a bracket assembly to hold a strut on an automobile in place. Welded to the auto-body frame, the bracket cups the strut and secures it to the frame via a single bolt with a lock washer. Proper alignment is necessary for both smooth installation during assembly and future performance. For mounting purposes the left-side hole, A, must be aligned on center with the right-side hole, B. If the

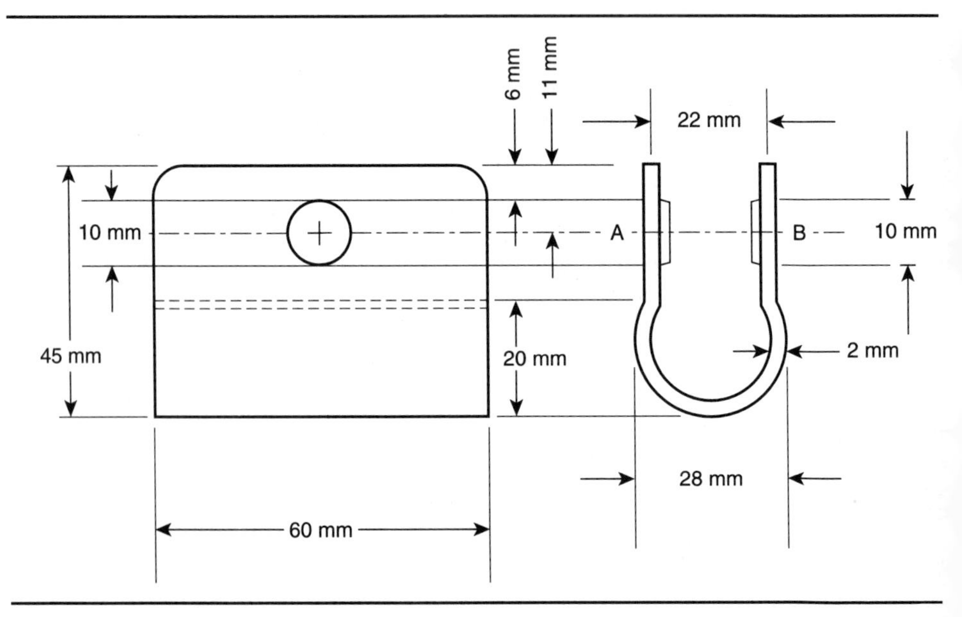

Figure 1 Bracket

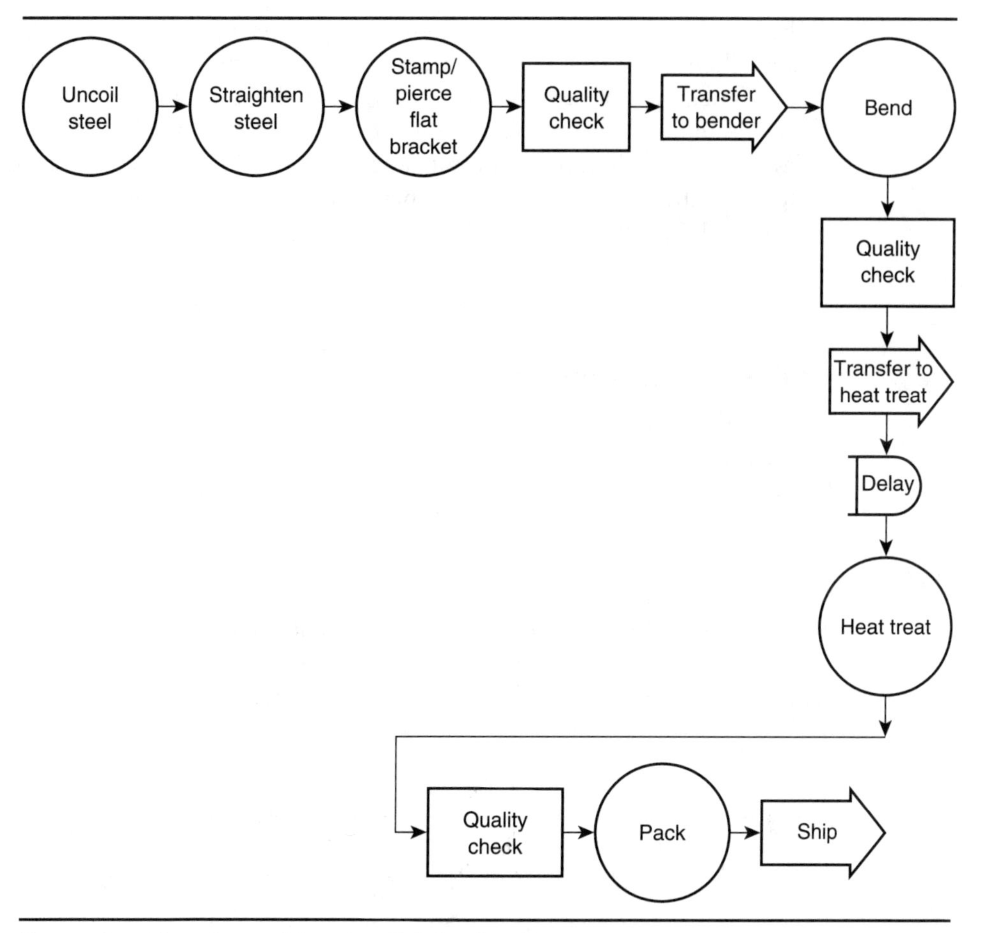

Figure 2 Flowchart of Bracket Fabrication Process

holes are centered directly opposite each other, in perfect alignment, then the angle between hole centers will measure 0 degrees. As the flowchart in Figure 2 shows, the bracket is created by passing coils of flat steel through a series of progressive dies. As the steel moves through the press, the bracket is stamped, pierced, and finally bent into the appropriate shape.

Recently customers have been complaining about having difficulty securing the bracket closed with the bolt and lock washer. The bolts have been difficult to slide through the holes and then tighten. Assemblers complain of stripped bolts and snug fittings. Unsure of where the problem is, WP Inc.'s management assembles a team consisting of representatives from process engineering, materials engineering, product design, and manufacturing.

Through the use of several cause-and-effect diagrams, the team determines that the most likely cause of the problems experienced by the customer is the alignment of the holes. At some stage in the formation process, the holes end up off-center. To confirm their suspicions, during the next production run, the bending press operator takes 20 subgroups of size 5 and measures the angle between the centers of the holes for each sample (Figure 3). The specification of the angle between insert hole A and insert hole B is 0.00 mm with a tolerance of ±0.30. The values recorded in Figure 3 represent the distance above (or when a minus sign is present, below) the nominal value of 0.00 mm.

Assignment

Utilize the data to create a histogram. Calculate the mean, median, mode, standard deviation, and range for all the data.

PART 2

Having a histogram and the accompanying statistical information tells the team a lot about the hole-alignment characteristics. Even more can be learned if these process data are compared with the specifications as set by the customers.

Assignment

Use the calculations for the area under the normal curve to determine the percentage of brackets created with misaligned holes. Remember the specifications for the angle between insert hole A and insert hole B are 0.00 mm with a tolerance of ±0.30. You will need to determine the percentage of brackets outside of each side of the tolerance. How are they doing?

PART 3

Using the problem-solving method described in Chapter 3 (and followed in Case 3.2), the team has determined that the fixture that holds the flat bracket in place during the bending operation needs to be replaced. One of the measures of performance that they created during step 4 of their problem-solving process is the percentage of the parts that are out of specification. We calculated this value in part 2 of this case. Now that the fixture has been replaced with a better one, the team would like to determine whether or not changing the fixture improved the process by removing a root cause of hole misalignment.

Sample \ Subgroup	1	2	3	4	5	6	7	8	9	10	11	12	13	14	15
1	0.31	0.27	0.30	0.30	0.25	0.18	0.26	0.15	0.30	0.31	0.18	0.22	0.19	0.14	0.29
2	0.29	0.23	0.30	0.20	0.20	0.26	0.27	0.21	0.24	0.25	0.16	0.30	0.28	0.27	0.23
3	0.30	0.31	0.28	0.21	0.19	0.18	0.12	0.24	0.26	0.25	0.21	0.21	0.26	0.25	0.27
4	0.28	0.23	0.24	0.23	0.26	0.24	0.20	0.27	0.27	0.28	0.29	0.24	0.29	0.28	0.24
5	0.23	0.29	0.32	0.25	0.25	0.17	0.23	0.30	0.26	0.25	0.27	0.26	0.24	0.16	0.23

Sample \ Subgroup	16	17	18	19	20
1	0.22	0.32	0.33	0.20	0.24
2	0.24	0.27	0.30	0.17	0.28
3	0.22	0.28	0.22	0.26	0.15
4	0.30	0.19	0.26	0.31	0.27
5	0.30	0.31	0.26	0.24	0.19

Figure 3 Hole A, B Alignment; Distance Above (+) or Below (−) Nominal

Sample \ Subgroup	1	2	3	4	5	6	7	8	9	10	11	12	13	14	15
1	0.03	0.06	0.06	0.03	−0.04	−0.02	−0.05	0.06	0.00	−0.02	−0.02	0.06	0.07	0.10	0.02
2	0.08	0.08	0.08	0.00	−0.07	0.06	−0.07	0.03	0.06	−0.01	0.06	0.02	−0.04	0.05	−0.05
3	−0.03	−0.01	0.05	0.05	0.00	0.02	0.11	0.04	0.02	0.00	−0.02	−0.01	0.08	0.03	0.04
4	0.07	0.08	−0.03	−0.01	−0.01	0.12	−0.03	0.03	−0.02	−0.10	0.02	−0.02	0.01	0.04	0.05
5	−0.02	0.02	0.03	0.07	−0.01	−0.07	0.03	0.01	0.00	−0.04	0.09	0.03	−0.04	0.06	−0.05

Sample \ Subgroup	16	17	18	19	20
1	−0.05	−0.06	−0.04	−0.06	−0.01
2	0.00	−0.02	−0.02	0.00	0.02
3	0.06	0.04	−0.01	0.00	0.05
4	−0.02	0.07	0.03	0.03	0.04
5	−0.01	−0.04	−0.02	0.04	0.03

Figure 4 **Hole A, B Alignment Following Process Improvement; Distance Above (+) or Below (−) Nominal**

Assignment

Utilize the data in Figure 4 to create a histogram. Calculate the mean, median, mode, standard deviation, and range for all new data. Use the calculations for the area under the normal curve to determine the new percentage of brackets created with misaligned holes. Remember the specifications for the angle between insert hole A and insert hole B are 0.00 mm with a tolerance of ±0.30. You will need to determine the percentage of brackets outside each side of the tolerance. How are they doing when you compare their measure of performance, percentage of the parts out of specification both before and after the fixture has changed?

5

Variable Control Charts

■ ***Learning Opportunities:***

1. To understand the concept of variation
2. To understand the difference between assignable causes and chance causes
3. To understand and learn the steps for control charting
4. To know how to construct a control chart for variables, either an $\overline{X}$ and R chart or an $\overline{X}$ and s chart
5. To understand the importance of the R and s charts when interpreting variables control charts
6. To recognize when a process is under control and when it is not
7. To know how to revise a control chart in which assignable causes have been identified and corrected ■

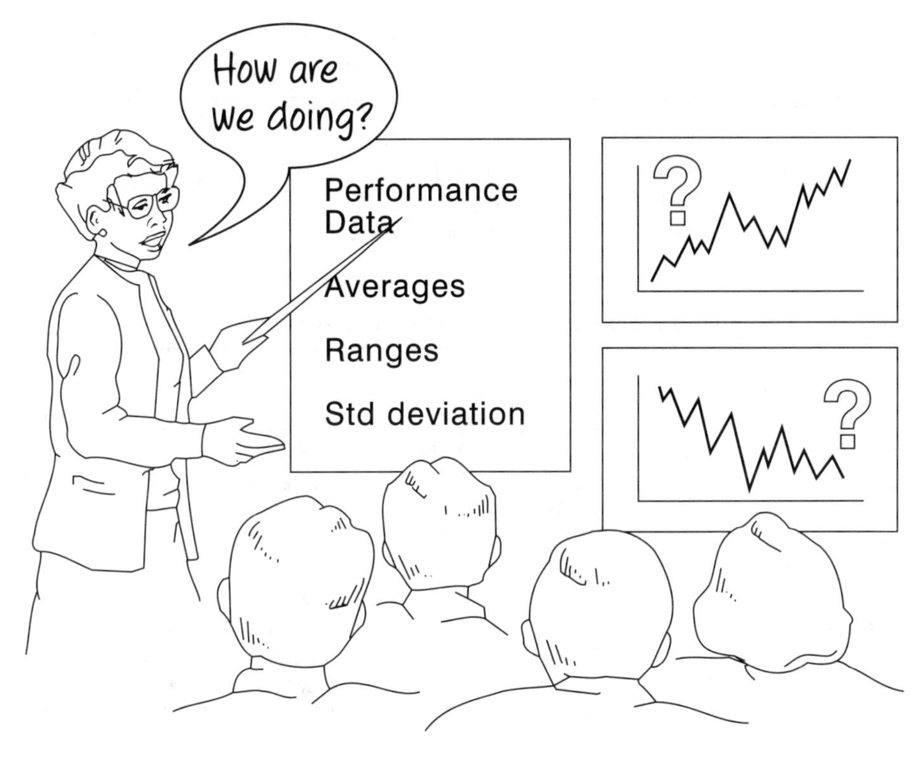

Profit Meeting

Is it possible for the people attending the meeting in the figure to answer the question? How would they do it? Can the performance data, average, range, and standard deviation be combined into a clear statement of their company's profit performance? The study of a process over time can be enhanced by the use of control charts in addition to the averages, ranges, and standard deviations calculated in the previous chapter. Process variation is recorded on **control charts,** *which are powerful aids to understanding the performance of a process over time.*

In Chapter 4, a group of engineers started working on a process that produces flat round plates. So far their analysis of the process has included constructing a histogram and calculating ranges, averages, and standard deviations. The only shortcoming in their analysis is its failure to show process performance over time. Let's take a closer look at the data in Table 4.6, reproduced here in Table 5.1, and see why that may be important. Figure 5.1*a* uses these data to create a simple chart that plots the average values calculated for each subgroup. When these averages are graphed in a

Table 5.1 Flat Round Plate Thickness: Sums and Averages

						ΣX_i	$\overline{X}$
Subgroup 1	0.0625	0.0626	0.0624	0.0625	0.0627	0.3127	0.0625
Subgroup 2	0.0624	0.0623	0.0624	0.0626	0.0625	0.3122	0.0624
Subgroup 3	0.0622	0.0625	0.0623	0.0625	0.0626	0.3121	0.0624
Subgroup 4	0.0624	0.0623	0.0620	0.0623	0.0624	0.3114	0.0623
Subgroup 5	0.0621	0.0621	0.0622	0.0625	0.0624	0.3113	0.0623
Subgroup 6	0.0628	0.0626	0.0625	0.0626	0.0627	0.3132	0.0626
Subgroup 7	0.0624	0.0627	0.0625	0.0624	0.0626	0.3126	0.0625
Subgroup 8	0.0624	0.0625	0.0625	0.0626	0.0626	0.3126	0.0625
Subgroup 9	0.0627	0.0628	0.0626	0.0625	0.0627	0.3133	0.0627
Subgroup 10	0.0625	0.0626	0.0628	0.0626	0.0627	0.3132	0.0626
Subgroup 11	0.0625	0.0624	0.0626	0.0626	0.0626	0.3127	0.0625
Subgroup 12	0.0630	0.0628	0.0627	0.0625	0.0627	0.3134	0.0627
Subgroup 13	0.0627	0.0626	0.0628	0.0627	0.0626	0.3137	0.0627
Subgroup 14	0.0626	0.0626	0.0625	0.0626	0.0627	0.3130	0.0626
Subgroup 15	0.0628	0.0627	0.0626	0.0625	0.0626	0.3132	0.0626
Subgroup 16	0.0625	0.0626	0.0625	0.0628	0.0627	0.3131	0.0626
Subgroup 17	0.0624	0.0626	0.0624	0.0625	0.0627	0.3126	0.0625
Subgroup 18	0.0628	0.0627	0.0628	0.0626	0.0630	0.3139	0.0627
Subgroup 19	0.0627	0.0626	0.0628	0.0625	0.0627	0.3133	0.0627
Subgroup 20	0.0626	0.0625	0.0626	0.0625	0.0627	0.3129	0.0626
Subgroup 21	0.0627	0.0626	0.0628	0.0625	0.0627	0.3133	0.0627
Subgroup 22	0.0625	0.0626	0.0628	0.0625	0.0627	0.3131	0.0626
Subgroup 23	0.0628	0.0626	0.0627	0.0630	0.0627	0.3138	0.0628
Subgroup 24	0.0625	0.0631	0.0630	0.0628	0.0627	0.3141	0.0628
Subgroup 25	0.0627	0.0630	0.0631	0.0628	0.0627	0.3143	0.0629
Subgroup 26	0.0632	0.0628	0.0631	0.0628	0.0627	0.3149	0.0630
Subgroup 27	0.0630	0.0628	0.0631	0.0628	0.0627	0.3144	0.0629
Subgroup 28	0.0632	0.0632	0.0628	0.0631	0.0630	0.3153	0.0631
Subgroup 29	0.0630	0.0628	0.0631	0.0632	0.0631	0.3152	0.0630
Subgroup 30	0.0632	0.0631	0.0630	0.0628	0.0628	0.3149	0.0630
						9.3997	

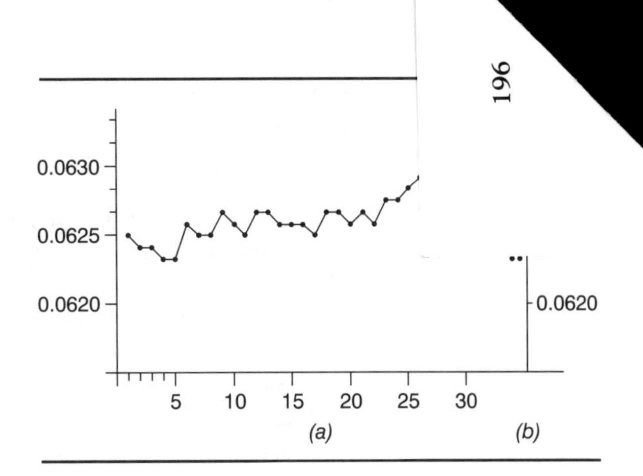

Figure 5.1 Chart with Histogram

histogram, the result closely resembles a normal curve (Figure 5.1*b*). In Chapter 4 we learned that a normal distribution is created whenever a large number of sample averages are taken. Graphing the averages by sample number, according to when they were produced, gives a different impression of the data (Figure 5.1*a*). From the chart, it appears that the thickness of the flat round plate is increasing as production continues. This was not evident during the creation of the histogram or the analysis of the average, range, and standard deviation.

Control charts enhance the analysis of a process by showing how that process is performing over time. By combining such a chart with an appropriate statistical summary, those studying a process can gain an understanding of what the process is currently producing. This will enable them to make decisions concerning future production. The axiom applicable to being lost—you can't know which way to go if you don't know where you are—also applies to producing a quality product: You can't produce quality product in the future if you don't know what you're producing now. Control charts describe where the process is in terms of current performance. Knowing where the process is permits those working with the process to make decisions to enhance the future quality of products and services. This chapter will introduce the statistical control charting concepts necessary to determine how the process is currently behaving.

CONTROL-CHART FUNCTIONS

Control charts serve two basic functions:

1. Control charts are decision-making tools. They provide an economic basis for making a decision as to whether to investigate for potential problems, to adjust the process, or to leave the process alone.
   a. First and most importantly, control charts provide information for timely decisions concerning recently produced items. If an out-of-control condition is shown by the control chart then a decision can be made about sorting or reworking the most recent production.

b. Control-chart information is used to determine the process capability, or the level of quality the process is capable of producing. Samples of completed product can be statistically compared with the process specifications. This comparison provides information concerning the process's ability to meet the specifications set by the product designer. Continual improvement in the production process can take place only if there is an understanding of what the process is currently capable of producing.

2. Control charts are problem-solving tools. They assist in the identification of problems in the process. They help to provide a basis on which to formulate improvement actions.
   a. Control chart information can be used to help locate and investigate the causes of the unacceptable or marginal quality. By observing the patterns on the chart the investigator can determine what adjustments need to be made. This type of coordinated problem solving utilizing statistical data leads to improved process quality.
   b. During daily production runs, the operator can monitor machine production and determine when to make the necessary adjustments to the process or when to leave the process alone to ensure quality production.

Control charts play a major role in successful decision making and problem solving. Statistical control charts, by pointing out the presence of process faults, can be used to observe process improvement opportunities. Data from the current process can be evaluated against expectations set by the designer or can be compared with past behavior. Through careful interpretation, control charts can be used to study changes made to the process.

VARIATION

In manufacturing and service industries, the goal of most processes is to produce products or provide services that exhibit little or no variation. ***Variation,*** *where no two items or services are exactly the same*, exists in all processes. Although it may take a very precise measuring instrument or a very astute consumer to notice the variation, any process in nature will exhibit variation. Even identical twins have some differences. Variation can be large, and therefore easily noticeable, or small and discernible only by high-precision measuring instruments.

EXAMPLE 5.1 Variation in a Process

The industrial engineering department is seeking to decrease the amount of time it takes to perform a printer assembly operation. An analysis of the methods used by the operators performing the assembly has revealed that one of the operators completes the assembly in 75 percent of the time it takes another operator performing the same assembly operation. Further investigation determines that the two operators use parts produced on different machines. The histograms in Figure 5.2 are based on measurements of the parts from the two different processes. In Figure 5.2*b* the spread of the process is considerably smaller, enabling the faster operator to assemble the parts much more quickly and with less effort. The slower

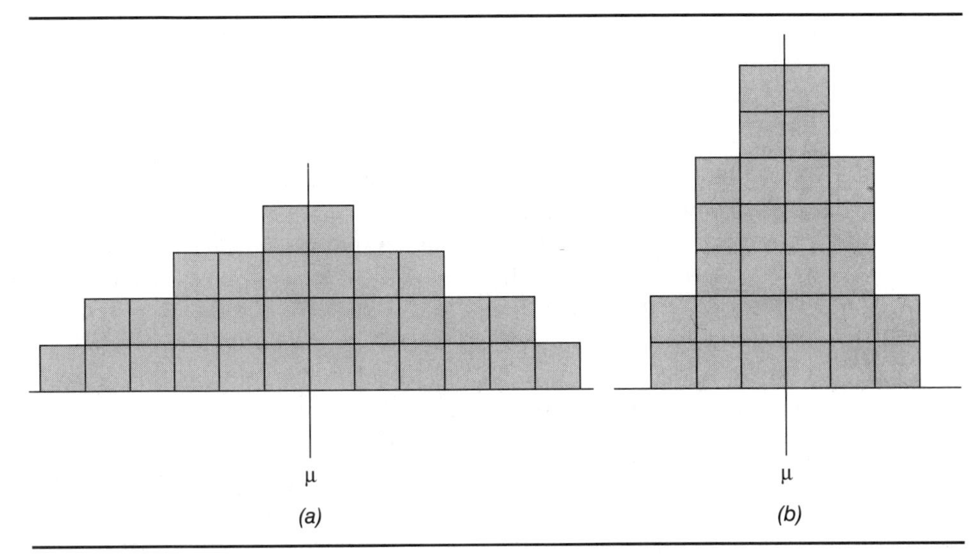

Figure 5.2 Histograms of Part Measurements from Two Machines

operator must try each part in the assembly, discard the part if it doesn't fit, and try another part. Repeating the operation when the parts do not fit has made the operator much less efficient. This operator's apparent lack of speed is actually caused by the process and is not the fault of the operator. Management intervention will be necessary to improve production at the previous operation and produce parts with less variation.

Several types of variation are tracked with statistical methods. These include:

1. Within-piece variation, or the variation within a single item or surface. For example, a single square yard of fabric may be examined to see if the color varies from one location to another.
2. Piece-to-piece variation, or the variation that occurs among pieces produced at approximately the same time. For example, in a production run filling gallon jugs with milk, when each of the milk jugs is checked after the filling station, the fill level from jug to jug will be slightly different.
3. Time-to-time variation, or the variation in the product produced at different times of the day—for example, the comparison of a part that has been stamped at the beginning of a production run with the part stamped at the end of a production run.

The variation in a process is studied by sampling the process. Samples are grouped into subgroups depending on whether or not the variation under study is piece-to-piece variation, within-piece variation, or time-to-time variation. The groupings of samples for piece-to-piece variation depend on the number of products analyzed during a particular time period. Within-piece variation samples are arranged in subgroups according to where in the process they are taken from. Samples for time-to-time variation are grouped according to the time of day that they were taken from the process.

Chance and Assignable Causes

Even in the most precise process no two parts are exactly alike. Whether these differences are due to chance causes in the process or to assignable causes needs to be determined. ***Chance,*** or ***common, causes*** *are small random changes in the process that cannot be avoided.* These small differences are due to the inherent variation present in all processes. They consistently affect the process and its performance day after day, every day. Variation of this type is only removable by making a change in the existing process. Changes of this type usually involve management intervention.

Assignable causes, on the other hand, *are large variations in the process that can be identified as having a specific cause.* Assignable causes are causes that are not part of the process on a regular basis. This type of variation arises because of specific circumstances. It is these circumstances that the quality-assurance analyst seeks to find. Examples of assignable causes include a size change in a part that occurs when chips build up around a work-holding device in a machining operation, changes in the thickness of incoming raw material, or a broken tool. Sources of variation can be found in the process itself, the materials used, the operator's actions, or the environment. Examples of factors that can contribute to process variation include tool wear, machine vibration, and work-holding devices. Changes in material thickness, composition, or hardness are sources of variation. Operator actions include overadjusting the machine, making an error during the inspection activity, changing the machine settings, or failing to properly align the part before machining. Environmental factors affecting variation include heat, light, radiation, and humidity.

To illustrate chance and assignable causes consider the following example.

EXAMPLE 5.2 Chance Causes and Assignable Causes

Picture a group of 25 students attending a lecture on quality control. The common causes affecting all of the students include room temperature, lighting, subject matter, and the professor's lecturing style. Special causes will affect only certain individuals. Special causes would include things like having stayed up late the night before to complete a difficult assignment, health problems, concentrating on preparing for a job interview, or a complete lack of interest in the subject. The professor needs to determine if the inattentiveness of a student or group of students is due to a common cause that affects the entire learning process or to a special cause that affects only one or two students. The manner in which the lack of attentiveness is dealt with will depend on the type of cause. If the room temperature is overly warm and the students are feeling drowsy, then the professor might choose to have a midclass break to allow the students to move around, get a drink of water, and wake up. If the professor notices a special cause affecting one or a few students, then she may ask those students after class what the difficulty is. The special cause can then be dealt with without affecting the remainder of the students.

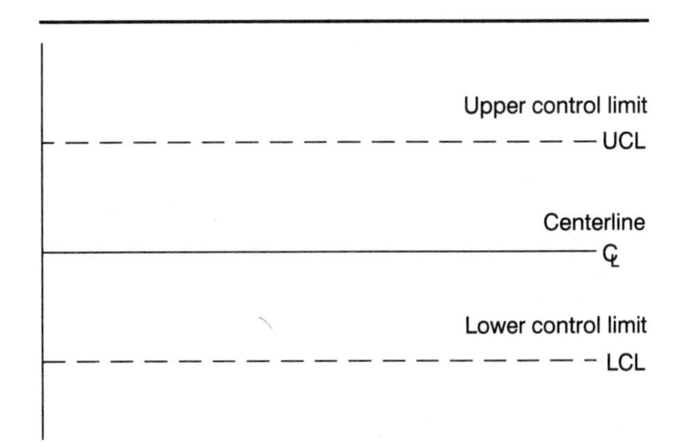

Figure 5.3 Control Chart Showing Centerline and Control Limits

CONTROL CHARTS

The type of variation monitored in a process is dependent on the particular process and the product or service being provided. Using statistical measures, process variation is recorded on control charts. Points on this chart are compared with each other to gain an understanding of what the process is capable of producing.

To create a control chart, individual data, arranged into subgroups, are sampled during the process. These measured values are plotted on the control chart. The ***centerline (℄)*** *of this chart shows where the process average is centered, or the central tendency of the data*. The ***upper control limit (UCL)*** and ***lower control limit (LCL)*** *describe the spread of the process* (Figure 5.3). Once the chart is constructed, it presents the investigator with a picture of what the process is currently capable of producing. In other words, we can expect future production to fall between these limits 99.73 percent of the time, providing the process does not change and is under control. Properly constructed, control charts will show changes in the distribution of measurements. The charts also help those closest to the process determine the causes of the changes.

Since control charts show changes in the process measurements, they allow for early detection of process changes. Instead of waiting until an entire production run is complete or until the product reaches the end of the assembly line, management can have the product checked and charted throughout the process. If a part or group of parts has been made incorrectly, production can be stopped, adjusted, or otherwise modified to produce parts correctly. This approach permits corrections to be made to the process before a large number of parts is produced or, in some cases, before the product exceeds the specifications. Early detection can avoid scrap, rework, unnecessary adjustments to the process, and/or production delays.

CONTROL CHARTS FOR VARIABLES

Variables *are the measurable characteristics of a product or service*. Examples of variables include the height, weight, or length of a part. Variables control charts take actual measurements and place them in chart form. One of the most commonly used

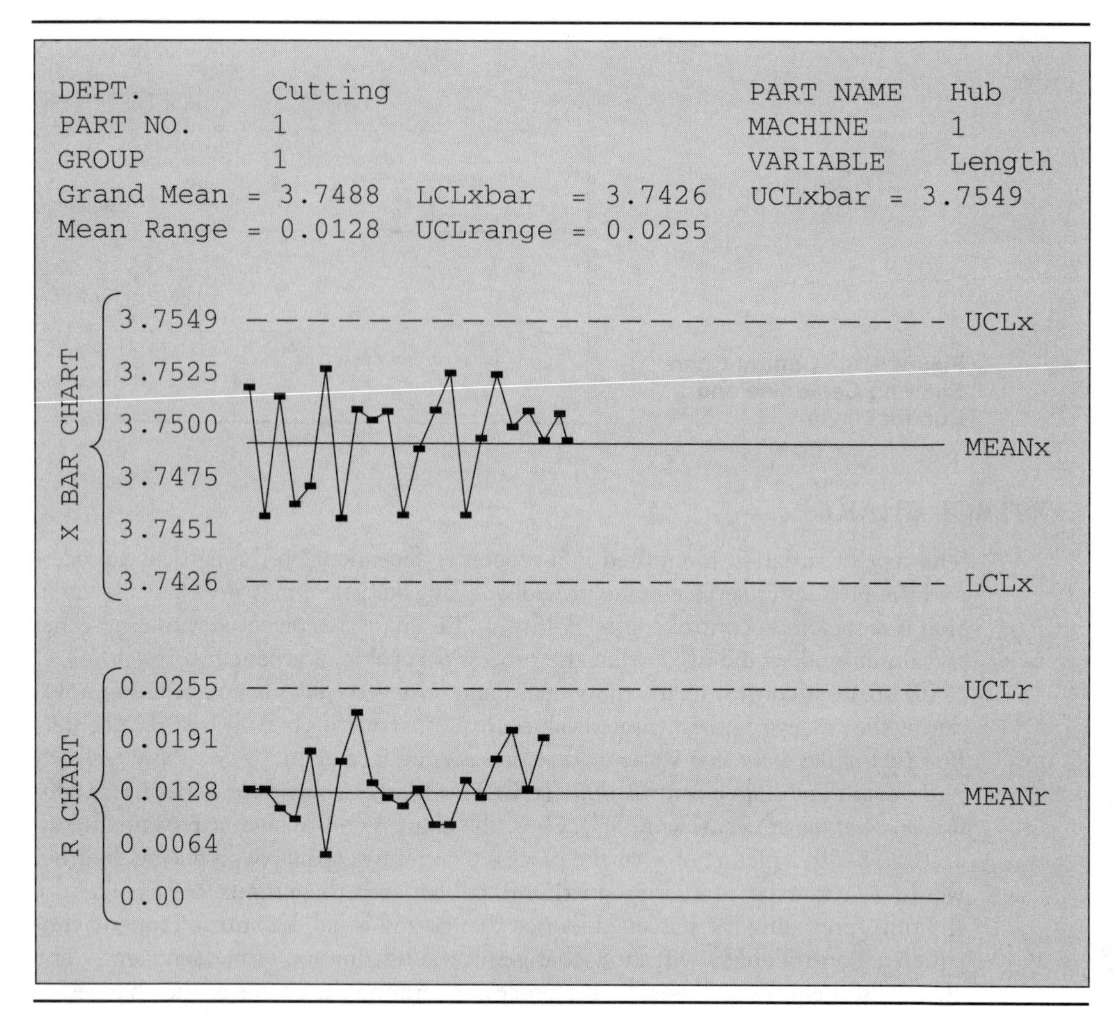

Figure 5.4 Typical $\overline{X}$ and R Chart

variables chart combinations in statistical process control is the $\overline{X}$ and R chart. A typical $\overline{X}$ and R chart utilized by an operator is shown in Figure 5.4. These charts are used together to determine the variation or distribution of the subgroup averages of sample measurements being produced by a process. The importance of using these two charts in conjunction with each other will become apparent shortly.

$\overline{X}$ and R Charts

The $\overline{X}$ ***chart*** *is used to monitor the variation of the subgroup averages that are calculated from the individual sampled data*. Averages rather than individual observations are used on control charts because average values will indicate a change in the amount of variation much faster than will individual values. Control limits on this chart are used to evaluate the variation from one subgroup to another.

The following steps and examples demonstrate the construction of an $\overline{X}$ chart.

1. Define the Problem

In any situation it is necessary to determine what the goal of monitoring a particular quality characteristic or group of characteristics is. To merely say, "Improve quality," is not enough. Nor is it sufficient to say, "We would like to see fewer parts out of specification." Out of which specification? Is it total product performance that is being affected or just one particular dimension? Sometimes several aspects of a part are critical for quality performance, occasionally only one is. It is more appropriate to say, "The length of these parts appears to be consistently below the lower specification limit. This causes the parts to mate incorrectly. Why are these parts below specification and how far below are they?" In the second statement we have isolated the length of the part as a critical dimension. From here, control charts can be placed on the process to help determine where the true source of the problem is located. Pareto analyses and cause-and-effect diagrams, as discussed in Chapter 3, are useful in establishing priorities and determining the potential causes of problems.

EXAMPLE 5.3 Defining the Problem

An assembly area has been experiencing serious delays in the construction of the computer printers. As quality assurance manager, you have been asked to determine the cause of these delays and fix the problems as soon as possible. To best utilize the limited time available, you convene a meeting involving those closest to the assembly problems. Representatives from production, supervision, manufacturing engineering, industrial engineering, quality assurance, and maintenance have been able to generate a variety of possible problems. From this meeting, a cause-and-effect diagram is created, showing the potential causes for the assembly difficulties (Figure 5.5). Discussions during the meeting reveal that the shaft on which the roller rests could be the major cause of assembly problems.

2. Select the Quality Characteristic to Be Measured

Variable control charts are based on measurements. Quality characteristics are the indicators used to measure system and process performance. Before creating control charts it is important to determine which quality characteristics will be studied. Choice of a characteristic to be measured depends on what is being investigated. Characteristic choice also depends on whether or not the process is being monitored for within-piece variation, piece-to-piece variation, or variation over time. Quality characteristics such as length, height, viscosity, color, temperature, and velocity are

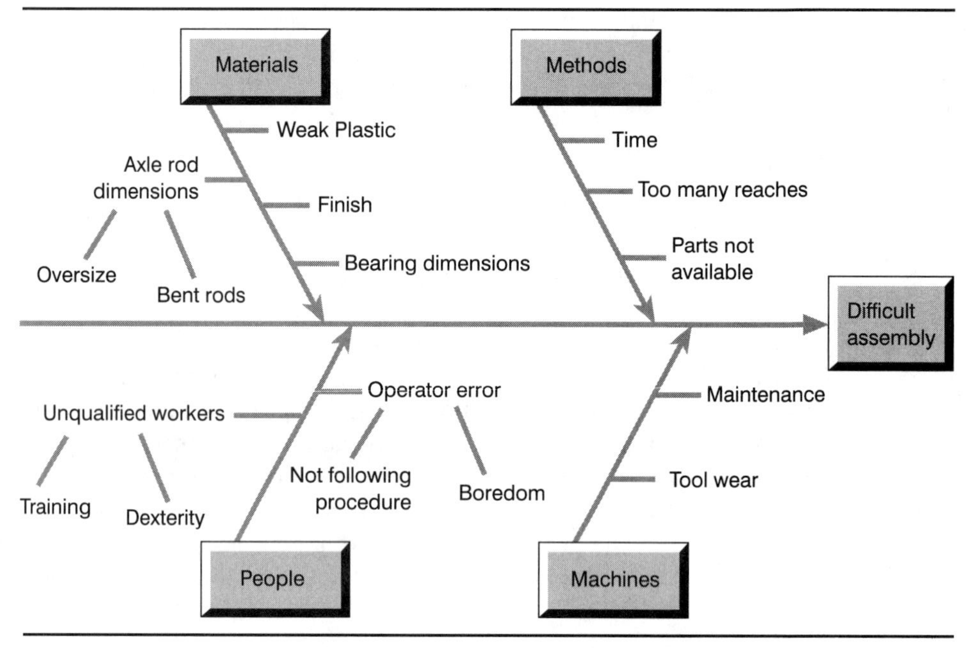

Figure 5.5 Cause-and-Effect Diagram

typically used in manufacturing settings. Number of errors, number of incorrect transactions, delivery times, checkout times, and service times are examples of quality characteristics chosen in a service industry. The characteristics selected should be ones that affect product or service performance. Quality characteristics affecting performance are found in many aspects of a product, including raw materials, components, subassemblies, and finished products. While it is crucial to identify important quality characteristics, avoid the tendency to try to establish control charts for all measurements.

EXAMPLE 5.4 Identifying the Quality Characteristic

As the troubleshooting meeting described in Example 5.3 continues, further discussions reveal that the length of the shaft is hindering assembly operations. The quality characteristic to measure has been identified as piece-to-piece variation in the length of the shafts. To begin to study the situation, measurements of the lengths of the shafts will be taken as they are produced. These will be compared with the specifications set for the assembly operations.

3. Choose a Rational Subgroup Size to Be Sampled

Subgroups, and the samples composing them, must be homogeneous. A ***homogeneous subgroup*** *will have been produced under the same conditions, by the same machine, the same operator, the same mold, and so on*. Homogeneous lots can also be designated by equal time intervals. Samples should be taken in an unbiased, random fashion. They should be representative of the entire population. Subgroup formation should reflect the type of variation under study. Subgroups in investigating piece-to-piece variation will not necessarily be constructed in the same manner as will subgroups formed to study time-to-time variation. The letter n is used to designate the number of samples taken within a subgroup. When constructing an $\overline{X}$ and R chart, keep the subgroup sample size constant for each subgroup taken.

Decisions concerning the specific size of the subgroup—n, or the number of samples—require judgment. Sampling should occur frequently enough to detect changes in the process. How often is the system expected to change? Examine the process and identify the factors causing change in the process. To be effective, sampling must occur as often as the system's most frequently changing factor. Once the number and frequency of sampling have been selected, they should not be changed unless the system itself has changed.

Realistically, sampling frequency must balance the value of the data obtained with the costs of taking the samples. Sampling is usually more frequent when control charts are first used to monitor the process. As process improvements are made and the process stabilizes, the frequency of sampling and subgroup size can be decreased.

When gathering sample data, it is important to have the following information in order to properly analyze the data:

1. *Who* will be collecting the data?
2. *What* aspect of the process is to be measured?
3. *Where* or at what point in the process will the sample be taken?
4. *When* or how frequently will the process be sampled?
5. *Why* is this particular sample being taken?
6. *How* will the data be collected?
7. *How many* samples will be taken (subgroup size)?

Some other guidelines to be followed include:

- The larger the subgroup size, the more sensitive the chart becomes to small variations in the process average. This will provide a better "picture" of the process since it allows the investigator to detect changes in the process quickly.
- While a larger subgroup size makes for a more sensitive chart, it also increases inspection costs.
- Destructive testing may make large subgroup sizes unfeasible. For example, it would not make sense for a fireworks manufacturer to test each and every one of its products.
- Subgroup sizes smaller than four do not create a representative distribution of subgroup averages. Subgroup averages are nearly normal for subgroups of four or more even when sampled from a nonnormal population.
- When the subgroup size exceeds 10, the standard deviation (s) chart, rather than the range (R) chart, should be used. For large subgroup sizes, the s chart gives a better representation of the true dispersion or true differences between the individuals sampled than does the R chart.

EXAMPLE 5.5 Selecting Subgroup Sample Size

The production from the machine making the shafts first looked at in Example 5.3 is consistent at 150 per hour. Since the process is currently exhibiting problems, your team has decided to take a sample of five measurements every 10 minutes from the production. The values for the day's production run are shown in Figure 5.6.

DEPT.	Roller	PART NAME	Shaft
PART NO.	1	MACHINE	1
GROUP	1	VARIABLE	length

Sample	1	2	3	4	5
Time	07:30	07:40	07:50	08:00	08:10
Date	07/02/95	07/02/95	07/02/95	07/02/95	07/02/95
1	11.95	12.03	12.01	11.97	12.00
2	12.00	12.02	12.00	11.98	12.01
3	12.03 (1)	11.96	11.97	12.00	12.02
4	11.98	12.00	11.98	12.03	12.03
5	12.01	11.98	12.00	11.99	12.02
$\overline{X}$	11.99 (2)	12.00	11.99	11.99	12.02
Range	0.08 (3)	0.07	0.04	0.06	0.03

Sample	6	7	8	9	10
Time	08:20	08:30	08:40	08:50	09:00
Date	07/02/95	07/02/95	07/02/95	07/02/95	07/02/95
1	11.98	12.00	12.00	12.00	12.02
2	11.98	12.01	12.01	12.02	12.00
3	12.00	12.03	12.04	11.96	11.97
4	12.01	12.00	12.00	12.00	12.05
5	11.99	11.98	12.02	11.98	12.00
$\overline{X}$	11.99	12.00	12.01	11.99	12.01
Range	0.03	0.05	0.04	0.06	0.08

Sample	11	12	13	14	15
Time	09:10	09:20	09:30	09:40	09:50
Date	07/02/95	07/02/95	07/02/95	07/02/95	07/02/95
1	11.98	11.92	11.93	11.99	12.00
2	11.97	11.95	11.95	11.93	11.98
3	11.96	11.92	11.98	11.94	11.99
4	11.95	11.94	11.94	11.95	11.95
5	12.00	11.96	11.96	11.96	11.93
$\overline{X}$	11.97	11.94	11.95	11.95	11.97
Range	0.05	0.04	0.05	0.06	0.07

Figure 5.6 Values for a Day's Production

Sample	16	17	18	19	20
Time	10:00	10:10	10:20	10:30	10:40
Date	07/02/95	07/02/95	07/02/95	07/02/95	07/02/95
1	12.00	12.02	12.00	11.97	11.99
2	11.98	11.98	12.01	12.03	12.01
3	11.99	11.97	12.02	12.00	12.02
4	11.96	11.98	12.01	12.01	12.00
5	11.97	11.99	11.99	11.99	12.01
$\overline{X}$	11.98	11.99	12.01	12.00	12.01
Range	0.04	0.05	0.03	0.06	0.03

Sample	21
Time	10:50
Date	07/02/95
1	12.00
2	11.98
3	11.99
4	11.99
5	12.02
$\overline{X}$	12.00
Range	0.04

(1) (2) $\frac{12.00 + 11.98 + 11.99 + 11.96 + 11.97}{5} = 11.98$ (3) $R = 0.05 = 12.02 - 11.97$

Figure 5.6 Continued

4. Collect the Data

To create a control chart, an amount of data sufficient to accurately reflect the statistical control of the process must be gathered. A minimum of 20 subgroups of sample size n = 4 is suggested. Each time a subgroup of sample size n is taken, an average is calculated for the subgroup. To do this, the individual values are recorded, summed, and then divided by the number of samples in the subgroup. This average, $\overline{X}_i$, is then plotted on the control chart.

EXAMPLE 5.6 Collecting Data

A sample of size n = 5 is taken at intervals from the process making shafts first introduced in Example 5.3. As shown in Figure 5.6, a total of 21 subgroups of sample size n = 5 have been taken. Each time a subgroup sample is taken, the individual values are recorded [Figure 5.6, (1)], summed, and then divided by the number of samples taken to get the average for the subgroup [Figure 5.6, (2)]. This subgroup average is then plotted on the control chart [Figure 5.7, (1)].

5. Determine the Trial Centerline for the $\overline{X}$ Chart

The centerline of the control chart is the process average. It would be the mean, μ, if the average of the population measurements for the entire process were known.

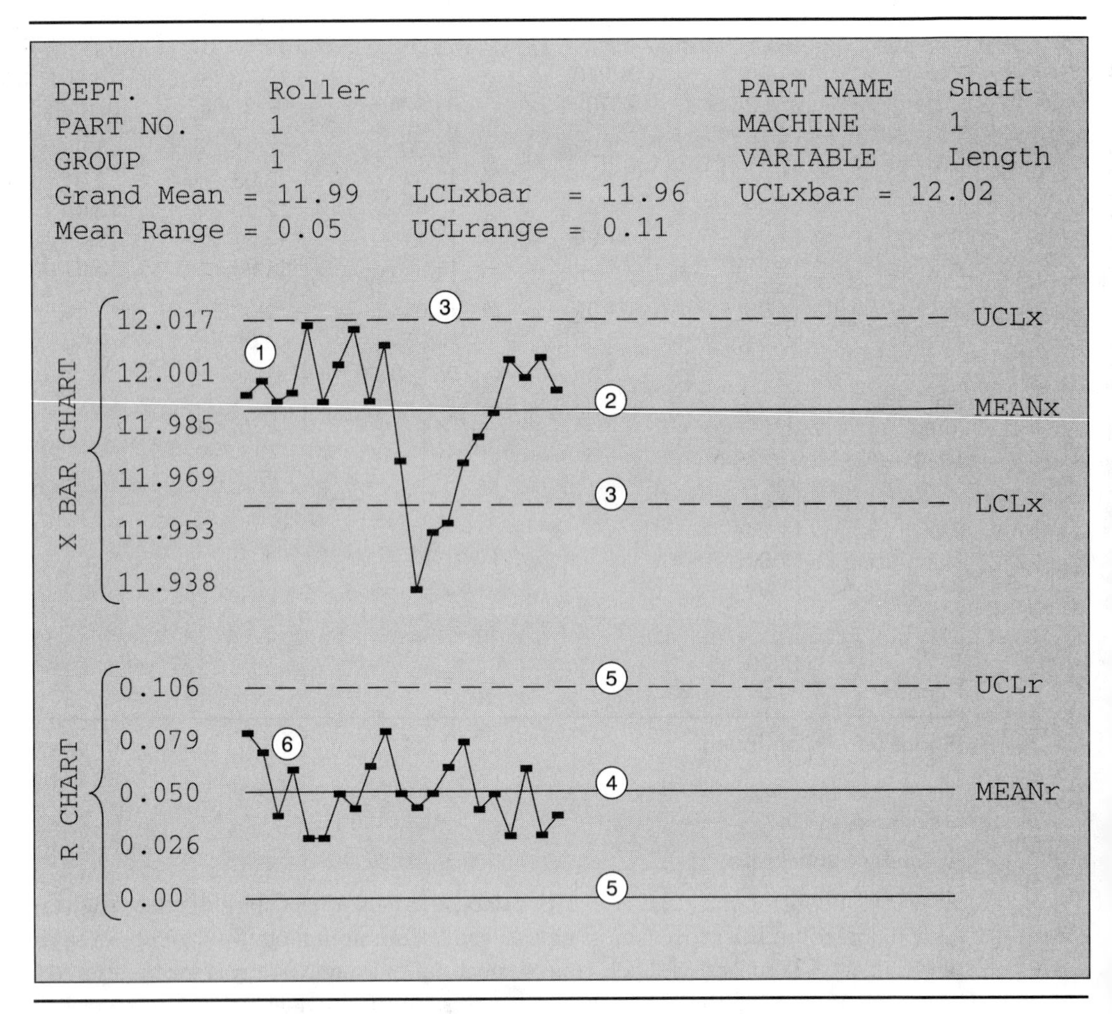

Figure 5.7 Control Chart

Since the value of the population mean μ cannot be determined unless all of the parts being produced are measured, in its place the grand average of the subgroup averages, $\overline{\overline{X}}$ (X double bar), is used. The grand average, or $\overline{\overline{X}}$, is calculated by summing all the subgroup averages and then dividing by the number of subgroups. This value is plotted as the centerline of the $\overline{X}$ chart:

$$\overline{\overline{X}} = \frac{\sum_{i=1}^{m} \overline{X}_i}{m}$$

where

$$\bar{\bar{X}} = \text{average of the subgroup averages}$$
$$\bar{X}_i = \text{average of the ith subgroup}$$
$$m = \text{number of subgroups}$$

6. Determine the Trial Control Limits for the $\bar{X}$ Chart

Figure 5.1 showed how a control chart is a time-dependent pictorial representation of a normal curve. The normal curve displays the distribution of the averages of the samples taken from the process. Just as the normal curve contains 99.73 percent of the data within $\pm 3\sigma$, 99.73 percent of the points plotted on a control chart will fall within the upper and lower control limits, which are set at ± 3 standard deviations from the grand average of the population. It is this statement that enables us to determine the capability of the process. Processes that are considered under control will have 99.73 percent of their graphed averages within the upper and lower control limits.

The control limits, which are a function of the subgroup averages, are established at ± 3 standard deviations [Figure 5.7 (3)] from the centerline for the process using the following formulas:

$$UCL_{\bar{X}} = \bar{\bar{X}} + 3\sigma_{\bar{x}}$$
$$LCL_{\bar{X}} = \bar{\bar{X}} - 3\sigma_{\bar{x}}$$

where

$$UCL = \text{upper control limit of the } \bar{X} \text{ chart}$$
$$LCL = \text{lower control limit of the } \bar{X} \text{ chart}$$
$$\sigma_{\bar{x}} = \text{population standard deviation of the subgroup averages}$$

The population standard deviation σ is needed to calculate the upper and lower control limits. Recalling Chapter 4, you'll note that calculating the $\sigma_{\bar{x}}$ for the population is not an easy task. A good approximation of $3\sigma_{\bar{x}}$ can be found by replacing $3\sigma_{\bar{x}}$ with the product of an A_2 factor multiplied by $\bar{R}$, the average of the ranges. The $A_2\bar{R}$ combination uses the sample data for its calculation. $\bar{R}$ is calculated by summing the values of the individual subgroup ranges and dividing by the number of subgroups m:

$$\bar{R} = \frac{\sum_{i=1}^{m} R_i}{m}$$

where

$$\overline{R} = \text{average of the ranges}$$
$$R_i = \text{individual range values for the sample}$$
$$m = \text{number of subgroups}$$

A_2 is a factor that allows the following approximation to be true: $A_2\overline{R} \approx 3\sigma_{\overline{X}}$; the selection of A_2 is made based on the subgroup sample size n. (See Appendix 2 for the A_2 factors.)

Upon replacement, the formulas for the upper and lower control limits become

$$UCL_{\overline{X}} = \overline{\overline{X}} + A_2\overline{R}$$
$$LCL_{\overline{X}} = \overline{\overline{X}} - A_2\overline{R}$$

After calculating the control limits, we place the centerline ($\overline{\overline{X}}$) and the upper and lower control limits (UCL and LCL, respectively) on the chart. As seen in Figure 5.7, the upper and lower control limits are shown by dashed lines. The grand average, or $\overline{\overline{X}}$, is shown by a solid line. The control limits on the $\overline{X}$ chart will be symmetrical about the central line. These control limits are used to evaluate the variation in quality from subgroup to subgroup.

EXAMPLE 5.7 Calculating the $\overline{X}$ Chart Centerline and Control Limits

Construction of an $\overline{X}$ chart begins with the calculation of the centerline, $\overline{\overline{X}}$. Using the 21 subgroups of sample size n = 5 provided in Figure 5.6, we calculate $\overline{\overline{X}}$ by summing all the subgroup averages from the individual samples taken and then dividing by the number of subgroups, m:

$$\overline{\overline{X}} = \frac{11.99 + 12.00 + 11.99 + \cdots + 12.00}{21}$$
$$= \frac{251.77}{21} = 11.99$$

This value is plotted as the centerline of the $\overline{X}$ chart [Figure 5.7, (2)].

$\overline{R}$ is calculated by summing the values of the individual subgroup ranges (Figure 5.6) and dividing by the number of subgroups, m:

$$\overline{R} = \frac{0.08 + 0.07 + 0.04 + \cdots + 0.04}{21}$$
$$= \frac{1.06}{21} = 0.05$$

The values for the upper and lower control limits of the $\overline{X}$ chart are calculated as follows:

$$UCL_{\overline{X}} = \overline{\overline{X}} + A_2\overline{R}$$
$$= 11.99 + 0.577(0.05) = 12.02$$
$$LCL_{\overline{X}} = \overline{\overline{X}} - A_2\overline{R}$$
$$= 11.99 - 0.577(0.05) = 11.96$$

The A_2 factor for a sample size of five is selected from the table in Appendix 2. Once calculated, the upper and lower control limits (UCL and LCL, respectively) are placed on the chart [Figure 5.7, (3)].

7. *Determine the Trial Control Limits for the* R *Chart*

When an $\overline{X}$ chart is used to evaluate the variation in quality from subgroup to subgroup, the range chart is one method of determining the amount of variation in the individual samples. The importance of the range chart is often overlooked. Without the range chart, or the standard deviation chart to be discussed later, it would not be possible to fully understand the process capability. Where the $\overline{X}$ chart shows the average of the individual subgroups, giving the viewer an understanding of where the process is centered, the range chart shows the spread or dispersion of the individual samples within the subgroup. The calculation of the spread of the measurements is necessary to determine whether the parts being produced are similar to one another or not. If the product displays a wide spread or a large range, then the individuals being produced are not similar to each other. Figure 5.8 shows that each of the subgroups has the same average; however, their samples spread differently. The optimal situation from a quality perspective is when the parts are grouped closely around the process average. This situation will yield a small value for both the range and the standard deviation, meaning that the measurements are very similar to each other.

The range chart limits are calculated in a manner similar to the $\overline{X}$ chart limits. Individual ranges are calculated for each of the subgroups by subtracting the highest value in the subgroup from the lowest value. These individual ranges are then summed and divided by the total number of subgroups to give $\overline{R}$, the centerline of the R chart:

$$\overline{R} = \frac{\sum_{i=1}^{m} R_i}{m}$$

$$UCL_R = \overline{R} + 3\sigma_R$$

$$LCL_R = \overline{R} - 3\sigma_R$$

where

UCL_R = upper control limit of the R chart
LCL_R = lower control limit of the R chart
σ_R = population standard deviation of the subgroup ranges

For the R chart, the average of the subgroup ranges multiplied by the D_3 and D_4 factors is used to estimate the standard deviation of the range σ_R:

$$UCL_R = D_4\overline{R}$$

$$LCL_R = D_3\overline{R}$$

Along with the value of A_2, the values of D_3 and D_4 are found in the table in Appendix 2. These values are selected on the basis of the subgroup sample size n.

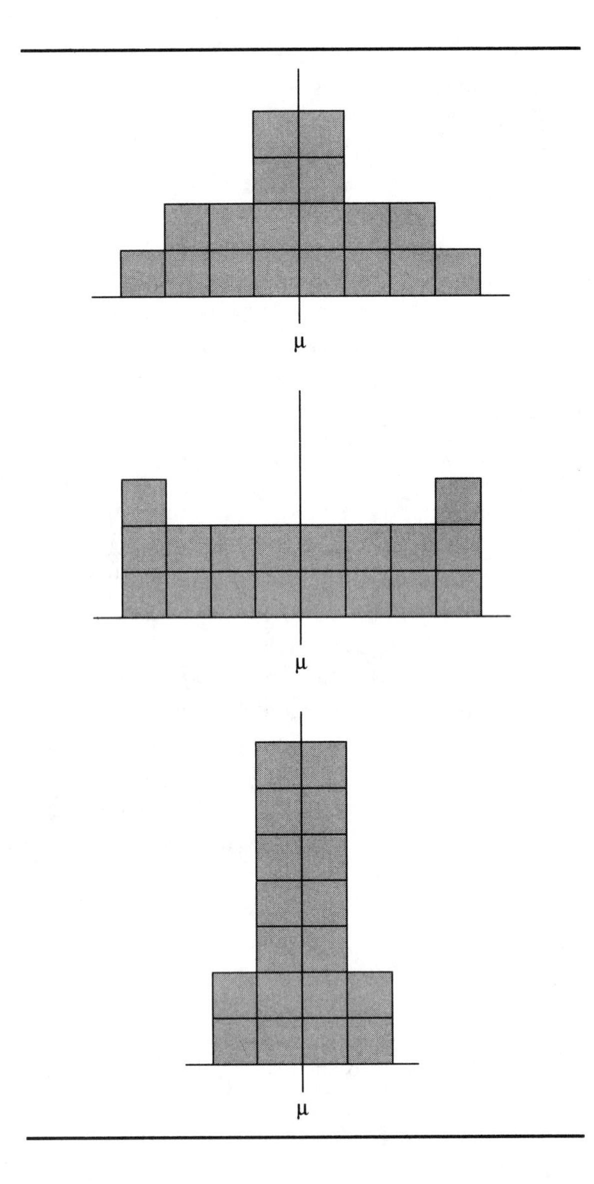

Figure 5.8 Histograms with Same Averages, Different Ranges

The control limits, when displayed on the R chart, should theoretically be symmetrical about the centerline ($\overline{R}$). However, because range values cannot be negative, a value of zero is given for the lower control limit with sample sizes of six or less. This results in an R chart that is asymmetrical. As with the $\overline{X}$ chart, control limits for the R chart are shown with a dashed line. The centerline is shown with a solid line.

Even with subgroup values larger than six, frequently the LCL of the range chart is kept at zero. This is in keeping with the desire to achieve zero variation between the parts being produced. However, this approach is not recommended because points below the calculated lower control limit would be identified as special variation. The values below the lower control limit indicate that reduced variation is present in the process. If the cause of this decrease in variation could be found and replicated, then the variation present in the process would be diminished.

EXAMPLE 5.8 Calculating the R-Chart Centerline and Control Limits

Constructing an R chart is similar to creating an $\overline{X}$ chart. To begin the process, individual range values are calculated for each of the subgroups by subtracting the highest value in the subgroup from the lowest value [Figure 5.6, (3)]. Once calculated, these individual range values (R_i) are plotted on the R chart [Figure 5.7, (6)].

To determine the centerline of the R chart, individual range (R_i) values are summed and divided by the total number of subgroups to give $\overline{R}$ [Figure 5.7, (4)].

$$\overline{R} = \frac{0.08 + 0.07 + 0.04 + \cdots + 0.04}{21}$$

$$= \frac{1.06}{21} = 0.05$$

The control limits for the R chart are calculated as follows:

$$UCL_R = D_4\overline{R}$$
$$= 2.114(0.05) = 0.11$$

$$LCL_R = D_3\overline{R}$$
$$= 0(0.05) = 0$$

With $n = 5$, the values of D_3 and D_4 are found in the table in Appendix 2. The control limits are placed on the R chart [Figure 5.7, (5)].

8. Examine the Process: Control-Chart Interpretation

Correct interpretation of control charts is essential to managing a process. Understanding the sources and potential causes of variation is critical to good management decisions. Managers must be able to determine if the variation present in a process is indicating a trend that must be dealt with or is merely random variation natural to the process. Misinterpretation can lead to a variety of losses, including

- Blaming people for problems that they cannot control
- Spending time and money looking for problems that do not exist
- Spending time and money on process adjustments or new equipment that are not necessary
- Taking action where no action is warranted
- Asking for worker-related improvements where process or equipment improvements need to be made first

If a process is understood and adjustments have been made to stabilize the process, then the benefits are many. Once the performance of a process is predictable, there is a sound basis for making plans and decisions concerning the process, the system, and its output. Costs to manufacture the product or provide the service become predictable. Quality levels and how quality compares with expectations can be determined. Productivity will be at a maximum for the process in its current state. The effects of changes made to the process can be measured and evaluated with greater accuracy and reliability.

EXAMPLE 5.9

Data is a key aspect of computer-integrated manufacturing at Whisks Electronics Corporation. Automated data-acquisition systems generate timely data about the product produced and the process producing it. Whisks believes that decision-making processes can be enhanced by using valid data that have been organized in an effective manner. For this reason, Whisks uses an integrated system of automated statistical process-control programming, data-collection devices, and programmable logic controllers (PLCs) to collect statistical information about its silicon wafer production. Utilizing the system relieves the process engineers from the burden of number crunching, freeing time for critical analysis of the data.

Recent quality problems in silicon wafer line B have necessitated creating the following $\overline{X}$ and R charts. Whisks begins by doing the following.

Step 1. Define the Problem. The production of integrated circuits by etching them onto silicon wafers requires silicone wafers of consistent thickness. Recently, Whisks' customers have raised questions about whether or not Whisks' process is capable of producing wafers of consistent thickness.

Step 2. Select the Quality Characteristic to Be Measured. Wafer thickness is the critical quality characteristic. The customer has established a target value for wafer thickness at 0.250 mm.

Step 3. Choose a Rational Subgroup Size to Be Sampled. Using micrometers linked to PLCs, every 15 minutes four wafers are randomly selected and measured in the order they are produced. These values are stored in a database that is accessed by the company's statistical process-control software. This subgroup size and sampling frequency were chosen based on the number of silicone wafers produced per hour.

Step 4. Collect the Data. The data collected by the micrometer and PLC are shown in Figure 5.9.

Step 5. Determine the Trial Centerline for the $\overline{X}$ Chart. The centerline of the $\overline{X}$ chart is created by using the following formula:

$$\overline{\overline{X}} = \frac{\Sigma \overline{X}_i}{m}$$

$$\overline{\overline{X}} = \frac{(0.2500 + 0.2503 + 0.2498 + \cdots + 0.2495)}{15}$$

$$\overline{\overline{X}} = 0.250 \text{ mm}$$

					$\overline{X}$	R
Subgroup 1	0.2500	0.2510	0.2490	0.2500	0.2500	0.002
Subgroup 2	0.2510	0.2490	0.2490	0.2520	0.2503	0.003
Subgroup 3	0.2510	0.2490	0.2510	0.2480	0.2498	0.003
Subgroup 4	0.2490	0.2470	0.2520	0.2480	0.2490	0.005
Subgroup 5	0.2500	0.2470	0.2500	0.2520	0.2498	0.005
Subgroup 6	0.2510	0.2520	0.2490	0.2510	0.2508	0.003
Subgroup 7	0.2510	0.2480	0.2500	0.2500	0.2498	0.003
Subgroup 8	0.2500	0.2490	0.2490	0.2520	0.2500	0.003
Subgroup 9	0.2500	0.2470	0.2500	0.2510	0.2495	0.004
Subgroup 10	0.2480	0.2480	0.2510	0.2530	0.2500	0.005
Subgroup 11	0.2500	0.2500	0.2500	0.2530	0.2508	0.003
Subgroup 12	0.2510	0.2490	0.2510	0.2540	0.2513	0.005
Subgroup 13	0.2500	0.2470	0.2500	0.2510	0.2495	0.004
Subgroup 14	0.2500	0.2500	0.2490	0.2520	0.2503	0.003
Subgroup 15	0.2500	0.2470	0.2500	0.2510	0.2495	0.004

Figure 5.9 Silicon Wafer Thickness

where

$\overline{\overline{X}}$ = the centerline of the $\overline{X}$ chart
$\overline{X}$ = the individual subgroup averages
m = the number of subgroups

Step 6. Determine the Trial Control Limits for the $\overline{X}$ Chart. The upper and lower control limits are based on the idea that 99.73 percent of the data should be located between ± 3 standard deviations of the mean. Rather than calculate the standard deviation for each subgroup, the $A_2\overline{R}$ value allows us to estimate $\pm 3\sigma$ using the average range ($\overline{R}$).

$$\overline{R} = \frac{\sum R_i}{m}$$

$$\overline{R} = \frac{(0.002 + 0.003 + 0.003 + \cdots + 0.004)}{15}$$

$$\overline{R} = 0.0037 \text{ mm}$$

Upper control limit, $\overline{X}$

$$UCL_{\overline{x}} = \overline{\overline{X}} + A_2\overline{R}$$
$$UCL_{\overline{x}} = 0.250 + 0.729(0.0037)$$
$$UCL_{\overline{x}} = 0.2527 \text{ mm}$$

Lower control limit, $\overline{X}$

$$LCL_{\bar{x}} = \overline{\overline{X}} - A_2\overline{R}$$
$$LCL_{\bar{x}} = 0.250 - 0.729(0.0037)$$
$$LCL_{\bar{x}} = 0.2473 \text{ mm}$$

Step 7. Determine the Trial Centerline and Control Limits for the R Chart.

$$\overline{R} = \frac{\sum R_i}{m}$$
$$\overline{R} = \frac{(0.002 + 0.003 + 0.003 + \cdots + 0.004)}{15}$$
$$\overline{R} = 0.0037 \text{ mm}$$

Upper control limit, R

$$UCL_R = D_4\overline{R}$$
$$UCL_R = 2.282(0.0037)$$
$$UCL_R = 0.0084 \text{ mm}$$

Lower control limit, R

$$LCL_R = D_3\overline{R}$$
$$LCL_R = 0(0.0037)$$
$$LCL_R = 0.0$$

Step 8. Interpret the Chart. $\overline{X}$ and R chart combinations provide information about process centering and process variation. If we study the $\overline{X}$ chart (Figure 5.10), it

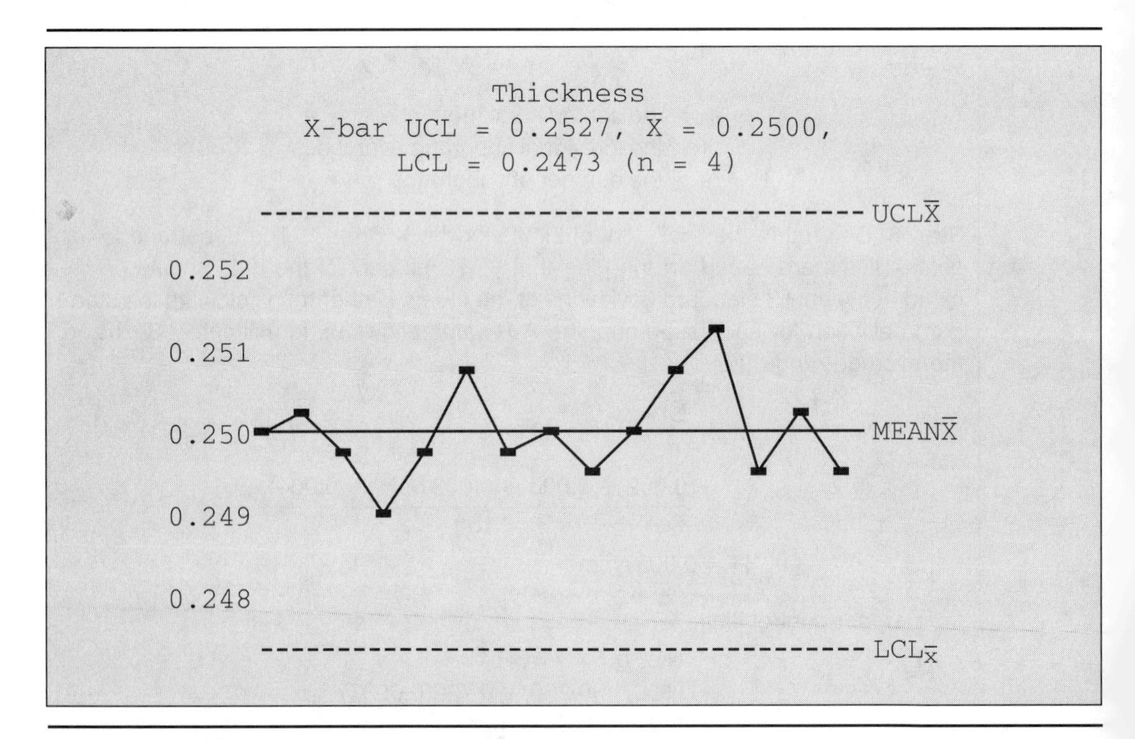

Figure 5.10 $\overline{X}$ Chart for Silicon Wafer Thickness

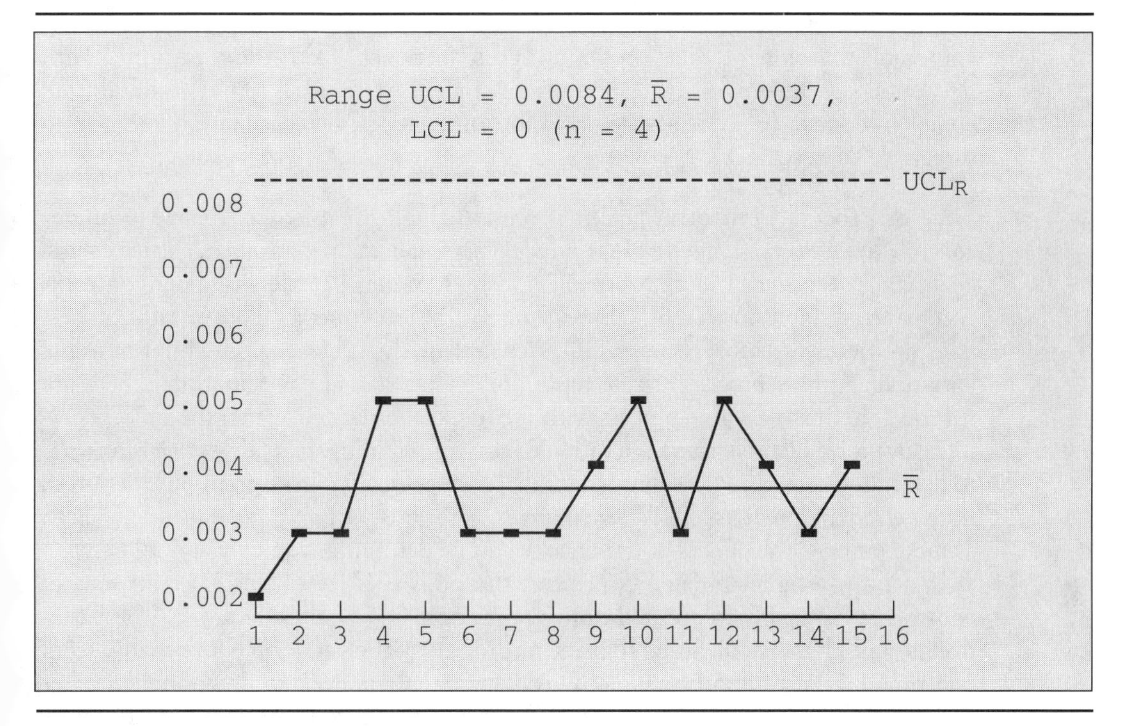

Figure 5.11 R Chart for Silicon Wafer Thickness

shows that the wafer thickness is centered around the target value of 0.250 mm. Studying the R chart (Figure 5.11) reveals that some variation is present in the process. Not all the wafers are uniform in their thickness. Whisks will have to determine the best method of removing variation from their silicon wafer production process.

Process Variation Because variation is present in all aspects of our lives, we already have developed an understanding of what is usual or unusual variation. For instance, on the basis of a six-month history of commuting, we may expect our commute to work to take 25 minutes, give or take a minute or two. We would be surprised if the commute took only 15 minutes. We would look for an assignable cause: Perhaps traffic was lighter because we left earlier. By the same token, we would be upset if the commute took 40 minutes and we would want to know why. A traffic accident could be the assignable cause that meant such an increase in commuting time. We continually make decisions on the basis of an interpretation of the amount of variation from the expected value we encounter. Included in these decisions are whether or not we think this is random (chance) variation or unusual (assignable) variation.

The patterns or variation on a control chart are not too different from the variation that exists in everyday life. Whether we are following weight gain or loss, household expenses from month to month, or the gas mileage of our cars, variation exists.

These values are never exactly the same from measurement to measurement. In the corporate environment, sales, profits, and costs are never exactly the same from month to month. At the university, test scores for students differ from test to test and from course to course. In a hospital, completion time for the same operation varies from surgery to surgery.

State of Process Control A *process is considered to be in a state of control, or* ***under control,*** *when the performance of the process falls within the statistically calculated control limits and exhibits only chance, or common, causes.* When a process is under control it is considered stable and the amount of future variation is predictable. A stable process does not necessarily meet the specifications set by the designer nor exhibit minimal variation; a stable process merely has a predictable amount of variation. There are several benefits to a stable process with predictable variation. When the process performance is predictable there is a rational basis for planning. It is fairly straightforward to determine costs associated with a stable process. Quality levels from time period to time period are predictable. When changes, additions, or improvements are made to a stable process, the effects of the change can be determined quickly and reliably.

When an assignable cause is present, the process is considered unstable, out of control, or beyond the expected normal variation. In an unstable process the variation is unpredictable, meaning that the magnitude of the variation could change from one time period to another. Quality-assurance analysts need to determine whether the variation that exists in a process is common or assignable. To treat an assignable cause as a chance cause could result in a disruption to a system or a process that is operating correctly except for the assignable cause. To treat chance causes as assignable causes is an ineffective use of resources because the variation is inherent in the process.

EXAMPLE 5.10 Common Variation in a Process

Consider the following information concerning work performed at a local doctor's office. The office management wanted to see a serious improvement in the amount of time it took to process the insurance forms. Without giving any specific instructions, the management encouraged the staff to work harder and smarter.

Process improvements were needed since a decrease in the time required to process and file insurance claim forms would help alleviate a backlog of work. Errors being made on the form also needed to be reduced. The individuals handling the insurance-form processing were a well-trained and experienced group who felt that they were all performing to the best of their ability, that this was a stable process. If it was a stable process, improvements would only come about with a change in the methods utilized to process and file the forms. To support this statement, the staff monitored their own performance, using control charts, for the next six months.

This control charting, using an $\overline{X}$ and R chart combination, showed that completing and processing each insurance form required an average of 23.875 minutes (Figure 5.12). The R chart is examined first in order to gain an understanding of the precision or lack of variation present in the process. As shown by the R chart, there was a slight variation in the time required to process the forms. Overall, the

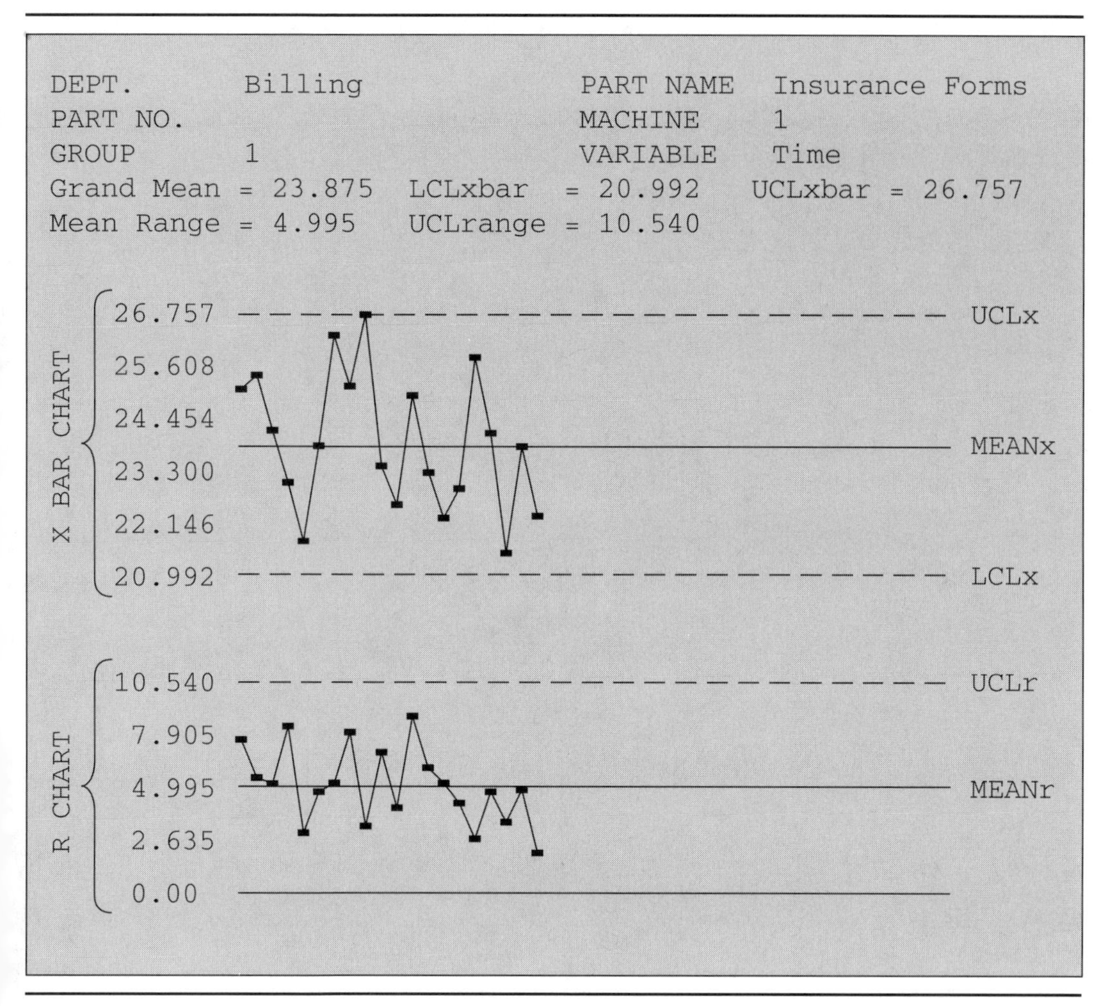

Figure 5.12 Control Chart with Grand Mean = 23.875

performance appeared to be very stable; no points went beyond the control limits and there were no unusual patterns. The process exhibited only chance causes of variation.

When presented with these data, management realized they had to do more than encourage the staff to work harder or work smarter. Beginning with the control chart, further investigation into the situation occurred. As a result of their investigations, management decided to change the insurance form and implement a new computerized billing/insurance processing system. Upon completion of operator training, this system was monitored through control charts, and as can be seen in Figure 5.13, the processing time was reduced to an average of 10.059 minutes. An increase in performance of this magnitude was possible only through intervention by management and a major change to the existing system.

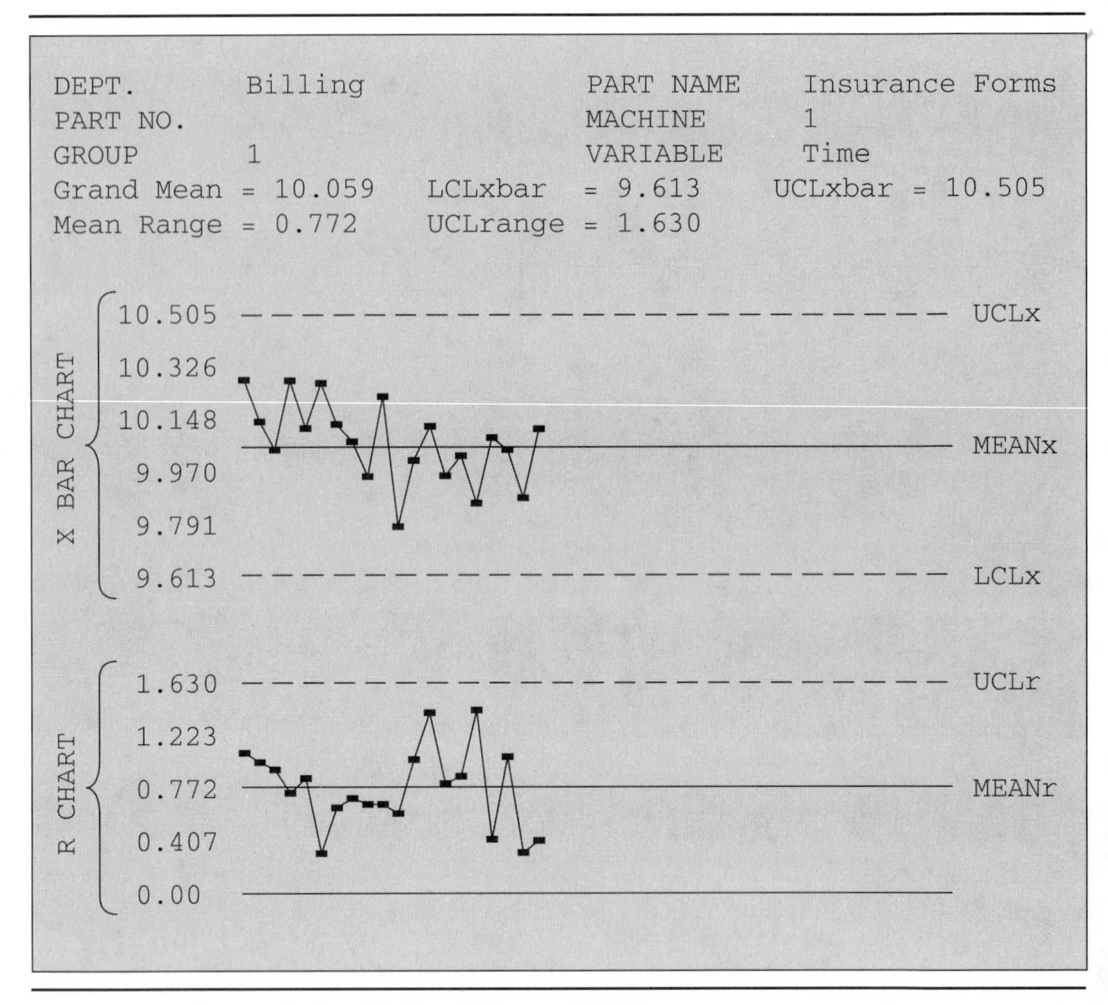

Figure 5.13 Control Chart with Grand Mean = 10.059

Once an understanding of process performance is created, it is possible to interpret the variation associated with the process and make decisions based on this interpretation. When charted, the variation of a quality characteristic present in the process is an indicator of how the system is operating.

When a system is subject to only chance causes of variation, 99.73 percent of the parts produced will fall within $\pm 3\sigma$. This means that if 1,000 subgroups are sampled, 997 of the subgroups will have values within the upper and lower control limits. Based on the normal curve, a control chart can be divided into three zones (Figure 5.14). Zone A is ± 1 standard deviation from the centerline and should contain approximately 68.3 percent of the calculated sample averages or ranges. Zone B is ± 2 standard

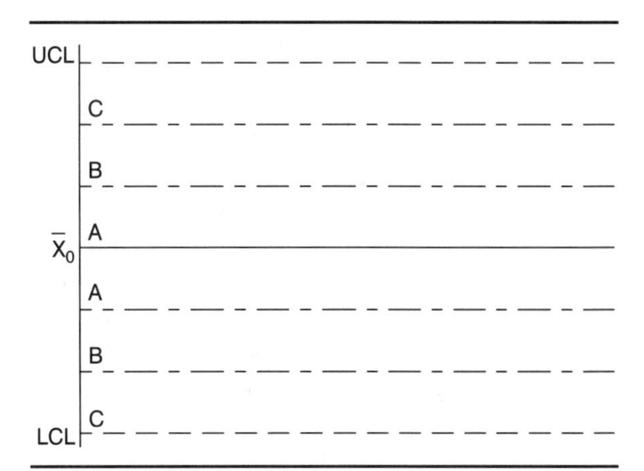

Figure 5.14 Zones on a Control Chart

deviations from the centerline and should contain 27.2 percent (95.5 percent − 68.3 percent) of the points. Zone C is ±3 standard deviations from the centerline and should contain only approximately 4.2 percent of the points (99.7 percent − 95.5 percent). With these zones as a guide, a control chart exhibits a state of control when:

1. Two-thirds of the points are near the center value.
2. A few of the points are close to the center value.
3. The points appear to float back and forth across the centerline.
4. The points are balanced (in roughly equal numbers) on both sides of the centerline.
5. There are no points beyond the control limits.
6. There are no patterns or trends on the chart.

While analyzing $\overline{X}$ and R charts, take a moment to study the scale of the range chart. The spread of the upper and lower control limits will reveal whether or not a significant amount of variation is present in the process. This clue to the amount of variation present may be overlooked if the R chart is checked only for patterns or out-of-control points.

The importance of having a process under control cannot be overemphasized. When a process is under control, a number of advantages can be found. The process capability, which tells how the current production compares with the specification limits, can be calculated. Once known, the process capability can be used in making decisions concerning the appropriateness of product specifications, of the amount of rework or scrap being created by the process, and of whether the parts should be used or shipped.

Identifying Patterns A process that is not under control or is unstable displays patterns of variation. Patterns signal the need to investigate the process and determine if an assignable cause can be found for the variation. Figures 5.15 through 5.26

display a variety of out-of-control conditions and give some reasons why those conditions may exist. The patterns in these figures and in Examples 5.11 through 5.15 have been exaggerated to make the patterns clear.

Trends or steady changes in level A trend is a steady, progressive change in where the data are centered on the chart. Figure 5.15 displays a downward trend. Note that the points were found primarily in the upper half of the control chart at the beginning of the process and on the lower half of the chart at the end. The key to identifying a trend or steady change in level is to recognize that the points are slowly and steadily working their way from one level of the chart to another.

A trend may appear on the $\overline{X}$ chart because of tool or die wear, a gradual deterioration of the equipment, a buildup of chips, a slowly loosening work-holding de-

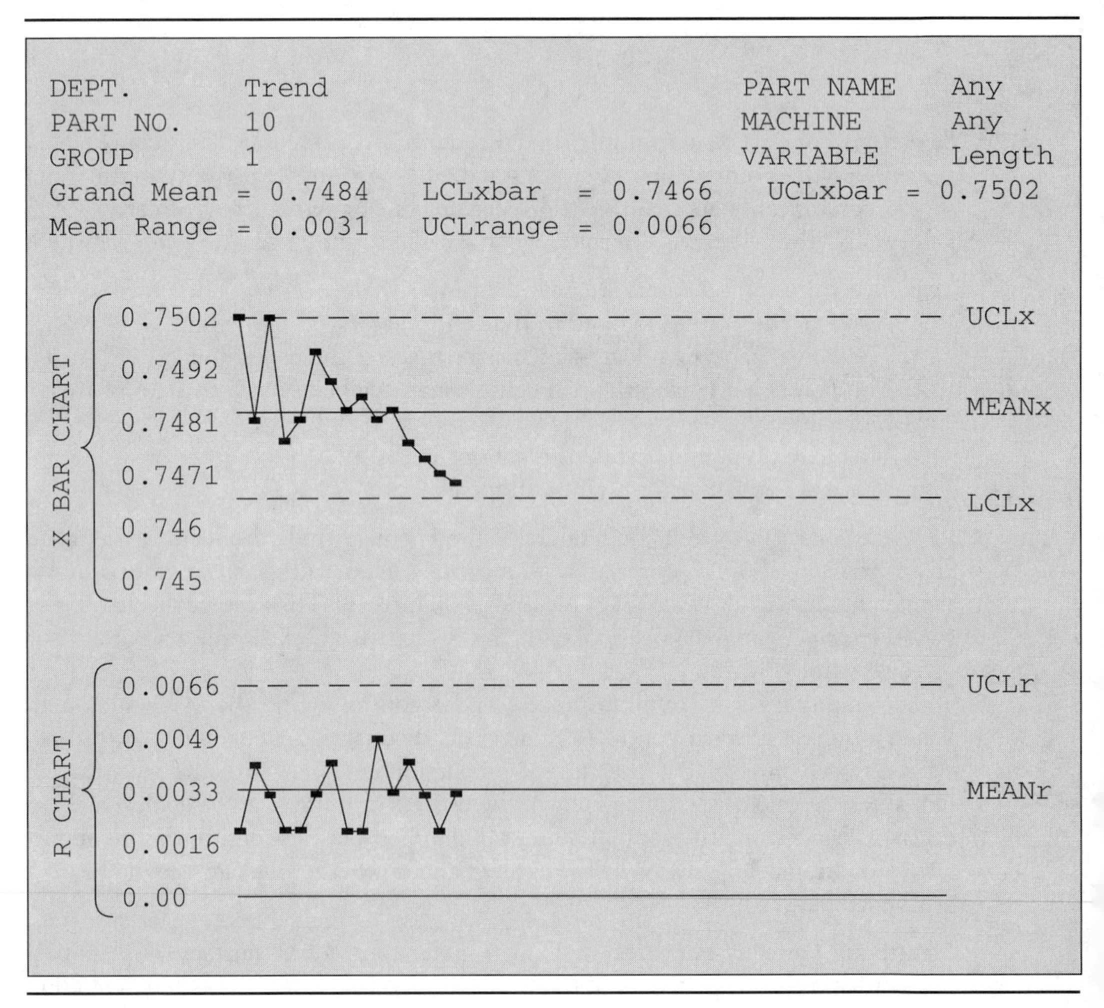

Figure 5.15 A Downward Trend

vice, a breakdown of the chemicals used in the process, or some other gradual change.

R-chart trends appear infrequently and could be due to changes in worker skills. Improvements would lead to less variation; increases in variation would reflect a decrease in skill or a change in the quality of the incoming material.

An oscillating trend would also need to be investigated (Figure 5.16). In this type of trend the points oscillate up and down for approximately 14 points or more. This could be due to a lack of homogeneity, perhaps a mixing of the output from two machines making the same product.

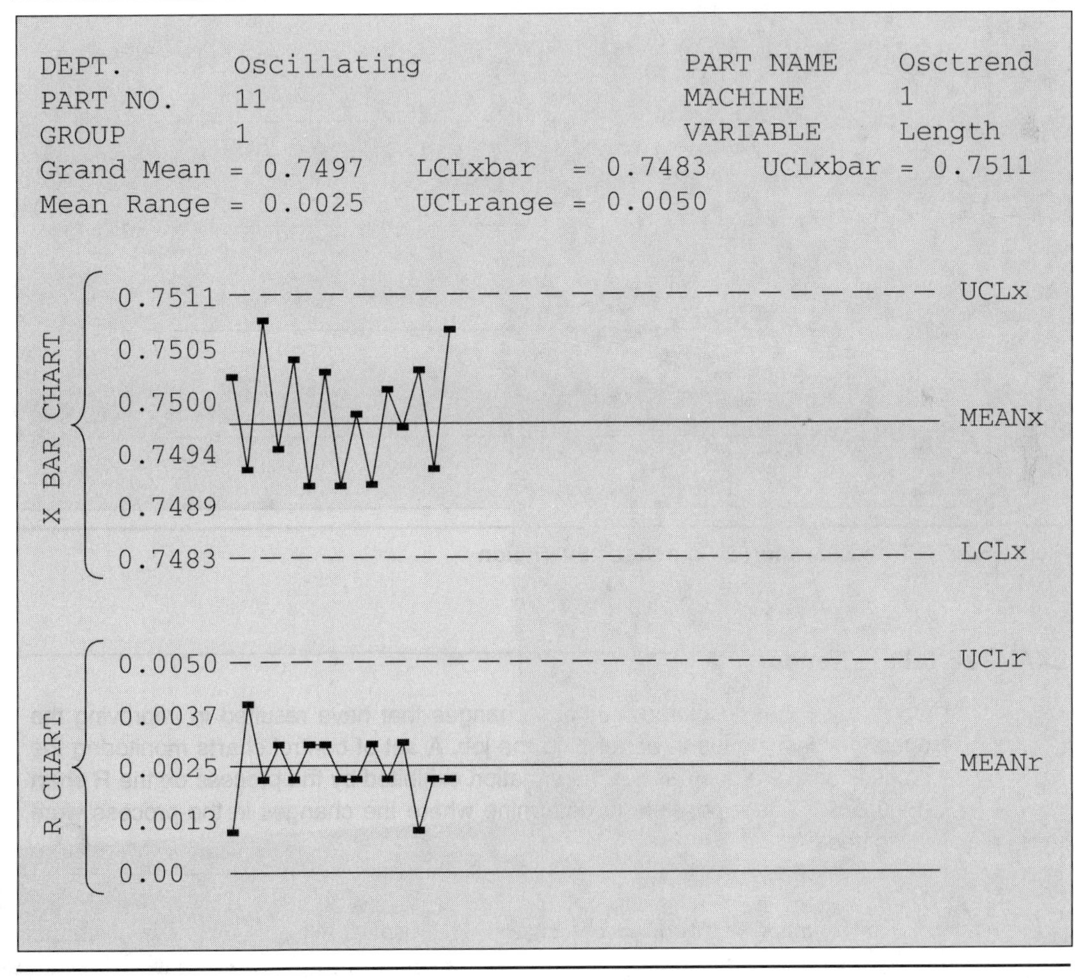

Figure 5.16 An Oscillating Trend

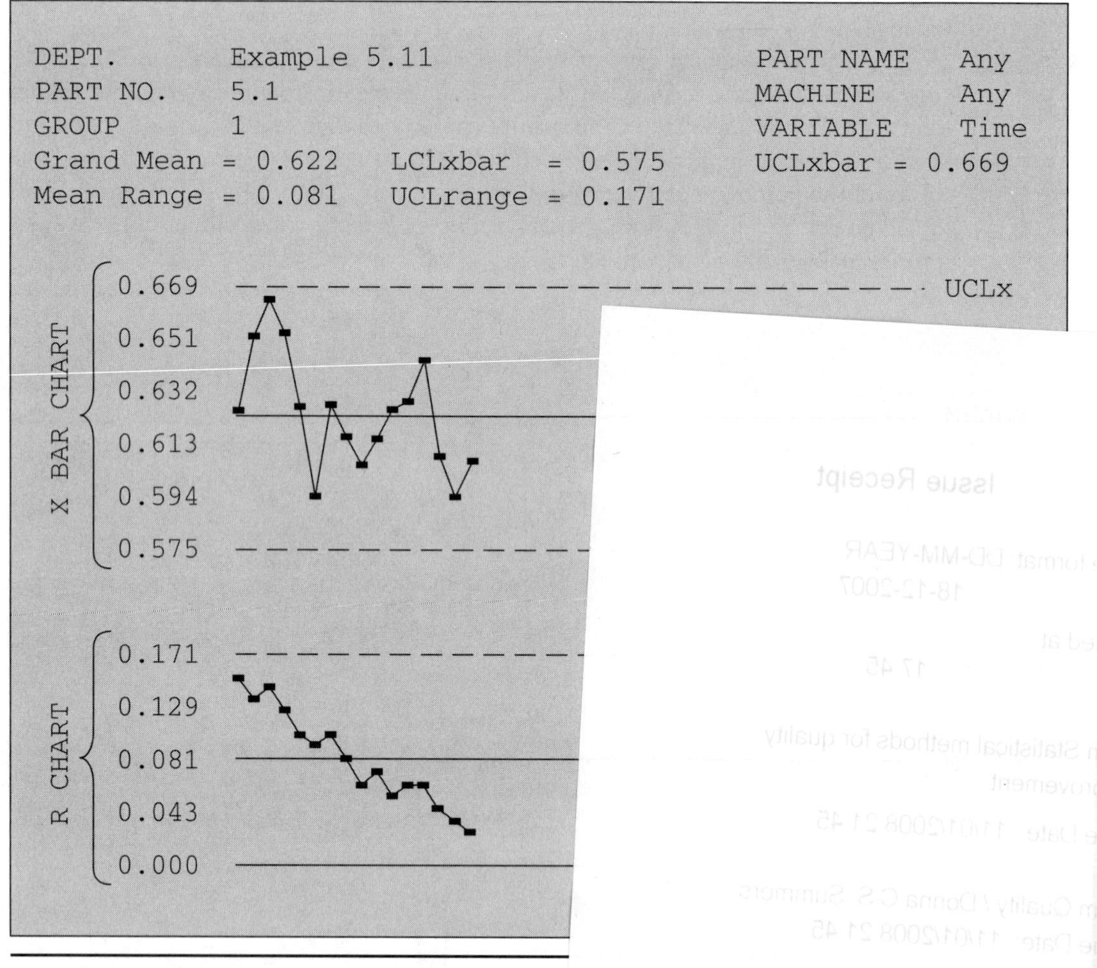

Figure 5.17 Trend Showing a Decrease in Variat

EXAMPLE 5.11 Trends

> Company B has instituted methods (
> operators' techniques in performing tl
> process shows a decrease in the vari
> (Figure 5.17). Is it possible to detern
> instigated?

Change, jump, or shift in level Figure 5.18 displays what is meant by a change, jump, or shift in level. Note that the process begins at one level (Figure 5.18*a*) and moves quickly to another level (Figure 5.18*b*) as the process continues to operate.

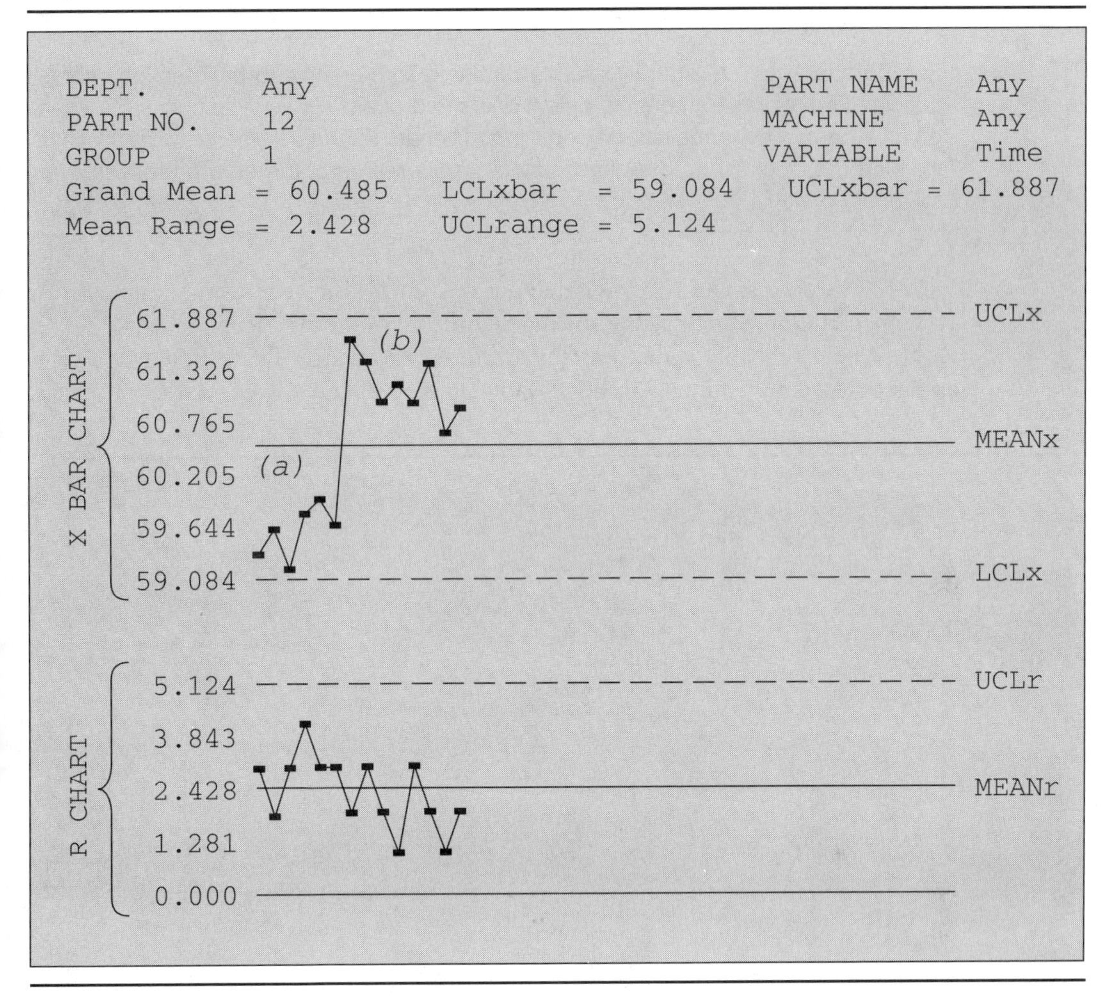

Figure 5.18 Change in Level

This change, jump, or shift in level is fairly abrupt, unlike a trend described above. A change, jump, or shift can occur either on the $\overline{X}$ or R chart or on both charts. Causes for sudden shifts in level tend to reflect some new and fairly significant difference in the process. When investigating a sudden shift or jump in level, look for significant changes that can be pinpointed to a specific moment in time. For the $\overline{X}$ chart, causes include new machines, dies, or tooling; minor failure of a machine part; new or inexperienced workers; new batches of raw material; new production methods; or changes to the process settings. For the R chart, potential sources of jumps or shifts in level causing a change in the process variability or spread include a new or inexperienced operator, a sudden increase in the play associated with gears or work-holding devices, or greater variation in incoming material.

EXAMPLE 5.12 Jump in Level

For the past week, Margaret has been training Tom on how to process loan applications. The average amount of time it takes to process six applications is recorded on an $\overline{X}$ chart. The range is recorded on the R chart. Figure 5.19 records the process for the past week. Is it possible to detect where Margaret turned the process over to Tom?

Runs A process can be considered out of control when there are unnatural runs present in the process. Imagine tossing a coin. If two heads occur in a row, the onlooker would probably agree that this occurred by chance. Even though the probability of the coin landing with heads showing is 50-50, no one expects coin tosses to

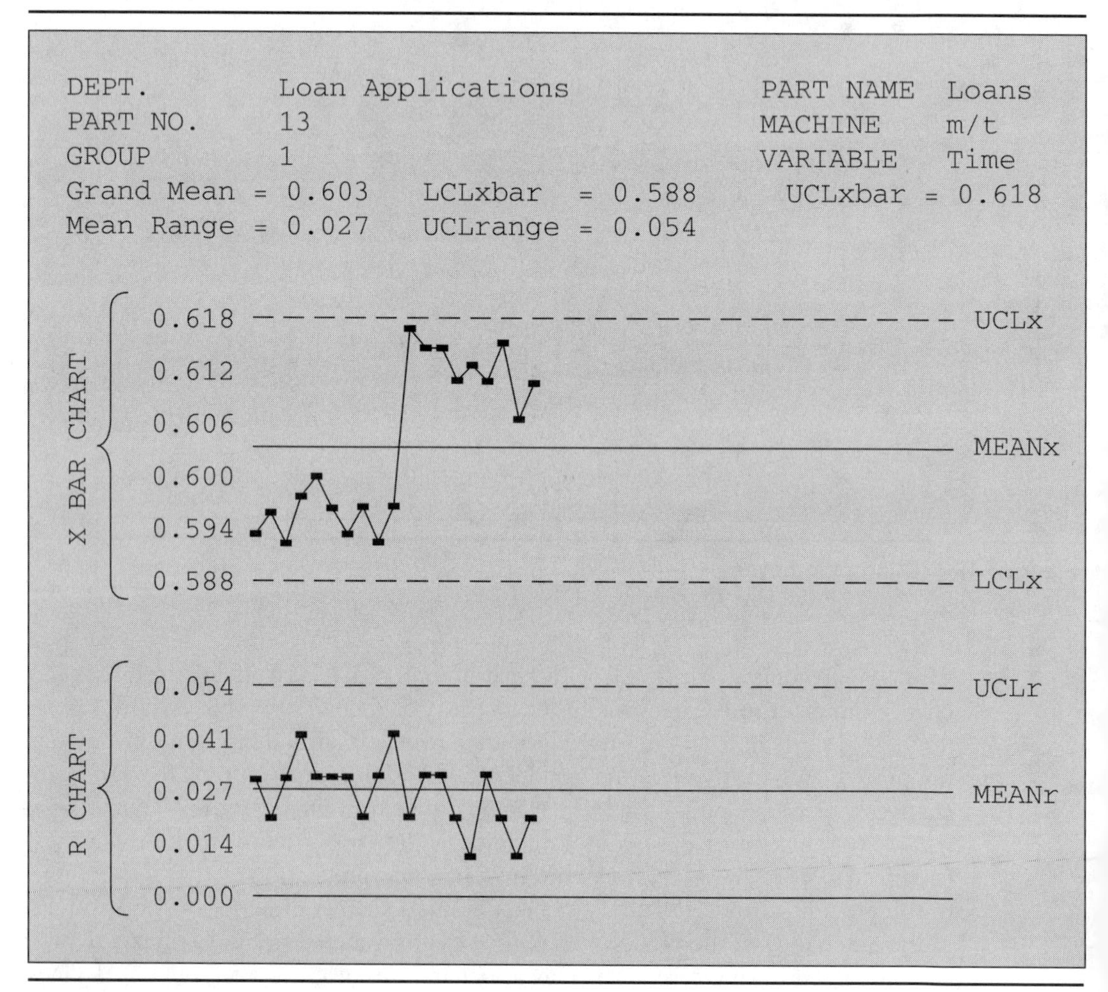

Figure 5.19 Change in Level

alternate between heads and tails. If, however, an onlooker saw someone toss six heads in a row, that onlooker would probably be suspicious that this set of events is not due to chance. The same principle applies to control charts. While the points on a control chart do not necessarily alternate above and below the centerline in a chart that is under control, the points are normally balanced above and below the centerline. A cluster of seven points in a row above or below the centerline would be improbable and would likely have an assignable cause. The same could be said for situations where 10 out of 11 points or 12 out of 14 points are located on one side or the other of the centerline (Figure 5.20*a*, *b*, *c*).

Runs on the $\overline{X}$ chart can be caused by temperature changes; tool or die wear; gradual deterioration of the process; or deterioration of the chemicals, oils, or cooling fluids used in the process. Runs on the R chart signal a change in the process variation. Causes for these R-chart runs could be a change in operator skill, either an improvement or a decrement, or a gradual improvement in the homogeneity of the process because of changes in the incoming material or changes to the process itself.

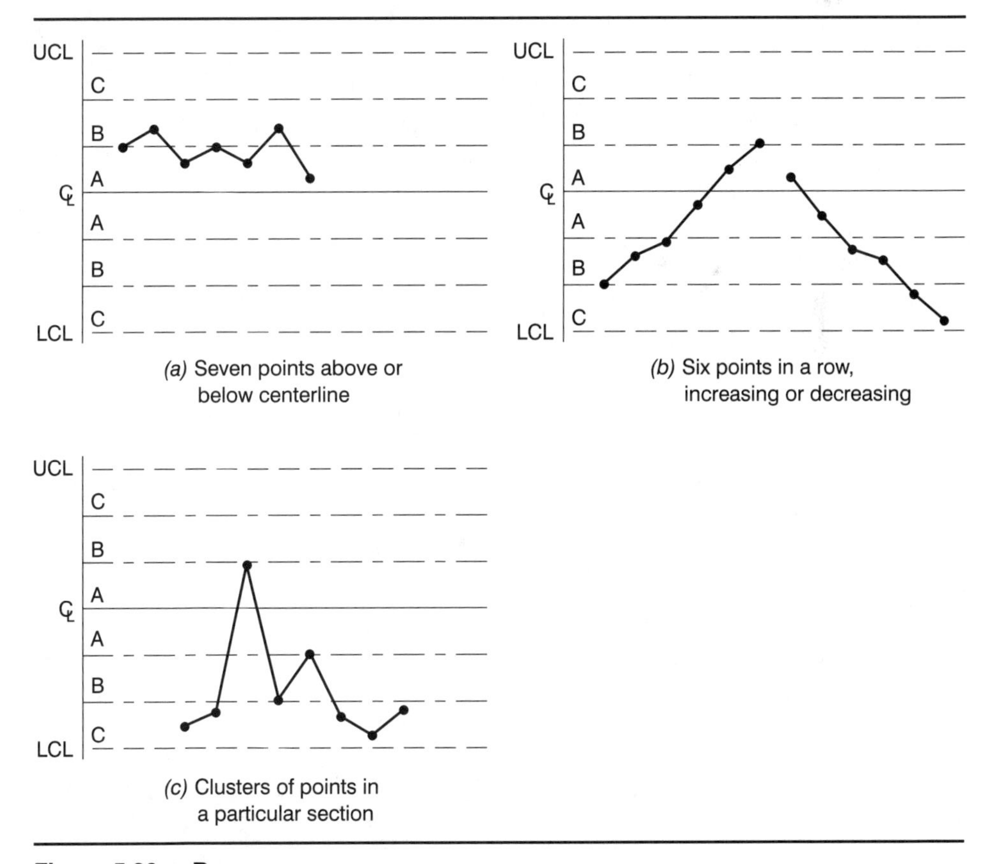

Figure 5.20 Runs

EXAMPLE 5.13 Runs

At the local burger-and-fries joint, the length of time that french fries are fried is based on the temperature of the oil. During the day, the settings are adjusted to reflect the oil temperature. The oil temperatures are sampled and recorded throughout the day. The correct oil temperature, combined with frying time, produces the best-tasting fries. Figure 5.21 shows a run of points on the R chart. The process will need to be studied to determine the cause of this run.

Recurring cycles Recurring cycles are caused by systematic changes related to the process. When investigating what appears to be cycles (Figure 5.22) on the chart, it is important to look for causes that will change, vary, or cycle over time. For the $\overline{X}$

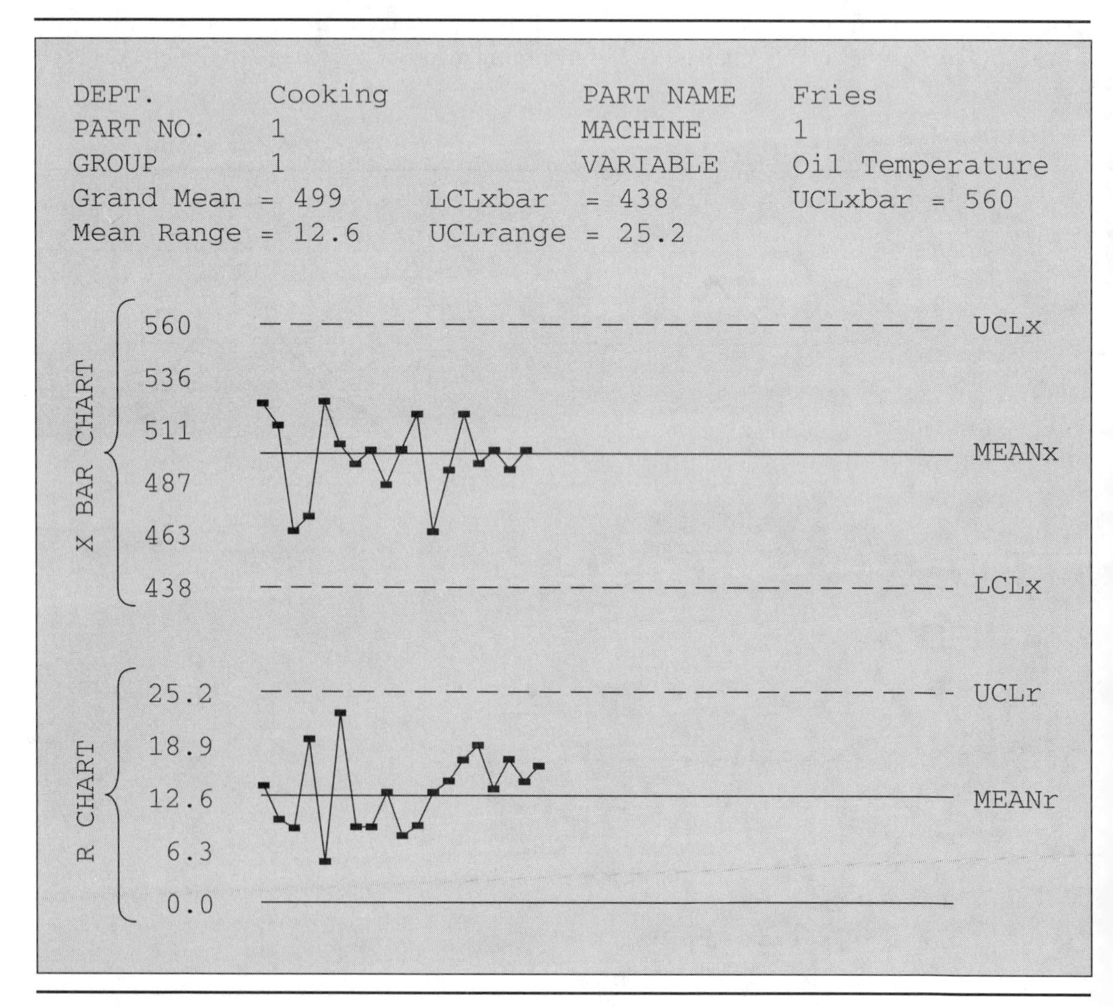

Figure 5.21 Run of Points

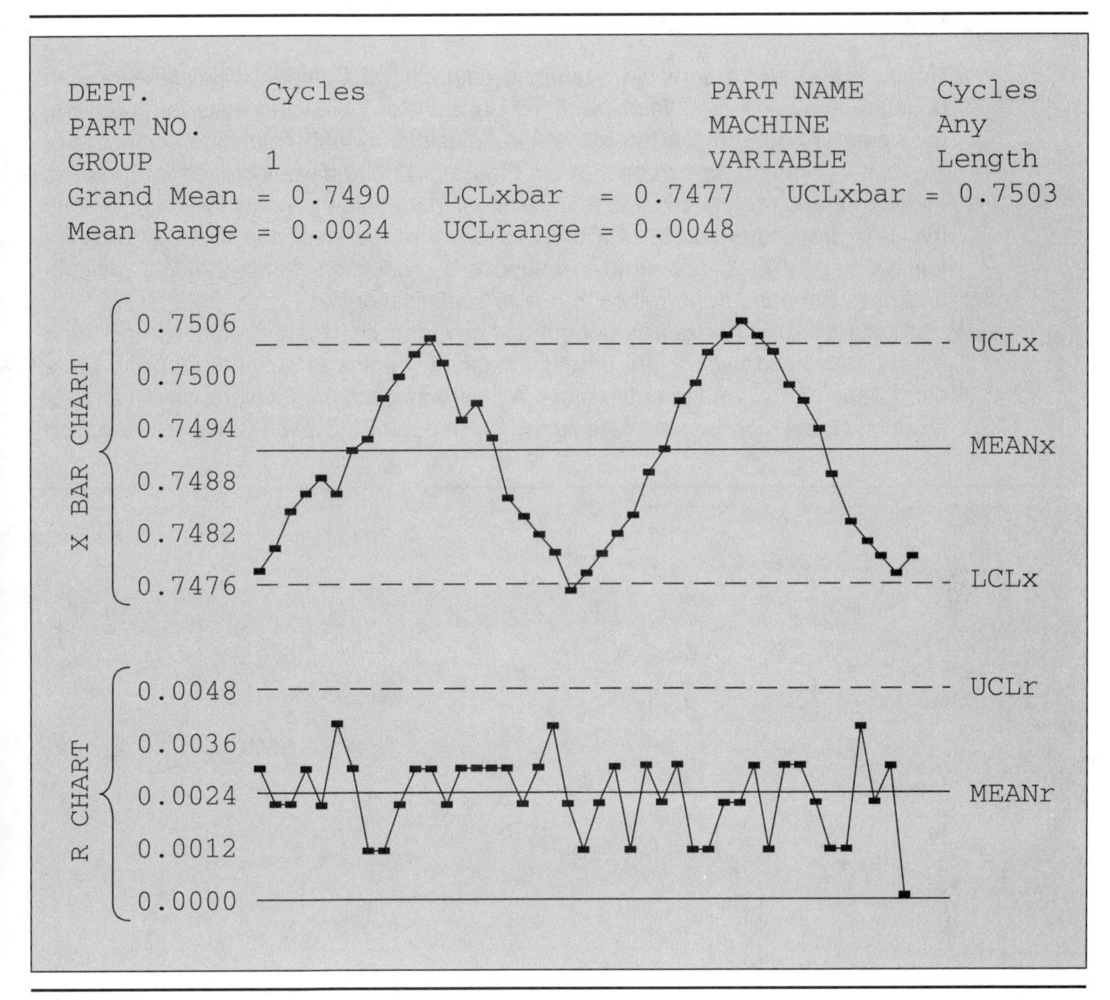

Figure 5.22 Cycles

chart, potential causes are tool or machine wear conditions, an accumulation of chips or other waste material around the tooling, maintenance schedules, periodic rotation of operators, worker fatigue, periodic replacement of cooling fluid or cutting oil, or changes in the process environment such as temperature or humidity. Cycles on an R chart are not as common; an R chart displays the variation or spread of the process, which usually does not cycle. Potential causes are related to lubrication cycles and operator fatigue.

Cycles can be difficult to locate because the entire cycle may not be present on a single chart. The frequency of inspection could potentially cause a cycle to be overlooked. For example if the cycle occurs every 15 minutes and samples are taken only every 30 minutes, then it is possible for the cycle to be overlooked.

EXAMPLE 5.14 Recurring Cycles

During the winter, a particular stamping operation at Company A experiences an unusual amount of machine downtime. Because of increasing material thickness, the operator has to make too many fine adjustments to the machine settings during daily operation. As can be seen in Figure 5.23, the parts increase in thickness as the morning progresses and then level out in the midafternoon. Investigation into the daily production and talks with the supplier of the steel do not reveal a cause for this trend. The charts from the supplier's production reflect that the material thickness throughout the roll conforms to specifications.

After reviewing her quality-control text covering chart interpretations, including runs, trends, and cycles, the process engineer begins to suspect that one day's chart does not reveal the entire story. A closer inspection of several days' worth of production laid side by side reveals the existence of a cycle. Further investigation

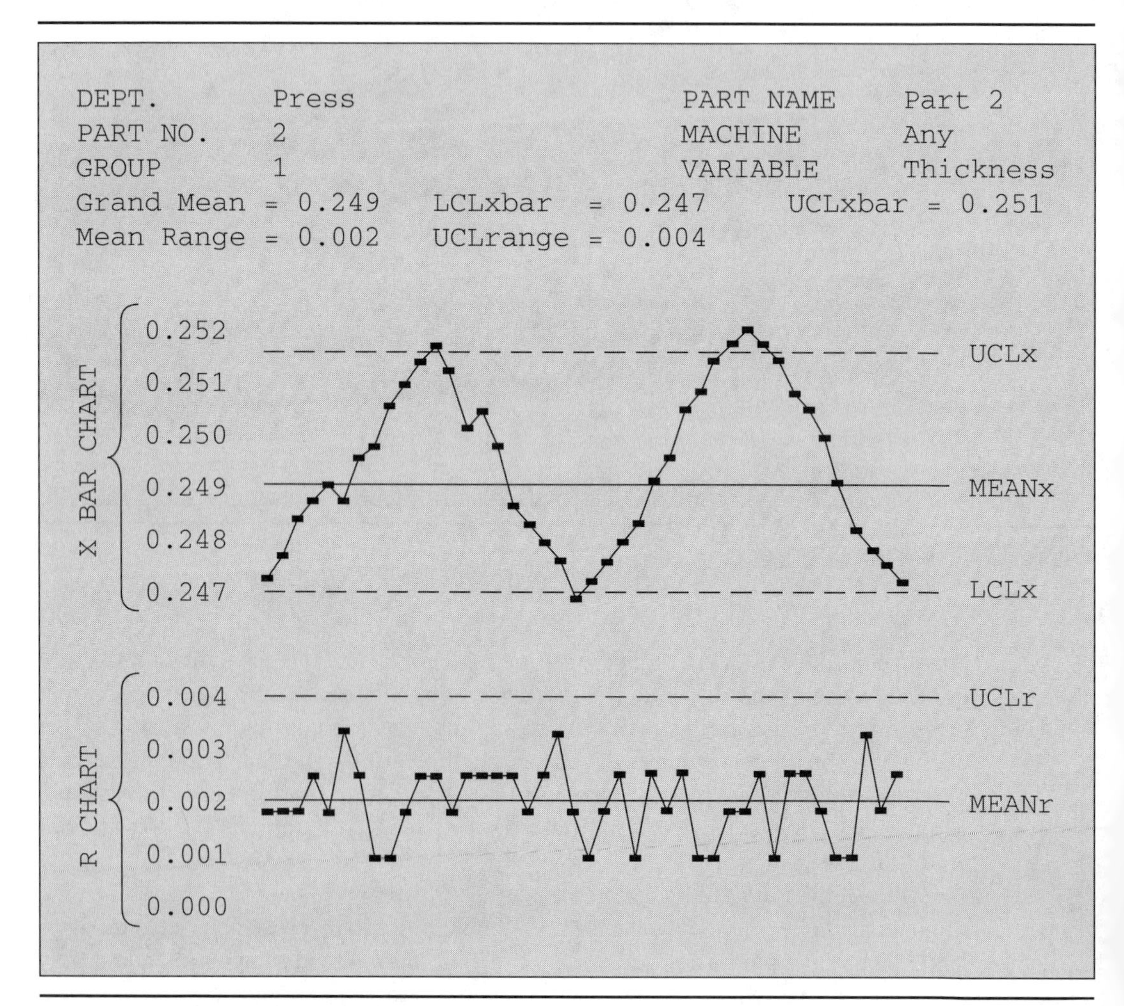

Figure 5.23 Cycle in Part Thickness

shows that this cycle is related to the temperature of the material during the day. Under the current method, the coils of steel are left on the loading dock until needed. Because of the winter weather, the material is quite cold when brought into the building for use each morning. As the coil warms up in the heated plant during the day, it expands, making it thicker. This increase in thickness is evident on the control chart. Since only one coil is used each day, the cycle was not apparent until the charts from day-to-day production were compared.

Two populations When a control chart is under control, the greatest number of the sample averages will fall closer to the centerline than the control limits. When a large number of the sample averages are plotted near or outside of the control limits, two populations of samples might exist (Figure 5.24). "Two populations" refers to the existence of two (or more) sources of production.

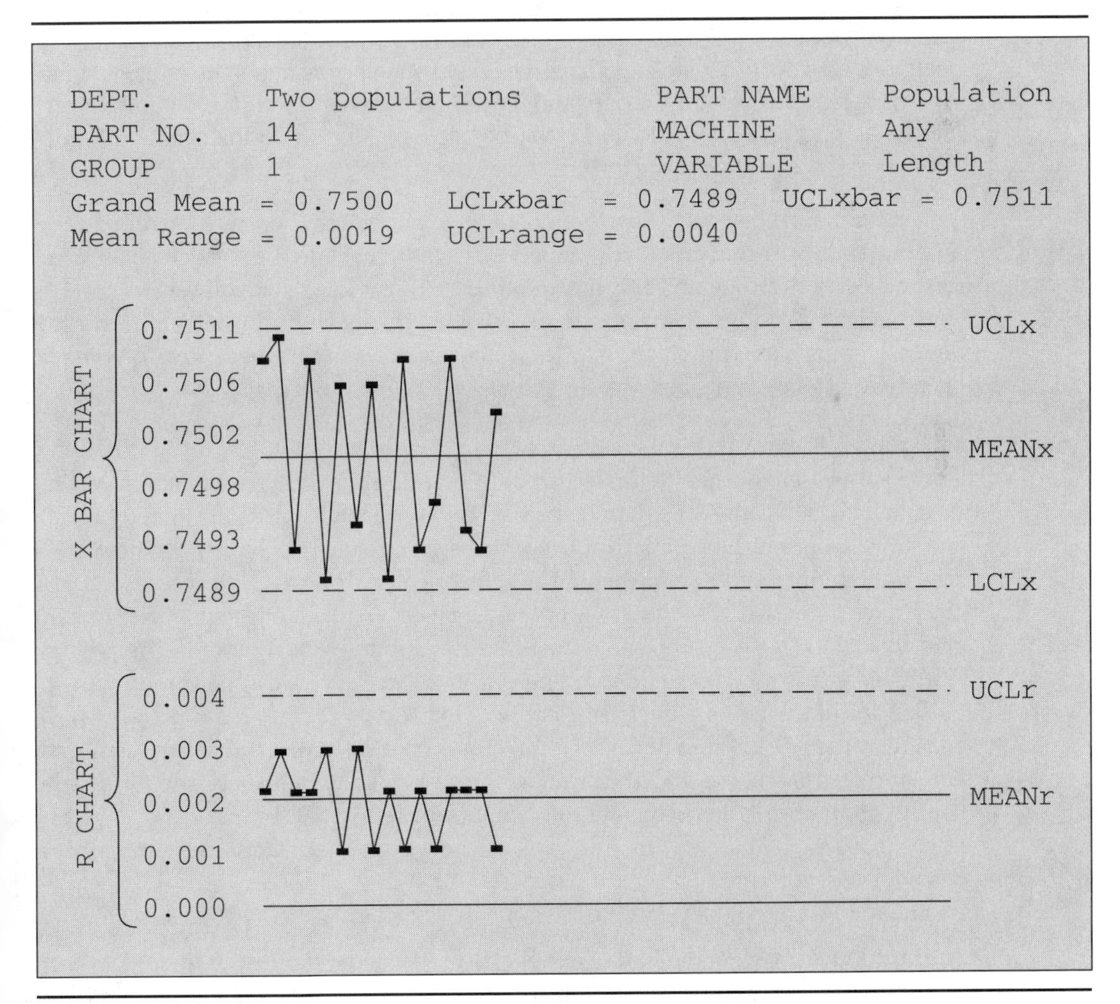

Figure 5.24 Two Populations

On an $\overline{X}$ chart, the different sources of production might be due to the output of two or more machines being combined before sampling takes place. It might also occur because the work of two different operators is combined or two different sources of raw materials are brought together in the process. A two-population situation means that the items being sampled are not homogeneous. Maintaining the homogeneity of the items being sampled is critical for creating and using control charts.

This type of pattern on an R chart signals that different workers are using the same chart or that the variation is due to the fact that raw materials are coming from different suppliers.

EXAMPLE 5.15 Two Populations

Don's Drive-In is keeping track of the amount of time it takes to serve a customer an order. Throughout each day the time it takes to take, make, and deliver a customer order is sampled, averaged, and recorded. The chart for this past week's worth of samples is given in Figure 5.25. To Don's son, a student taking a quality-control class, the chart looks unusual. Upon questioning his father, he finds out that the samples are taken, averaged, and placed on the chart regardless of the server filling the order. This has created a two-population situation on the chart.

Mistakes The "mistakes" category combines all the other out-of-control patterns that can exist on a control chart (Figure 5.26). Mistakes can arise from calculation errors, calibration errors, and testing-equipment errors. They generally show up on a control chart as a lone point or a small group of points above or below the control limits. Finding these problems takes effort but is necessary if the process is going to operate to the best of its ability.

9. *Revise the Charts*

There are two circumstances under which the control chart is revised and new limits calculated. Existing calculations can be revised if a chart exhibits good control and any changes made to improve the process are permanent. When the new operating conditions become routine and no out-of-control signals have been seen, the chart may be revised. The revisions provide a better estimate of the population standard deviation, representing the spread of all of the individual parts in the process. With this value, a better understanding of the entire process can be gained.

Control limits are also revised if patterns exist, provided that the patterns have been identified and eliminated. Once the causes have been determined, investigated, and corrected in such a way that they will not affect the process in the future, the control chart can be revised. The new limits will reflect the changes and improvements made to the process. In both cases the new limits are used to judge the process behavior in the future.

The following four steps are taken to revise the charts.

A. Interpret the Original Charts The R chart reflects the stability of the process and should be analyzed first. A lack of control on the R chart shows that the process is not

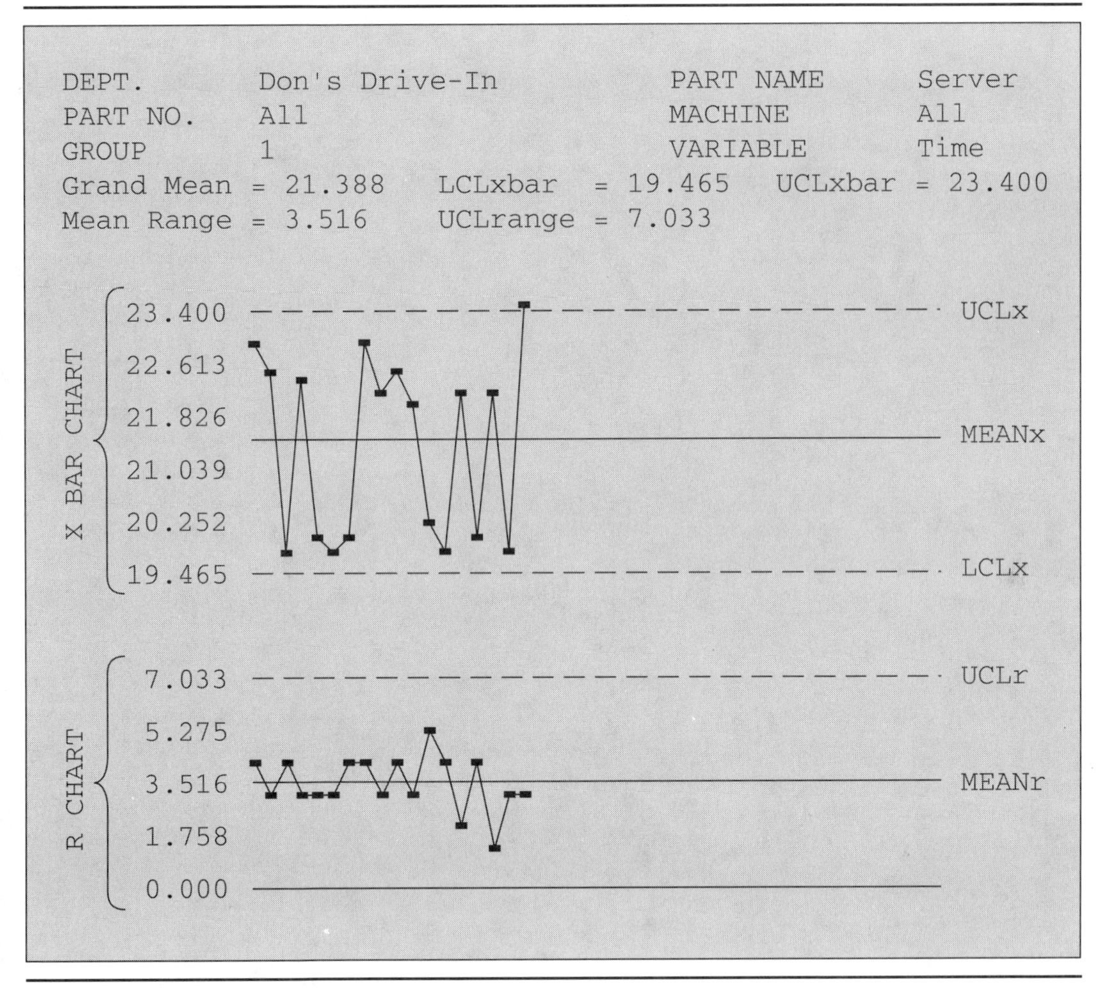

Figure 5.25 Don's Drive-In: Two Populations

producing parts that are very similar to each other. The process is not precise. Look at the chart and determine if cycles, trends, runs, two populations, mistakes, or other examples of lack of control exist. If both the $\overline{X}$ and R charts are exhibiting good control, proceed to step D. If the charts display out-of-control conditions, then continue to step B.

EXAMPLE 5.16 Examining the Control Charts for Example 5.3

Begin an examination of the $\overline{X}$ and R charts combination created for the situation in Example 5.3 by investigating the R chart, which displays the variation present in the process. Excessive variation indicates that the process is not producing consistent product. The R chart from Example 5.3 (Figure 5.7) exhibits good control.

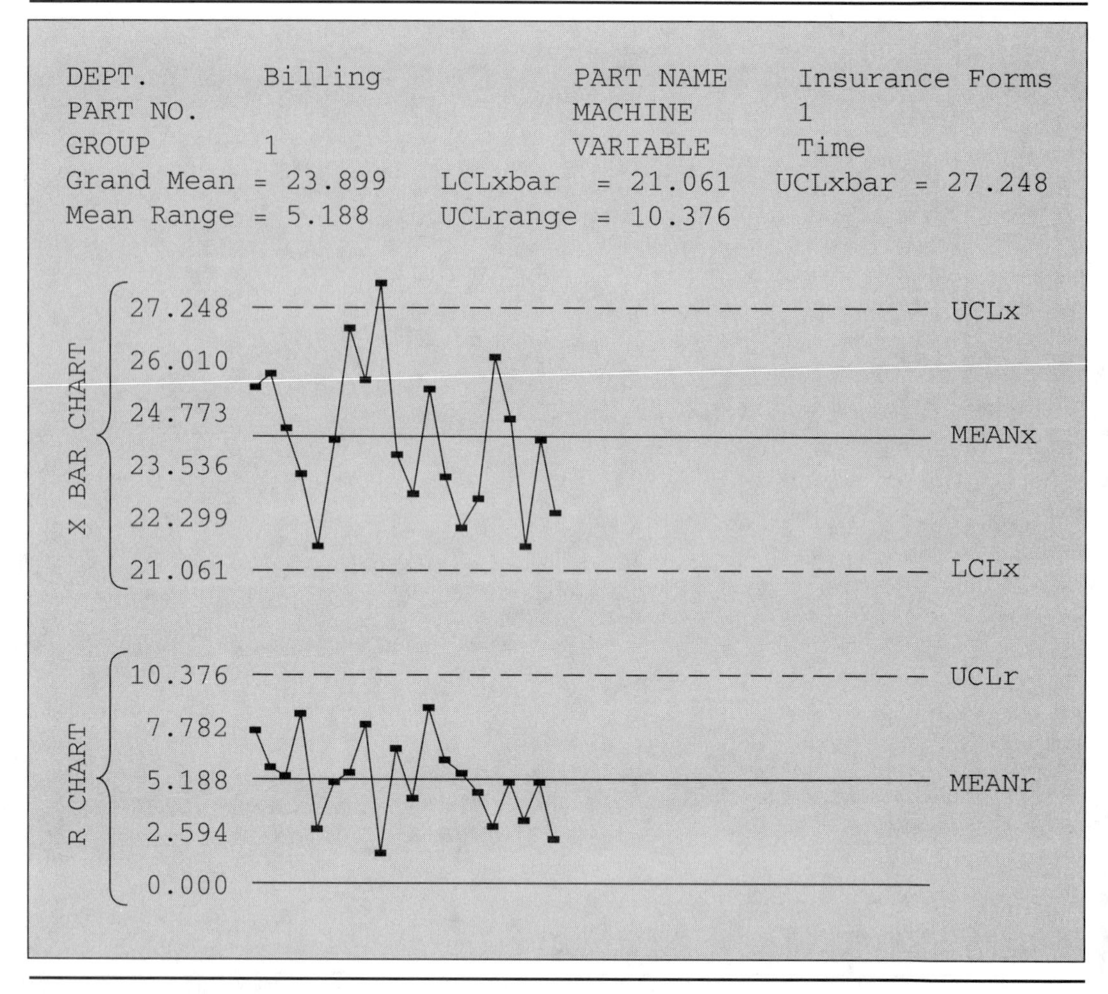

Figure 5.26 Mistakes

The points are evenly spaced on both sides of the centerline and there are no points beyond the control limits. There are no unusual patterns or trends in the data. Given these observations, it can be said that the process is producing parts of similar dimensions.

Next the $\overline{X}$ chart is examined. An inspection of the $\overline{X}$ chart reveals an unusual pattern occurring at points 12, 13, and 14. These measurements are all below the lower control limit. When compared with other samples throughout the day's production, the parts produced during the time when samples 12, 13, and 14 were taken were much shorter than parts produced during other times in the production run. A glance at the R chart reveals that the range of the individual measurements taken in the samples is small, meaning that the parts produced during samples 12, 13, and 14 are all similar in size. An investigation into the cause of the production of undersized parts needs to take place.

B. Isolate the Cause If either the $\overline{X}$ or R chart is not exhibiting good statistical control, find the cause of the problem. Problems shown on the control chart may be removed only if the causes of those problems have been isolated and steps have been taken to eliminate them.

C. Take Corrective Action Take the necessary steps to correct the causes associated with the problems exhibited on the chart. Once the causes of variation have been removed from the process, these points may be removed from the control chart and the calculations revised.

EXAMPLE 5.17 A Further Examination of the Control Charts

An investigation into the differences in shaft lengths has been conducted. The $\overline{X}$ and R chart combination created for the situation in Example 5.3 aids the investigators by allowing them to isolate when the differences were first noticed. Since the R chart (Figure 5.7) exhibits good control, the investigators are able to concentrate their attention on possible causes for a consistent change in shaft length for those three subgroups. Their investigation reveals that the machine settings have been bumped during the loading of the machine. For the time being, operators are being asked to check the control panel settings after loading the machine. To take care of the problem for the long term, manufacturing engineers are looking into possible design changes to protect the controls against accidental manipulation.

D. Revise the Chart To determine the new limits against which the process will be judged in the future, it is necessary to remove any points from the calculations that have been corrected. In the case of charts that are exhibiting good statistical control, the points removed will equal zero and the calculations will continue from there. The criteria for removing points are based on finding the cause behind the out-of-control condition. If no cause can be found and corrected, then the points *cannot* be removed from the chart. Two methods can be used to discard the data. When it is necessary to remove a subgroup from the calculations, it can be removed from only the out-of-control chart or it can be removed from both charts. In one case, when an $\overline{X}$ value must be removed from the $\overline{X}$ control chart, its corresponding R value is *not* removed from the R chart and vice versa. In this text, the points are removed from *both* charts. This approach has been chosen because the values on both charts are interrelated. The R-chart values describe the spread of the data on the $\overline{X}$ chart. Removing data from one chart or the other negates this relationship. Groups of points, runs, trends, and other patterns can be removed in the same manner as removing individual points.

The formulas for revising both the $\overline{X}$ and R charts are as follows:

$$\overline{\overline{X}}_{new} = \frac{\Sigma\overline{X} - \overline{X}_d}{m - m_d}$$

$$\overline{R}_{new} = \frac{\Sigma R - R_d}{m - m_d}$$

where

$$\overline{X}_d = \text{discarded subgroup averages}$$
$$m_d = \text{number of discarded subgroups}$$
$$R_d = \text{discarded subgroup ranges}$$

The newly calculated values of $\overline{\overline{X}}$ and $\overline{R}$ are used to establish updated values for the centerline and control limits on the chart. These new limits reflect that improvements have been made to the process and future production should be capable of meeting these new limits. The formulas for the revised limits are

$$\overline{\overline{X}}_{new} = \overline{X}_0 \qquad \overline{R}_{new} = R_0$$

$$\sigma_0 = R_0/d_2$$

$$UCL_{\overline{X}} = \overline{X}_0 + A\sigma_0$$
$$LCL_{\overline{X}} = \overline{X}_0 - A\sigma_0$$

$$UCL_R = D_2\sigma_0$$
$$LCL_R = D_1\sigma_0$$

where A, D_1, and D_2 are factors from the table in Appendix 2.

EXAMPLE 5.18 Revising the Control Limits

Since a cause for the undersized parts has been determined for the machine of Example 5.3, the values for 12, 13, and 14 can be removed from the calculations for the $\overline{X}$ and R chart. The new or revised limits will be used to monitor future production. The new limits will extend from the old limits, as shown in Figure 5.27.

Revising the calculations is performed as follows:

$$\overline{\overline{X}}_{new} = \overline{X}_0 = \frac{\sum_{i=1}^{m} \overline{X} - \overline{X}_d}{m - m_d} =$$
$$= \frac{251.77 - 11.94 - 11.95 - 11.95}{21 - 3}$$
$$= 12.00$$

$$\overline{R}_{new} = R_0 = \frac{\sum_{i=1}^{m} R - R_d}{m - m_d}$$
$$= \frac{1.06 - 0.04 - 0.05 - 0.06}{21 - 3}$$
$$= 0.05$$

Calculating the σ_0 for the process, when n = 5,

$$\sigma_0 = \frac{R_0}{d_2} = \frac{0.05}{2.326} = 0.02$$

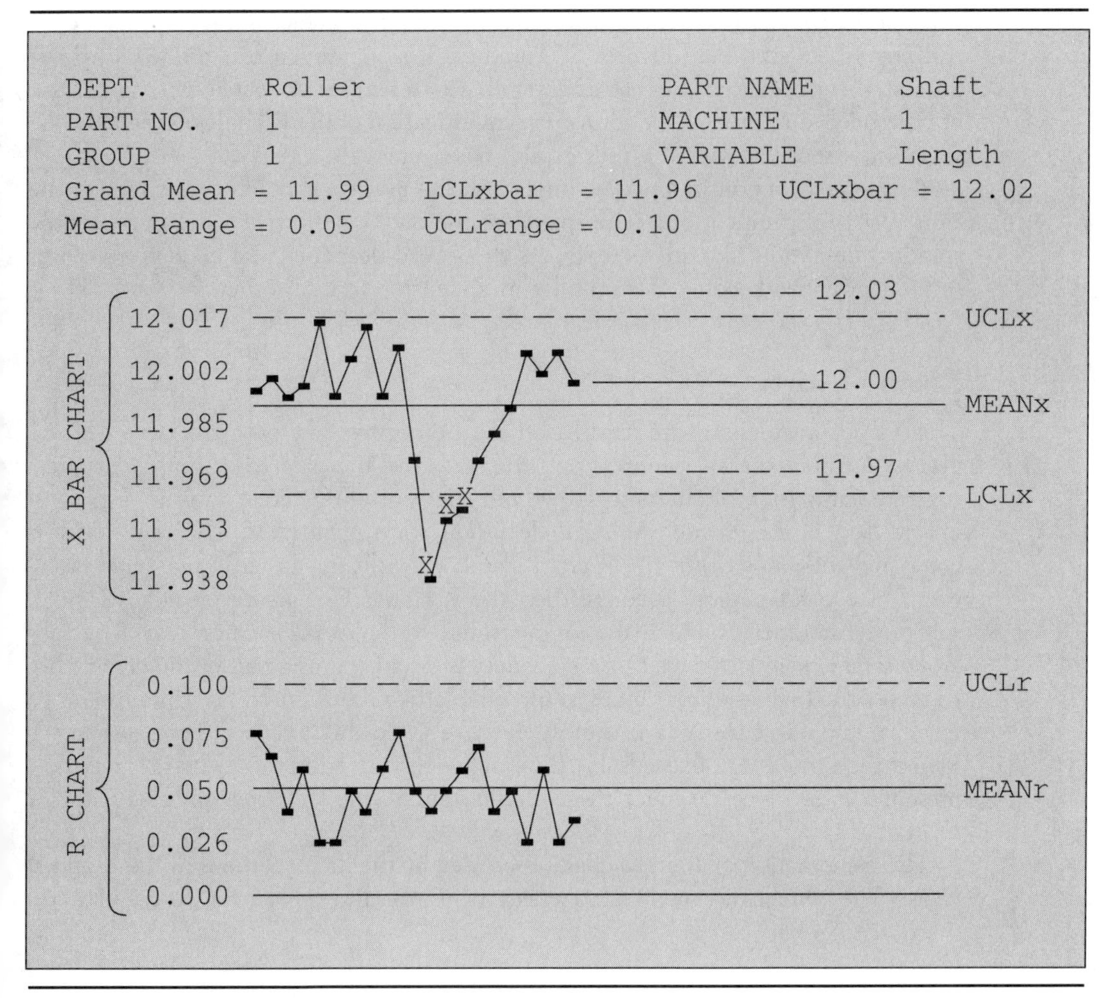

Figure 5.27 Extension of Limits on a Chart

$$UCL_{\bar{X}} = 12.00 + 1.342(0.02) = 12.03$$
$$LCL_{\bar{X}} = 12.00 - 1.342(0.02) = 11.97$$

$$UCL_R = 4.918(0.02) = 0.10$$
$$LCL_R = 0(0.02) = 0$$

10. Achieve the Purpose

Users of control charts are endeavoring to decrease the variation inherent in a process over time. Once established, control charts enable the user to understand where the process is currently centered and what the distribution of that process is. To know this information and not utilize it to improve the process defeats the purpose of control charts. As shown in Figures 5.12 and 5.13, as improvements are made to the process, the process becomes more centered, with a significant decrease in the amount of time needed to process the forms. An investigation of the R chart reveals that the variation associated with the process also has decreased.

As the process continues to be improved, the average should come closer to the center of the specifications. The spread of the data, as shown by the range or the standard deviation, should decrease. As the spread decreases, the parts produced or services provided become more similar in dimension to each other.

$\overline{X}$ and s Charts

The $\overline{X}$ and range charts are used together in order to show both the center of the process measurements (accuracy) and the spread of the data (precision). An alternative combination of charts to show both the central tendency and the dispersion of the data is the $\overline{X}$ and ***standard deviation,*** or ***s, chart.*** When an R chart is compared with an s chart, the R chart stands out as being easier to compute. However, the s chart is more accurate than the R chart. This greater accuracy is a result of the manner in which the sample standard deviation is calculated. The subgroup sample standard deviation, or s value, is calculated using all of the data rather than just the high and low values in the sample like the R chart. So while the range chart is simple to construct, it is most effective when the sample size is less than 10. When the sample size exceeds 10, the range does not truly represent the variation present in the process. Under these circumstances, the s chart is used with the $\overline{X}$ chart.

The combination of $\overline{X}$ and s charts is created by the same methods as the $\overline{X}$ and R charts. The formulas are modified to reflect the use of s instead of R for the calculations.

For the $\overline{X}$ chart,

$$\overline{\overline{X}} = \frac{\sum_{i=1}^{m} \overline{X}_i}{m}$$

$$UCL_{\overline{X}} = \overline{\overline{X}} + A_3\bar{s}$$

$$LCL_{\overline{X}} = \overline{\overline{X}} - A_3\bar{s}$$

For the s chart,

$$\bar{s} = \frac{\sum_{i=1}^{m} s_i}{m}$$

$$UCL_s = B_4\bar{s}$$

$$LCL_s = B_3\bar{s}$$

where

s_i = standard deviation of the subgroup values
$\bar{s}_i$ = average of the subgroup sample standard deviations
A_3, B_3, B_4 = factors used for calculating 3σ control limits for $\overline{X}$ and s charts using the average sample standard deviation and Appendix 2

As in the $\overline{X}$ and R chart, the revised control limits can be calculated using the following formulas:

$$\overline{X}_0 = \overline{\overline{X}}_{new} = \frac{\Sigma\overline{X} - \overline{X}_d}{m - m_d}$$

$$s_0 = \bar{s}_{new} = \frac{\Sigma s - s_d}{m - m_d}$$

and

$$\sigma = s_0/c_4$$

$$UCL_{\overline{X}} = \overline{X}_0 + A\sigma_0$$
$$LCL_{\overline{X}} = \overline{X}_0 - A\sigma_0$$

$$UCL_s = B_6\sigma_0$$
$$LCL_s = B_5\sigma_0$$

where

s_d = sample standard deviation of the discarded subgroup
c_4 = factor found in Appendix 2 for computing σ_0 from $\bar{s}$
A, B_5, B_6 = factors found in Appendix 2 for computing the 3σ process control limits for $\overline{X}$ and s charts

EXAMPLE 5.19 $\overline{X}$ and s Charts

This example uses the same data as Example 5.5 except the range has been replaced by s, the standard deviation of the sample. A sample of size n = 5 is taken at intervals from the process making shafts (Figure 5.28). A total of 21 subgroups of measurements is taken. Each time a sample is taken, the individual values are recorded [Figure 5.28, (1)], summed, and then divided by the number of samples taken to get the average [Figure 5.28, (2)]. This average is then plotted on the control chart [Figure 5.29, (1)]. Note that in this example, the values for $\overline{X}$, s have been calculated to three decimal places for clarity.

Using the 21 samples provided in Figure 5.28, we can calculate $\overline{\overline{X}}$ by summing all the averages from the individual samples taken and then dividing by the number of subgroups:

$$\overline{\overline{X}} = \frac{251.76}{21} = 11.99$$

This value is plotted as the centerline of the $\overline{X}$ chart [Figure 5.29, (2)].

$\overline{s}$, the grand standard deviation average, is calculated by summing the values of the sample standard deviations (Figure 5.28) and dividing by the number of subgroups m:

$$\overline{s} = \frac{0.414}{21} = 0.02$$

DEPT.	Roller	PART NAME	Shaft
PART NO.	1	MACHINE	1
GROUP	1	VARIABLE	length

Sample	1	2	3	4	5
Time	07:30	07:40	07:50	08:00	08:10
Date	07/02/95	07/02/95	07/02/95	07/02/95	07/02/95
1	11.95	12.03	12.01	11.97	12.00
2	12.00	12.02	12.00	11.98	12.01
3	12.03 (1)	11.96	11.97	12.00	12.02
4	11.98	12.00	11.98	12.03	12.03
5	12.01	11.98	12.00	11.99	12.02
$\overline{X}$	11.99 (2)	12.00	11.99	11.99	12.02
s	0.031 (3)	0.029	0.016	0.023	0.011

Sample	6	7	8	9	10
Time	08:20	08:30	08:40	08:50	09:00
Date	07/02/95	07/02/95	07/02/95	07/02/95	07/02/95
1	11.98	12.00	12.00	12.00	12.02
2	11.98	12.01	12.01	12.02	12.00
3	12.00	12.03	12.04	11.96	11.97
4	12.01	12.00	12.00	12.00	12.05
5	11.99	11.98	12.02	11.98	12.00
$\overline{X}$	11.99	12.00	12.01	11.99	12.01
s	0.013	0.018	0.017	0.023	0.030

Sample	11	12	13	14	15
Time	09:10	09:20	09:30	09:40	09:50
Date	07/02/95	07/02/95	07/02/95	07/02/95	07/02/95
1	11.98	11.92	11.93	11.99	12.00
2	11.97	11.95	11.95	11.93	11.98
3	11.96	11.92	11.98	11.94	11.99
4	11.95	11.94	11.94	11.95	11.95
5	12.00	11.96	11.96	11.96	11.93
$\overline{X}$	11.97	11.94	11.95	11.95	11.97
s	0.019	0.018	0.019	0.023	0.029

Figure 5.28 Shafts, Averages, and Standard Deviations

Sample	16	17	18	19	20
Time	10:00	10:10	10:20	10:30	10:40
Date	07/02/95	07/02/95	07/02/95	07/02/95	07/02/95
1	12.00	12.02	12.00	11.97	11.99
2	11.98	11.98	12.01	12.03	12.01
3	11.99	11.97	12.02	12.00	12.02
4	11.96	11.98	12.01	12.01	12.00
5	11.97	11.99	11.99	11.99	12.01
$\overline{X}$	11.98	11.999	12.01	12.00	12.01
s	0.016	0.019	0.011	0.022	0.011

Sample	21
Time	10:50
Date	07/02/95
1	12.00
2	11.98
3	11.99
4	11.99
5	12.02
$\overline{X}$	12.00
s	0.015

1

2: $$\frac{12.00 + 11.98 + 11.99 + 11.96 + 11.97}{5} = 11.98$$

3: $$\sqrt{\frac{(12 - 11.98)^2 + (11.98 - 11.98)^2 + (11.99 - 11.98)^2 + (11.96 - 11.98)^2 + (11.97 - 11.98)^2}{5 - 1}} = 0.016$$

Figure 5.28 Continued

The values for the upper and lower control limits of the $\overline{X}$ chart are calculated as follows:

$$\begin{aligned} UCL_{\overline{X}} &= \overline{\overline{X}} + A_3\overline{s} \\ &= 11.99 + 1.427(0.02) = 12.02 \end{aligned}$$

$$\begin{aligned} LCL_{\overline{X}} &= \overline{\overline{X}} - A_3\overline{s} \\ &= 11.99 - 1.427(0.02) = 11.96 \end{aligned}$$

The A_3 factor for a sample size of five is selected from the table in Appendix 2. Once calculated, the upper and lower control limits (UCL and LCL, respectively) are placed on the chart [Figure 5.29, (3)].

Individual standard deviations are calculated for each of the subgroups by utilizing the formula for calculating standard deviations, as presented in Chapter 4. Once calculated, these values are plotted on the s chart [Figure 5.29, (6)].

To determine the centerline of the s chart, individual standard deviation values are summed and divided by the total number of subgroups to give $\overline{s}$ [Figure 5.29, (4)]:

$$\overline{s} = \frac{0.414}{21} = 0.02$$

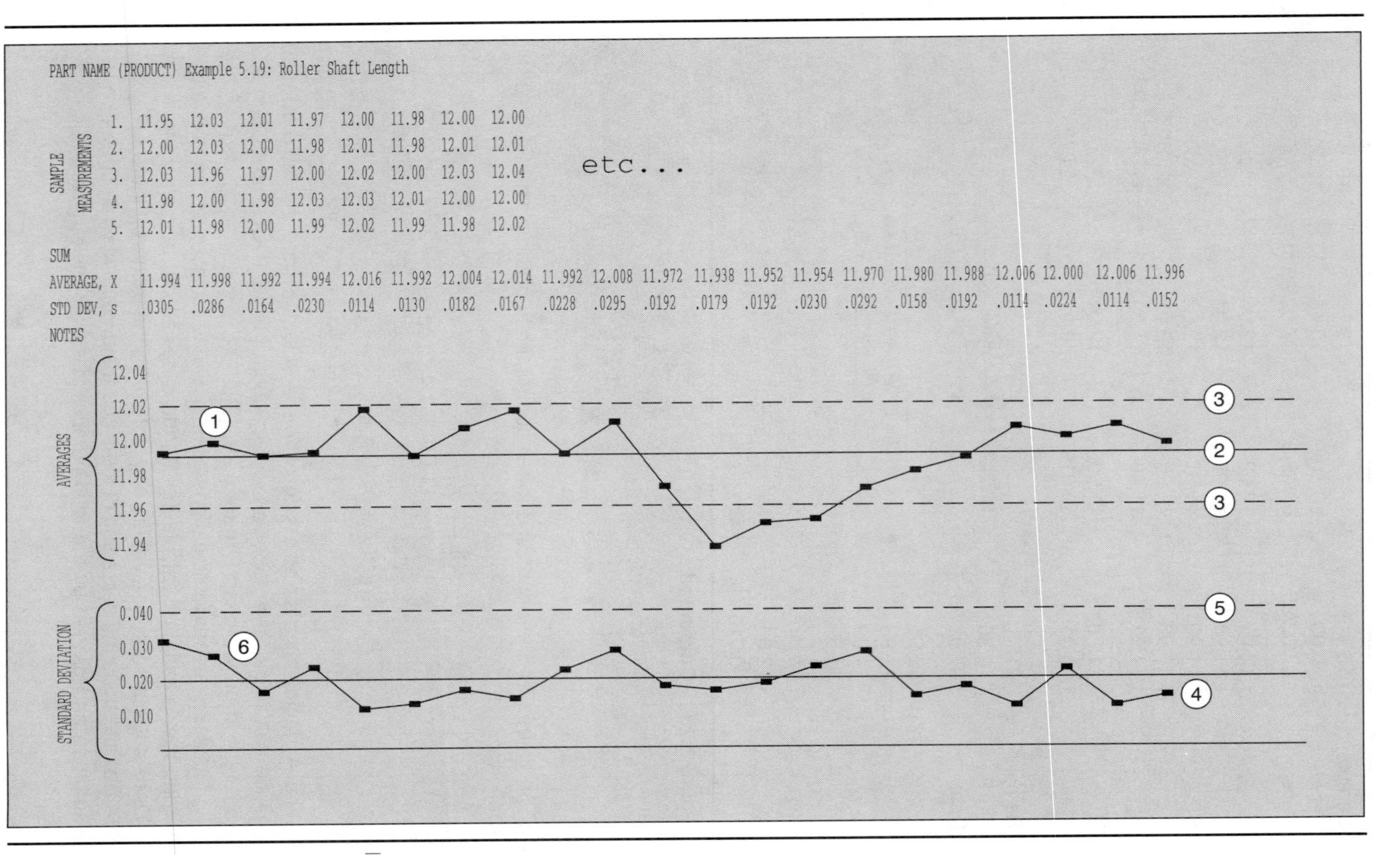

Figure 5.29 Control Charts for $\overline{X}$ and s

The control limits for the s chart are calculated as follows:

$$UCL_s = B_4\bar{s}$$
$$= 2.089(0.02) = 0.04$$

$$LCL_s = B_3\bar{s}$$
$$= 0(0.02) = 0$$

With $n = 5$, the values of B_3 and B_4 are found in the table in Appendix 2. The control limits are placed on the s chart [Figure 5.29, (5)].

Similar to the interpretation of the R chart in Example 5.3, the s chart (Figure 5.29) exhibits good control. The points are evenly spaced on both sides of the centerline and there are no points beyond the control limits. There are no unusual patterns or trends in the data. Given these observations, it can be said that the process is producing parts of similar dimensions.

Next the $\overline{X}$ chart is examined. Once again, an inspection of the $\overline{X}$ chart reveals an unusual pattern occurring at points 12, 13, and 14. These measurements are all below the lower control limit. When compared with other samples throughout the day's production, the parts produced during the time that samples 12, 13, and 14 were taken were much smaller than parts produced during other times in the production run. As before, this signals that an investigation into the cause of the production of undersized parts needs to take place.

Since a cause for the undersized parts was determined in previous examples, these values can be removed from the calculations for the $\overline{X}$ chart and s chart. The new or revised limits will be used to monitor future production. The new limits will extend from the old limits, as shown in Figure 5.30.

Revising the calculations is performed as follows:

$$\overline{\overline{X}}_{new} = \overline{X}_0 = \frac{\Sigma\overline{X} - \overline{X}_d}{m - m_d}$$
$$= \frac{251.76 - 11.94 - 11.95 - 11.95}{21 - 3}$$
$$= 12.00$$

$$\bar{s}_{new} = s_0 = \frac{\Sigma s - s_d}{m - m_d}$$
$$= \frac{0.414 - 0.0179 - 0.0192 - 0.0230}{21 - 3}$$
$$= 0.02$$

Calculating σ_0,

$$\sigma_0 = \frac{s_0}{c_4} = \frac{0.02}{0.9400} = 0.02$$

$$UCL_{\overline{X}} = 12.00 + 1.342(0.02) = 12.03$$
$$LCL_{\overline{X}} = 12.00 - 1.342(0.02) = 11.97$$

$$UCL_s = 1.964(0.02) = 0.04$$
$$LCL_s = 0(0.02) = 0$$

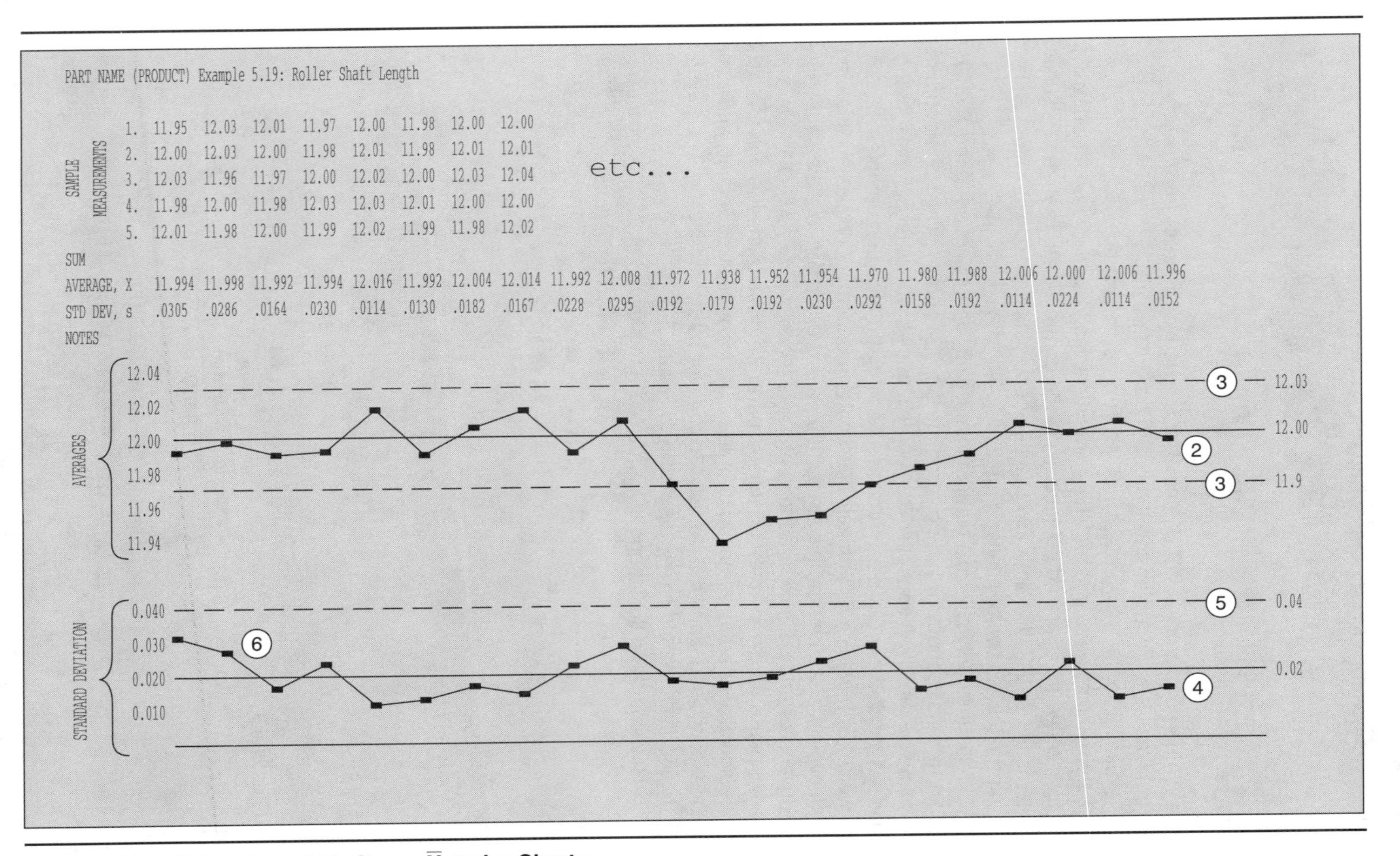

Figure 5.30 **Extension of Limits on $\overline{X}$ and s Charts**

SUMMARY

Control charts are easy to construct and use in studying a process, whether that process is in a manufacturing or service environment. Control charts indicate areas for improvements. Knowing the difference between assignable and chance causes helps focus problem-solving efforts on the true problems in a process. Once identified, changes can be proposed and tested, and improvements can be monitored through the use of control charts. Consider the situation described in Example 5.9. The doctors and office manager were encouraging the people working in billing and insurance to work harder. When control charts revealed that this was the best the process was going to perform, major changes were made to the process. Changes that would not have otherwise been made resulted in increased productivity. The billing is being completed faster with fewer errors. Faster completion times and reduced errors mean a decrease in cost and an improvement in worker satisfaction and worker-management relations. The cost of refiling the claims is significantly reduced and payments are reaching the office faster. One chart led to major process changes that have been justified by all the gains achieved.

Through the use of control charts, similar gains can be realized in the manufacturing sector. Users of control charts report savings in scrap, including material and labor; lower rework costs; reduced inspection; higher product quality; more consistent part characteristics; greater operator confidence; lower troubleshooting costs; reduced completion times; faster deliveries; and other gains.

Lessons Learned

1. Control charts enhance the analysis of a process by showing how that process performs over time. Control charts allow for early detection of process changes.
2. Control charts serve two basic functions: They provide an economic basis for making a decision as to whether to investigate for potential problems, adjust the process, or leave the process alone; and they assist in the identification of problems in the process.
3. Variation, differences between items, exists in all processes. Variation can be within-piece, piece-to-piece, and time-to-time.
4. The $\overline{X}$ chart is used to monitor the variation in the average value of the measurements of groups of samples. Averages rather than individual observations are used on control charts because average values will indicate a change in the amount of variation much faster than individual values will.
5. The $\overline{X}$ chart, showing the central tendency of the data, is always used in conjunction with either a range or a standard deviation chart. The R and s charts show the spread or dispersion of the data.

6. The centerline of a control chart shows where the process is centered. The upper and lower control limits describe the spread of the process.
7. A homogeneous subgroup is essential to the proper study of a process. Certain guidelines can be applied in choosing a rational subgroup.
8. Common, or chance, causes are small random changes in the process that cannot be avoided. Assignable causes are large variations in the process that can be identified as having a specific cause.
9. Losses due to confusing assignable causes and common causes are large.
10. A process is considered to be in a state of control, or under control, when the performance of the process falls within the statistically calculated control limits and exhibits only common, or chance, causes. Certain guidelines can be applied for determining by control chart when a process is under control.
11. Patterns on a control chart indicate a lack of statistical control. Patterns may take the form of changes or jumps in level, runs, trends, or cycles or may reflect the existence of two populations or mistakes.
12. The steps for revising a control chart are (a) examine the chart for out-of-control conditions; (b) isolate the causes of the out-of-control condition; (c) eliminate the cause of the out-of-control condition; and (d) revise the chart, using the formulas presented in the chapter. Revisions to the control chart can take place only when the assignable causes have been determined and eliminated. ■

Formulas

Average and Range Charts

$\overline{X}$ chart:

$$\overline{\overline{X}} = \frac{\sum_{i=1}^{m} \overline{X}_i}{m}$$

$$UCL_{\overline{X}} = \overline{\overline{X}} + A_2\overline{R}$$
$$LCL_{\overline{X}} = \overline{\overline{X}} - A_2\overline{R}$$

R chart:

$$\overline{R} = \frac{\sum_{i=1}^{m} R_i}{m}$$

$$UCL_R = D_4\overline{R}$$
$$LCL_R = D_3\overline{R}$$

Revising the charts:

$$\overline{X} = \overline{\overline{X}}_{new} = \frac{\sum_{i=1}^{m} \overline{X} - \overline{X}_d}{m - m_d}$$

$$UCL_{\overline{X}} = \overline{X}_0 + A\sigma_0$$
$$LCL_{\overline{X}} = \overline{X}_0 - A\sigma_0$$

$$\sigma_0 = R_0/d_2$$

$$\overline{R}_{new} = \frac{\sum_{i=1}^{m} R - R_d}{m - m_d}$$

$$UCL_R = D_2\sigma_0$$
$$LCL_R = D_1\sigma_0$$

Average and Standard Deviation Charts

$\overline{X}$ chart:

$$\overline{\overline{X}} = \frac{\sum_{i=1}^{m} X_i}{m}$$

$$UCL_{\overline{X}} = \overline{\overline{X}} + A_3\bar{s}$$
$$LCL_{\overline{X}} = \overline{\overline{X}} - A_3\bar{s}$$

s chart:

$$\bar{s} = \frac{\sum_{i=1}^{m} s_i}{m}$$
$$UCL_s = B_4\bar{s}$$
$$LCL_s = B_3\bar{s}$$

Revising the charts:

$$X_0 = \overline{\overline{X}}_{new} = \frac{\sum_{i=1}^{m} \overline{X} - \overline{X}_d}{m - m_d}$$

$$s_0 = \bar{s}_{new} = \frac{\sum_{i=1}^{m} s - s_d}{m - m_d}$$

and

$$\sigma_0 = s_0/c_4$$

$$UCL_{\overline{X}} = \overline{X}_0 + A\sigma_0$$
$$LCL_{\overline{X}} = \overline{X}_0 - A\sigma_0$$

$$UCL_s = B_6\sigma_0$$
$$LCL_s = B_5\sigma_0$$

Chapter Problems

1. Describe the difference between chance and assignable causes.
2. How would you use variation to manage a group of people? Why should a manager be aware of assignable and chance causes?

$\overline{X}$ and R Charts

3. A large bank establishes $\overline{X}$ and R charts for the time required to process applications for its charge cards. A sample of five applications is taken each day. The first four weeks (20 days) of data give

$$\overline{\overline{X}} = 16\,\text{min} \qquad \overline{s} = 3\,\text{min} \qquad \overline{R} = 7\,\text{min}$$

   Based on the values given, calculate the centerline and control limits for the $\overline{X}$ and R charts.

4. The data below are $\overline{X}$ and R values for 25 samples of size n = 4 taken from a process filling bags of fertilizer. The measurements are made on the fill weight of the bags in pounds.

Sample Number	$\overline{X}$	*Range*
1	50.3	0.73
2	49.6	0.75
3	50.8	0.79
4	50.9	0.74
5	49.8	0.72
6	50.5	0.73
7	50.2	0.71
8	49.9	0.70
9	50.0	0.65
10	50.1	0.67
11	50.2	0.65
12	50.5	0.67
13	50.4	0.68
14	50.8	0.70
15	50.0	0.65

Sample Number	$\overline{X}$	*Range*
16	49.9	0.66
17	50.4	0.67
18	50.5	0.68
19	50.7	0.70
20	50.2	0.65
21	49.9	0.60
22	50.1	0.64
23	49.5	0.60
24	50.0	0.62
25	50.3	0.60

Set up an $\overline{X}$ and R chart on this process. Interpret the chart. Does the process seem to be in control? If necessary, assume assignable causes and revise the trial control limits. If the average fill of the bags is to be 50.0 pounds, how does this process compare?

5. The data below are $\overline{X}$ and R values for 12 samples of size n = 5. They were taken from a process producing bearings. The measurements are made on the inside diameter of the bearing. The data have been coded from 0.50; in other words, a measurement of 0.50345 has been recorded as 345. Range values are coded from 0.0000; that is, 0.00003 is recorded as 3.

Sample Number	$\overline{X}$	*Range*
1	345	13
2	342	14
3	356	12
4	366	19
5	350	15
6	341	16
7	339	14
8	338	13
9	348	12
10	356	15
11	352	13
12	355	16

Set up the $\overline{X}$ and R charts on this process. Does the process seem to be in control? Why or why not? If necessary, assume assignable causes and revise the trial control limits.

6. What is meant by the statement, "The process is in a state of statistical control"?

7. Describe how both an $\overline{X}$ and R or s chart would look if they were under normal statistical control.
8. $\overline{X}$ charts describe the accuracy of a process, and R and s charts describe the precision. How would accuracy be recognized on an $\overline{X}$ chart? How would precision be recognized on either an R or s chart?
9. Why is the use and interpretation of an R or s chart so critical when examining an $\overline{X}$ chart?
10. Create an $\overline{X}$ and R chart for the information in Figure 5.1. You will need to calculate the range values for each subgroup. Calculate the control limits and centerline for each chart. Graph the data with the calculated values. Beginning with the R chart, how does the process look?
11. RM Manufacturing makes thermometers for use in the medical field. These thermometers, which read in degrees Celsius, are able to measure temperatures to a level of precision of two decimal places. Each hour, RM Manufacturing tests eight randomly selected thermometers in a solution that is known to be at a temperature of 3°C. Use the following data to create and interpret an X and R chart. Based on the desired thermometer reading of 3°, interpret the results of your plotted averages and ranges.

Subgroup	*Average Temperature*	*Range*
1	3.06	0.10
2	3.03	0.09
3	3.10	0.12
4	3.05	0.07
5	2.98	0.08
6	3.00	0.10
7	3.01	0.15
8	3.04	0.09
9	3.00	0.09
10	3.03	0.14
11	2.96	0.07
12	2.99	0.11
13	3.01	0.09
14	2.98	0.13
15	3.02	0.08

12. Interpret the $\overline{X}$ and R charts in Figure 1.
13. Interpret the $\overline{X}$ and R charts in Figure 2.
14. The variables control chart seen in Figure 3 is monitoring the main score residual for a peanut canister pull top. The data are coded from 0.00 (in

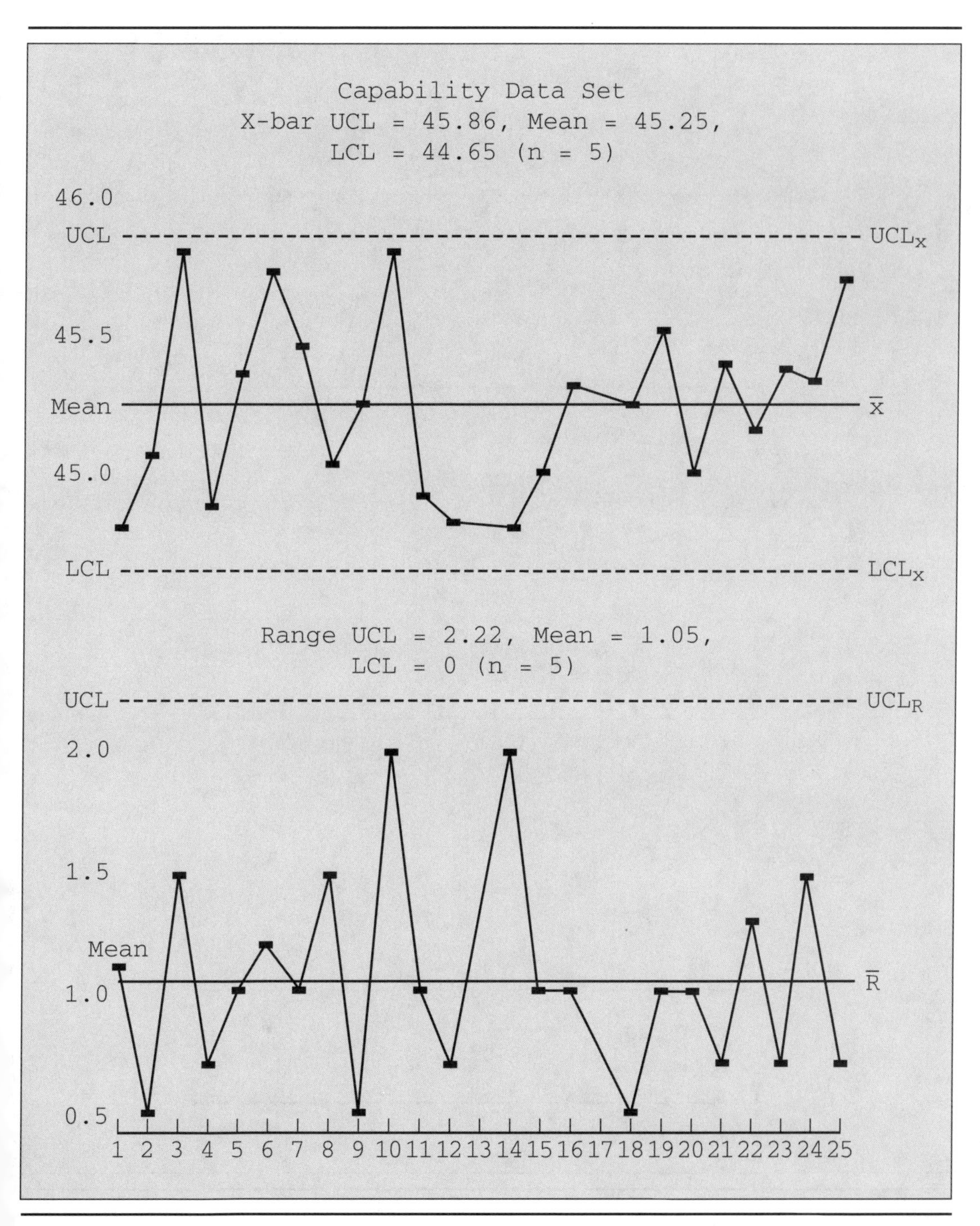
Capability Data Set
X-bar UCL = 45.86, Mean = 45.25,
LCL = 44.65 (n = 5)
46.0
UCL
UCLx
45.5
Mean
x̄
45.0
LCL
LCLx
Range UCL = 2.22, Mean = 1.05,
LCL = 0 (n = 5)
UCL
UCLR
2.0
1.5
Mean
1.0
R̄
0.5
1 2 3 4 5 6 7 8 9 10 11 12 13 14 15 16 17 18 19 20 21 22 23 24 25

Problem 12

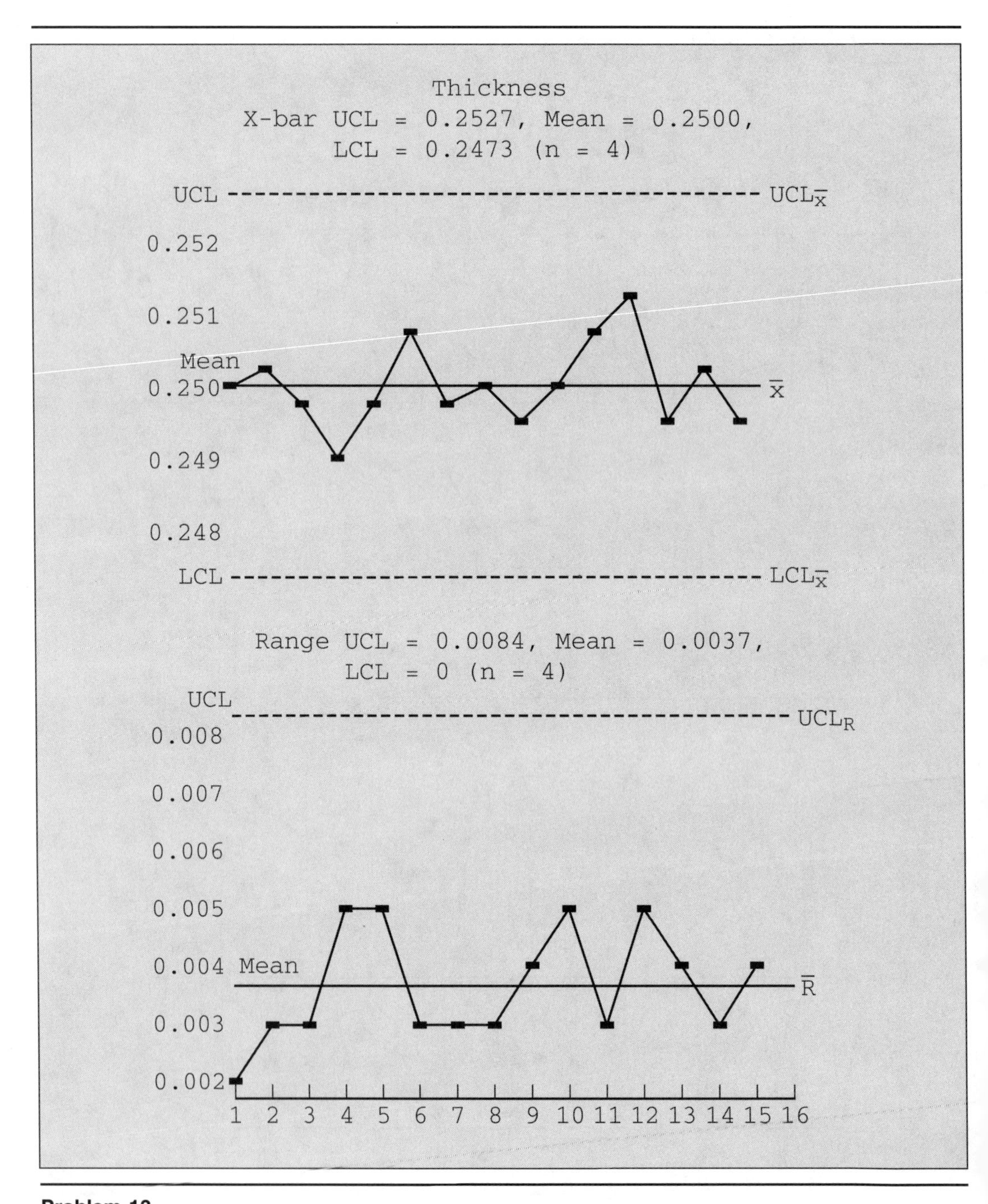

Thickness
X-bar UCL = 0.2527, Mean = 0.2500,
LCL = 0.2473 (n = 4)
UCL
UCL$_{\bar{x}}$
0.252
0.251
Mean
0.250
$\bar{x}$
0.249
0.248
LCL
LCL$_{\bar{x}}$
Range UCL = 0.0084, Mean = 0.0037,
LCL = 0 (n = 4)
UCL
UCL$_R$
0.008
0.007
0.006
0.005
0.004 Mean
$\bar{R}$
0.003
0.002
1 2 3 4 5 6 7 8 9 10 11 12 13 14 15 16

Problem 13

VARIABLES CONTROL CHART

PART NAME (PRODUCT) Pull Up Tab OPERATION (PROCESS) Main Score Residual PART NUMBER SPECIFICATION LIMITS .0028 +/- 0.0005

OPERATOR You MACHINE Press #34 GAGE 1 UNIT OF MEASURE 0.0001* ZERO EQUALS

DATE	9/21					9/22					
TIME	1:00	2:00	2:30	3:00	3:30	1:00	1:30	2:00	2:30	3:00	3:30
SAMPLE MEASUREMENTS 1.	28	27	28	26	25	28	28	26	30	27	27
2.	36	35	35	33	33	34	34	27	35	29	31
3.	32	33	32	31	31	30	30	31	36	33	29
4.	28	27	27	27	26	38	24	33	33	33	26
SUM	124	122	122	117	115	120	116	117	134	122	113
AVERAGE, X	31	31	31	29	29	30	29	29	34	31	28
RANGE, R	8	8	8	7	8	6	10	7	6	6	5

DATE	9/23						9/24					
TIME	1:00	1:30	2:00	2:30	3:00	3:30	1:00	1:30	2:00	2:30	3:00	3:30
SAMPLE MEASUREMENTS 1.	25	25	23	24	24	25	24	25	26	25	25	23
2.	30	28	27	23	23	30	29	27	28	26	26	27
3.	28	25	26	27	27	29	27	25	26	28	28	25
4.	24	22	22	28	24	24	24	24	24	27	28	24
SUM	107	100	98	102	98	108						
AVERAGE, X	27	25	25	26	25	27						
RANGE, R	6	6	5	5	4	6						

Problem 14

other words, a value of 26 in the chart is actually 0.0026). Finish the calculations for the sum, averages, and range. Create an $\overline{X}$ and R chart, calculate the limits, plot the points, and interpret the chart.

$\overline{X}$ and s Charts

15. Create an $\overline{X}$ and s chart for the information in Figure 5.1. You will need to calculate the standard deviation values for each subgroup. Calculate the control limits and centerline for each chart. Graph the data. Beginning with the s chart, how does the process look?
16. Create an $\overline{X}$ and s chart for the information given in Problem 14.

CASE STUDY 5.1
Quality Control for Variables

PART 1

This case study provides some details about the activities of the Whisk Wheel Company, which is currently in the process of applying statistical quality control and problem-solving techniques to their wheel hub operation. Whisk Wheel supplies hubs and wheels to a variety of bicycle manufacturers. The wheel hubs under discussion in this case fit on an all-terrain model bicycle.

Background

The Whisk Wheel Company has just been notified by its largest customer, Rosewood Bicycle Inc., that Whisk Wheel will need to dramatically improve the quality level associated with the hub operation. Currently the operation is unable to meet the specification limits set by the customer. Rosewood has been sorting the parts on the production line before assembly, but they want to end this procedure. Beginning immediately, Whisk Wheel will be required to provide detailed statistical information about each lot of products they produce. (A lot is considered one day's worth of production.) At the end of each day, the lot produced is shipped to Rosewood just-in-time for their production run.

The Product

Figure 1 diagrams the product in question, a wheel hub. The hub shaft is made of chrome-moly steel, is 0.750 inch in diameter and 3.750 inches long. The dimension in question is the length. The specifications for the length are 3.750 $\pm$ 0.005 inches.

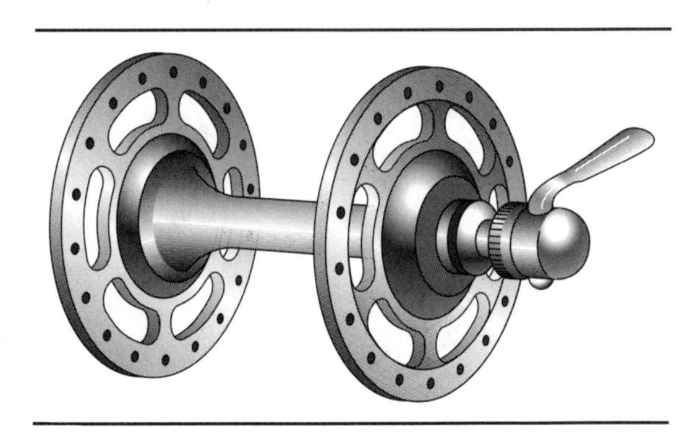

Figure 1 Hub Assembly

The Process

Twelve-foot-long chrome-moly steel shafts are purchased from a supplier. The shafts are straightened and then cut to the 3.750-inch length. Several different machines perform the cutting operation. The data presented here are for the production off one machine only.

Management Strategy

On the basis of new customer requirements, until greater control can be placed on the process, management has decided to intensify product inspection. This will allow the staff engineers to complete their study of the problem and recommend an action plan. Each piece produced will be inserted in a go/no-go gauge to determine if it meets specifications. This will work fairly well by preventing improperly sized shafts from going to the customer. Several managers want to make this a permanent arrangement, but some of the more forward-thinking managers feel that this will not get at the root cause of the problem. There is also the concern that 100 percent inspection is costly and not effective in the long term.

In the meantime, the staff engineers (including you) are continuing to study the problem more carefully. The following information is from today's production run. From the finished parts, an operator samples six hubs 24 times during the day.

Assignment

On the computer, create a histogram from today's data (Table 1). Mark on the histogram where the specifications set by the designer fall in relation to the process information. Write a summary of the results. Use the estimated sigma and the Z tables to calculate the percentage of parts produced above and below the specification limits.

Table 1 A Day's Data

DATA Day 1

DEPT.	Machine	PART NAME	Whisk Wheel
PART NO.	01-92	MACHINE	Cutoff
GROUP	Quality Control	VARIABLE	

Sample	1	2	3	4	5
Time	07:00:00	07:18:00	07:36:00	07:54:00	08:12:00
Date	01/30/96	01/30/96	01/30/96	01/30/96	01/30/96
1	3.757	3.753	3.744	3.755	3.757
2	3.749	3.739	3.745	3.753	3.760
3	3.751	3.747	3.740	3.753	3.751
4	3.755	3.751	3.741	3.749	3.754
5	3.749	3.751	3.742	3.739	3.750
6	3.759	3.744	3.743	3.747	3.753

Table 1 Continued

DATA Day 1

DEPT.	Machine	PART NAME	Whisk Wheel
PART NO.	01-92	MACHINE	Cutoff
GROUP	Quality Control	VARIABLE	

Sample	6	7	8	9	10
Time	08:30:00	08:48:00	09:06:00	09:24:00	09:42:00
Date	01/30/96	01/30/96	01/30/96	01/30/96	01/30/96
1	3.741	3.746	3.746	3.760	3.741
2	3.749	3.743	3.753	3.755	3.745
3	3.745	3.753	3.747	3.757	3.751
4	3.742	3.751	3.755	3.757	3.741
5	3.743	3.747	3.758	3.749	3.740
6	3.742	3.751	3.756	3.741	3.741

Sample	11	12	13	14	15
Time	10:00:00	10:18:00	10:36:00	10:54:00	11:12:00
Date	01/30/96	01/30/96	01/30/96	01/30/96	01/30/96
1	3.749	3.746	3.743	3.755	3.745
2	3.751	3.749	3.751	3.744	3.751
3	3.757	3.744	3.745	3.753	3.747
4	3.754	3.757	3.739	3.755	3.751
5	3.755	3.737	3.750	3.754	3.744
6	3.753	3.749	3.747	3.766	3.745

Sample	16	17	18	19	20
Time	11:30:00	11:48:00	12:06:00	01:06:00	01:24:00
Date	01/30/96	01/30/96	01/30/96	01/30/96	01/30/96
1	3.748	3.757	3.740	3.756	3.742
2	3.746	3.747	3.739	3.757	3.753
3	3.755	3.756	3.752	3.749	3.754
4	3.755	3.759	3.744	3.755	3.743
5	3.749	3.751	3.745	3.744	3.741
6	3.749	3.756	3.757	3.741	3.748

Sample	21	22	23	24
Time	01:42:00	02:00:00	02:18:00	02:36:00
Date	01/30/96	01/30/96	01/30/96	01/30/96
1	3.752	3.746	3.745	3.752
2	3.751	3.753	3.762	3.755
3	3.749	3.741	3.753	3.753
4	3.753	3.746	3.750	3.749
5	3.755	3.743	3.744	3.754
6	3.750	3.744	3.750	3.756

PART 2

Although process capability calculations have not been made, on the basis of the histogram, the process does not appear to be capable. It is apparent from the histogram that a large proportion of the process does not meet the individual length specification. The data appear to be a reasonable approximation of a normal distribution.

During a rare quiet moment in your day, you telephone a good friend from your quality-control class to reflect on the events so far. You also remember some of the comments made by your SQC professor about appropriate sampling and measuring techniques. After listening to your story, your friend brings up several key concerns.

1. Product Control Basically, management has devised a stop-gap procedure to prevent poor quality products from reaching the customer. This work—screening, sorting, and selectively shipping parts—is a strategy consistent with the "detect and sort" approach to quality control. Management has not really attempted to determine the root cause of the problem.
2. The Engineering Approach While a little more on track, the focus of engineering on the process capability was purely from the "conformance to specifications" point of view. Appropriate process capability calculations should be based on a process that is under statistical control. No information has yet been gathered on this particular process to determine if the process is under statistical control. In this situation, process capability was calculated without determining if the process was in a state of statistical control—something you now remember your quality professor cautioning against.

Another consideration deals with the statistical significance of the sample using the "best operator and the best machine." Few or no details have been given about the sampling techniques used or the training level of the operator.

A Different Approach

After much discussion, you and your friend come up with a different approach to solving this problem. You gather together your fellow team members and plan a course of action. The goal of the group is to determine the source of variation in the process of producing wheel hubs. A process flowchart is created to carefully define the complete sequence of processing steps: raw material handling, straightening, cutting, and finish polish (Figure 2). Creating a process flowchart has helped all members of the team to better understand what is happening during the manufacture of the hub.

At each step along the way, your team discusses all the factors that could be contributing to the variation in the final product. To aid and guide the discussion, the group creates a cause-and-effect diagram, which helps keep the group discussions focused and allows the team to discuss all the possible sources of variation. There are several of these, including the raw materials (their properties and preparation), the methods (procedures for setup and machine operation at each of the three operations),

FLOW PROCESS CHART

Part Charted:	Whisk wheel hub	Chart No.	92-05
Drawing No.	D-92-05	Chart of Method	Present
Chart Begins	Raw material storage	Charted by:	
Chart Ends	Hub assembly bin	Date	01/27/92 1 of 1

Distance, ft	Unit Time, min	Symbols	Process Description
		▽	In stock storage
40	0.14	⇨	Hand cart to straightener
	0.35	○	Straighten stock
	0.02	○	Load to cart
7	0.10	⇨	Cart to cutoff
	0.92	○	Cut stock to six pc. 3.750
2	0.12	○	Finish ends
1	0.02	○	Load on conveyer
20	1.12	⇨	Conveyer to hub assembly bin
			02/92

Figure 2 Process Flowchart

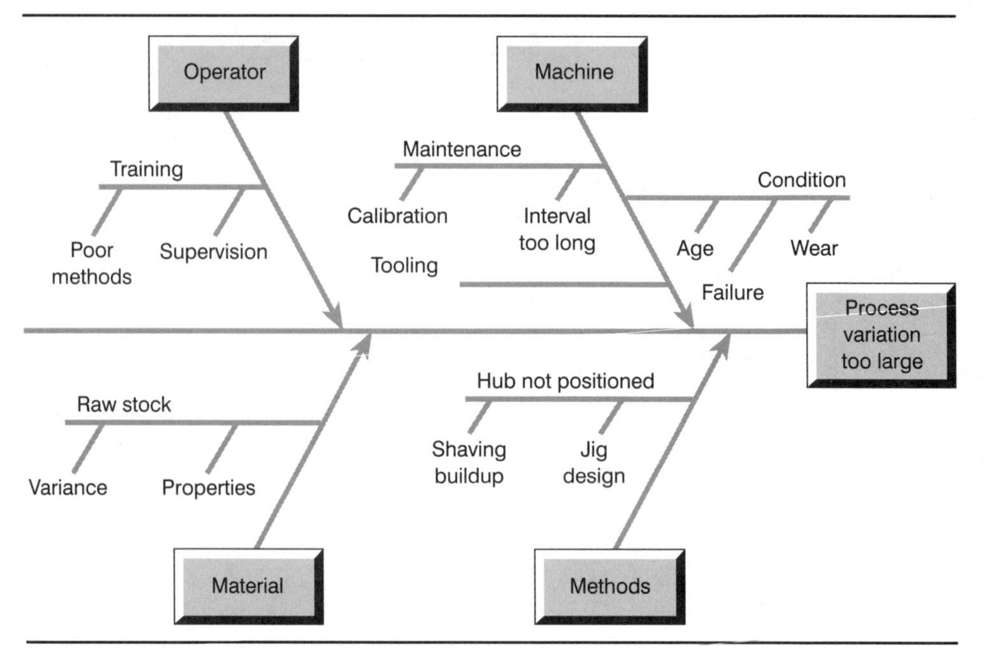

Figure 3 Cause-and-Effect Diagram

the machine conditions (operating settings, maintenance conditions), and the operator (training, supervision, techniques). The diagram created is shown in Figure 3.

The team originally focuses on the inherent equipment capability as the key problem. This approach leads too quickly to the conclusion that new machines should be purchased. This approach does not enable team members to learn to use the equipment, processes, and people already available to their fullest potential.

After studying and discussing the complete process flow, the team decides that they do not know enough about the process to suggest solutions. They assign team members to more fully investigate the four areas (raw materials, straightening, cutting, and finishing). Close contact of the team members will ensure that the discoveries in one area are quickly shared with other related areas. After all, in this process of making a wheel hub, no one area can function without the others.

You and your partner have been assigned to the cutting area. To discover the source of variation, the two of you decide to run $\overline{X}$ and R charts on the data from the preceding day as well as the data from this day (an additional 24 subgroups of sample size six).

Assignment

Add the new data (Table 2) for today to your previous file. Create an $\overline{X}$ and R chart containing both days' data and discuss what the charts look like. Use all the

Table 2 Data for Day 2

DATA Day 2

DEPT.	Machine	PART NAME	Whisk Wheel
PART NO.	01-92	MACHINE	Cutoff
GROUP	Quality Control	VARIABLE	

Sample	1	2	3	4	5
Time	07:00:00	07:18:00	07:36:00	07:54:00	08:12:00
Date	01/31/96	01/31/96	01/31/96	01/31/96	01/31/96
1	3.744	3.741	3.749	3.745	3.751
2	3.753	3.744	3.747	3.750	3.737
3	3.757	3.745	3.747	3.749	3.755
4	3.749	3.754	3.757	3.745	3.751
5	3.757	3.745	3.750	3.749	3.751
6	3.749	3.747	3.755	3.741	3.739

Sample	6	7	8	9	10
Time	08:30:00	08:48:00	09:06:00	09:24:00	09:42:00
Date	01/31/96	01/31/96	01/31/96	01/31/96	01/31/96
1	3.751	3.747	3.749	3.751	3.747
2	3.755	3.753	3.755	3.743	3.757
3	3.749	3.737	3.743	3.753	3.747
4	3.753	3.745	3.749	3.757	3.745
5	3.755	3.749	3.752	3.749	3.751
6	3.751	3.744	3.749	3.744	3.756

Sample	11	12	13	14	15
Time	10:00:00	10:18:00	10:36:00	10:54:00	11:12:00
Date	01/31/96	01/31/96	01/31/96	01/31/96	01/31/96
1	3.747	3.755	3.747	3.754	3.741
2	3.741	3.742	3.749	3.747	3.739
3	3.751	3.749	3.755	3.749	3.751
4	3.750	3.755	3.749	3.751	3.753
5	3.747	3.745	3.754	3.756	3.749
6	3.740	3.745	3.746	3.754	3.743

Sample	16	17	18	19	20
Time	11:30:00	11:48:00	12:06:00	01:06:00	01:24:00
Date	01/31/96	01/31/96	01/31/96	01/31/96	01/31/96
1	3.751	3.758	3.745	3.755	3.747
2	3.745	3.748	3.744	3.753	3.754
3	3.757	3.743	3.752	3.750	3.741
4	3.747	3.751	3.743	3.753	3.745
5	3.749	3.748	3.750	3.751	3.759
6	3.745	3.759	3.746	3.752	3.747

(continued)

Table 2 Continued

DATA Day 2

DEPT.	Machine		PART NAME	Whisk Wheel
PART NO.	01-92		MACHINE	Cutoff
GROUP	Quality Control		VARIABLE	

Sample	21	22	23	24
Time	01:42:00	02:00:00	02:18:00	02:36:00
Date	01/31/96	01/31/96	01/31/96	01/31/96
1	3.753	3.755	3.747	3.751
2	3.755	3.749	3.749	3.753
3	3.747	3.752	3.752	3.753
4	3.747	3.750	3.745	3.752
5	3.751	3.753	3.750	3.757
6	3.747	3.755	3.743	3.751

information available to create a histogram. Draw the specification limits on the histogram by hand. Use the estimated sigma and the Z tables to calculate the percentage of parts produced above and below the specification limits.

PART 3

In order to determine the root causes of variation, you and your partner spend the remainder of day 2 studying the cutting operation and the operator. You randomly select a machine and an operator to watch as he performs the operation and measures the parts. You note several actions taken by the operator that could be sources of variation.

Investigation reveals that the operator runs the process in the following manner. Every 18 minutes, he measures the length of six hubs with a micrometer. The length values for the six consecutively produced hubs are averaged, and the average is plotted on a piece of charting paper. Periodically, the operator reviews the evolving data and makes a decision as to whether or not the process mean (the hub length) needs to be adjusted. These adjustments can be accomplished by stopping the machine, loosening some clamps, and jogging the cutting device back or forth depending on the adjustment the operator feels is necessary. This process takes about five minutes and appears to occur fairly often.

Based on what you have learned about process control in SQC class, it is obvious to you that the operator is adding variation to the process! He appears to be overcontrolling (overadjusting) the process because he cannot distinguish between common-cause variation and special-cause variation. The operator has been reacting to patterns in the data that may be inherent (common) to the process. The consequences of this mistake are devastating to a control chart. Each time an adjustment is made when it is not necessary, variation is introduced to the process that would not be there otherwise. Not only is quality essentially decreased (made more variable) with each adjustment, but production time is unnecessarily lost.

A glance at the histogram created the first day shows that overadjustment is indeed occurring, resulting in the bimodal distribution. Control charts can be used to help distinguish between the presence of common and special causes of variation. Removing this source of variation will allow the process to operate more consistently. Removing this obstacle can also help uncover the root cause of the variation.

The data from day 3 have been gathered and reflect the suggested change. The operator has been told not to adjust the process at all during the day. If the process goes beyond the previous day's limits and out of control, the operator is to contact you.

Assignment

Create an $\overline{X}$ and R chart for *only the new data from day 3* (Table 3). Compare the new chart with the charts from the two previous days. Draw the previous day's limits on the new chart for day 3 by hand. Using only the data from day 3, create a histogram and hand-draw the specification limits on the histogram. Use the estimated sigma and the Z tables to calculate the percentage of parts produced above and below the specification limits.

The new chart should allow you to better distinguish between the presence of common and special causes of variation. Compare all of your mathematical and graphical results. What conclusions can you and your partner draw?

PART 4

With one source of variation identified and removed, quality and productivity on the line have improved. The process has been stabilized because no unnecessary

Table 3 Data for Day 3

DATA Day 3

DEPT.	Machine		PART NAME	Whisk Wheel	
PART NO.	92-01		MACHINE	Cutoff	
GROUP	Quality Control		VARIABLE		

Sample	1	2	3	4	5
Time	07:00:00	07:18:00	07:36:00	07:54:00	08:12:00
Date	02/06/96	02/06/96	02/06/96	02/06/96	02/06/96
1	3.749	3.751	3.747	3.753	3.755
2	3.752	3.742	3.749	3.750	3.754
3	3.751	3.745	3.755	3.744	3.751
4	3.746	3.739	3.753	3.745	3.753
5	3.745	3.751	3.752	3.741	3.749
6	3.749	3.751	3.740	3.748	3.752

(continued)

Table 3 Continued

DATA Day 3

DEPT.	Machine	PART NAME	Whisk Wheel
PART NO.	92-01	MACHINE	Cutoff
GROUP	Quality Control	VARIABLE	

Sample	6	7	8	9	10
Time	08:30:00	08:48:00	09:06:00	09:24:00	09:42:00
Date	02/06/96	02/06/96	02/06/96	02/06/96	02/06/96
1	3.752	3.746	3.749	3.750	3.745
2	3.752	3.750	3.753	3.746	3.747
3	3.750	3.749	3.752	3.750	3.745
4	3.748	3.747	3.757	3.754	3.754
5	3.751	3.753	3.747	3.753	3.748
6	3.748	3.751	3.747	3.749	3.743

Sample	11	12	13	14	15
Time	10:00:00	10:18:00	10:36:00	10:54:00	11:12:00
Date	02/06/96	02/06/96	02/06/96	02/06/96	02/06/96
1	3.751	3.751	3.749	3.746	3.741
2	3.751	3.753	3.753	3.751	3.748
3	3.748	3.753	3.755	3.747	3.744
4	3.748	3.747	3.749	3.749	3.749
5	3.755	3.741	3.745	3.755	3.751
6	3.750	3.741	3.759	3.747	3.744

Sample	16	17	18	19	20
Time	11:30:00	11:48:00	12:06:00	01:06:00	01:24:00
Date	02/06/96	02/06/96	02/06/96	02/06/96	02/06/96
1	3.749	3.745	3.743	3.744	3.743
2	3.749	3.746	3.749	3.748	3.745
3	3.745	3.753	3.745	3.754	3.749
4	3.745	3.753	3.747	3.753	3.749
5	3.749	3.751	3.755	3.755	3.749
6	3.745	3.751	3.753	3.749	3.748

Sample	21	22	23	24
Time	01:42:00	02:00:00	02:18:00	02:36:00
Date	02/06/96	02/06/96	02/06/96	02/06/96
1	3.748	3.747	3.752	3.747
2	3.752	3.750	3.751	3.745
3	3.752	3.748	3.753	3.751
4	3.755	3.751	3.749	3.749
5	3.750	3.755	3.751	3.753
6	3.753	3.753	3.755	3.740

adjustments have been made. The method of overcontrol has proved costly from both a quality (inconsistent product) and a productivity (machine downtime, higher scrap) point of view. The search continues for other sources of variation.

During day 3, you and your partner watched the methods the operator used to measure the hub. Neither of you feel that this technique is very good. Today you replace the old method and tool with a new measuring tool, and the operator is carefully trained to use the new tool.

Assignment

Continue day 3's control chart to record the data for day 4 (Table 4). How do the data look overall? Are there any trends or patterns? Comment on the tighter control limits as compared with days 1 and 2. Create a histogram with the data from only days 3 and 4 combined. How does the overall spread of the process look when compared with the specifications (Z table calculations)? What conclusions can you and your partner draw?

PART 5

Now that two unusual causes of variation have been removed from the process, you and your partner are able to spend the fourth day studying the resulting stable process. You are able to determine that the design of the jig used by the operation is causing a buildup of chips. Each time a part is cut, a small amount of chips builds up in the back of the jig. Unless the operator clears these chips away before inserting the new stock into the jig, they build up. The presence or absence of chips is causing variation in the length of the hub.

Table 4 Data for Day 4

DATA Day 4

DEPT.	Machine	PART NAME	Whisk Wheel	
PART NO.	92-01	MACHINE	Cutoff	
GROUP	Quality Control	VARIABLE		

Sample	1	2	3	4	5
Time	07:00:00	07:18:00	07:36:00	07:54:00	08:12:00
Date	02/10/96	02/10/96	02/10/96	02/10/96	02/10/96
1	3.745	3.753	3.751	3.747	3.749
2	3.745	3.756	3.755	3.753	3.753
3	3.751	3.750	3.750	3.744	3.749
4	3.749	3.751	3.749	3.748	3.748
5	3.753	3.749	3.751	3.750	3.745
6	3.740	3.749	3.752	3.747	3.750

(continued)

Table 4 Continued

DATA Day 4

DEPT.	Machine	PART NAME	Whisk Wheel
PART NO.	92-01	MACHINE	Cutoff
GROUP	Quality Control	VARIABLE	

Sample	6	7	8	9	10
Time	08:30:00	08:48:00	09:06:00	09:24:00	09:42:00
Date	02/10/96	02/10/96	02/10/96	02/10/96	02/10/96
1	3.752	3.755	3.751	3.751	3.755
2	3.749	3.749	3.750	3.750	3.750
3	3.753	3.753	3.749	3.749	3.750
4	3.749	3.752	3.749	3.751	3.751
5	3.750	3.749	3.749	3.750	3.751
6	3.751	3.753	3.751	3.752	3.752

Sample	11	12	13	14	15
Time	10:00:00	10:18:00	10:36:00	10:54:00	11:12:00
Date	02/10/96	02/10/96	02/10/96	02/10/96	02/10/96
1	3.749	3.750	3.752	3.751	3.747
2	3.748	3.749	3.750	3.752	3.751
3	3.752	3.749	3.751	3.751	3.748
4	3.749	3.751	3.752	3.746	3.746
5	3.749	3.750	3.751	3.749	3.749
6	3.746	3.750	3.751	3.750	3.753

Sample	16	17	18	19	20
Time	11:30:00	11:48:00	01:10:00	01:28:00	01:46:00
Date	02/10/96	02/10/96	02/10/96	02/10/96	02/10/96
1	3.753	3.751	3.747	3.750	3.751
2	3.747	3.750	3.745	3.749	3.750
3	3.754	3.749	3.749	3.748	3.753
4	3.754	3.748	3.748	3.749	3.750
5	3.749	3.749	3.750	3.750	3.749
6	3.750	3.750	3.750	3.749	3.754

Sample	21	22	23	24
Time	02:04:00	02:22:00	02:40:00	02:58:00
Date	02/10/96	02/10/96	02/10/96	02/10/96
1	3.750	3.749	3.753	3.755
2	3.755	3.751	3.749	3.748
3	3.746	3.755	3.745	3.749
4	3.750	3.751	3.748	3.748
5	3.751	3.752	3.749	3.748
6	3.750	3.751	3.749	3.750

To correct this, during the night maintenance shift a slot is placed in the back of the jig, allowing the chips to drop out of the jig. Additionally, a solvent flush system is added to the fixture to wash the chips clear of the jig.

Assignment

Create an $\overline{X}$ and R chart for just day 5's data (Table 5). Compare the new chart with the charts from days 3 and 4. Have the control limits changed? How? How is the process doing now (compare Z table calculations)? Overall, how would you view the process?

PART 6

When reviewing the chart produced on day 5, you see that the process seems to have settled down with the three improvements made.

Table 5 Data for Day 5

DATA Day 5

DEPT.	Cutting	PART NAME	Whisk Wheel
PART NO.	92-01	MACHINE	Cutoff
GROUP	Quality Control	VARIABLE	

Sample	1	2	3	4	5
Time	07:00:00	07:18:00	07:36:00	07:54:00	08:12:00
Date	02/12/96	02/12/96	02/12/96	02/12/96	02/12/96
1	3.750	3.752	3.752	3.750	3.751
2	3.749	3.750	3.751	3.751	3.752
3	3.750	3.749	3.750	3.750	3.752
4	3.750	3.750	3.749	3.751	3.750
5	3.750	3.751	3.750	3.749	3.750
6	3.751	3.752	3.750	3.750	3.750

Sample	6	7	8	9	10
Time	08:30:00	08:48:00	09:06:00	09:24:00	09:42:00
Date	02/12/96	02/12/96	02/12/96	02/12/96	02/12/96
1	3.749	3.750	3.750	3.751	3.751
2	3.750	3.752	3.750	3.750	3.752
3	3.750	3.750	3.750	3.750	3.751
4	3.750	3.750	3.750	3.751	3.750
5	3.751	3.750	3.751	3.752	3.749
6	3.750	3.751	3.750	3.750	3.750

(continued)

Table 5 Continued

DATA Day 5

DEPT.	Cutting	PART NAME	Whisk Wheel
PART NO.	92-01	MACHINE	Cutoff
GROUP	Quality Control	VARIABLE	

Sample	11	12	13	14	15
Time	10:00:00	10:18:00	10:36:00	10:54:00	11:12:00
Date	02/12/96	02/12/96	02/12/96	02/12/96	02/12/96
1	3.751	3.750	3.750	3.751	3.749
2	3.750	3.751	3.751	3.750	3.750
3	3.750	3.750	3.751	3.751	3.751
4	3.749	3.749	3.750	3.751	3.750
5	3.751	3.750	3.750	3.750	3.750
6	3.750	3.750	3.750	3.750	3.752
Sample	16	17	18	19	20
Time	11:30:00	11:48:00	12:06:00	12:24:00	12:42:00
Date	02/12/96	02/12/96	02/12/96	02/12/96	02/12/96
1	3.750	3.750	3.750	3.751	3.752
2	3.750	3.750	3.752	3.750	3.750
3	3.751	3.749	3.750	3.751	3.751
4	3.750	3.750	3.751	3.752	3.750
5	3.749	3.751	3.749	3.750	3.749
6	3.750	3.750	3.750	3.751	3.750
Sample	21	22	23	24	
Time	13:00:00	13:18:00	13:36:00	13:54:00	
Date	02/12/96	02/12/96	02/12/96	02/12/96	
1	3.750	3.750	3.750	3.751	
2	3.750	3.752	3.751	3.751	
3	3.749	3.751	3.749	3.750	
4	3.749	3.750	3.751	3.750	
5	3.751	3.750	3.750	3.750	
6	3.751	3.749	3.750	3.750	

Assignment

Create a control chart for only the sixth day's data (Table 6). How is the process behaving? What will you recommend to management that they tell the customer? How will you support your recommendation?

Table 6 Data for Day 6

DATA Day 6

DEPT.	Machine	PART NAME	Whisk Wheel
PART NO.	92-01	MACHINE	Cutoff
GROUP	Quality Control	VARIABLE	

Sample	1	2	3	4	5
Time	07:00:00	07:18:00	07:36:00	07:54:00	08:12:00
Date	02/20/96	02/20/96	02/20/96	02/20/96	02/20/96
1	3.751	3.749	3.751	3.751	3.750
2	3.751	3.751	3.751	3.752	3.751
3	3.750	3.750	3.751	3.750	3.751
4	3.751	3.751	3.751	3.751	3.750
5	3.750	3.750	3.750	3.750	3.751
6	3.750	3.750	3.750	3.750	3.750
Sample	6	7	8	9	10
Time	08:30:00	08:48:00	09:06:00	09:24:00	09:42:00
Date	02/20/96	02/20/96	02/20/96	02/20/96	02/20/96
1	3.751	3.750	3.750	3.750	3.751
2	3.751	3.751	3.751	3.750	3.751
3	3.750	3.750	3.750	3.751	3.751
4	3.750	3.749	3.750	3.752	3.751
5	3.750	3.750	3.751	3.751	3.750
6	3.750	3.750	3.750	3.751	3.750
Sample	11	12	13	14	15
Time	10:00:00	10:18:00	10:36:00	10:54:00	11:12:00
Date	02/20/96	02/20/96	02/20/96	02/20/96	02/20/96
1	3.751	3.750	3.750	3.751	3.750
2	3.751	3.751	3.751	3.750	3.751
3	3.750	3.751	3.751	3.750	3.750
4	3.750	3.751	3.752	3.750	3.751
5	3.750	3.751	3.750	3.750	3.751
6	3.750	3.751	3.752	3.750	3.751
Sample	16	17	18	19	20
Time	11:30:00	11:48:00	01:10:00	01:28:00	01:46:00
Date	02/20/96	02/20/96	02/20/96	02/20/96	02/20/96
1	3.750	3.751	3.751	3.750	3.752
2	3.751	3.750	3.752	3.750	3.751
3	3.751	3.750	3.752	3.750	3.750
4	3.752	3.752	3.751	3.750	3.750
5	3.751	3.750	3.750	3.751	3.750
6	3.751	3.750	3.751	3.751	3.750

(continued)

Table 6 Continued

DATA Day 6

DEPT.	Machine	PART NAME	Whisk Wheel
PART NO.	92-01	MACHINE	Cutoff
GROUP	Quality Control	VARIABLE	

Sample	21	22	23	24
Time	02:04:00	02:22:00	02:40:00	02:58:00
Date	02/20/96	02/20/96	02/20/96	02/20/96
1	3.751	3.751	3.752	3.751
2	3.750	3.750	3.750	3.751
3	3.750	3.750	3.751	3.750
4	3.750	3.749	3.751	3.750
5	3.750	3.751	3.750	3.750
6	3.750	3.751	3.750	3.750

CASE STUDY 5.2
Process Improvement

This case is the third in a four-part series of cases involving process improvement. The other cases are found at the end of Chapters 3, 4, and 6. Data and calculations for this case establish the foundation for the future cases; however, it is not necessary to complete this case in order to complete and understand the cases in Chapters 3, 4, and 6. Completing this case will provide insight into the use of $\overline{X}$ and R charts in process improvement. The case can be worked by hand or with the software provided.

PART 1

Figure 1 provides the details of a bracket assembly to hold a strut on an automobile in place. Welded to the auto-body frame, the bracket cups the strut and secures it to the frame via a single bolt with a lock washer. Proper alignment is necessary for both smooth installation during assembly and future performance. For mounting purposes the left-side hole, A, must be aligned on center with the right-side hole, B. If the holes are centered directly opposite each other in perfect alignment, then the angle between hole centers will measure 0 degrees. As the flowchart in Figure 2 shows, the bracket is created by passing coils of flat steel through a series of progressive dies. As

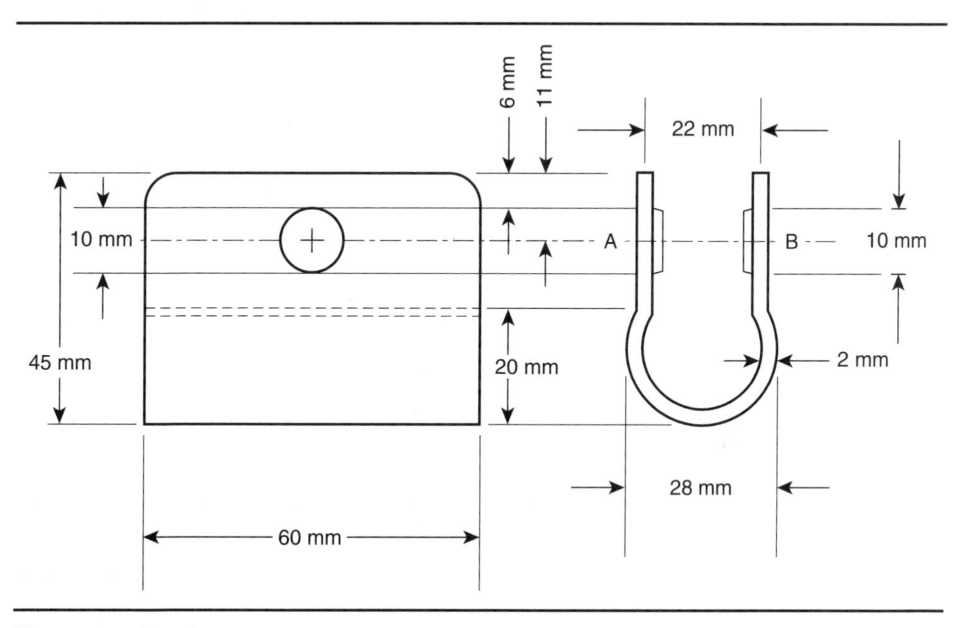

Figure 1 Bracket

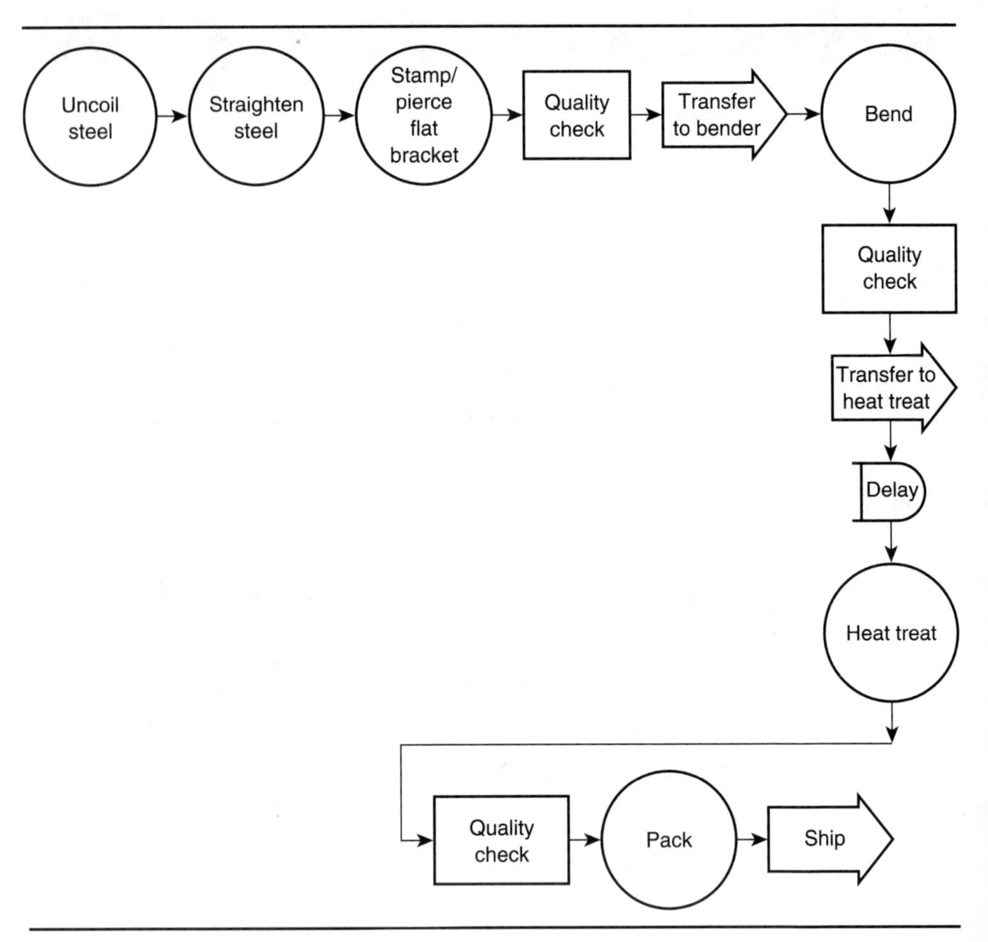

Figure 2 Flowchart of Bracket Fabrication Process

the steel moves through the press, the bracket is stamped, pierced, and finally bent into the appropriate shape.

Recently customers have been complaining about having difficulty securing the bracket closed with the bolt and lock washer. The bolts have been difficult to slide through the holes and then tighten. Assemblers complain of stripped bolts and snug fittings. Unsure of where the problem is, WP Inc.'s management assembles a team consisting of representatives from process engineering, materials engineering, product design, and manufacturing.

Through the use of several cause-and-effect diagrams, the team determines that the most likely cause of the problems experienced by the customer is the alignment of the holes. At some stage in the formation process, the holes end up off-center. To confirm their suspicions, during the next production run, the bending press operator takes 20 subgroups of size 5 and measures the angle between the centers of the holes for each sample (Figure 3). The specification for the angle between insert hole A and insert hole

Sample \ Subgroup	1	2	3	4	5	6	7	8	9	10	11	12	13	14	15
1	0.31	0.27	0.30	0.30	0.25	0.18	0.26	0.15	0.30	0.31	0.18	0.22	0.19	0.14	0.29
2	0.29	0.23	0.30	0.20	0.20	0.26	0.27	0.21	0.24	0.25	0.16	0.30	0.28	0.27	0.23
3	0.30	0.31	0.28	0.21	0.19	0.18	0.12	0.24	0.26	0.25	0.21	0.21	0.26	0.25	0.27
4	0.28	0.23	0.24	0.23	0.26	0.24	0.20	0.27	0.27	0.28	0.29	0.24	0.29	0.28	0.24
5	0.23	0.29	0.32	0.25	0.25	0.17	0.23	0.30	0.26	0.25	0.27	0.26	0.24	0.16	0.23

Sample \ Subgroup	16	17	18	19	20
1	0.22	0.32	0.33	0.20	0.24
2	0.24	0.27	0.30	0.17	0.28
3	0.22	0.28	0.22	0.26	0.15
4	0.30	0.19	0.26	0.31	0.27
5	0.30	0.31	0.26	0.24	0.19

Figure 3 Hole A, B Alignment; Distance Above (+) or Below (−) Nominal

B is 0.00 mm with a tolerance of ±0.30. The values recorded in Figure 3 represent the distance above (or when a minus sign is present, below) the nominal value of 0.00 mm.

Assignment

Follow the steps outlined in the chapter and utilize the data to create a set of $\overline{X}$ and R charts. You will need to calculate the mean and range for each subgroup. Describe the performance of the process.

PART 2

Having $\overline{X}$ and R charts and the accompanying statistical information tells the team a lot about the hole-alignment characteristics. Using the problem-solving method described in Chapter 3 (and followed in Case 3.2), the team has determined that the fixture that holds the flat bracket in place during the bending operation needs to be replaced. One of the measures of performance that they created during step 4 of their problem-solving process is the percentage of the parts out of specification. Now that the fixture has been replaced with a better one, the team would like to determine whether or not changing the fixture improved the process by removing a root cause of hole misalignment.

Assignment

Utilize the data in Figure 4 to create another set of $\overline{X}$ and R charts. Has their change to the fixtures resulted in a process improvement? How do you know? How are they doing when you compare their measure of performance, percentage of the parts out of specification, both before and after the fixture is changed?

Sample \ Subgroup	1	2	3	4	5	6	7	8	9	10	11	12	13	14	15
1	0.03	0.06	0.06	0.03	−0.04	−0.02	−0.05	0.06	0.00	−0.02	−0.02	0.06	0.07	0.10	0.02
2	0.08	0.08	0.08	0.00	−0.07	0.06	−0.07	0.03	0.06	−0.01	0.06	0.02	−0.04	0.05	−0.05
3	−0.03	−0.01	0.05	0.05	0.00	0.02	0.11	0.04	0.02	0.00	−0.02	−0.01	0.08	0.03	0.04
4	0.07	0.08	−0.03	−0.01	−0.01	0.12	−0.03	0.03	−0.02	−0.10	0.02	−0.02	0.01	0.04	0.05
5	−0.02	0.02	0.03	0.07	−0.01	−0.07	0.03	0.01	0.00	−0.04	0.09	0.03	−0.04	0.06	−0.05

Sample \ Subgroup	16	17	18	19	20
1	−0.05	−0.06	−0.04	−0.06	−0.01
2	0.00	−0.02	−0.02	0.00	0.02
3	0.06	0.04	−0.01	0.00	0.05
4	−0.02	0.07	0.03	0.03	0.04
5	−0.01	−0.04	−0.02	0.04	0.03

Figure 4 Hole A, B Alignment Following Process Improvement; Distance Above (+) or Below (−) Nominal

CASE STUDY 5.3
Sample-Size Considerations

PART 1

WT Corporation manufactures crankshafts for 2-L automotive engines. In order to attach a crankshaft to a flywheel, six holes are drilled in the flange end of the crankshaft. These holes are to be drilled 0.3750 ($\frac{3}{8}$) inch in diameter. The holes are not threaded and go all the way through the flange.

All six holes are drilled simultaneously. Every hour, the operator inspects four cranks resulting from four consecutive cycles of the drill press. All six holes on each of four crankshafts are measured and the values are recorded. The values are provided in Figure 1.

HOUR 1

	Crank			
Hole	1	2	3	4
1	3751	3752	3750	3750
2	3752	3751	3750	3752
3	3747	3752	3752	3749
4	3745	3745	3741	3745
5	3752	3751	3750	3752
6	3753	3750	3752	3750
$\overline{X}$-bar	3750	3750	3749	3750
Range	8	7	11	7

HOUR 2

	Crank			
Hole	1	2	3	4
1	3750	3751	3752	3753
2	3749	3752	3754	3752
3	3748	3748	3753	3751
4	3745	3744	3745	3746
5	3750	3754	3753	3750
6	3751	3750	3752	3753
	3749	3750	3752	3751
	6	10	9	7

HOUR 3

	Crank			
Hole	1	2	3	4
1	3751	3749	3752	3753
2	3748	3752	3751	3753
3	3749	3749	3753	3752
4	3745	3744	3744	3743
5	3750	3751	3752	3750
6	3752	3749	3750	3753

HOUR 4

	Crank			
Hole	1	2	3	4
1	3751	3753	3752	3750
2	3750	3751	3751	3751
3	3749	3750	3751	3752
4	3741	3745	3744	3745
5	3752	3755	3751	3750
6	3753	3752	3754	3753

Figure 1

HOUR 5

Hole	Crank 1	2	3	4
1	3751	3752	3754	3753
2	3754	3750	3751	3752
3	3752	3753	3752	3751
4	3745	3746	3747	3746
5	3751	3751	3753	3754
6	3750	3752	3753	3751

HOUR 6

Hole	Crank 1	2	3	4
1	3752	3750	3751	3750
2	3751	3750	3752	3750
3	3753	3750	3753	3750
4	3744	3745	3746	3744
5	3751	3750	3751	3751
6	3750	3751	3750	3750

HOUR 7

Hole	Crank 1	2	3	4
1	3751	3749	3751	3750
2	3752	3750	3754	3751
3	3753	3750	3752	3750
4	3744	3742	3754	3745
5	3750	3750	3750	3750
6	3751	3749	3751	3750

HOUR 8

Hole	Crank 1	2	3	4
1	3752	3751	3753	3750
2	3751	3752	3753	3750
3	3753	3753	3750	3751
4	3744	3746	3745	3744
5	3751	3751	3752	3750
6	3750	3750	3752	3750

HOUR 9

Hole	Crank 1	2	3	4
1	3750	3752	3751	3750
2	3751	3750	3751	3750
3	3752	3750	3750	3749
4	3741	3742	3740	3742
5	3751	3752	3750	3750
6	3752	3754	3750	3754

HOUR 10

Hole	Crank 1	2	3	4
1	3750	3752	3751	3750
2	3750	3751	3752	3750
3	3750	3752	3751	3750
4	3745	3744	3746	3745
5	3750	3752	3752	3751
6	3750	3751	3752	3751

HOUR 11

Hole	Crank 1	2	3	4
1	3750	3750	3751	3750
2	3750	3749	3751	3750
3	3751	3752	3750	3751
4	3742	3744	3743	3744
5	3750	3750	3751	3752
6	3750	3749	3750	3751

HOUR 12

Hole	Crank 1	2	3	4
1	3750	3750	3749	3750
2	3750	3751	3750	3749
3	3750	3750	3751	3751
4	3741	3746	3745	3744
5	3751	3750	3749	3750
6	3750	3750	3751	3750

Figure 1 Continued

Assignment

Following the example set with the first two subgroups, use the data in Figure 1 to calculate $\overline{X}$ and R values based on the crank. Make $\overline{X}$ and R charts using the data. Interpret the control charts.

PART 2

Based on the $\overline{X}$ and R charts, the process appears to be performing smoothly. Two-thirds of the points are near the centerline, and there are no patterns or points out of control.

After shipping this group of crankshafts, you receive a call from your customer. They are very disturbed. Apparently the crankshafts are not of the quality expected. The customer feels that the hole diameters on each crank are not consistent. You point out that the process is under control, as verified by the control charts. In response, your customer suggests that you take another look at the data.

Assignment

Referring to the sections in Chapter 5 discussing variation and choosing a rational subgroup size to be sampled, what information is being provided by the control charts? How were the averages arrived at? What data were combined? What information is the R chart providing?

PART 3

After taking a close look at how the data are organized, you should have discovered that the average is based on summing up the diameters for all six holes on the crank. The average is the average value of the diameter of all the holes on one particular crank. The cycle-to-cycle or crank-to-crank differences can be monitored by comparing one $\overline{X}$ value with another. If the data are separated into the hours in which they were produced, then the hour-to-hour differences can be seen. Both measurements are between-subgroup measurements.

To study within-subgroup measurements, the range chart is used. The range chart displays the variation present between the different hole diameters for a single crank. Reexamine the charts. Discuss the range chart in light of this information. What does the range tell you about the variation present in the average values?

Assignment

Since the study of the data from this point of view has not yielded an answer to the customer's concerns, perhaps it would be beneficial to determine: What question are we trying to answer? What aspect of product performance is important to the customer?

PART 4

In their earlier discussion, the customer mentioned that the hole diameters on each crank are not consistent. Although the Range chart reveals that significant within-crank variation is present, the charts you created do not clearly show the variation caused by each individual drill. The data combine information from six holes, which are drilled simultaneously. The subgroups in Figure 1 have been arranged to study the variation present in each crank, not the variation present in each individual drill. To study crank-to-crank and hole-to-hole differences between subgroups, the hole-size data must be reorganized.

Assignment

Using Figure 2 and the first hour's calculations as a guide, recreate the control charts. Interpret the between-subgroup and within-subgroup variation present.

HOUR 1

Hole	Crank 1	Crank 2	Crank 3	Crank 4	AVE	R
1	3751	3752	3750	3750	3751	2
2	3752	3751	3750	3752	3751	2
3	3747	3752	3752	3749	3750	5
4	3745	3745	3741	3745	3744	4
5	3752	3751	3750	3752	3751	2
6	3753	3750	3752	3750	3751	3

HOUR 2

Hole	Crank 1	Crank 2	Crank 3	Crank 4
1	3750	3751	3752	3753
2	3749	3752	3754	3752
3	3748	3748	3753	3751
4	3745	3744	3745	3746
5	3750	3754	3753	3750
6	3751	3750	3752	3753

HOUR 3

Hole	Crank 1	Crank 2	Crank 3	Crank 4
1	3751	3749	3752	3753
2	3748	3752	3751	3753
3	3749	3749	3753	3752
4	3745	3744	3744	3743
5	3750	3751	3752	3750
6	3752	3749	3750	3753

HOUR 4

Hole	Crank 1	Crank 2	Crank 3	Crank 4
1	3751	3753	3752	3750
2	3750	3751	3751	3751
3	3749	3750	3751	3752
4	3741	3745	3744	3745
5	3752	3755	3751	3750
6	3753	3752	3754	3753

HOUR 5

Hole	Crank 1	Crank 2	Crank 3	Crank 4
1	3751	3752	3754	3753
2	3754	3750	3751	3752
3	3752	3753	3752	3751
4	3745	3746	3747	3746
5	3751	3751	3753	3754
6	3750	3752	3753	3751

HOUR 6

Hole	Crank 1	Crank 2	Crank 3	Crank 4
1	3752	3750	3751	3750
2	3751	3750	3752	3750
3	3753	3750	3753	3750
4	3744	3745	3746	3744
5	3751	3750	3751	3751
6	3750	3751	3750	3750

Figure 2

HOUR 7

Hole	Crank 1	Crank 2	Crank 3	Crank 4
1	3751	3749	3751	3750
2	3752	3750	3754	3751
3	3753	3750	3752	3750
4	3744	3742	3754	3745
5	3750	3750	3750	3750
6	3751	3749	3751	3750

HOUR 8

Hole	Crank 1	Crank 2	Crank 3	Crank 4
1	3752	3751	3753	3750
2	3751	3752	3753	3750
3	3753	3753	3750	3751
4	3744	3746	3745	3744
5	3751	3751	3752	3750
6	3750	3750	3752	3750

HOUR 9

Hole	Crank 1	Crank 2	Crank 3	Crank 4
1	3750	3752	3751	3750
2	3751	3750	3751	3750
3	3752	3750	3750	3749
4	3741	3742	3740	3742
5	3751	3752	3750	3750
6	3752	3754	3750	3754

HOUR 10

Hole	Crank 1	Crank 2	Crank 3	Crank 4
1	3750	3752	3751	3750
2	3750	3751	3752	3750
3	3750	3752	3751	3750
4	3745	3744	3746	3745
5	3750	3752	3752	3751
6	3750	3751	3752	3751

HOUR 11

Hole	Crank 1	Crank 2	Crank 3	Crank 4
1	3750	3750	3751	3750
2	3750	3749	3751	3750
3	3751	3752	3750	3751
4	3742	3744	3743	3744
5	3750	3750	3751	3752
6	3750	3749	3750	3751

HOUR 12

Hole	Crank 1	Crank 2	Crank 3	Crank 4
1	3750	3750	3749	3750
2	3750	3751	3750	3749
3	3750	3750	3751	3751
4	3741	3746	3745	3744
5	3751	3750	3749	3750
6	3750	3750	3751	3750

Figure 2 Continued

PART 5

An investigation of the new $\overline{X}$ and R charts reveals that there are significant differences between the sizes of holes being drilled on the crank. Unlike the first charts created with the same data, these charts display a definite pattern. The differences between drill bits can now be easily seen. Hole 4 is consistently being drilled undersize, creating a very noticeable pattern on the $\overline{X}$ chart. It should also be noted that based on a nominal dimension of 0.3750, the other holes are being drilled slightly oversized.

Rearranging the data has clarified the between-subgroup differences that can now be seen in the $\overline{X}$ chart. By focusing on the aspect of the product critical to the customer, the control chart more clearly answers the customer's questions. In this case, the customer is interested in whether or not systematic differences exist between the holes drilled. Investigated from this point of view, we can now understand why the customer is upset. The inconsistent hole sizes cause assembly problems, making it difficult to assemble the cranks and flywheels. Oversize and undersize holes lead to fit problems. To improve the process, each of the drills will need to be adjusted to bring the averages closer to the nominal dimension of 0.3750 inch.

The customer is also interested in whether or not there are cycle-to-cycle differences. In other words, are each of the six individual drills drilling holes of consistent size from cycle to cycle? These within-subgroup differences can be seen on the R chart. The variation present in the hole sizes is relatively consistent. Further reductions in the variation present in the process will decrease assembly difficulties later.

Assignment

Creating control charts separating data for each drill bit will enhance understanding the process by increasing the sensitivity of the charts. When the data are separated into six charts, the users of the charts will be able to determine quickly how each drill is operating. Placing data for all six drill bits on different charts unclutters the data and more clearly shows the process centering and variation present for each drill.

Reuse your calculations of $\overline{X}$ and R from Figure 2 to create and interpret six separate control charts. Using the same values, separate them according to which drill they are from. Describe the process centering and spread present for each drill.

6

Process Capability

■ *Learning Opportunities:*

1. To gain an understanding of the relationship of individual values and their averages
2. To understand the difference between specification limits and control limits
3. To learn to calculate and interpret the process capability indices: C_p, C_r, and C_{pk} ■

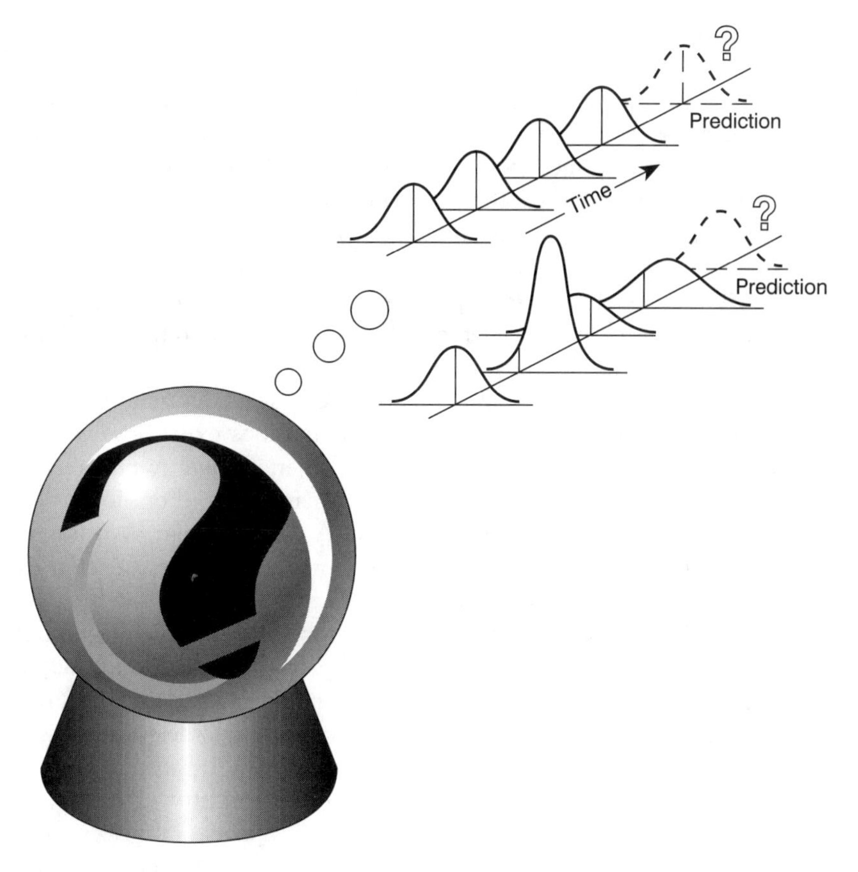

Crystal Ball

It certainly would be interesting to have some insight into the future. We would know which questions were to be on a test, whether a traffic jam existed on the road we wished to take, which investments would make us the greatest return on our money. When manufacturing a product or providing a service, decision makers would like to know if the product or service provided in the future is capable of meeting specifications. Control charts and their calculations, while they can't replace a crystal ball, can give those who use them insight into the capability of a process and what future production might be like.

Process capability *refers to the ability of a process to meet the specifications set by the customer or designer.* As discussed in the last chapter, variation affects a process and may prevent the process from producing products or services that meet customer specifications. Knowing process capability gives insight into whether or not the process will be able to meet future demands placed on it. We have all encountered limitations in process capability. Consider what happens when a student is given several major assignments from different classes that are all due on the same date. Since there are only so many hours in the day, the capability of the process (the student) will be severely affected by the overload of assignments. The student may choose to complete each of the assignments as best he can, resulting in a group of assignments of average work. Or the student may choose to perform superior work on one project and mediocre work on the remainder. Or the student may decide to turn some of the projects in late. Any number of combinations exist; however, completion of all of the assignments on the due date with perfect quality probably won't happen. The process is incapable of meeting the demands placed on it.

In industry, the concept is similar. A customer may ask for part tolerances so fine that the machines are not capable of producing to that level of exactness. In assembly, it is difficult to assemble products that vary from the high side of the specification limits to the low side of the specifications. An undersized part A may not mate correctly with an oversized part B. A patron at a restaurant may wish to be served a full lunch within ten minutes of arriving during the noon-hour rush. If no seats are available and the kitchen is behind in cooking orders, then the process will not be capable of meeting the customer's demands.

Determining the process capability aids industry in meeting their customer demands. Manufacturers of products and providers of services who know the process capability can pass this information on to the customer. It can then be used to assist in decisions concerning product or process specifications, appropriate production methods, equipment to be used, and time commitments. Process capability studies can help determine the product uniformity around a target. Reducing process variability and creating consistent quality increase the viability of predictions of future process performance (Figure 6.1).

INDIVIDUAL VALUES COMPARED WITH AVERAGES

Process capability is based on the performance of individual products or services against specifications. In quality assurance, rather than execute the time-consuming task of measuring or checking each product produced or service provided, we take samples to study the process. Analysts must use information from the samples to determine the behavior of individuals in a process. To do this, the analyst needs an understanding of the relationship between individual values and their averages.

Subgroup sample averages are composed of individual values. Table 6.1 repeats the tally of individual and average values of flat round plate thickness data from Table 5.1. With this actual production line data, two frequency diagrams have been created in Figure 6.2. One frequency diagram is constructed of individual values, the other is made up of subgroup averages. Both distributions are approximately normal. When the two diagrams are compared, the averages are grouped closer to the center

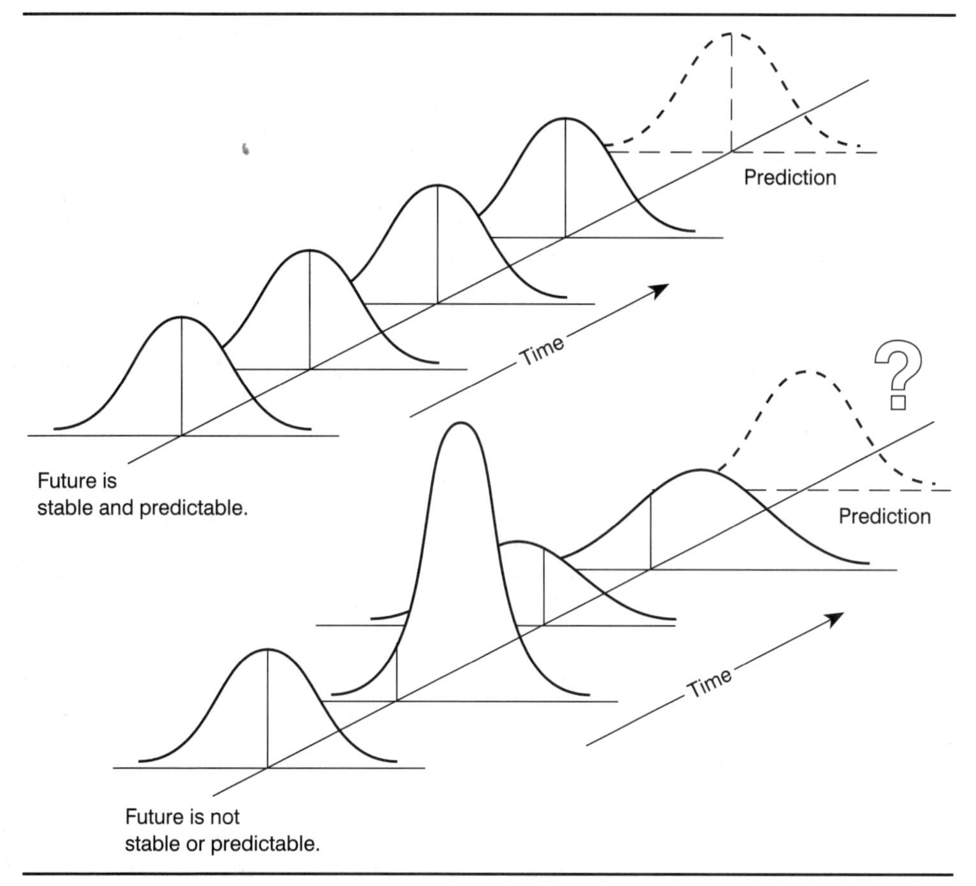

Figure 6.1 Future Predictions

value than are the individual values, as described by the central limit theorem. Average values smooth out the highs and lows associated with individuals. The important difference to note is that individual values spread much more widely than their averages. This comparison will be important to keep in mind when comparing the behavior of averages within control limits and that of individual values with specification limits (Figure 6.3).

ESTIMATION OF POPULATION σ FROM SAMPLE DATA

Sample values and their averages can give insight into the behavior of an entire population. It is possible to estimate the spread of the population of individuals using the sample data. The relationship between the population standard deviation (σ) and the standard deviation of the subgroup averages ($\sigma_{\overline{X}}$ is given by the formula

$$\sigma_{\overline{X}} = \frac{\sigma}{\sqrt{n}}$$

Table 6.1 Flat Round Plate Thickness: Sums and Averages

						ΣX_i	$\bar{X}$	$\bar{R}$
Sample 1	0.0625	0.0626	0.0624	0.0625	0.0627	0.3127	0.0625	0.0003
Sample 2	0.0624	0.0623	0.0624	0.0626	0.0625	0.3122	0.0624	0.0003
Sample 3	0.0622	0.0625	0.0623	0.0625	0.0626	0.3121	0.0624	0.0004
Sample 4	0.0624	0.0623	0.0620	0.0623	0.0624	0.3114	0.0623	0.0004
Sample 5	0.0621	0.0621	0.0622	0.0625	0.0624	0.3113	0.0623	0.0004
Sample 6	0.0628	0.0626	0.0625	0.0626	0.0627	0.3132	0.0626	0.0003
Sample 7	0.0624	0.0627	0.0625	0.0624	0.0626	0.3126	0.0625	0.0003
Sample 8	0.0624	0.0625	0.0625	0.0626	0.0626	0.3126	0.0625	0.0002
Sample 9	0.0627	0.0628	0.0626	0.0625	0.0627	0.3133	0.0627	0.0003
Sample 10	0.0625	0.0626	0.0628	0.0626	0.0627	0.3132	0.0626	0.0003
Sample 11	0.0625	0.0624	0.0626	0.0626	0.0626	0.3127	0.0625	0.0002
Sample 12	0.0630	0.0628	0.0627	0.0625	0.0627	0.3134	0.0627	0.0005
Sample 13	0.0627	0.0626	0.0628	0.0627	0.0626	0.3137	0.0627	0.0002
Sample 14	0.0626	0.0626	0.0625	0.0626	0.0627	0.3130	0.0626	0.0002
Sample 15	0.0628	0.0627	0.0626	0.0625	0.0626	0.3132	0.0626	0.0003
Sample 16	0.0625	0.0626	0.0625	0.0628	0.0627	0.3131	0.0626	0.0003
Sample 17	0.0624	0.0626	0.0624	0.0625	0.0627	0.3126	0.0625	0.0003
Sample 18	0.0628	0.0627	0.0628	0.0626	0.0630	0.3139	0.0627	0.0004
Sample 19	0.0627	0.0626	0.0628	0.0625	0.0627	0.3133	0.0627	0.0003
Sample 20	0.0626	0.0625	0.0626	0.0625	0.0627	0.3129	0.0626	0.0002
Sample 21	0.0627	0.0626	0.0628	0.0625	0.0627	0.3133	0.0627	0.0003
Sample 22	0.0625	0.0626	0.0628	0.0625	0.0627	0.3131	0.0626	0.0003
Sample 23	0.0628	0.0626	0.0627	0.0630	0.0627	0.3138	0.0628	0.0004
Sample 24	0.0625	0.0631	0.0630	0.0628	0.0627	0.3141	0.0628	0.0006
Sample 25	0.0627	0.0630	0.0631	0.0628	0.0627	0.3143	0.0629	0.0004
Sample 26	0.0632	0.0628	0.0631	0.0628	0.0627	0.3149	0.0630	0.0005
Sample 27	0.0630	0.0628	0.0631	0.0628	0.0627	0.3144	0.0629	0.0004
Sample 28	0.0632	0.0632	0.0628	0.0631	0.0630	0.3153	0.0631	0.0004
Sample 29	0.0630	0.0628	0.0631	0.0632	0.0631	0.3152	0.0630	0.0004
Sample 30	0.0632	0.0631	0.0630	0.0628	0.0628	0.3149	0.0630	0.0004
						9.3997		

where

$\sigma_{\bar{X}}$ = standard deviation of subgroup averages
σ = population standard deviation of the lot from which the sample was drawn
n = subgroup sample size

The population standard deviation, σ, is determined by measuring the individuals. This necessitates measuring every value. To avoid complicated calculations, if

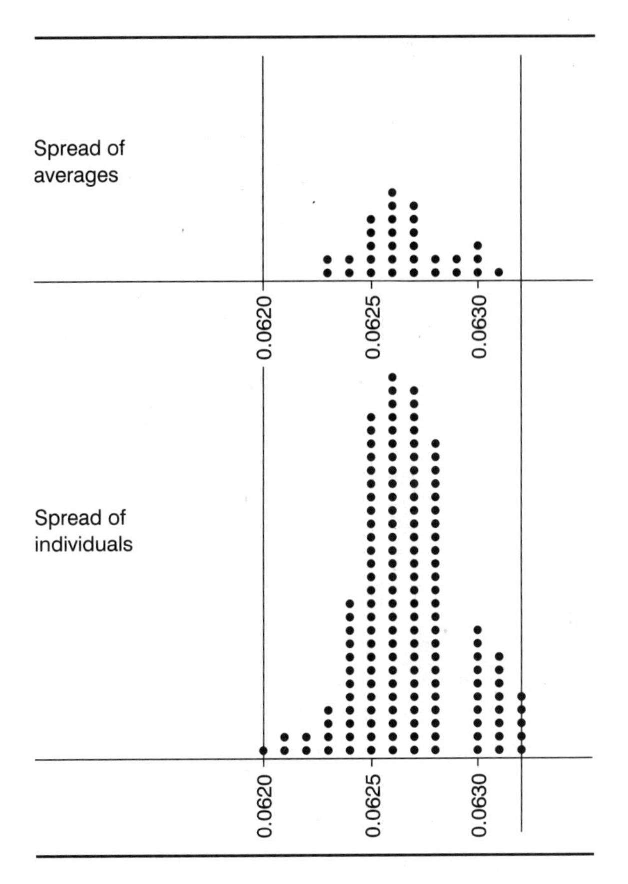

Figure 6.2 Normal Curves for Individuals and Averages

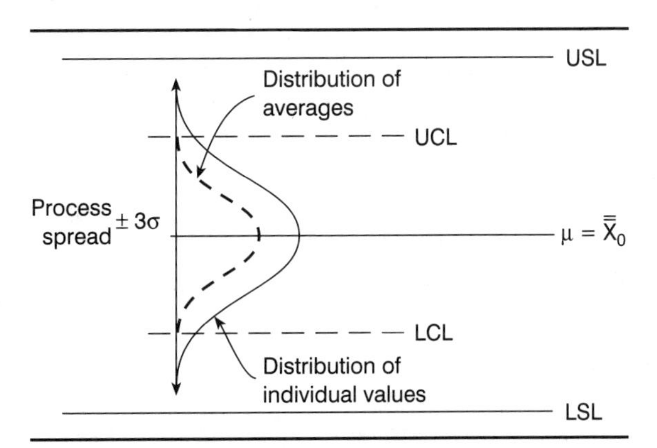

Figure 6.3 Comparison of the Spread of Individual Values with Averages

the process can be assumed to be normal, the population standard deviation can be estimated from either the standard deviation associated with the sample standard deviation (s) or the range (R):

$$\hat{\sigma} = \frac{\bar{s}}{c_4} \quad \text{or} \quad \hat{\sigma} = \frac{\overline{R}}{d_2}$$

where

$\hat{\sigma}$ = estimate of population standard deviation
$\bar{s}$ = sample standard deviation calculated from process
$\overline{R}$ = average range of subgroups
c_4 as found in Appendix 2
d_2 as found in Appendix 2

Because of the estimators (c_4 and d_2), these two formulas will yield similar but not identical values for $\hat{\sigma}$.

CONTROL LIMITS AND SPECIFICATION LIMITS

A process performing within statistical control is said to be in control or under control. A process is in control only when its process centering remains constant over time. The amount of variation present in the process must also remain constant over time. If both are true, then the behavior of the process will be predictable. As discussed in Chapter 5, it exhibits the following characteristics:

1. Two-thirds of the points are near the center value.
2. A few of the points are close to the center value.
3. The points appear to float back and forth across the centerline.
4. The points are balanced (in roughly equal numbers) on both sides of the centerline.
5. There are no points beyond the control limits.
6. There are no patterns or trends on the chart.

A process under control meets these guidelines and can be expected to perform in the same fashion in the near future. The control limits found on the charts represent what the process is capable of producing. The control limits on the $\overline{X}$ chart represent process centering. The R chart limits represent the amount of variation present in the process.

It is important to note that a process in statistical control will not necessarily meet specifications as established by the customer. There is a difference between a process conforming to specifications and a process performing within statistical control. Control limits and specification limits are two separate concepts, which may, at first, seem difficult to separate. The importance of understanding the difference between the two cannot be overestimated. Established during the design process or from customer requests, specifications communicate what the customers expect, want, or need from the process. Specifications can be considered the voice of the customer. Control limits are the voice of the process. Based on averages, control limits are determined by using either s or R, the measures of the variation present in the current process. Control limits are a prediction of the variation that the process will exhibit in the near future. The difference between specifications and control limits is that specifications relay wishes and control limits tell of reality.

It is important to note that a process can be out of control and within specification or under control and out of specification. For explanatory purposes, both control limits and specification limits appear on the charts in Figure 6.4, a practice not to be followed in industry. The variation in Figure 6.4*a* exceeds control limits marking the expected process variation but not the specification limits. While the process is out of control, for the time being the customer's needs are being met. The process in Figure 6.4*b* is under control and within the control limits, but the specification limits do not correspond with the control limits. This situation reveals that the process is performing to the best of its abilities, but not well enough to meet the specifications set by the customer or designer. In this case (Figure 6.4*b*), it may be possible to shift the process centering to meet the specifications.

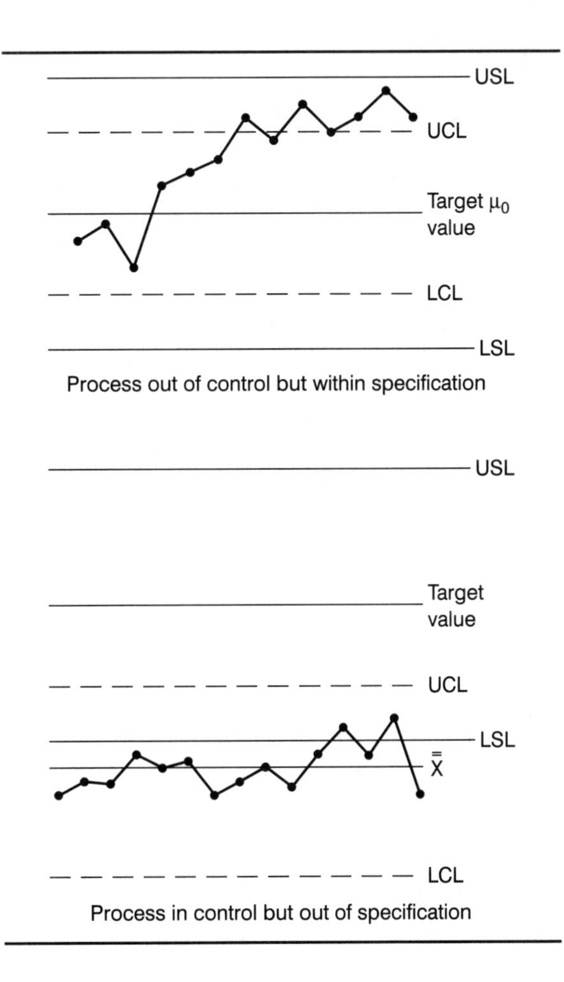

Figure 6.4 Control Limits versus Specification Limits

As we learned in Chapter 4 during the discussion of the central limit theorem and in Figure 6.2, the spread of individual values is wider than the spread of the averages. For this reason, control limits can not be compared directly with specification limits. An $\overline{X}$ chart does not reflect how widely the individual values composing the plotted averages spread. This is one reason why an R or s chart is always used in conjunction with the $\overline{X}$ chart. The spread of the individual data can be seen only by observing what is happening on the R or s chart. If the values on the R or s chart are large, then the variation associated with the average is large. Figure 6.5 shows a control chart created using the concepts from Chapter 5 and the values in Table 6.1. Individual values are also plotted on this chart. Note where the individual values fall in relation to the control limits established for the process. The individuals spread more widely than the averages and follow the pattern established by the R chart. Overlaying the R chart pattern on the $\overline{X}$ chart can significantly increase the understanding of how the process is performing.

EXAMPLE 6.1 Using the $\overline{X}$ and R Charts to Assess the Process

Engineers in a materials testing lab have been studying the results of a tensile strength test. To better understand the process performance and the spread of the individual values that compose the averages, they have overlaid the R chart pattern on the $\overline{X}$ chart. To do this easily, they divided the $\overline{X}$ chart into four sections (Figure 6.5) chosen on the basis of how the data on the $\overline{X}$ chart appear to have grouped. $\overline{X}$ values in section A are centered at the mean and are very similar. In section B, the $\overline{X}$ values are above the mean and more spread out. Section C values have a slight downward shift in level. In section D, the $\overline{X}$ values are below the centerline.

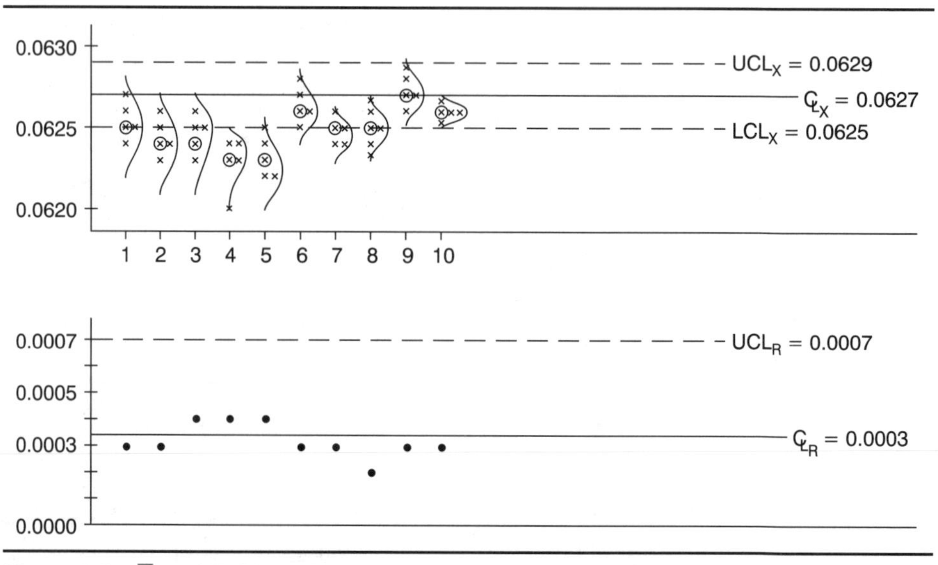

Figure 6.5 $\overline{X}$ and R Chart Showing Averages and Individuals

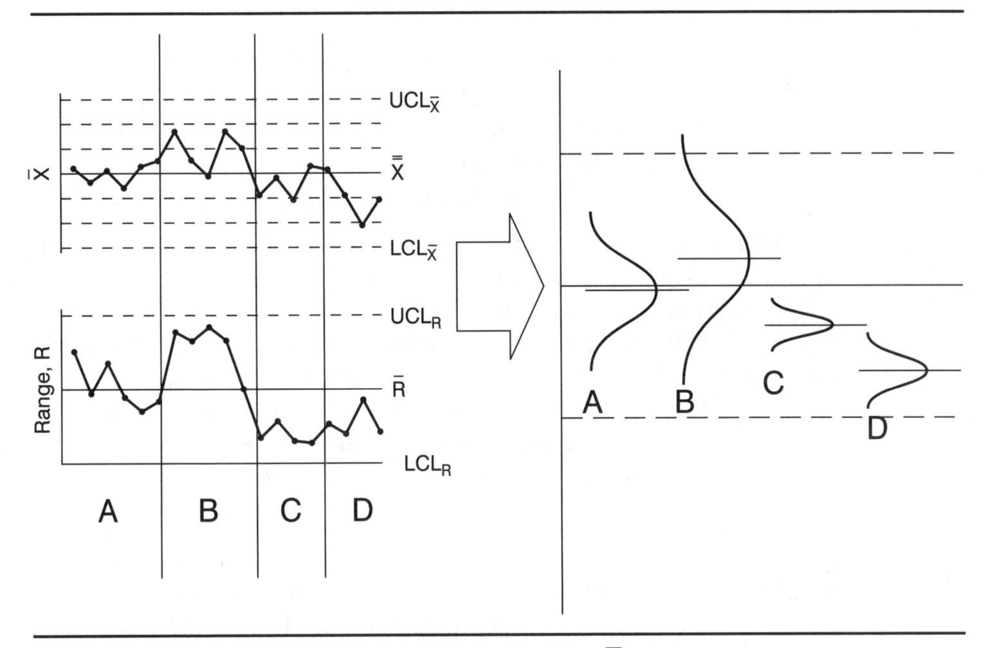

Figure 6.6 Overlaying the R Chart Pattern on the $\bar{X}$ Chart

Studying the sections on the R chart reveals that the spread of the data is changing (Figure 6.6). The values in section A have an average amount of variation, denoted by the normal curve corresponding to section A. Variation increases in section B, resulting in a much broader spread on the normal curve (B). The significant decrease in variation in sections C and D is shown by narrow, peaked distributions. Note that the location of the center of these distributions on the graph varies according to the behavior of the averages on the $\bar{X}$ chart. This is because the $\bar{X}$ chart shows the location of the center of the product produced, or in this case, tested. The R chart describes the spread of the individuals on the $\bar{X}$ chart.

Before making a comparison between control limits and specification limits, it is necessary to estimate the spread of the individual values.

THE 6σ SPREAD VERSUS SPECIFICATION LIMITS

The spread of the individuals in a process, 6σ, is the measure used to compare the realities of production with the desires of the customers. The process standard deviation is based on either s or R from control chart data:

$$\hat{\sigma} = \frac{\bar{s}}{c_4} \quad \text{or} \quad \hat{\sigma} = \frac{\bar{R}}{d_2}$$

where

$\hat{\sigma}$ = estimate of production standard deviation
$\bar{s}$ = sample standard deviation calculated from process
$\bar{R}$ = average range of subgroups calculated from process
c_4 as found in Appendix 2
d_2 as found in Appendix 2

Remember, because of the estimators (c_4 and d_2), these two formulas will yield similar but not identical values for $\hat{\sigma}$.

Specification limits, the allowable spread of the individuals, are compared with the 6σ spread of the process to determine how capable the process is of meeting the specifications. Three different situations can exist when specifications and 6σ are compared: (1) the 6σ spread can be less than the spread of the specification limits; (2) the 6σ spread can be equal to the spread of the specification limits; (3) the 6σ spread can be greater than the spread of the specification limits.

Case I: 6σ < USL − LSL. This is the most desirable case. Figure 6.7 illustrates this relationship. The control limits have been placed on the diagram, as well as the spread of the process averages (dotted line). The 6σ spread of the process individuals is shown by the solid line. As expected, the spread of the individual values is greater than the spread of the averages; however, the values are still within the specification limits. The 6σ spread of the individuals is less than the spread of the specifications. This allows for more room for process shifts while staying within the specifications. Notice that even if the process drifts out of control (Figure 6.7*b*), the change must be dramatic before the parts are considered out of specification.

Case II: 6σ = USL − LSL. In this situation, 6σ is equal to the tolerance (Figure 6.8*a*). As long as the process remains in control and centered, with no change in

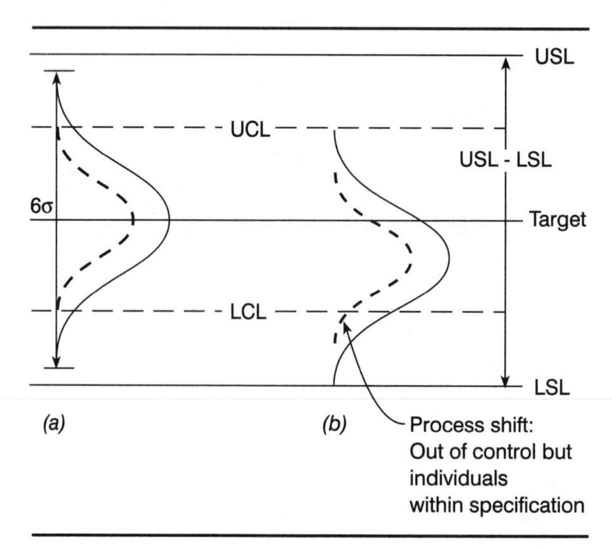

Figure 6.7 Case I: 6σ < USL − LSL

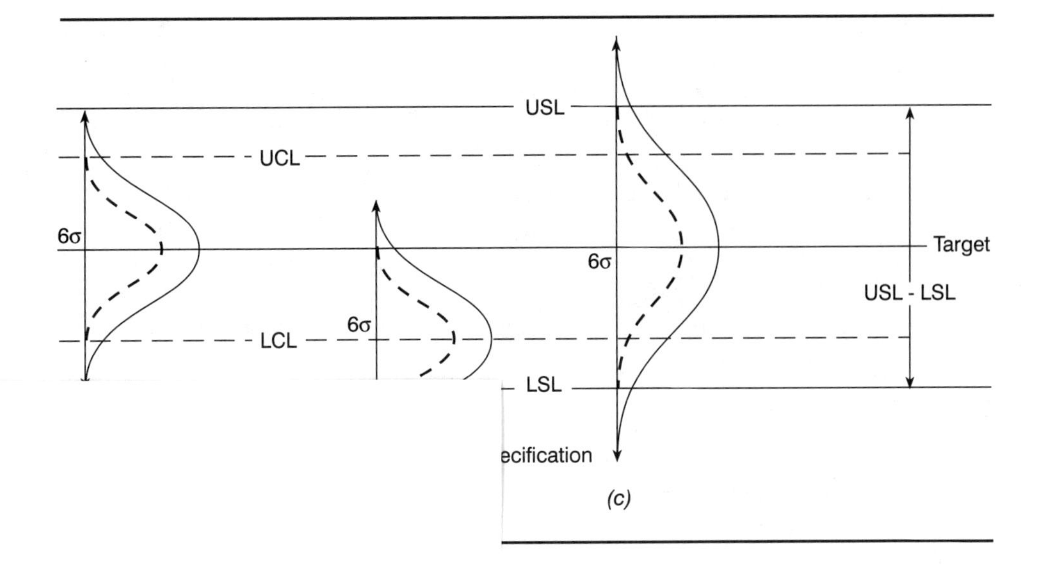

ll be within specification. However, a shift in
ılt in the production of parts that are out of
on present in the process also creates an out-

ne that the 6σ spread is greater than the tol-
exists (Figure 6.9*a*). Even though the process
riation, it is incapable of meeting the specifi-
this problem, management intervention will
ɔcess to decrease the variation or to recenter

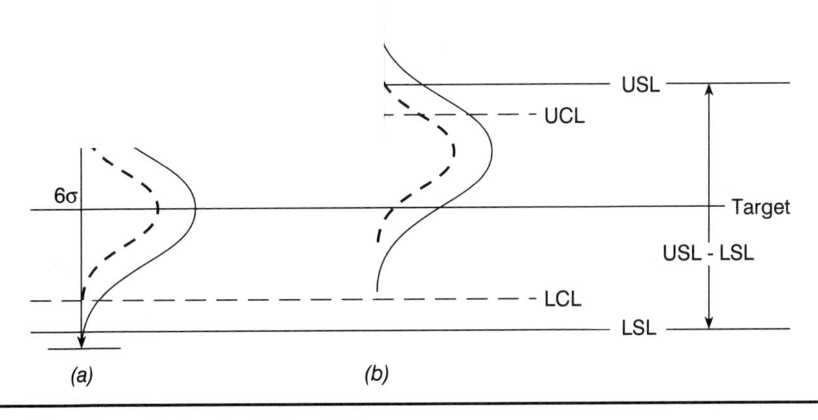

Figure 6.9 Case III: 6σ > USL − LSL

the process if necessary. The capability of the process cannot be improved without changing the existing process. To achieve a substantial reduction in the standard deviation or spread of the data, management will have to authorize the utilization of different materials, the overhaul of the machine, the purchase of a new machine, the retraining of the operator, or other significant changes to the process. Other, less desirable approaches to dealing with this problem are to perform 100 percent inspection on the product, increase the specification limits, or shift the process average so that all of the nonconforming products occur at one end of the distribution (Figure 6.9*b*). In certain cases, shifting the process average can eliminate scrap and increase the amount of rework, thus saving scrap costs by increasing rework costs.

CALCULATING PROCESS CAPABILITY INDICES

Process capability indices are mathematical ratios that quantify the ability of a process to produce products within the specifications. The capability indices compare the spread of the individuals created by the process with the specification limits set by the customer or designer. The 6σ spread of the individuals can be calculated for a new process that has not produced a significant number of parts or for a process currently under operation. In either case, a true 6σ value cannot be determined until the process has achieved stability, as described by the $\overline{X}$ and R charts or $\overline{X}$ and s charts. If the process is not stable, the calculated values may or may not be representative of the true process capability.

Calculating $6\hat{\sigma}$

Assuming the process is under statistical control, we use the following method to calculate $6\hat{\sigma}$ of a new process:

1. Take at least 20 subgroups of sample size 4 for a total of 80 measurements.
2. Calculate the sample standard deviation, s_i, for each subgroup.
3. Calculate the average sample standard deviation, $\bar{s}$:

$$\bar{s} = \frac{\sum_{i=1}^{m} s_i}{m}$$

where

s_i = standard deviation for each subgroup
m = number of subgroups

4. Calculate the estimate of the population standard deviation:

$$\hat{\sigma} = \frac{\bar{s}}{c_4}$$

where c_4 is obtained from Appendix 2.

5. Multiply the population standard deviation by 6.

EXAMPLE 6.2 Calculating $6\hat{\sigma}$

The individuals monitoring the process making shafts for printers wish to calculate $6\hat{\sigma}$. They intend to use the data gathered during a recent production run (Table 6.2). They have 21 subgroups of sample size 5 for a total of 105 measurements, more than the 80 recommended.

They calculate the sample standard deviation, s_i, for each subgroup (Table 6.2). From these values they calculate the average sample standard deviation, $\bar{s}$:

$$\bar{s} = \frac{\sum_{i=1}^{m} s_i}{m} = \frac{0.031 + 0.029 + \cdots + 0.015}{21}$$
$$= \frac{0.414}{21} = 0.02$$

The next step involves calculating the estimate of the population standard deviation:

$$\hat{\sigma} = \frac{\bar{s}}{c_4} = \frac{0.02}{0.9400} = 0.021$$

The value of c_4 is obtained from Appendix 2 and is based on a sample size of 5.

Table 6.2 $\bar{X}$ and s Values of Shafts

Sample Number	X_i					$\bar{X}$	s
1	11.950	12.000	12.030	11.980	12.010	11.994	0.031
2	12.030	12.020	11.960	12.000	11.980	11.998	0.029
3	12.010	12.000	11.970	11.980	12.000	11.992	0.016
4	11.970	11.980	12.000	12.030	11.990	11.994	0.023
5	12.000	12.010	12.020	12.030	12.020	12.016	0.011
6	11.980	11.980	12.000	12.010	11.990	11.992	0.013
7	12.000	12.010	12.030	12.000	11.980	12.004	0.018
8	12.000	12.010	12.040	12.000	12.020	12.014	0.017
9	12.000	12.020	11.960	12.000	11.980	11.992	0.023
10	12.020	12.000	11.970	12.050	12.000	12.008	0.030
11	11.980	11.970	11.960	11.950	12.000	11.972	0.019
12	11.920	11.950	11.920	11.940	11.960	11.938	0.018
13	11.980	11.930	11.940	11.950	11.960	11.952	0.019
14	11.990	11.930	11.940	11.950	11.960	11.954	0.023
15	12.000	11.980	11.990	11.950	11.930	11.970	0.029
16	12.000	11.980	11.970	11.960	11.990	11.980	0.016
17	12.020	11.980	11.970	11.980	11.990	11.988	0.019
18	12.000	12.010	12.020	12.010	11.990	12.006	0.011
19	11.970	12.030	12.000	12.010	11.990	12.000	0.022
20	11.990	12.010	12.020	12.000	12.010	12.006	0.011
21	12.000	11.980	11.990	11.990	12.020	11.996	0.015

As the final step they multiply the population standard deviation by 6:

$$6\hat{\sigma} = 6(0.021) = 0.126$$

This value can be compared with the spread of the specifications to determine how the individual products produced by the process compare with the specifications set by the designer.

A second method of calculating $6\hat{\sigma}$ is to use the data from a control chart. Once again, it is assumed that the process is under statistical control, with no unusual patterns of variation.

1. Take the past 10 subgroups, sample size of 4 or more.
2. Calculate the range, R, for each subgroup.
3. Calculate the average range, $\overline{R}$:

$$\overline{R} = \frac{\sum_{i=1}^{m} R_i}{m}$$

where

R_i = individual range values for the subgroups
m = number of subgroups

4. Calculate the estimate of the population standard deviation, $\hat{\sigma}$:

$$\hat{\sigma} = \frac{\overline{R}}{d_2}$$

where d_2 is obtained from Appendix 2.

5. Multiply the population standard deviation by 6.

Using more than 10 subgroups will improve the accuracy of the calculations.

EXAMPLE 6.3 Calculating 6σ for the Flat Round Plate

The engineers use the data in Table 6.1 to calculate $6\hat{\sigma}$ for the flat round plate. Thirty subgroups of sample size 5 and their ranges are used to calculate the average range, $\overline{R}$:

$$\overline{R} = \frac{\sum_{i=1}^{m} R_i}{m} = \frac{0.0003 + 0.0003 + \cdots + 0.0004}{21}$$
$$= 0.0003$$

Next, the engineers calculate the estimate of the population standard deviation, $\hat{\sigma}$:

$$\hat{\sigma} = \frac{\overline{R}}{d_2} = \frac{0.0003}{2.326} = 0.0001$$

Using a sample size of 5, they take the value for d_2 from Appendix 2.

To determining $6\hat{\sigma}$, they multiply the population standard deviation by 6:

$$6\hat{\sigma} = 6(0.0001) = 0.0006$$

They now compare this value with the specification limits to determine how well the process is performing.

The Capability Index

Once calculated, the σ values can be used to determine several indices related to process capability. The ***capability index*** C_p *is the ratio of tolerance (USL − LSL) and* $6\hat{\sigma}$:

$$C_p = \frac{USL - LSL}{6\hat{\sigma}}$$

where

C_p = capability index
USL − LSL = upper specification limit − lower specification limit, or tolerance

The capability index is interpreted as follows: If the capability index is larger than 1.00, a Case I situation exists (Figure 6.7*a*). This is desirable. The greater this value, the better. If the capability index is equal to 1.00, then a Case II situation exists (Figure 6.8*a*). This is not optimal, but it is feasible. If the capability index is less than 1.00, then a Case III situation exists (Figure 6.9*a*). Values of less than 1 are undesirable and reflect the process's inability to meet the specifications.

EXAMPLE 6.4 Finding the Capability Index I

The engineers in Example 6.3 are working with specification limits of 0.0625 ± 0.0001. The upper specification limit is 0.0626 and the lower specification limit is 0.0624. They now calculate C_p:

$$C_p = \frac{USL - LSL}{6\hat{\sigma}} = \frac{0.0626 - 0.0624}{0.0006}$$
$$= 0.3333$$

A low value like this means that the process is not capable of meeting the demands placed on it by the customer's specifications. Major changes will need to occur to improve the process performance and meet specifications.

EXAMPLE 6.5 Finding the Capability Index II

Those monitoring the process in Example 6.2 want to calculate its C_p. They use specification limits of USL = 12.05, LSL = 11.95:

$$C_p = \frac{USL - LSL}{6\hat{\sigma}} = \frac{12.05 - 11.95}{0.126}$$
$$= 0.794$$

This process is not capable of meeting the demands placed on it.

The Capability Ratio

Another indicator of process capability is called the ***capability ratio.*** This ratio is similar to the capability index, though it reverses the numerator and the denominator. It is defined as

$$C_r = \frac{6\hat{\sigma}}{USL - LSL}$$

A capability ratio less than 1 is the most desirable situation. The larger the ratio, the less capable the process is of meeting specifications. Be aware that it is easy to confuse the two indices. The most commonly used index is the capability index.

C_{pk}

The centering of the process is also an important concept when discussing process capability. As shown by the Taguchi loss function described in Chapter 2, a process operating in the center of the specifications set by the designer is usually more desirable than one that is consistently producing parts to the high or low side of the specification limits. In Figure 6.10, all three distributions have the same C_p index value of 1.3. In the first situation, the process is centered as well as capable. In the second, a further upward shift in the process would result in an out-of-specification situation. The reverse holds true in the third situation. Though each of these processes have the same capability index, they represent three different scenarios. The capability indices introduced so far do not take into account the centering of the process. *The ratio that reflects how the process is performing in terms of a nominal, center, or target value is* $\boldsymbol{C_{pk}}$. C_{pk} can be calculated using the following formula:

$$C_{pk} = \frac{Z(\min)}{3}$$

where Z(min) is the smaller of

$$Z(USL) = \frac{USL - \overline{X}}{\hat{\sigma}}$$

$$\text{or } Z(LSL) = \frac{\overline{X} - LSL}{\hat{\sigma}}$$

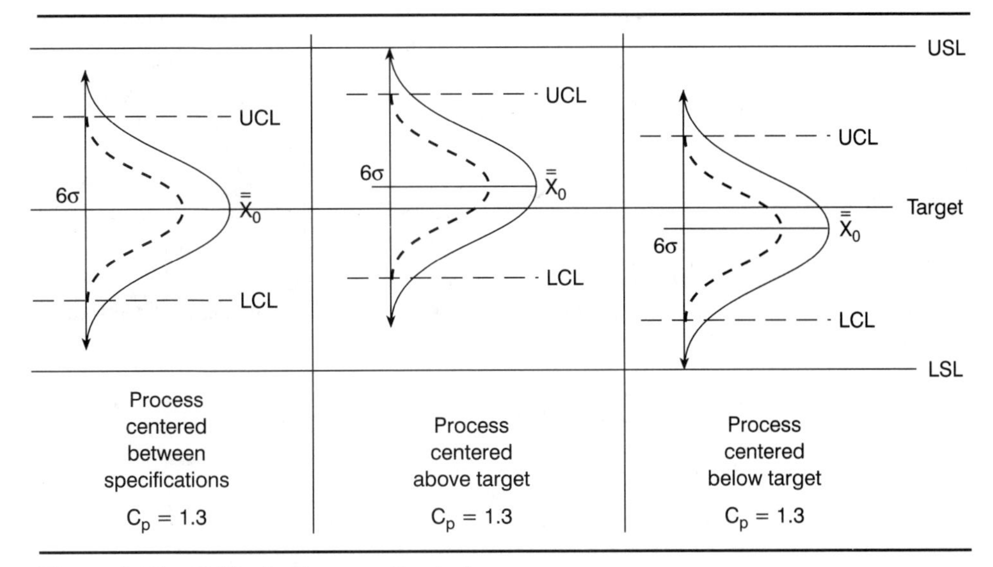

Figure 6.10 Shifts in Process Centering

When $C_{pk} = C_p$ the process is centered. Figure 6.11 illustrates C_p and C_{pk} values for a process that is centered and one that is off center. The relationships between C_p and C_{pk} are as follows:

1. The C_p value does not reflect process centering.
2. When the process is centered, $C_p = C_{pk}$.
3. C_{pk} is always less than or equal to C_p.
4. When C_{pk} has a value of 1.00, it indicates the process is producing product that conforms to specifications.
5. When C_{pk} has a value less than 1.00, it indicates the process is producing product that does not conform to specifications.
6. A C_p value of less than 1.00 indicates that the process is not capable.
7. A C_{pk} value of zero indicates the process average is equal to one of the specification limits.
8. A negative C_{pk} value indicates that the average is outside the specification limits.

EXAMPLE 6.6 Finding C_{pk}

Determine the C_{pk} for Example 6.2. The average, $\bar{X}$, is equal to 11.990.

$$C_{pk} = \frac{Z(\min)}{3}$$

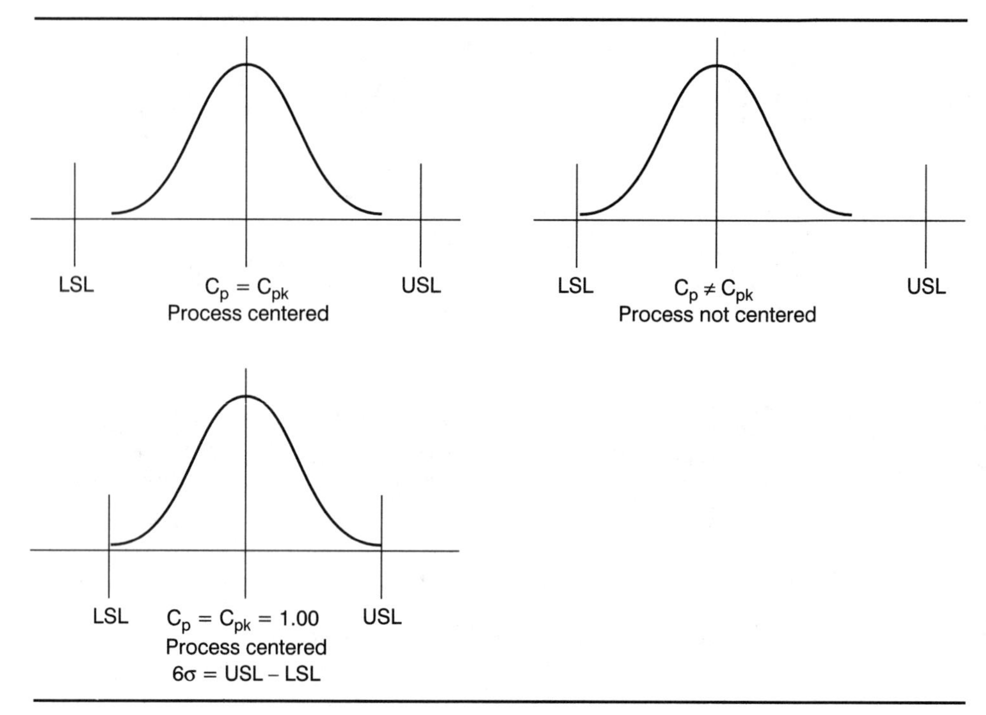

Figure 6.11 Process Centering: C_p versus C_{pk}

where

$$Z(\min) = \text{smaller of } \frac{(USL - \overline{X})}{\hat{\sigma}} \text{ or } \frac{(\overline{X} - LSL)}{\hat{\sigma}}$$

$$Z(USL) = \frac{(12.050 - 11.990)}{0.021} = 2.857$$

$$Z(LSL) = \frac{(11.990 - 11.950)}{0.021} = 1.905$$

$$C_{pk} = \frac{1.905}{3} = 0.635$$

A C_{pk} value of less than 1 means that the process is not capable. Because the C_p value (0.794, from Example 6.5) and the C_{pk} value (0.635) are not equal, the process is not centered between the specification limits.

EXAMPLE 6.7 Is the Process Capable?

The production of integrated circuits by etching them onto silicon wafers requires silicon wafers of consistent thickness. Recall from Example 5.9 that recent quality problems in Whisks' silicon wafer line B necessitated creating an $\overline{X}$ and R chart with the data in Figure 6.12. Though Whisks' control charts (Figures 6.13, 6.14) exhibit good

					X-bar	*R*
Subgroup 1	0.2500	0.2510	0.2490	0.2500	0.2500	0.002
Subgroup 2	0.2510	0.2490	0.2490	0.2520	0.2503	0.003
Subgroup 3	0.2510	0.2490	0.2510	0.2480	0.2498	0.003
Subgroup 4	0.2490	0.2470	0.2520	0.2480	0.2490	0.005
Subgroup 5	0.2500	0.2470	0.2500	0.2520	0.2498	0.005
Subgroup 6	0.2510	0.2520	0.2490	0.2510	0.2508	0.003
Subgroup 7	0.2510	0.2480	0.2500	0.2500	0.2498	0.003
Subgroup 8	0.2500	0.2490	0.2490	0.2520	0.2500	0.003
Subgroup 9	0.2500	0.2470	0.2500	0.2510	0.2495	0.004
Subgroup 10	0.2480	0.2480	0.2510	0.2530	0.2500	0.005
Subgroup 11	0.2500	0.2500	0.2500	0.2530	0.2508	0.003
Subgroup 12	0.2510	0.2490	0.2510	0.2540	0.2513	0.005
Subgroup 13	0.2500	0.2470	0.2500	0.2510	0.2495	0.004
Subgroup 14	0.2500	0.2500	0.2490	0.2520	0.2503	0.003
Subgroup 15	0.2500	0.2470	0.2500	0.2510	0.2495	0.004

Figure 6.12 Silicon Wafer Thickness

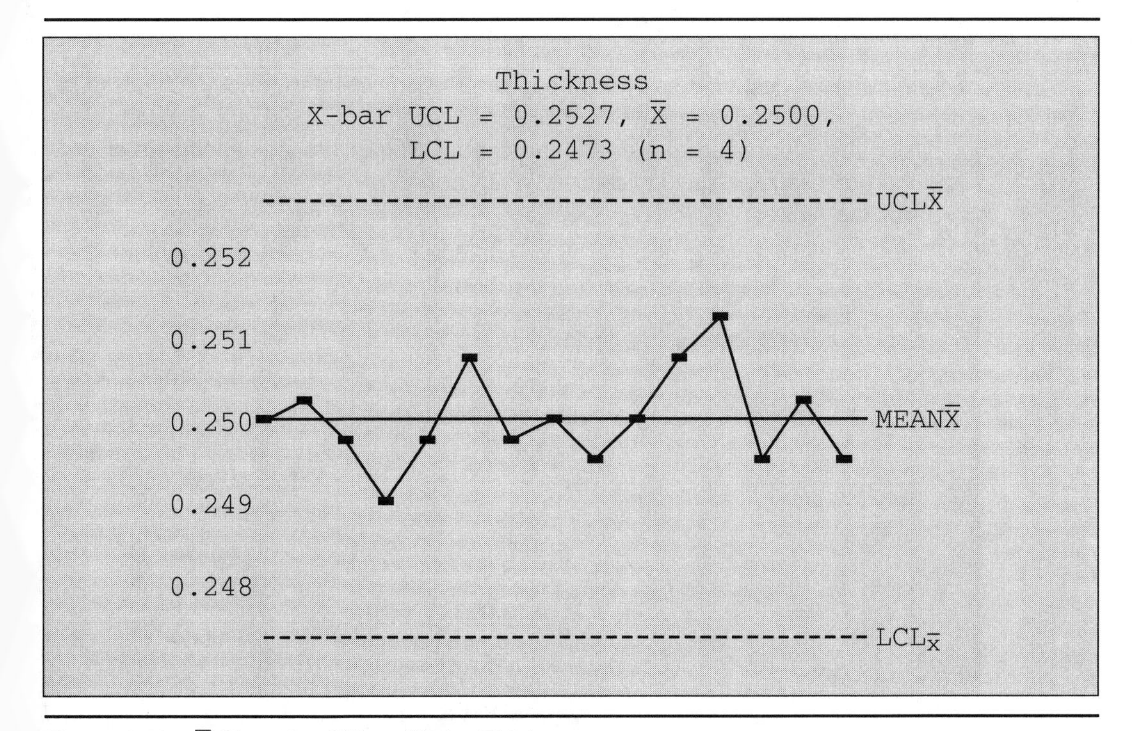

Figure 6.13 $\overline{X}$ Chart for Silicon Wafer Thickness

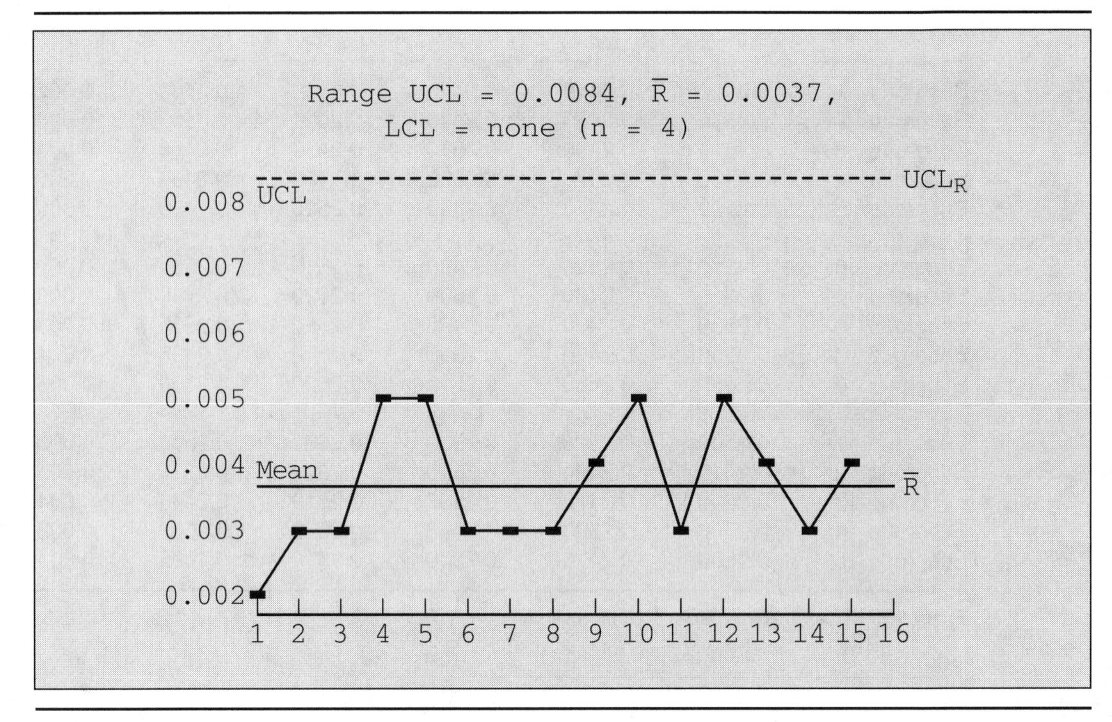

Figure 6.14 R Chart for Silicon Wafer Thickness

control, Whisks' customer has raised questions about whether or not the process is capable of producing wafers within their specifications of 0.2500 mm $\pm$ 0.0050.

Using the information provided in the charts, Whisks' process engineers calculate the process capability indicators 6σ, C_p, and C_{pk}.

From the figures:

$$\bar{\bar{X}} = 0.2500$$
$$\bar{R} = 0.0037$$
$$UCL_{\bar{X}} = 0.2527$$
$$LCL_{\bar{X}} = 0.2473$$
$$USL_{\bar{X}} = 0.2550$$
$$LSL_{\bar{X}} = 0.2450$$
$$n = 4$$

To calculate 6σ:

$$\sigma = \frac{\bar{R}}{d_2}$$
$$\sigma = \frac{0.0037}{2.059}$$
$$\sigma = 0.0018$$
$$6\sigma = 0.0108$$

To calculate C_p:

$$C_p = \frac{USL - LSL}{6\sigma}$$

$$C_p = \frac{0.2550 - 0.2450}{0.0108}$$

$$C_p = 0.9259$$

To calculate C_{pk}:

$$C_{pk} = \frac{Z(\min)}{3}$$

Calculating Z(min) as the smaller of Z(USL) and Z(LSL) gives

$$Z(USL) = \frac{(USL - \overline{\overline{X}})}{\sigma}$$

$$Z(USL) = \frac{(0.2550 - 0.2500)}{0.0018}$$

$$Z(USL) = 2.78$$

$$Z(LSL) = \frac{(\overline{\overline{X}} - LSL)}{\sigma}$$

$$Z(LSL) = \frac{(0.2500 - 0.2450)}{0.0018}$$

$$Z(LSL) = 2.78$$

Therefore,

$$C_{pk} = \frac{2.78}{3}$$

$$C_{pk} = 0.9267$$

Interpretation

The $\overline{X}$ and R charts in Figures 6.13 and 6.14 are within statistical control according to the guidelines discussed in this chapter. If we study process centering, the $\overline{X}$ chart (Figure 6.13) shows that the wafer thickness is centered around the target value of 0.2500 mm. Studying process variation, the R chart (Figure 6.14) reveals that some variation is present in the process; not all the wafers are uniform in their thickness.

As we noted earlier in the chapter, a process within statistical control is not necessarily capable of meeting the specifications set by the customer. The control charts do not tell the entire story. Whisks' process capability index, C_p, reveals that the process is currently producing wafers that do not meet the specifications. The C_{pk} value is nearly equal to the C_p value; that is, the process is close to being centered within the specifications. Essentially, the process is producing product as shown in Figure 6.15. As we suspected from earlier examination of the R chart, too much variation is present in the process.

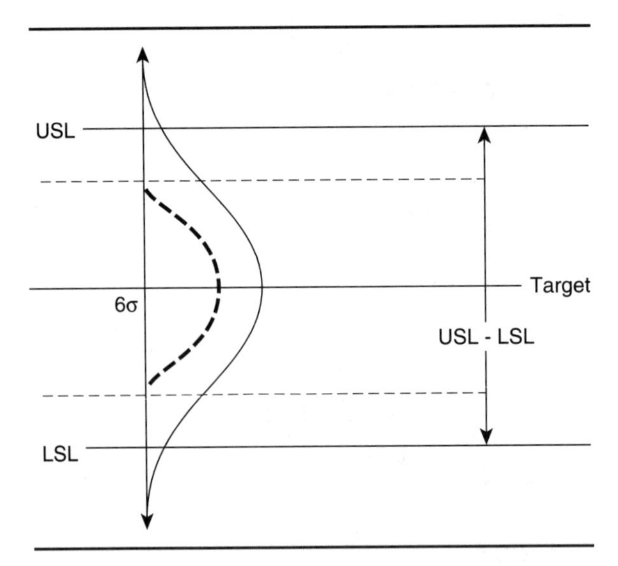

Figure 6.15 Spread of Averages, 6σ versus Specifications

SUMMARY

Manufacturing a product or providing a service is more than meeting specifications. As reducing the variation associated with a process has become a greater concern, the use of process capability indices has increased. Used to judge how consistently the process is performing, these indices provide a great deal of information concerning process centering and the ability of the process to meet specifications. Process capability indices can guide the improvement process toward uniformity about a target value.

■ *Lessons Learned*

1. Process capability refers to the ability of a process to meet the specifications set by the customer or designer.
2. Individuals in a process have a greater spread around the center than do the averages.
3. Specification limits are set by the designer or customer. Control limits are determined by the current process.
4. $6\hat{\sigma}$ is the spread of the process or process capability.
5. C_p, the capability index, is the ratio of the tolerance (USL − LSL) and the process capability ($6\hat{\sigma}$).
6. C_r, the capability ratio, is the ratio of the process capability (6σ) to the tolerance (USL − LSL).
7. C_{pk} is the ratio that reflects how the process is performing in relation to a nominal, center, or target value. ■

Formulas

$$\hat{\sigma} = \frac{\bar{s}}{c_4}$$

$$\hat{\sigma} = \frac{\bar{R}}{d_2}$$

Capability Indices

$$C_p = \frac{USL - LSL}{6\hat{\sigma}}$$

$$C_r = \frac{6\hat{\sigma}}{USL - LSL}$$

$$C_{pk} = \frac{Z(\min)}{3}$$

where Z(min) is the smaller of $Z(USL) = (USL - \bar{X})/\hat{\sigma}$ or $Z(LSL) = (\bar{X} - LSL)/\hat{\sigma}$.

Chapter Problems

1. What do control limits represent? What do specification limits represent? Describe the three cases that compare specification limits to control limits.
2. Describe the relationship between averages and individuals in terms of specification limits and control limits.
3. Why can a process be in control but not be capable of meeting specifications?
4. A hospital is using $\bar{X}$ and R charts to record the time it takes to process patient account information. A sample of five applications is taken each day. The first four weeks' (20 days') data give the following values:

   $$\bar{\bar{X}} = 16 \text{ min} \quad \bar{R} = 7 \text{ min}$$

   If the upper and lower specifications are 21 minutes and 13 minutes, respectively, calculate $6\hat{\sigma}$, C_p, and C_{pk}. Interpret the indices.
5. Two television manufacturers operate plants with differing quality philosophies. One plant works to target and the other works to specification limit. Describe the difference between these two philosophies.
6. For the data in Problem 4 of Chapter 5, calculate $6\hat{\sigma}$, C_p, and C_{pk}. Interpret the indices. The specification limits are 50 ± 0.5.
7. For the data in Problem 5 of Chapter 5, calculate $6\hat{\sigma}$, C_p, and C_{pk}. Interpret the indices. The specification limits are 0.50345 ± 0.0020 mm.
8. From the information given in Figure 5.1 and the calculations in Problem 10 of Chapter 5, calculate $6\hat{\sigma}$, C_p, and C_{pk}. Interpret the indices. The specification limits are 0.0625 ± 0.0005.

9. From the information in Problem 11 of Chapter 5, calculate $6\hat{\sigma}$, C_p, and C_{pk}. Interpret the indices. The specification limits are 0.0625 ± 0.0005.
10. For the data in Problem 12 of Chapter 5, use $\overline{R}/d_2$ to calculate $6\hat{\sigma}$, C_p, and C_{pk}. Interpret the indices. The specification limits are 0.0028 ± 0.0005.
11. For the data in Problem 12 of Chapter 5 and the chart of Problem 13 of Chapter 5, use $\overline{s}/c_4$ to calculate $6\hat{\sigma}$, C_p, and C_{pk}. Interpret the indices.
12. For the case study in Chapter 5, calculate $6\hat{\sigma}$, C_p, and C_{pk} for each day. How do the indices change? Do you feel comfortable using the data from day 6 as the true process capability? Why or why not?
13. A quality analyst is checking the process capability associated with the production of struts, specifically the amount of torque used to tighten a fastener. Twenty-five samples of size 4 have been taken. These were used to create $\overline{X}$ and R charts. The values for these charts are as follows. The upper and lower control limits for the $\overline{X}$ chart are 74.80 Nm and 72.37 Nm, respectively. $\overline{\overline{X}}$ is 73.58 Nm. $\overline{R}$ is 1.66. The specification limits are 80 Nm ± 10. Calculate $6\hat{\sigma}$, C_p, and C_{pk}. Interpret the values.
14. Complete the following table by calculating $6\hat{\sigma}$, C_p, and C_{pk} for the data from Case 5.1, parts 1–5. Specifications: 3.750 ± 0.005 inch. When interpreting the completed table, be sure to compare and contrast the values to each other as well as comment about their changes over the five days. Note that you do not have to complete the cases to answer this question.

Day	1	2	3	4	5
$\overline{\overline{X}}$	3.4794	3.7493	3.7492	3.7496	3.7503
$\hat{\sigma}$	0.0048	0.0046	0.0036	0.0029	0.0008
$6\hat{\sigma}$	0.0288	0.0276			
C_p	0.3471	0.3623			
C_{pk}	0.3056	0.3116			

CASE STUDY 6.1
Process Capability

PART 1

Hop Scotch Drive-In Restaurant is considering a new marketing idea. At Hop Scotch, diners have the choice of dining inside or staying in their cars to eat. To serve those dining in their cars, carhops take the orders at the car and bring the food to the car when it's ready. Many people want fast service when they pull into the drive-in restaurant, but they also want more than a drive-through restaurant can offer. To try to meet their customers' expectations, Hop Scotch Drive-In is proposing to promise to have a server at your car in two minutes or less after your arrival at their drive-in. Before making this promise, the owners of Hop Scotch want to see how long it takes their servers to reach the cars now. They have gathered the information seen in Table 1.

Assignment

Calculate the range for each of the subgroups containing five server times. Since the owners of Hop Scotch Drive-In are interested in the combined performance of their servers and not the performance of any particular server, the subgroups contain times from each server. Use these calculated values to determine the process capability and the C_p and C_{pk} values. How well is the process performing when compared with the specifications of 2.0 min + 0.0/ − 1.0?

PART 2

Judging by the time measurements, it now takes longer to reach the cars than the proposed two minutes. Hop Scotch probably won't be able to keep up with their advertisement of two minutes or less without some changes in the process. The process capability index is too low. The C_{pk} shows that the process is not centered around the center of the specification. In fact, the C_{pk} value tells us that the average for this distribution is outside the specifications.

Management has decided to explain the proposed advertisement to their servers. They hope that by doing this, the servers will put more effort into getting to the cars in two minutes or less. Management carefully explains how moving faster to get to the cars will have a positive effect on increasing business because the customers will be happier with faster service. At first skeptical, the servers decide it might be possible to increase business by serving the customers faster. They agree to do their best

Table 1 Server-to-Car Times in Minutes

Subgroup Number	*Server 1*	*Server 2*	*Server 3*	*Server 4*	*Server 5*	*Range*
1	2.5	2.4	2.3	2.5	2.1	0.4
2	2.5	2.2	2.4	2.3	2.8	0.6
3	2.4	2.5	2.7	2.5	2.5	0.3
4	2.3	2.4	2.5	2.9	2.9	0.6
5	2.3	2.6	2.4	2.7	2.2	
6	2.6	2.4	2.6	2.5	2.5	
7	2.7	2.6	2.4	2.4	2.5	
8	2.4	2.3	2.5	2.6	2.4	
9	2.5	2.2	2.3	2.4	2.7	
10	2.4	2.4	2.3	2.7	2.4	
11	2.5	2.4	2.5	2.5	2.3	
12	2.4	2.3	2.4	2.2	2.1	
13	2.5	2.5	2.4	2.5	2.7	
14	2.5	2.4	2.3	2.3	2.6	
15	2.4	2.3	2.3	2.4	2.8	
16	2.6	2.7	2.4	2.6	2.9	
17	2.0	2.3	2.2	2.7	2.5	
18	2.2	2.4	2.1	2.9	2.6	
19	2.3	2.5	2.6	2.4	2.3	
20	2.3	2.5	2.6	2.6	2.5	

to get to the cars as quickly as possible. The time measurements seen in Table 2 were made after the meeting with the servers.

Assignment

Use the new data in Table 2 to calculate the process capability, C_p, and C_{pk}. How is the process performing now?

PART 3

Based on what they learned from the previous day's study, the owners of Hop Scotch have decided to supply their servers with roller skates. The servers are obviously working very hard but are still unable to reach the cars quickly. Management hopes that after the servers become comfortable with their skates, they will be able to move faster around the restaurant parking lot. Table 3 shows the results of adding roller skates.

Table 2 Server-to-Car Times in Minutes

Subgroup Number	*Server 1*	*Server 2*	*Server 3*	*Server 4*	*Server 5*
1	2.3	2.2	2.0	2.4	2.2
2	2.2	2.1	2.0	2.2	2.3
3	2.2	2.1	2.0	2.3	2.0
4	2.0	2.3	2.3	2.2	2.1
5	2.0	2.3	2.3	2.1	1.9
6	2.2	2.5	2.2	2.3	2.1
7	2.2	2.2	2.3	2.4	2.0
8	2.4	2.6	2.4	2.5	2.2
9	2.5	2.2	2.4	1.9	2.2
10	2.2	2.0	2.2	2.5	2.0
11	2.0	2.3	2.1	2.2	2.1
12	2.3	2.2	2.1	2.3	2.2
13	2.1	2.2	2.1	2.1	2.0
14	2.1	2.3	2.2	2.4	2.0
15	2.2	2.1	2.1	2.3	2.0
16	2.0	2.1	2.2	2.3	2.1
17	2.1	2.0	2.3	2.4	1.9
18	2.0	2.0	2.2	1.9	2.2
19	2.3	2.1	2.4	2.3	2.3
20	2.0	2.3	2.1	1.9	2.0

Assignment

Calculate the process capability, C_p, and C_{pk}. Are the roller skates a big help?

PART 4

Without doubt, the roller skates have improved the overall performance of the servers dramatically. As an added bonus, the drive-in has become more popular because the customers really like seeing the servers skate. For some of them, it reminds them of cruising drive-ins in the fifties; for others, they like to see the special skating skills of some of the servers. The servers seem very happy with the new skates. Several have said that their job is more fun now.

To make further improvements, the owners of Hop Scotch have decided to change how the servers serve customers. In the past, each server was assigned a section of

Table 3 Server-to-Car Times in Minutes

Subgroup Number	*Server 1*	*Server 2*	*Server 3*	*Server 4*	*Server 5*
1	1.9	1.9	2.0	2.0	1.9
2	1.9	2.0	2.0	2.0	1.9
3	1.9	2.0	1.9	1.9	2.0
4	1.9	2.0	1.8	1.8	2.0
5	2.0	2.1	2.1	2.0	2.0
6	2.0	2.0	2.1	2.1	2.1
7	2.0	1.8	1.9	2.0	2.0
8	2.0	2.0	2.0	2.0	1.9
9	2.0	2.0	1.9	1.8	1.9
10	1.9	1.8	1.9	1.8	1.7
11	1.8	1.8	1.8	1.7	1.6
12	1.9	1.7	1.7	1.8	1.7
13	1.7	1.7	1.7	1.6	1.6
14	1.8	1.6	1.6	1.6	1.6
15	1.9	1.9	1.8	1.9	1.8
16	1.8	1.8	2.0	2.0	1.9
17	1.5	1.6	1.6	1.6	1.9
18	1.6	1.5	1.6	1.8	1.7
19	1.4	1.6	1.7	1.8	1.8
20	1.8	1.8	1.9	1.7	1.7

the parking lot. But now, observing the activities in the restaurant parking lot, the owners see that the lot does not fill up evenly; some sections fill up first, especially when a group comes in to be served all at once. This unevenness often overwhelms a particular server while another server has very few cars to wait on. Under the new program, servers will handle any area of the parking lot, serving cars on a first-come, first-served basis.

Assignment

Use the data in Table 4 to calculate the process capability, C_p, and C_{pk}. Do the new rules help?

Table 4 Server-to-Car Times in Minutes

Subgroup Number	*Server 1*	*Server 2*	*Server 3*	*Server 4*	*Server 5*
1	1.1	1.1	1.0	1.1	1.1
2	0.9	1.0	1.0	1.1	1.1
3	1.0	1.0	1.0	1.0	1.2
4	1.1	1.1	1.0	1.1	1.0
5	0.7	0.9	0.8	0.9	0.8
6	0.8	0.8	0.8	1.1	1.0
7	0.9	1.0	1.0	1.0	0.8
8	1.0	1.1	1.1	1.0	1.2
9	0.8	0.9	0.8	0.8	0.9
10	0.8	0.7	0.8	0.7	0.7
11	0.8	0.8	0.8	0.9	1.0
12	0.8	0.9	0.8	0.8	0.9
13	0.7	0.7	0.7	0.7	0.7
14	0.7	0.7	0.7	0.8	0.9
15	0.9	0.9	0.9	1.0	0.9
16	0.9	0.9	0.8	1.0	1.0
17	0.9	0.9	0.9	0.9	0.9
18	0.7	0.8	0.7	0.8	0.7
19	0.5	0.6	0.5	0.7	0.6
20	0.6	0.7	0.6	0.7	0.6

CASE STUDY 6.2
Process Improvement

This case is the fourth in a four-part series of cases involving process improvement. The other cases are found at the end of Chapters 3, 4, and 5. Data and calculations for this case establish the foundation for the future cases; however, it is not necessary to have completed the cases in Chapters 3, 4, and 5 in order to complete and understand this case. Portions of the case in Chapter 5 are repeated here and must be completed in order to calculate process capability. Completing this case will provide insight into process capability and process improvement. The case can be worked by hand or with the software provided.

PART 1

Figure 1 provides the details of a bracket assembly to hold a strut on an automobile in place. Welded to the auto-body frame, the bracket cups the strut and secures it to the frame via a single bolt with a lock washer. Proper alignment is necessary for both smooth installation during assembly and future performance. For mounting purposes the left-side hole, A, must be aligned on center with the right-side hole, B. If the holes are centered directly opposite each other, in perfect alignment, then the angle between hole centers will measure 0 degrees. As the flowchart in Figure 2 shows, the bracket is created by passing coils of flat steel through a series of progressive dies. As

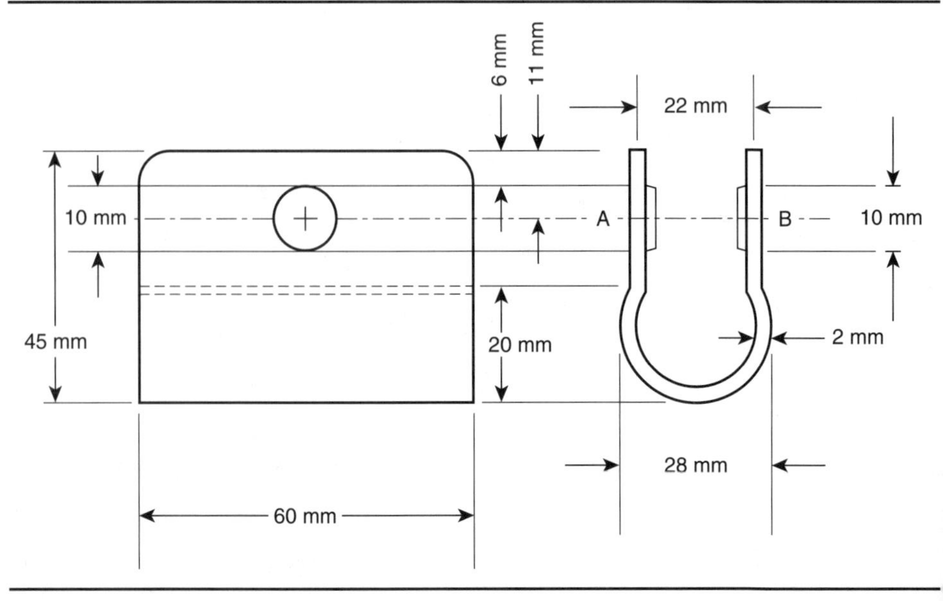

Figure 1 Bracket

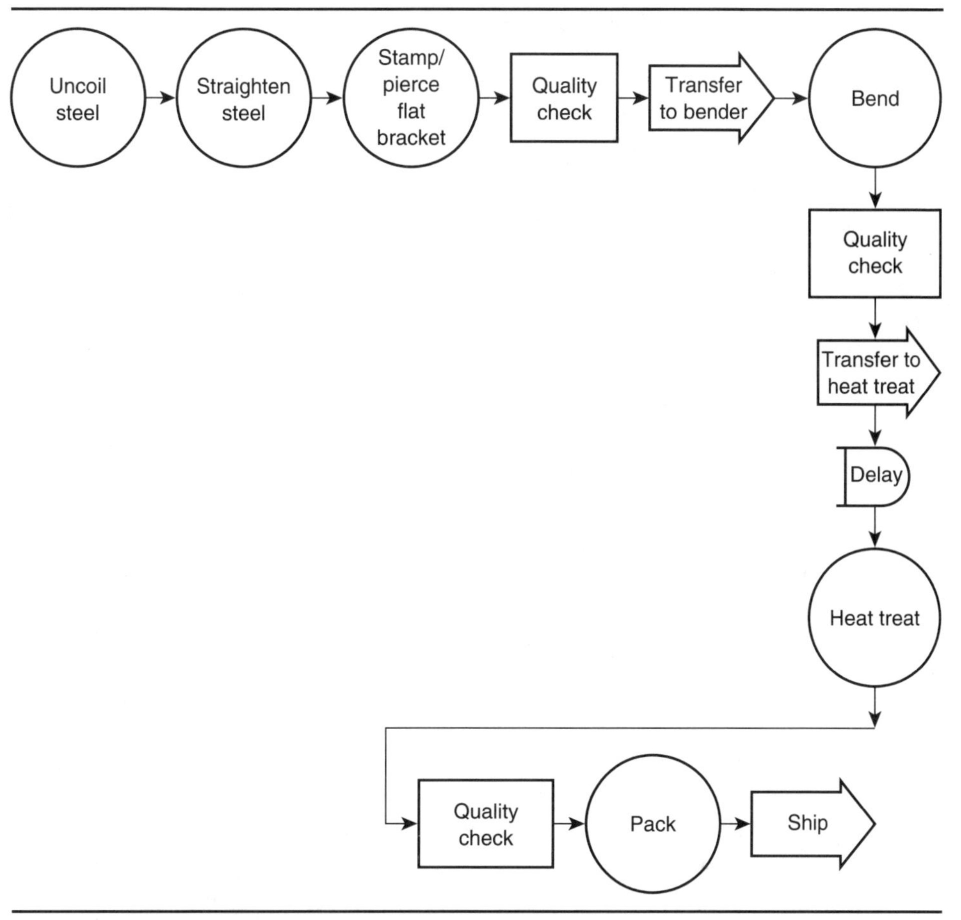

Figure 2 Flowchart of Bracket Fabrication Process

the steel moves through the press, the bracket is stamped, pierced, and finally bent into appropriate shape.

Recently customers have been complaining about having difficulty securing the bracket closed with the bolt and lock washer. The bolts have been difficult to slide through the holes and then tighten. Assemblers complain of stripped bolts and snug fittings. Unsure of where the problem is, WP Inc.'s management assembled a team consisting of representatives from process engineering, materials engineering, product design, and manufacturing.

Through the use of several cause-and-effect diagrams, the team determines that the most likely cause of the problems experienced by the customer is the alignment of the holes. At some stage in the formation process, the holes end up off-center. To confirm their suspicions, during the next production run, the bending press operator takes 20 subgroups of size 5 and measures the angle between the centers of the holes for each sample (Figure 3). The specification for the angle between insert hole A and insert hole B is 0.00 mm with a tolerance of ± 0.30. The values recorded in

Sample \ Subgroup	1	2	3	4	5	6	7	8	9	10	11	12	13	14	15
1	0.31	0.27	0.30	0.30	0.25	0.18	0.26	0.15	0.30	0.31	0.18	0.22	0.19	0.14	0.29
2	0.29	0.23	0.30	0.20	0.20	0.26	0.27	0.21	0.24	0.25	0.16	0.30	0.28	0.27	0.23
3	0.30	0.31	0.28	0.21	0.19	0.18	0.12	0.24	0.26	0.25	0.21	0.21	0.26	0.25	0.27
4	0.28	0.23	0.24	0.23	0.26	0.24	0.20	0.27	0.27	0.28	0.29	0.24	0.29	0.28	0.24
5	0.23	0.29	0.32	0.25	0.25	0.17	0.23	0.30	0.26	0.25	0.27	0.26	0.24	0.16	0.23

Sample \ Subgroup	16	17	18	19	20
1	0.22	0.32	0.33	0.20	0.24
2	0.24	0.27	0.30	0.17	0.28
3	0.22	0.28	0.22	0.26	0.15
4	0.30	0.19	0.26	0.31	0.27
5	0.30	0.31	0.26	0.24	0.19

Figure 3 Hole A, B Alignment; Distance Above (+) or Below (−) Nominal

Figure 3 represent the distance above (or when a minus sign is present, below) the nominal value of 0.00 mm.

Assignment

Follow the steps outlined in Chapter 5 and utilize the data to create a set of $\overline{X}$ and R charts. You will need to calculate the mean and range for each subgroup. Describe the performance of the process. Using $\sigma = \overline{R}/d_2$, calculate $6\hat{\sigma}$, C_p, and C_{pk}. Interpret the charts and these values. Do they tell you the same thing? Is the process capable of meeting specifications?

PART 2

Having $\overline{X}$ and R charts and the accompanying statistical information tells the team a lot about the hole-alignment characteristics. Using the problem-solving method described in Chapter 3 (and followed in Case 3.2), the team has determined that the fixture that holds the flat bracket in place during the bending operation needs to be replaced. One of the measures of performance that they created during Step 4 of their problem-solving process is percentage of the parts out of specification. Now that the fixture has been replaced with a better one, the team would like to determine whether or not changing the fixture improved the process by removing a root cause of hole misalignment.

Assignment

Utilize the data in Figure 4 to create another set of $\overline{X}$ and R charts. Has their change to the fixtures resulted in a process improvement? Using $\sigma = \overline{R}/d_2$, calculate 6σ, C_p, and C_{pk}. Interpret the charts and these values. Do they tell you the same thing? Is the process capable of meeting specifications? How do you know? Compare the calculations from part 1 of this case with part 2. Describe the changes. How are they doing when you compare their measure of performance, percentage of the parts out of specification, both before and after the fixture is changed?

Sample \ Subgroup	1	2	3	4	5	6	7	8	9	10	11	12	13	14	15
1	0.03	0.06	0.06	0.03	−0.04	−0.02	−0.05	0.06	0.00	−0.02	−0.02	0.06	0.07	0.10	0.02
2	0.08	0.08	0.08	0.00	−0.07	0.06	−0.07	0.03	0.06	−0.01	0.06	0.02	−0.04	0.05	−0.05
3	−0.03	−0.01	0.05	0.05	0.00	0.02	0.11	0.04	0.02	0.00	−0.02	−0.01	0.08	0.03	0.04
4	0.07	0.08	−0.03	−0.01	−0.01	0.12	−0.03	0.03	−0.02	−0.10	0.02	−0.02	0.01	0.04	0.05
5	−0.02	0.02	0.03	0.07	−0.01	−0.07	0.03	0.01	0.00	−0.04	0.09	0.03	−0.04	0.06	−0.05

Sample \ Subgroup	16	17	18	19	20
1	−0.05	−0.06	−0.04	−0.06	−0.01
2	0.00	−0.02	−0.02	0.00	0.02
3	0.06	0.04	−0.01	0.00	0.05
4	−0.02	0.07	0.03	0.03	0.04
5	−0.01	−0.04	−0.02	0.04	0.03

Figure 4 Hole A, B Alignment Following Process Improvement; Distance Above (+) or Below (−) Nominal

7

Other Variable Control Charts

■ *Learning Opportunities:*

1. To introduce other control charts, including charts for individuals, median charts, moving-average/moving-range charts, and run charts
2. To introduce the concept of precontrol
3. To familiarize the student with short-run production charts ■

What about the Short Run?

$\overline{X}$ *and* R *or* s *charts track the performance of processes that have long production runs or repeated services. What happens when there is an insufficient number of sample measurements to create a traditional* $\overline{X}$ *and* R *chart? How do statistical control ideas apply to new processes or short production runs? Can processes be tracked with other methods? What happens when only one sample is taken from a process? This chapter covers the variety of different statistical charts that deal with situations where traditional* $\overline{X}$ *and* R *or* s *charts cannot be applied.*

INDIVIDUALS AND MOVING-RANGE CHARTS

Charts for individuals with their accompanying moving-range charts are often used when the data collection occurs either once a day or on a week-to-week or month-to-month basis. As their title suggests, *charts for* ***individuals and moving range*** *are created when the measurements are single values or when the number of products produced is too small to form traditional* $\overline{X}$ *and* R *charts*.

To construct an individual values ($\overline{X}_i$) and moving-range chart, the individual X_i measurements are taken and plotted on the $\overline{X}_i$ chart. To find the centerline of the chart, the individual X_i values are summed and divided by the total number of X_i readings:

$$\overline{X}_i = \frac{\Sigma X_i}{m}$$

Moving ranges *are calculated by measuring the value-to-value difference of the individual data*. Two consecutive individual data-point values are compared and the magnitude or absolute value of their difference is recorded on the moving-range chart. This reading is usually placed on the chart between the space on the R chart designated for a value and its preceding value (Figure 7.1). This shows the reader of the chart that the two individual values were compared to determine the range value. An alternative method of plotting R is to place the first R value in the space designated for the second reading, leaving the first space blank.

$\overline{R}$, the centerline, is equal to the summation of the individual R values divided by the total number of Rs calculated. The total number of Rs calculated will be 1 less than the total number of individual $\overline{X}_i$ values:

$$\overline{R} = \frac{\Sigma R_i}{m - 1}$$

Once $\overline{R}$ has been determined, control limits for both charts are calculated by adding $\pm 3\sigma$ to the centerline:

$$UCL_{Xi} = \overline{X}_i + 3\sigma$$
$$LCL_{Xi} = \overline{X}_i - 3\sigma$$

where

$$\sigma = \frac{\overline{R}}{d_2}$$

To approximate $\pm 3\sigma$, divide 3 by the d_2 value for $n = 2$ found in the table in Appendix 2. The formulas become

$$UCL_X = \overline{X}_i + 2.66(\overline{R})$$
$$LCL_X = \overline{X}_i - 2.66(\overline{R})$$

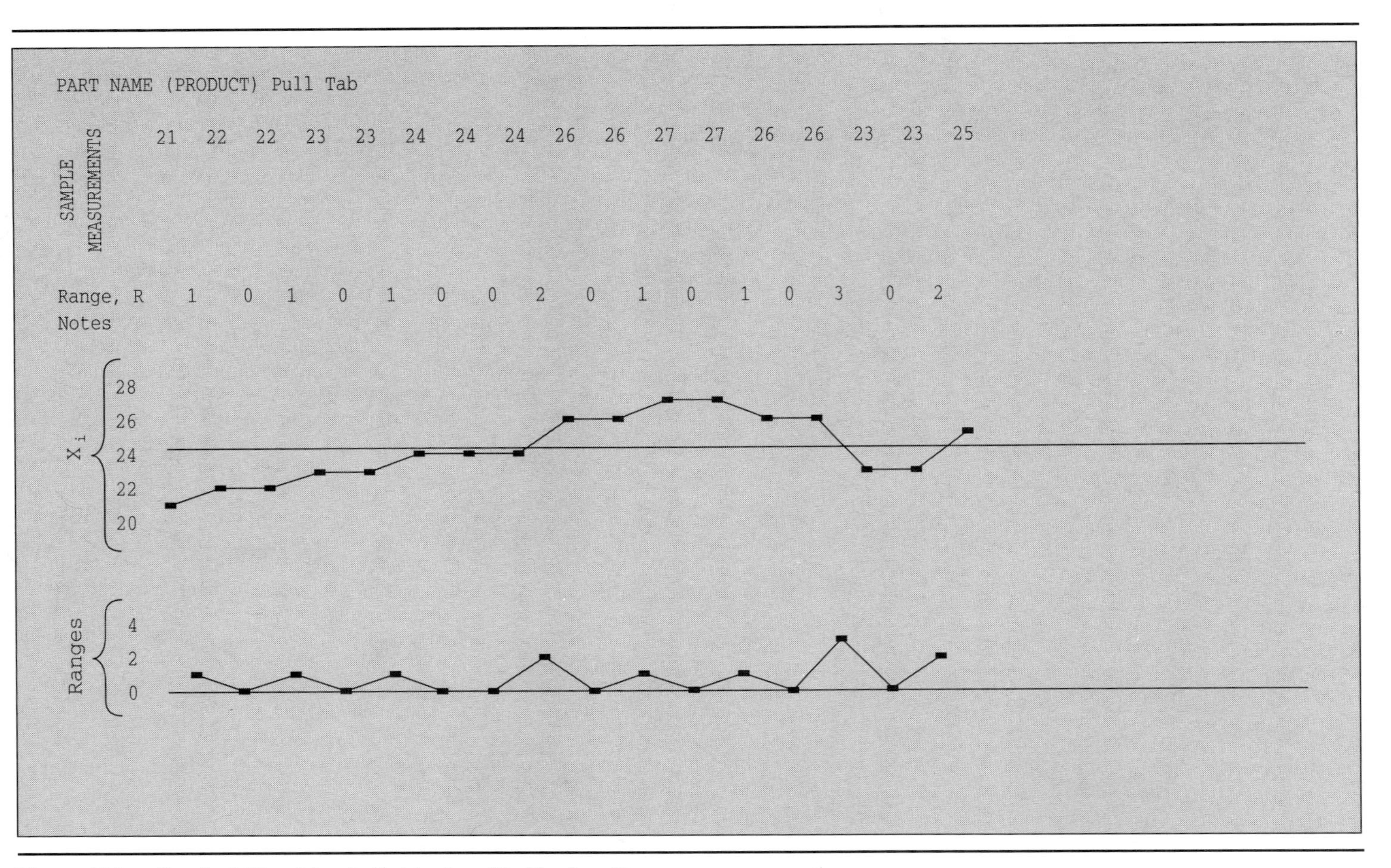

Figure 7.1 Control Chart for Individuals with Moving Range

The upper control limit for the R chart is found by multiplying D_4 by $\overline{R}$. From Appendix 2, $D_4 = 3.27$ when n = 2. The lower control limit for the R chart is equal to zero:

$$UCL_R = D_4 \times \overline{R}$$
$$= 3.27 \times \overline{R}$$

Interpreting individuals and moving-range charts is similar to the interpretation of $\overline{X}$ and R charts, as presented in previous chapters. When studying the charts, look for trends, shifts, or changes in level that indicate patterns in the process. Study the range chart to determine the amount of variation present in the process. Look for cycles or patterns in the data. On both charts, the data should randomly flow back and forth across the centerline. Data should not be concentrated at either control limit, which would indicate a skewed distribution. (See Chapter 5 for a complete review of control-chart interpretation. Once the process is considered in control, process capability can be determined using the methods discussed in Chapter 6.) As with traditional variables control charts, a large number of subgroups of sample size n = 1 ensures that the distribution is approximately normal. Individual and moving-range charts are more reliable when the number of samples taken exceeds 80.

EXAMPLE 7.1 Creating and Interpreting Individuals and Moving-Range Charts

In a test lab, two different Rockwell hardness (R_c) testers are operated. One Rockwell hardness tester has a dial indicator; the other has a digital indicator. Both of the hardness testers are calibrated on a daily basis. For the data from the C scale for the first 10 days of February, create and interpret charts for individuals and the moving range for each of the Rockwell hardness testers.

For the Dial Indicator Tester

Step 1. To construct individual $(\overline{X}_i)$ and moving-range charts, first measure the individual values $(\overline{X}_i)$ and record them (Table 7.1).

Step 2. To find the centerline for the process charts, sum the individual X_i values and divide by the total number of X_i readings:

$$\overline{X}_i = \frac{\Sigma X_i}{m} = \frac{46.1 + 45.9 + 46.0 + \cdots + 45.9}{10}$$
$$= 45.9$$

Scale the chart and place the centerline on it (Figure 7.2).

Step 3. Plot the individual X_i values on the chart (Figure 7.2).

Step 4. Calculate the range values for the data, comparing each measurement with the one preceding it (Table 7.1).

Step 5. Calculate the average of the range values using the following formula:

$$\overline{R} = \frac{\Sigma R_i}{m - 1} = \frac{0.2 + 0.1 + 0.0 + 0.0 + \cdots + 0.1}{9}$$
$$= 0.16$$

Table 7.1 Recorded Values

Dial: Individual R_c Measures	Dial: Range	Digital: Individual R_c Measures	Digital: Range
46.1		46.1	
	0.2		0.1
45.9		46.0	
	0.1		0.1
46.0		45.9	
	0.0		0.6
46.0		46.5	
	0.0		0.5
46.0		46.0	
	0.3		0.1
45.7		46.1	
	0.1		0.0
45.8		46.1	
	0.3		0.1
46.1		46.2	
	0.3		0.8
45.8		45.4	
	0.1		0.8
45.9		46.2	

Note that the value for m for the $\overline{R}$ calculation is one less than the value of m for the $\overline{X}_i$ calculation. $\overline{R}$ is the centerline of the R chart.

Step 6. The upper control limit for the $\overline{X}_i$ chart for a sample size of two is calculated using the following formulas:

$$UCL_X = \overline{X}_i + 2.66(\overline{R})$$
$$= 45.9 + 2.66(0.16) = 46.3$$

$$LCL_X = \overline{X}_i - 2.66(\overline{R})$$
$$= 45.9 - 2.66(0.16) = 45.5$$

Record the control limits on the individuals chart (Figure 7.2).

Step 7. Find the upper and lower control limits for the R chart using the following formulas:

$$UCL_R = 3.27 \times \overline{R} = 3.27 \times 0.16$$
$$= 0.52$$
$$LCL_R = 0$$

Step 8. Record the centerline, control limits, and range values on the moving-range chart (Figure 7.2).

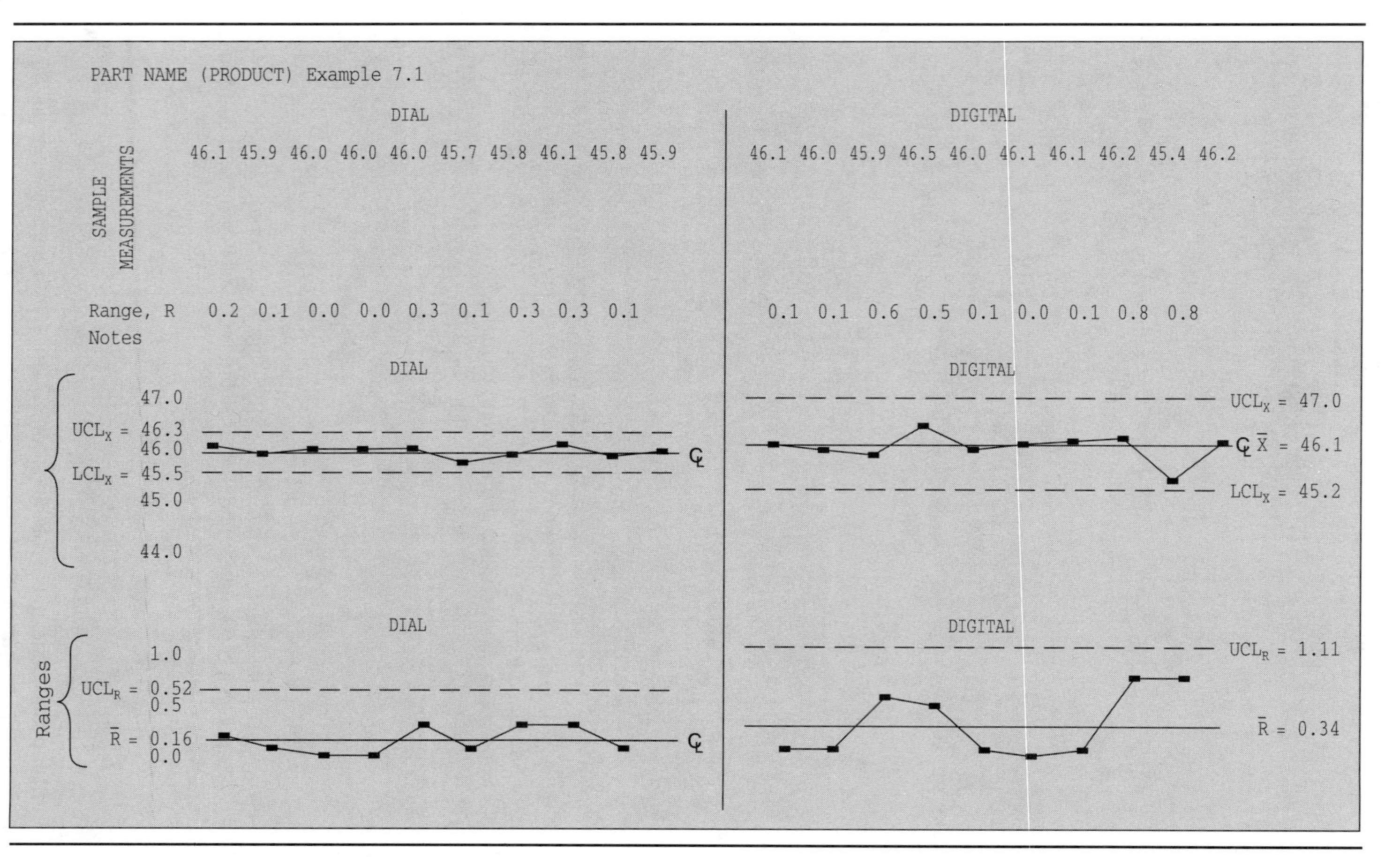

Figure 7.2 Control Chart for Individuals with Moving Range

For the Digital Indicator Tester

Repeating the same steps as taken for the dial indicator,

$$\overline{X}_i = \frac{\Sigma X_i}{m} = \frac{46.1 + 46.0 + 45.9 + \cdots + 46.2}{10}$$
$$= 46.1$$

$$\overline{R} = \frac{\Sigma R_i}{m - 1} = \frac{0.1 + 0.1 + 0.6 + 0.5 + \cdots + 0.8}{9}$$
$$= 0.34$$

$$UCL_X = \overline{X}_i + 2.66(\overline{R})$$
$$= 46.1 + 2.66(0.34) = 47.0$$

$$LCL_X = \overline{X}_i - 2.66(\overline{R})$$
$$= 46.1 - 2.66(0.34) = 45.2$$

$$UCL_R = 3.27 \times \overline{R} = 3.27 \times 0.34$$
$$= 1.11$$
$$LCL_R = 0$$

Step 9. Interpret the charts. In this case the variation present in the digital indicator's chart is significantly greater than the variation seen in the dial indicator's chart. This significant difference is unexpected since readings taken from a digital indicator are usually more precise than those of a dial indicator (where the operator may interpolate). The results of investigating these charts reveal a calibration problem with the digital indicator. Once repaired, thevariation for the digital indicator should decrease dramatically.

MOVING-AVERAGE AND MOVING-RANGE CHARTS

Rather than plot each individual reading, we can use *moving-average and moving-range charts to combine* n *number of individual values to create an average*. When a new individual reading is taken, the oldest value forming the previous average is discarded. The new reading is combined with the remaining values from the previous average and the newest value to form a new average, thus the term "moving average." This average is plotted on the chart and the limits are calculated using the formulas for the $\overline{X}$ and R charts presented in Chapter 5. Capability indices can also be calculated using the methods presented in Chapter 6. The number of individual values grouped together to form a subgroup will be the size of n used to select the A_2, D_4, and D_3 values.

By combining individual values produced over time, moving averages smooth out short-term variation and allow users to study the underlying trends in the data. For this reason, moving-average charts are frequently used for seasonal products. Moving averages, because of the nature of their construction, will always lag behind changes in the process, making them less sensitive to such changes. The larger the size of the moving subgroup, the less sensitivity to change. To quickly detect process changes, use the individuals and moving-range combination of charts. Moving-average charts are best used when the process changes slowly.

EXAMPLE 7.2 Creating Moving-Average and Moving-Range Charts

A manufacturer of skis is interested in making moving-average and moving-range charts of the sales figures for the past year (Table 7.2). Because of the seasonal nature of the business, they have decided to combine the individual sales values for three months. This is the average to be plotted on the chart.

Their first step is to combine the values from January, February, and March to obtain the first moving average. These values are compared and the lowest value is subtracted from the highest value to generate the range associated with the three points.

Combining these values, they find

January	75,000
February	71,000
March	68,000

$$\text{Moving average 1} = \frac{75{,}000 + 71{,}000 + 68{,}000}{3} = 71{,}333$$

$$\text{Moving range 1} = 75{,}000 - 68{,}000 = 7{,}000$$

These values are plotted on the chart. Subsequent values are found by dropping off the oldest measurement and replacing it with the next consecutive value. A new

Year 1	
January	75,000
February	71,000
March	68,000
April	59,800
May	55,100
June	54,000
July	50,750
August	50,200
September	55,000
October	67,000
November	76,700
December	84,000

Table 7.2 Ski Sales for the Past Year

average and range are then calculated and plotted. Adding the month of April's figures, they find

April 59,800

$$\text{Moving average 2} = \frac{71{,}000 + 68{,}000 + 59{,}800}{3} = 66{,}267$$

$$\text{Moving range 2} = 71{,}000 - 59{,}800 = 11{,}200$$

Then May:

May 55,100

$$\text{Moving average 3} = \frac{68{,}000 + 59{,}800 + 55{,}100}{3} = 60{,}967$$

$$\text{Moving range 3} = 68{,}000 - 55{,}100 = 12{,}900$$

and so on.

To calculate the limits associated with a moving-average and moving-range chart combination, they use the traditional $\overline{X}$ and R chart formulas:

$$\text{Centerline } \overline{\overline{X}} = \frac{\Sigma X_i}{m} = \frac{71{,}333 + 66{,}267 + \cdots + 75{,}900}{10}$$
$$= 61{,}132$$

$$\overline{R} = \frac{\Sigma R_i}{m} = \frac{7{,}000 + 11{,}200 + 12{,}900 + \cdots + 17{,}000}{10}$$
$$= 10{,}535$$

$$UCL_{\overline{X}} = \overline{\overline{X}} + A_2\overline{R}$$
$$= 61{,}132 + 1.023(10{,}535) = 71{,}909$$

$$LCL_{\overline{X}} = \overline{\overline{X}} - A_2\overline{R}$$
$$= 61{,}132 - 1.023(10{,}535) = 50{,}355$$

$$UCL_R = D_4\overline{R}$$
$$= 2{,}574(10{,}535) = 27{,}117$$

$$LCL_R = D_3\overline{R}$$
$$= 0$$

Figure 7.3 shows the completed chart. An investigation of the chart reveals that the sales of skis decreases dramatically from the winter months to the summer. Sales are fairly constant during the summer months, where the R chart exhibits very little variation. When sales begin to increase, there is a good deal of variation from month to month. Company sales appear to be strong, with sales later in the year increasing much more dramatically than they decreased in the earlier season.

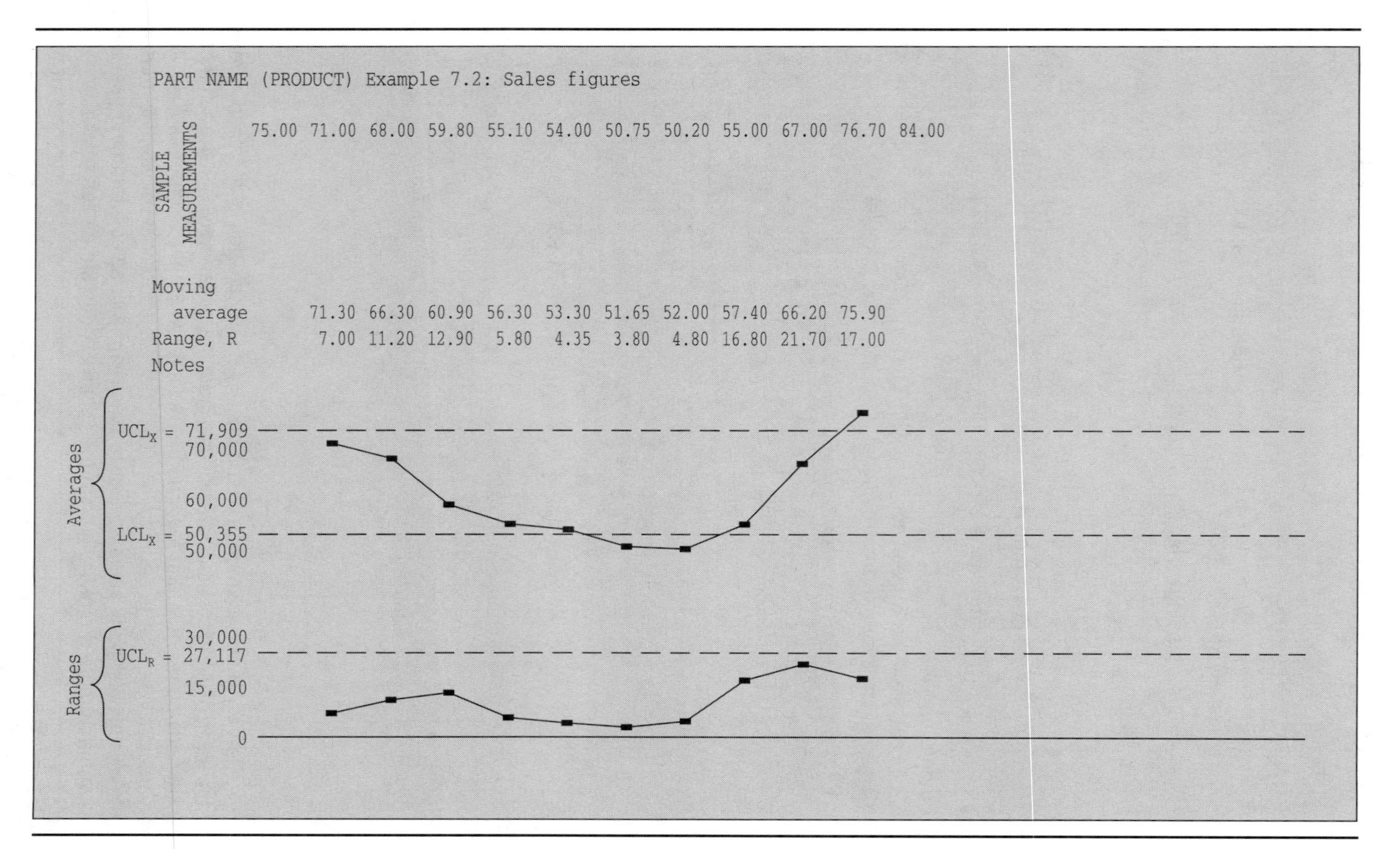

Figure 7.3 Control Chart for Moving Average with Moving Range

A CHART PLOTTING ALL INDIVIDUAL VALUES

Control charts that plot each individual value found in a subgroup (Figure 7.4), while cluttered looking, are useful when explaining the concept of variation. As discussed in Chapters 5 and 6, the points plotted on an $\overline{X}$ chart are averages made up of individual values. The range chart accompanying the $\overline{X}$ chart gives the user an understanding of the spread of the data. Sometimes, however, a picture is worth a thousand words, or in this case, all the values read during a subgroup present a more representative picture than does a single point on the R chart.

On an individual values chart, the individual values from the subgroup are represented by a small mark. The average of those values is represented by a circle. If an individual value and the average are the same, the mark is placed inside the circle. By looking at a control chart that plots all of the individual readings from a subgroup, those using the chart can gain a clearer understanding of how the data are

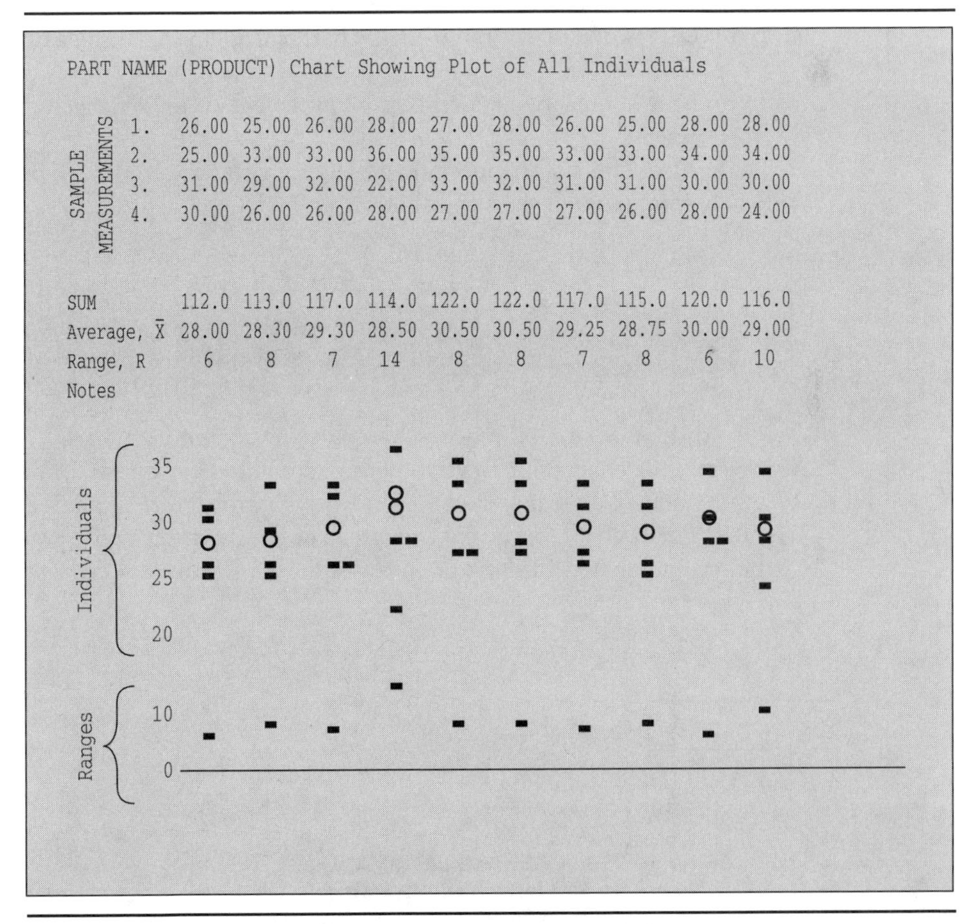

Figure 7.4 Control Chart Plotting All Individual Values

spreading with respect to the average value (Figure 7.4). In most circumstances, because of its cluttered appearance, this chart is reserved for training people on interpreting the values of R or s charts.

MEDIAN AND RANGE CHARTS

*When a **median chart** is used, the median of the data is calculated and charted rather than the value for the average*. Just like variables control charts, median charts can be used to study the variation of process, although ease of calculation is a trade-off with some loss of sensitivity. Constructed and interpreted in a manner similar to $\overline{X}$ and R charts, median and range charts are created through the following steps:

1. Calculate and record the sample measurements.
2. Arrange the sample measurements in each subgroup in order from highest to lowest.
3. Calculate the median and the range for each subgroup.
4. To find the centerline of the median chart, calculate the average of the subgroup medians.
5. To find the centerline of the range chart, calculate the average of the subgroup ranges.
6. The control limits of the median and range charts are determined by the following formulas and the values shown in Table 7.3:

$$UCL_{Md} = \overline{X}_{Md} + A_6\overline{R}_{Md}$$
$$LCL_{Md} = \overline{X}_{Md} - A_6\overline{R}_{Md}$$

$$UCL_R = D_4\overline{R}_{Md}$$
$$LCL_R = D_3\overline{R}_{Md}$$

7. Record the median and range values on their appropriate charts.
8. Interpret the chart using the information presented in Chapters 5 and 6. Patterns, trends, shifts, or points beyond the control limits should be investigated. The R chart will reveal the spread of the variation in the process, and the median chart will show the average median of the process.

n	A_6
2	1.88
3	1.19
4	0.80
5	0.69
6	0.55
7	0.51

Table 7.3 Values of A_6 Used in Median Charts Constructed with Averages of Subgroup Medians

Table 7.4 Recorded Values: Die Roll Measurements in Inches

Subgroups					
1	2	3	4	5	6
0.0059	0.0054	0.0052	0.0054	0.0059	0.0060
0.0059	0.0060	0.0058	0.0054	0.0058	0.0055
0.0052	0.0057	0.0057	0.0054	0.0057	0.0056
Subgroups					
7	8	9	10	11	12
0.0052	0.0057	0.0058	0.0056	0.0058	0.0059
0.0051	0.0055	0.0059	0.0055	0.0057	0.0057
0.0046	0.0054	0.0060	0.0054	0.0057	0.0058

EXAMPLE 7.3 Creating a Median Chart I

WP Corporation produces surgical instruments. After one particular instrument is stamped, a channel is machined into the part. Because of this channel, part thickness plays an important role in the quality of the finished product. Since die roll can effectively reduce part thickness, it is a key quality characteristic to monitor after the stamping operation. ("Die roll" is a by-product of the stamping process and refers to the feathered edge created by the downward force used to shear the metal.) In order to monitor the amount of die roll on the stamped part, WP has decided to use a median chart.

Step 1. Using a sample size of three, the sample measurements are taken and recorded (Table 7.4).

Step 2. To create the median chart, the sample measurements for each subgroup are arranged in order from highest to lowest (Table 7.5).

Step 3. The median and range for each subgroup are calculated (Figure 7.5).

Step 4. The average of the subgroup medians is calculated to find the centerline of the median chart. This value is placed on the chart (Figure 7.5).

Medians of the subgroups

$$\overline{X}_{Md} = \frac{0.0059 + 0.0057 + 0.0057 + \cdots + 0.0058}{12}$$

$$\overline{X}_{Md} = 0.0056$$

Step 5. To find the centerline of the range chart, WP calculates the average of the subgroup ranges and places this value on the chart (Figure 7.5):

Subgroup ranges

$$\overline{R}_{Md} = \frac{0.0007 + 0.0006 + 0.0006 + \cdots + 0.0002}{12}$$

$$\overline{R}_{Md} = 0.0003$$

Step 6. They determine the control limits of the median chart using the following formulas and the values in Table 7.3:

$$UCL_{Md} = \overline{X}_{Md} + A_6\overline{R}_{Md}$$
$$= 0.0056 + 1.19(0.0003) = 0.0060$$

$$LCL_{Md} = \overline{X}_{Md} - A_6\overline{R}_{Md}$$
$$= 0.0056 - 1.19(0.0003) = 0.0052$$

$$UCL_R = D_4\overline{R}_{Md}$$
$$= 2.575(0.0003) = 0.0008$$

$$LCL_R = D_3\overline{R}_{Md}$$
$$= 0(0.0003) = 0$$

where D_3 and D_4 are found in Appendix 2 using $n = 3$.

Step 7. They record the median and range values on their appropriate charts (Figure 7.5).

Step 8. In this example, the median chart contains a point that is out of control. The R chart also should be watched to see if the run of points below the centerline continues.

An alternative to using the average of the subgroup medians for the centerline of the median chart is to use the median of the subgroup medians. The centerline for

Table 7.5 Recorded Values Arranged in Order: Die Roll Measurements in Inches

Subgroups						
	1	2	3	4	5	6
	0.0059	0.0060	0.0058	0.0054	0.0059	0.0060
Median	0.0059	0.0057	0.0057	0.0054	0.0058	0.0056
	0.0052	0.0054	0.0052	0.0054	0.0057	0.0055
Range	0.0007	0.0006	0.0006	0.0000	0.0002	0.0004

Subgroups						
	7	8	9	10	11	12
	0.0052	0.0057	0.0058	0.0056	0.0058	0.0059
Median	0.0051	0.0055	0.0059	0.0055	0.0057	0.0058
	0.0046	0.0054	0.0060	0.0054	0.0057	0.0057
Range	0.0006	0.0003	0.0002	0.0002	0.0001	0.0002

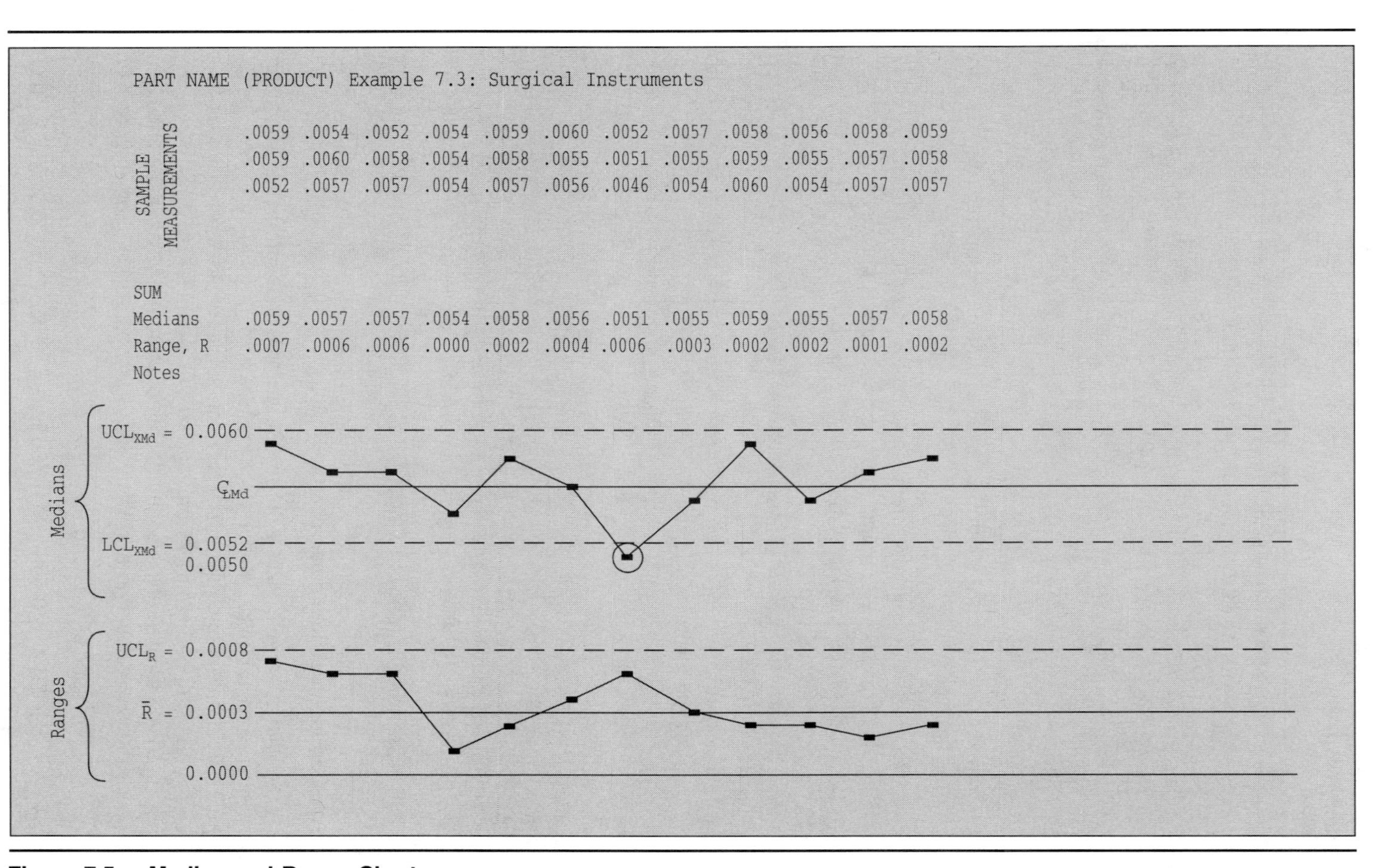
PART NAME (PRODUCT) Example 7.3: Surgical Instruments

SAMPLE MEASUREMENTS	.0059	.0054	.0052	.0054	.0059	.0060	.0052	.0057	.0058	.0056	.0058	.0059
	.0059	.0060	.0058	.0054	.0058	.0055	.0051	.0055	.0059	.0055	.0057	.0058
	.0052	.0057	.0057	.0054	.0057	.0056	.0046	.0054	.0060	.0054	.0057	.0057
SUM												
Medians	.0059	.0057	.0057	.0054	.0058	.0056	.0051	.0055	.0059	.0055	.0057	.0058
Range, R	.0007	.0006	.0006	.0000	.0002	.0004	.0006	.0003	.0002	.0002	.0001	.0002
Notes												

Figure 7.5 Median and Range Charts

Subgroup Size	A_5	D_5	D_6	d_3
2	2.224	0	3.865	0.954
3	1.265	0	2.745	1.588
4	0.829	0	2.375	1.978
5	0.712	0	2.179	2.257
6	0.562	0	2.055	2.472
7	0.520	0.078	1.967	2.645

Table 7.6 Median Chart Values A_5, D_5, D_6, and d_3 Used in Median-Chart Construction with Median of Medians

the range chart is the median range. The chart-creation process and formulas are modified as follows:

1. Record the sample measurements on the control-chart form.
2. Calculate the median and the range for each subgroup.
3. Calculate the median of the subgroup medians. This will be the centerline of the median chart.
4. Calculate the median of the subgroup ranges. This will be the centerline of the range chart.
5. The control limits of the median and range charts are determined from the following formulas and the values in Table 7.6:

$$UCL_{Md} = Md_{Md} + A_5R_{Md}$$
$$LCL_{Md} = Md_{Md} - A_5R_{Md}$$

$$UCL_R = D_6R_{Md}$$
$$LCL_R = D_5R_{Md}$$

6. Record the median and range values on their appropriate charts.
7. Interpret the chart using the information presented in Chapters 5 and 6.

EXAMPLE 7.4 Creating a Median Chart II

Rework Example 7.3 for WP Corporation's surgical instruments using the second method for creating median charts.

1. With a sample size of three, the sample measurements are taken and recorded (Table 7.4).
2. To create the median chart for each subgroup, the sample measurements are arranged in order from highest to lowest (Table 7.5).
3. To find the centerline of the median chart, the median of the subgroup medians is calculated and this value is placed on the chart (Figure 7.6).

   Medians of the subgroups:

0.0059	0.0057	0.0057	0.0054	0.0058	0.0056
0.0051	0.0055	0.0059	0.0055	0.0057	0.0058

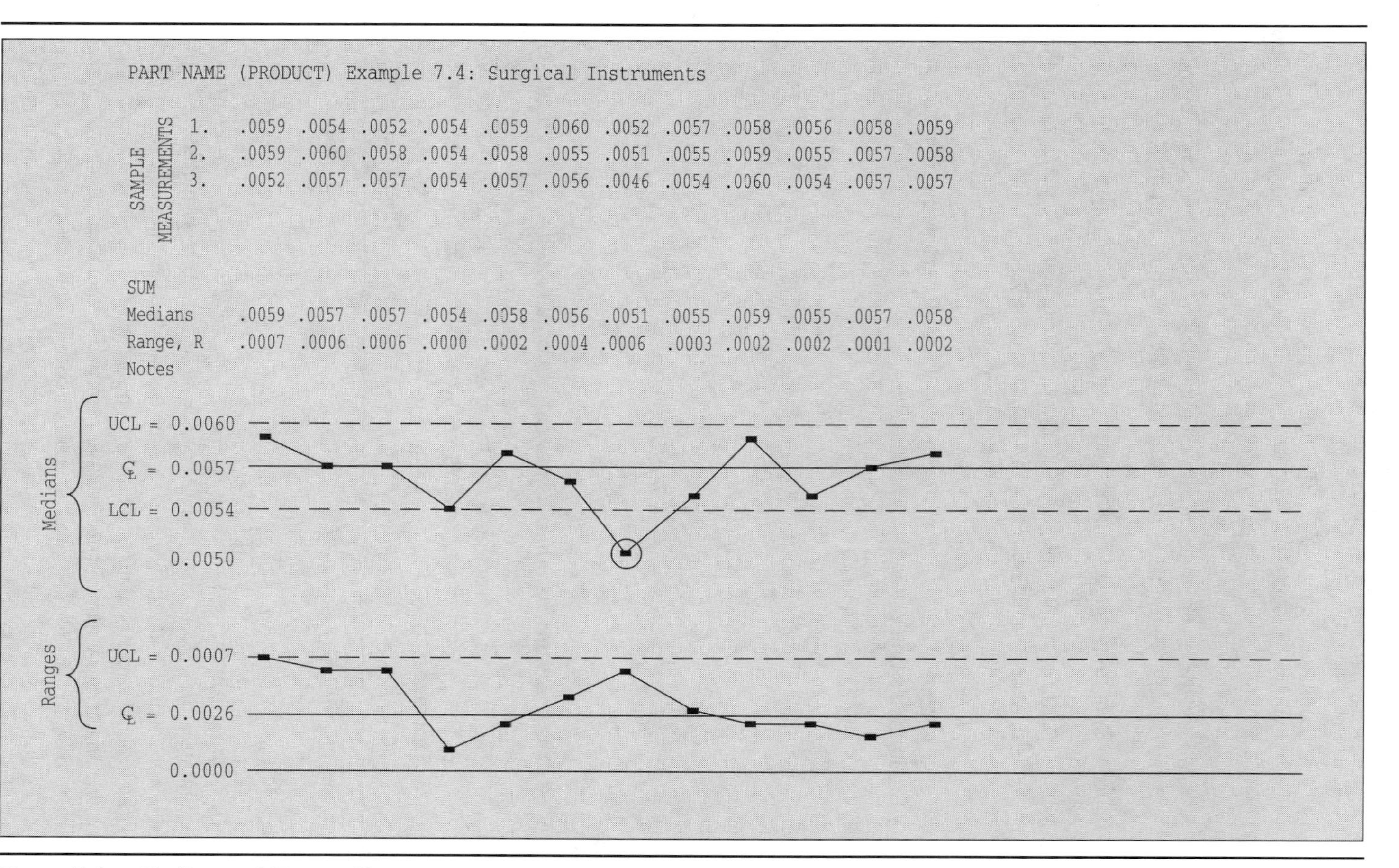

Figure 7.6 Median and Range Charts

Arranged in order from highest to lowest:

0.0059	0.0059	0.0058	0.0058	0.0057	0.0057
0.0057	0.0056	0.0055	0.0055	0.0054	0.0051

$$Md_{Md} = 0.0057$$

4. To find the centerline of the range chart, the median of the subgroup ranges is calculated and this value is placed on the chart (Figure 7.6):

Subgroup ranges in order from highest to lowest:

0.0007	0.0006	0.0006	0.0006	0.0004	0.0003
0.0002	0.0002	0.0002	0.0002	0.0001	0.0000

$$R_{Md} = 0.00025$$

5. The control limits of the median chart are determined using the following formulas and the values in Table 7.6:

$$\begin{aligned} UCL_{Md} &= Md_{Md} + A_5R_{Md} \\ &= 0.0057 + 1.265(0.00025) = 0.0060 \\ LCL_{Md} &= Md_{Md} - A_5R_{Md} \\ &= 0.0057 - 1.265(0.00025) = 0.0054 \end{aligned}$$

$$\begin{aligned} UCL_R &= D_6R_{Md} \\ &= 2.745(0.00025) = 0.0007 \\ LCL_R &= D_5R_{Md} \\ &= 0(0.00025) = 0 \end{aligned}$$

6. The median and range values are recorded on their appropriate charts (Figure 7.6).
7. The chart is interpreted using the information presented in Chapters 5 and 6. The median chart contains a single point below the lower control limit. The range chart displays a slight upward trend in the center of the chart. Both charts will be watched to see if anything of interest develops.

RUN CHARTS

Run charts *can be used to monitor process changes associated with a particular characteristic over time*. Run charts are versatile and can be constructed with either variables or attributes data. These data can be gathered in many forms, including individual measurements, counts, or subgroup averages. Time is displayed on the x axis of the chart; the value of the variable or attribute being investigated is recorded on the y axis. Cycles, trends, runs, and other patterns are easily spotted on a run chart. As Figure 7.7 shows, financial performance over time is often displayed in the form of a run chart.

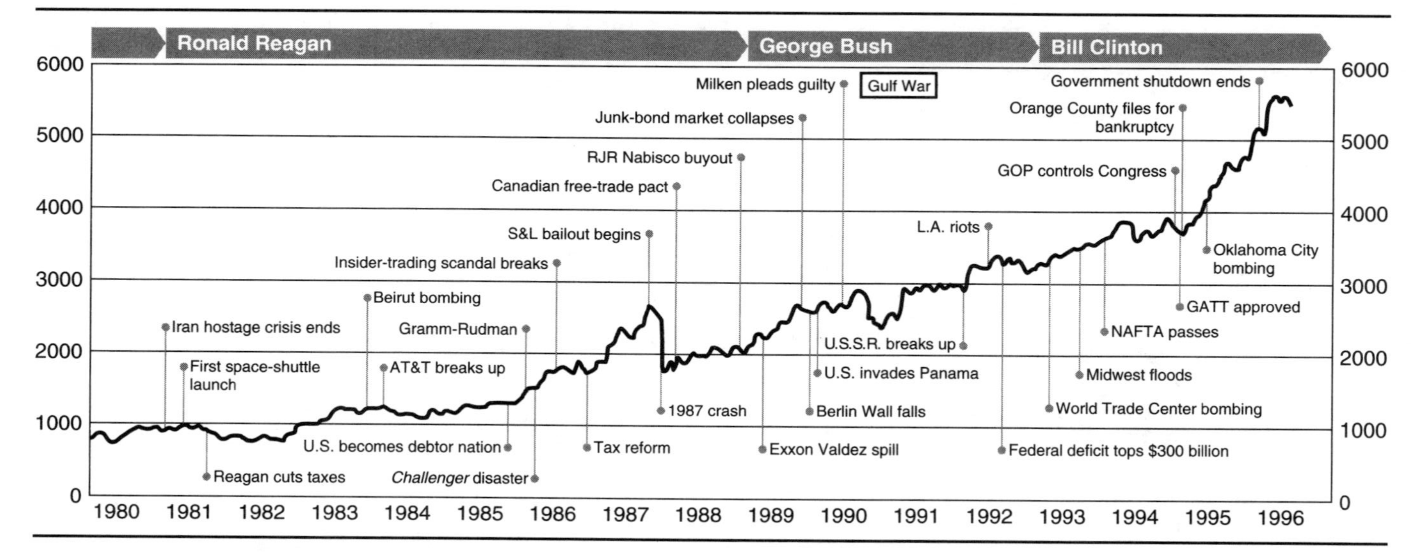

Figure 7.7 **Run Chart Showing Stock Market Performance, 1980–1996**

A run chart is created in five steps:

1. Determine the time increments necessary to properly study the process. These may be based on the rate at which the product is produced, on how often an event occurs, or according to any other time frame associated with the process under study. Mark these time increments on the x axis of the chart.
2. Scale the y axis to reflect the values that the measurements or attributes data will take.
3. Collect the data.
4. Record the data on the chart as they occur.
5. Interpret the chart. Since there are no control limits, the interpretation of a run chart is limited to looking for patterns in the data. A point that appears to be high or low cannot be judged as a special cause without control limits. It may be the worst data point in a series of points but that does not equate with being out of control.

EXAMPLE 7.5 Creating a Run Chart

At a large vegetable-processing plant, completed cans of product are placed in boxes containing 24 cans. These boxes are then placed on skids. When a skid is filled, the entire skid is shrink-wrapped. Concerns have been raised about the amount of shrink-wrap plastic used in this process. A run chart was suggested as a way to keep track of usage. The company followed the procedure for creating a run chart:

1. Determine the time increments necessary to properly study the process. The company decided to count the number of shrink-wrap rolls used per day. Time increments based on the days of the week were marked on the x axis of the chart.
2. The y axis was scaled to reflect the number of rolls of shrink wrap used per day.
3. The data were collected (Figure 7.8).
4. The data were recorded on the chart as they were gathered (Figure 7.9).
5. The chart, when interpreted, showed that on two occasions the use of shrink wrap increased. Investigation into the first occurrence revealed that shipments of canned tomato sauce increased because of a special one-time order (a big spaghetti cook-off). On the second occasion a new operator had not been trained on how much shrink wrap was appropriate to safely protect a skid. The operator received additional training, and new training methods were instituted. The problem was not expected to reoccur.

A CHART FOR VARIABLE SUBGROUP SIZE

Traditional variables control charts are created using a constant subgroup sample size. There are rare occurrences when, in the process of gathering data, the subgroup size varies. When this occurs it is necessary to recalculate the control limits to reflect the

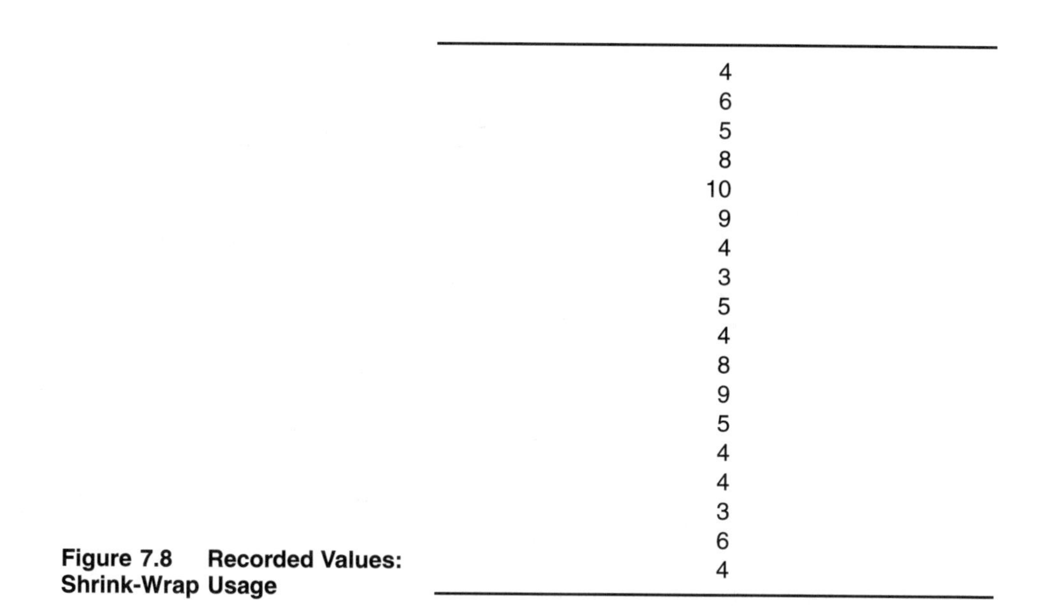

4
6
5
8
10
9
4
3
5
4
8
9
5
4
4
3
6
4

Figure 7.8 Recorded Values: Shrink-Wrap Usage

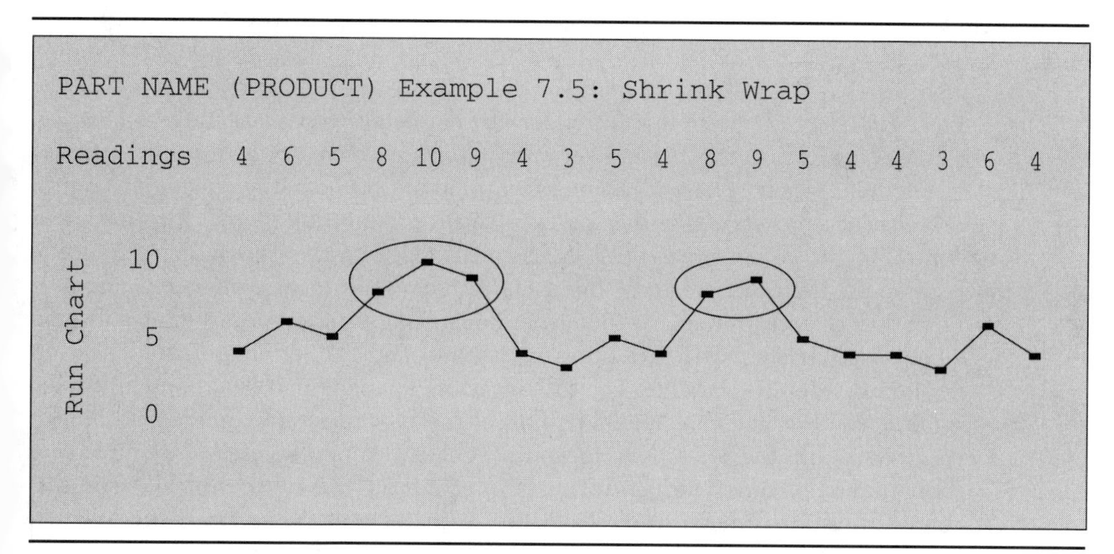

Figure 7.9 Run Chart for Shrink-Wrap Usage

change in the A_2, D_4, and D_3 values caused by the different number of samples taken. For example, for a subgroup size n = 5, the A_2 value is 0.577; if n were to equal 3, then the A_2 value would be 1.023. Each subgroup with a different sample size will have its own control limits plotted on the chart, as shown in Figure 7.10. As discussed in previous chapters, as the subgroup size increases, the control limits will come closer to the centerline. The numerous calculations required with changing subgroup sizes limit the usefulness of this chart.

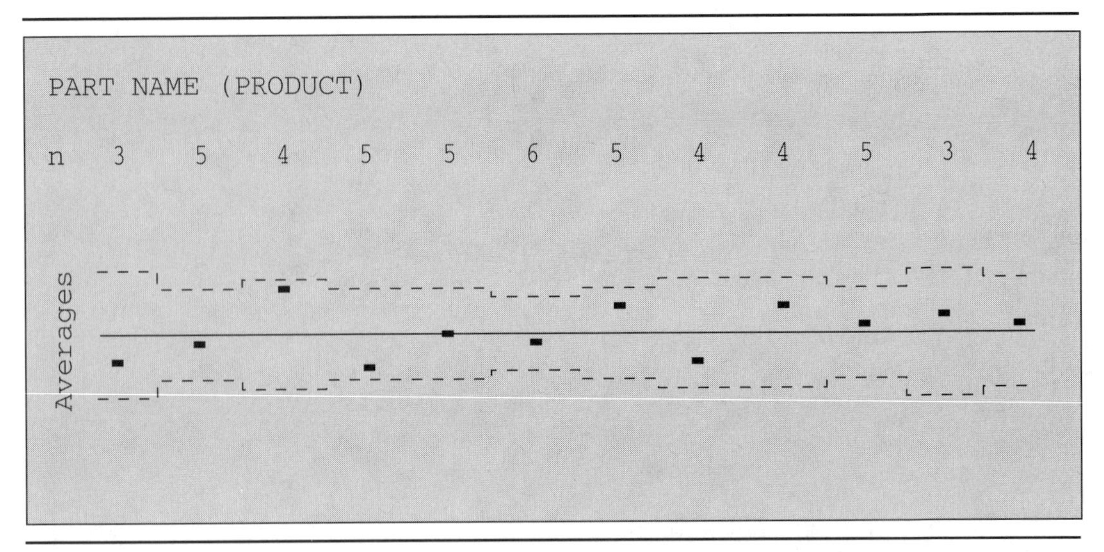

Figure 7.10 Variables Control Chart for Variable Subgroup Size

PRECONTROL CHARTS

Precontrol concepts were first introduced by Frank Satherwaite at Rath and Strong, a consulting firm, in the 1950s. ***Precontrol charts*** *study and compare product produced with tolerance limits*. The underlying assumption associated with applying precontrol concepts to a process is that the process is capable of meeting the specifications. Precontrol charts do not use control limits calculated from the data gathered from the process; the limits are created using specifications. This reliance on specifications or tolerances can result in charts that generate more false alarms or missed signals than a control chart does. Still, precontrol charts are simple to set up and run. They can be used with either variables or attribute data. They are particularly useful during setup operations and can help determine if the process setup is producing product centered within the tolerances. Precontrol charts can also be used to monitor very short production runs. Like control charts, precontrol charts can identify if the process center has shifted; they can also indicate an increase in the spread of the process. Unfortunately, precontrol charts are not as powerful as traditional control charts. They reveal little about the actual process performance. Unlike control charts, they cannot be used for problem solving, nor can they be used for calculating process capability.

Keeping in mind the differences between the spread of the individuals versus the spread of the averages, some users of precontrol charts recommend using only a portion of the tolerance spread to ensure that the product produced will be within specification. How much of the tolerance to use depends on the process capability desired. If the process capability index is expected to be 1.2, then 100 percent divided by 120 percent, or 83 percent, of the tolerance should be used. Similarly, if the process capability index is expected to be 1.15, then 0.85 should be multiplied by the tolerance spread. A process capability index of 1.10 requires that 90 percent of the tolerance be used.

Creating a precontrol chart is a three-step process:

1. *Create the zones.*
   a. Place the upper and lower specification limits on the chart.
   b. Determine the center of the specification; this becomes the centerline on the chart.
   c. Create the zones by finding the center of area between the specification limits and the center of the tolerance. To do this, subtract the centerline from the upper specification limit, divide this value in half, and add the result to the centerline.
   d. To divide the lower half of the precontrol chart, subtract the lower specification limit from the centerline, divide this value in half, and subtract the result from the centerline.

These steps create four equal zones, as shown in Figure 7.11. The center two zones, one above and one below the centerline, are combined to create the green, or "go," section of the chart. The green zone centers on the process average and is the most desirable location for the measurements. The two sections nearest the upper or lower specification limit are colored yellow for caution. Part measurements in this area are farther away from the target value and approach the specification limits and are thus less desirable. The areas above and below the upper and lower specification limits, respectively, are colored in red. Points falling in these zones are undesirable. If the measurement falls in the red zones, the process should be stopped and adjusted.

2. *Take measurements and apply setup rules.* Once the zones are established, measurements are plotted on the precontrol chart without further calculations. Set up the job and, beginning with the first piece, measure each piece as it is produced.

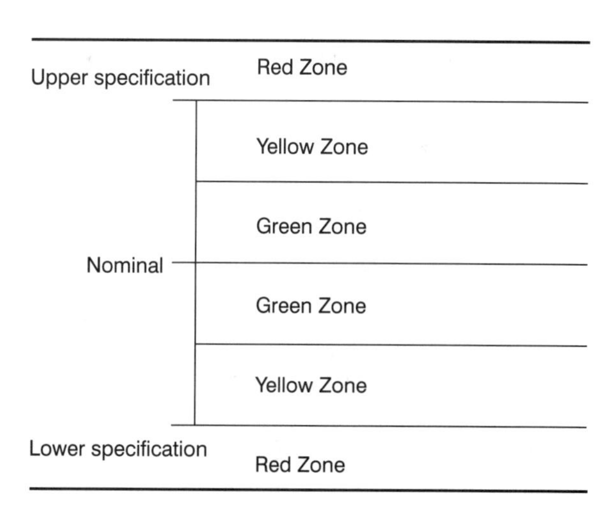

Figure 7.11 Precontrol Chart Showing Zones

Behavior of Chart	Action to Be Taken
Point in green zone	Continue.
Point in yellow zone	Check another piece.
Two points in a row in same yellow zone	Adjust the process average.
Two points in a row in opposite yellow zones	Stop and adjust process to remove variation.
Point in red zone	Stop and adjust process to remove variation.

Table 7.7 Summary of Precontrol Setup Rules

Record the consecutive measurements on the chart. Use the setup rules described below and shown in Table 7.7 to monitor and adjust the process as necessary.

a. If the measured pieces are in the green zone, continue running.
b. If one piece is inside the specification limits, but outside the green limits, check the next piece.
c. If the second piece is also outside the green limits, but still inside the specification limits, reset the process.
d. If a piece is found to be outside the specification limits, stop, make corrections, and reset the process.
e. If two successive pieces fall outside the green zone, one on the high side and one on the low side, those operating the process must immediately take steps to reduce the variation in the process.
f. Whenever a process or machine is reset, five successive pieces inside the green zone must occur before the operator implements any type of sampling plan.

3. *Apply the precontrol sampling plan.* When five pieces in a row fall in the green zone, it is okay to begin running the job. Use the run rules (Table 7.8), randomly

Behavior of Chart	Action to Be Taken
Both points in green zone	Continue.
One point in yellow zone and one in green zone	Continue.
Two points in same yellow zone	Adjust the process average.
Two points in opposite yellow zones	Stop and adjust process to remove variation.
Point in red zone	Stop and adjust process to remove variation; begin again at precontrol setup rules.

Table 7.8 Summary of Precontrol Run Rules

sampling *two* pieces at intervals, to monitor the process. The two consecutive pieces must be selected randomly, yet with enough frequency to be representative of the process. Sampling may be based on time intervals—for instance, sampling two parts every 15 minutes. Some users of precontrol charts suggest sampling a minimum of 25 pairs between setups.

If the machine is reset for any reason, the precontrol setup rules should again be used to set up the process.

EXAMPLE 7.6 Creating a Precontrol Chart

At WP Corporation, which manufactures surgical instruments, parts are stamped from steel that arrives at the plant in coils, which are straightened before being stamped. During the setup of the straightening machine, the setup operator uses a precontrol chart to ensure that the setup operation has been performed properly. The designer has set the specifications as 0.0000 to 0.0150 inch in every 4 inches of uncoiled steel. The operator uses the following steps.

Step 1. Create the Zones. The upper and lower specification limits to be placed on the chart are 0.0000 and 0.0150. From this the operator is able to determine the center of the specification, 0.0075, which is the centerline of the chart. After placing the specifications and centerline on the chart (Figure 7.12), the operator creates the zones by finding the center of the area between the specification limits and the centerline. He divides the upper half of the precontrol chart:

$$0.0150 - 0.0075 = 0.0075$$
$$0.0075 \div 2 = 0.00375$$
$$0.0075 + 0.00375 = 0.01125$$

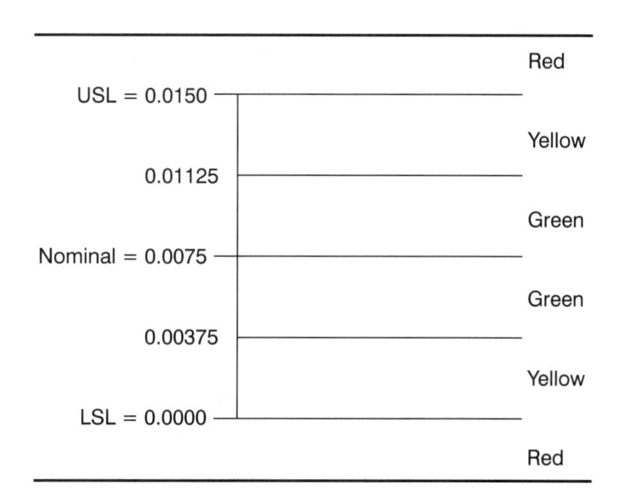

Figure 7.12 Precontrol Chart Showing Zones

Setup Measurements

0.0110
0.0141
0.0132
0.0210
0.0111
0.0090
0.0083
0.0065
0.0071

Run Measurements (2 taken every 10 min)

0.0082
0.0064

0.0080
0.0073

0.0075
0.0068

0.0124
0.0110

0.0131
0.0140

Figure 7.13 Recorded Values: Flatness

He then divides the lower half of the precontrol chart:

$$0.0075 - 0.0000 = 0.0075$$
$$0.0075 \div 2 = 0.00375$$
$$0.0000 + 0.00375 = 0.00375$$

He thus creates four equal zones and labels them as shown in Figure 7.12.

Step 2. Take Measurements and Apply the Setup Rules. Next the operator sets up the job, measures the coil as it is straightened and records the measurements (Figure 7.13). As the first pieces are straightened, the operator uses the setup rules (Table 7.7) to monitor and adjust the machine as necessary.

For this example, as the consecutive measurements are placed on the precontrol chart (Figure 7.14), the following scenario unfolds. The first measurement of flatness is 0.0110, which is inside the green, or "run," zone of the precontrol chart. The second measurement is 0.0141, which is in the yellow zone. According to the rules, the next measurement must be taken: 0.0132. Since this measurement and the last one are both in the yellow zone, the process must be adjusted.

After making the adjustment, the operator takes another measurement: 0.0210. This value is in the red zone and unacceptable. The adjustment the operator made to the straightener has not worked out. Once again he must adjust the machine. The next reading is 0.0111, which is in the green zone so no further adjustments are needed. The next four pieces—0.0090, 0.0083, 0.0065,

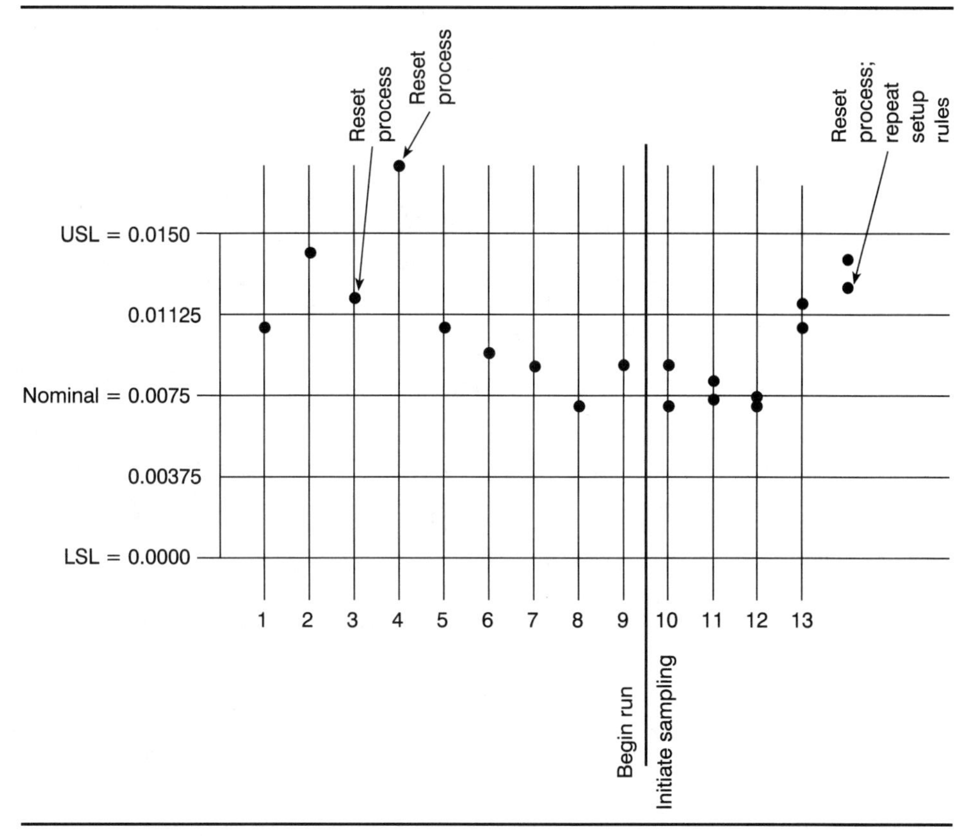

Figure 7.14 Completed Precontrol Chart

0.0071—are also in the green zone. Since five measurements in a row are in the green zone, the machine is now ready to run, and sampling can begin.

Step 3. Apply the Precontrol Sampling Plan. Having completed the setup and produced five pieces in a row in the green zone, the operator begins running the job. Using the run decision rules (Table 7.8), he randomly samples two pieces at intervals to monitor the process. The sampling plan for this process calls for taking two measurements every 10 minutes.

If the sets of measurements are in the green zone, the straightener should continue to run. In this instance, the first three sets of data—0.0082, 0.0064; 0.0080, 0.0073; 0.0075, 0.0068—are all within the green zone (Figure 7.14). Trouble begins to occur in the fourth subgroup—0.0124, 0.0110. The first measurement is in the yellow zone, signaling a change in the process; the second measurement is in the green zone, so no adjustment is made at this time. The next set of readings—0.0131, 0.0140—confirms that the process has shifted. Both of these measurements are in the yellow zone. At this point the operator must adjust the process. After any changes are made to the process, monitoring must begin again with the precontrol setup rules.

SHORT-RUN CHARTS

Traditional variables control charts used in statistical process control work most effectively with long, continuous production runs. But in today's competitive environment, manufacturers find it more economical to switch from one production run to another as needed. As the production runs get shorter and the number of products produced during each run decreases, applying traditional variables control charts becomes more difficult. Short-run charts have been developed to support statistical process control in this new production environment.

Several different methods have been developed to monitor shorter production runs. One possibility, studying the first and last pieces of the run, does not tell the investigator anything about the parts produced in between. Another method, 100 percent inspection, is costly and time-consuming and can be inaccurate. Traditional variables control charts, with separate charts for each part number and each different run of each part number, can be used. This creates an abundance of charts, yet very little information. Usually there is not enough information from a single run to calculate the control limits. The separate charts do not allow investigators to see time-related changes in the process associated with the equipment.

Unlike traditional control charts, which only have one part number per chart, short-run charts include multiple part numbers on the same chart. Because they display multiple part numbers, short-run control charts enable users to view the impact of variation on both the process and the part numbers. This focus on the process actually aids improvement efforts because the effects of changes can be seen as they relate to different part numbers.

Nominal $\overline{X}$ and R Charts

When creating a short-run control chart, the data are coded so that all the data, regardless of the part number, are scaled to a common denominator. This creates a common distribution and set of control limits. *The* ***nominal*** $\overline{X}$ ***and*** **R** ***charts*** *use coded measurements based on the nominal print dimension*. For example, if the print dimension is 3.750 ± 0.005, then the nominal dimension used for coding purposes is 3.750. Coding the measurements allows all part numbers produced by a given process to be plotted on the same chart. Because all products produced on a single machine are plotted on the same chart, the nominal $\overline{X}$ and R chart combination shows process centering and process spread. A critical assumption associated with this chart is that the process variation, as seen on the range chart, is similar for each of the part numbers being graphed on the charts. To ensure this, the chart should be constructed using parts from the same operator, machine, material, methods, and measurement techniques. If the process variation of a particular part number is more than 1.3 times greater than the total $\overline{R}$ calculated, then it must be charted on a separate control chart.

Use the following steps to create a nominal $\overline{X}$ and R chart:

1. Determine which parts will be monitored with the same control chart. Pay careful attention to select parts made by the same operator using the same machine, methods, materials, and measurement techniques.

2. Determine the nominal specification for each part number.
3. Begin the chart by collecting the data. The subgroup sample size should be the same for all part numbers.
4. Once the measurements have been taken, subtract the nominal value for the appropriate part number from the measurements. $\overline{X}$ is then calculated for each subgroup.
5. Plot the coded average measurements, $\overline{X}$'s, from step 4 on the chart. The coded values will show the difference between the nominal dimension, represented by a zero on the control chart, and the average measurement.
6. Continue to calculate, code, and plot measurements for the entire run of this particular part number.
7. When another part number is to be run, repeat the above steps and plot the points on the chart.
8. When 20 subgroups have been plotted from any combination of parts, the control limits can be calculated with a modified version of the traditional variables control chart formulas:

Nominal $\overline{X}$ chart

$$\text{Centerline} = \frac{\sum \text{coded } \overline{X}}{m}$$

$$UCL_X = \text{centerline} + A_2\overline{R}$$

$$LCL_X = \text{centerline} - A_2\overline{R}$$

Nominal range chart

$$\overline{R} = \frac{\Sigma R_i}{m}$$

$$UCL_R = D_4\overline{R}$$

$$LCL_R = D_3\overline{R}$$

9. Draw the centerlines and control limits on the chart.
10. Interpret the chart.

EXAMPLE 7.7 Creating a Nominal $\overline{X}$ and R Chart

The Special Garden Tools Corporation has created a variety of garden tools designed using ergonomic (human factors) concepts. One of their product lines, pruning shears, offers three different sizes. A user will choose the size according to the operation the shears will perform. The smallest tool is used for light pruning; the largest tool is used for larger jobs, such as pruning tree limbs. All the pruners are handheld and are designed so as not to place undue force requirements on the user's hands. To ensure this, the shears undergo a clasping force test before being shipped to retail outlets for sale to customers. This inspection tests the amount of force it takes to close the tool during typical operations. Since this is a just-in-time operation, the production runs tend to be small. Management wants to create a nominal $\overline{X}$ and R chart that contains information from three different runs of pruning shears.

1. Those monitoring the process have determined that measurements from three parts will be recorded on the same chart. Careful attention has been paid to ensure that the parts have been made using the same machine, methods, materials, and measurement techniques.
2. They determine the nominal specification for each part number:

   - Hand pruners, the smallest shears, have a force specification of 300 N $\pm$ 20.
   - Pruning shears, the midsized shears, have a force specification of 400 N $\pm$ 30.
   - Lopping shears, the heavy-duty shears, have a force specification of 500 N $\pm$ 40.

3. They now collect the data. The sample size n is 3. Table 7.9 contains the data for the hand pruner samples.
4. They subtract the nominal value for the appropriate part number from the measurements (Table 7.10). $\overline{X}_1$ is calculated by adding 4, -1, and -2, and dividing by 3.

$$\overline{X}_1 = \frac{4 + (-1) + (-2)}{3} = 0.33$$

5. They plot the coded average measurements from step 4 on the chart (Figure 7.15).
6. They then continue to calculate, code, and plot measurements for the entire run of the hand pruner shears.
7. Repeating the above steps, they add lopping shears (Table 7.11) and pruning shears (Table 7.12) to the chart.
8. Now that 20 subgroups have been plotted for the three types of pruners, they calculate the control limits with the modified version of the traditional $\overline{X}$ and R chart formulas:

Nominal $\overline{X}$ chart

$$\text{Centerline} = \frac{\sum \text{coded } \overline{X}}{m} = \frac{0.33 + 6.33 + \cdots + 10 + 7 + \cdots + 1.67}{20}$$

$$= \frac{86.67}{20} = 4.33$$

$$UCL_X = \text{centerline} + A_2\overline{R}$$
$$= 4.33 + 1.023(5.55) = 10$$

$$LCL_X = \text{centerline} - A_2\overline{R}$$
$$= 4.33 - 1.023(5.55) = -1.35$$

Nominal range chart

$$\overline{R} = \frac{\sum R_i}{m} = \frac{111}{20} = 5.55$$

Table 7.9 Recorded Values: Hand Pruners, with a Force Specification of 300 N ± 20

	Part
	304
1	299
	298
	306
2	308
	305
	300
3	298
	303
	306
4	300
	307
	305
5	301
	295
	298
6	306
	302
	304
7	301
	297

Table 7.10 Recorded Values and Differences: Hand Pruners

	Part	*Nominal*	*Difference*
	304	300	4
1	299	300	−1
	298	300	−2
	306	300	6
2	308	300	8
	305	300	5
	300	300	0
3	298	300	−2
	303	300	3
	306	300	6
4	300	300	0
	307	300	7
	305	300	5
5	301	300	1
	295	300	−5
	298	300	−2
6	306	300	6
	302	300	2
	304	300	4
7	301	300	1
	297	300	−3

$$UCL_R = D_4\overline{R}$$
$$= 2.574(5.55) = 14.29$$
$$LCL_R = D_3R = 0$$

9. They draw the centerlines and control limits on the chart (Figure 7.16).
10. And finally they interpret the chart, which reveals that the hand pruner shears require lower forces than the target value to operate. For the lopping shears, there is less variation between shears; however, they all consistently need above-average force to operate. As seen on the R chart, the pruning shears exhibit an increasing amount of variation in the force required. In general, the pruning shears require less force than the target amount to operate.

Nominal $\overline{X}$ and R charts are the most useful when the subgroup size (n) sampled is the same for all part numbers. This type of chart is best used when the nominal or center of the specifications is the most appropriate target value for all part numbers.

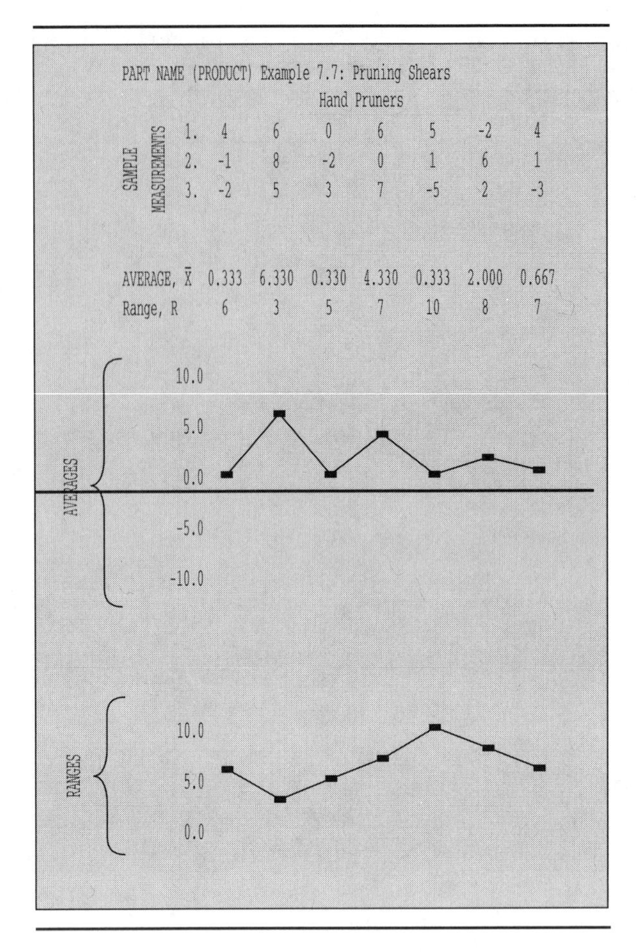

Figure 7.15 Short-Run Chart: Hand Pruners

SUMMARY

Different control charts are applicable to different situations. With such a wide variety of charts available, creators of charts to monitor processes should determine what aspects of the process they wish to study and which chart will meet their needs.

■ *Lessons Learned*

1. Individuals and moving-range charts are used to monitor processes that do not produce enough data to construct traditional variables control charts.
2. Moving-average and moving-range charts are also used when individual readings are taken. Once a subgroup size is chosen, the oldest measurement is removed from calculations as each successive measurement is taken. These charts are less sensitive to changes in the process.

Table 7.11 Recorded Values: Lopping Shears with a Force Specification of 500 N ± 40

	Part	*Nominal*	*Difference*
	510	500	10
1	512	500	12
	508	500	8
	507	500	7
2	509	500	9
	505	500	5
	509	500	9
3	510	500	10
	512	500	12
	507	500	7
4	510	500	10
	512	500	12
	507	500	7
5	509	500	9
	511	500	11
	512	500	12
6	508	500	8
	509	500	9
	507	500	7
7	505	500	5
	506	500	6
	506	500	6
8	510	500	10
	504	500	4

Table 7.12 Recorded Values: Pruning Shears with a Force Specification of 400 N ± 30

	Part	*Nominal*	*Difference*
	406	400	6
1	401	400	1
	399	400	−1
	404	400	4
2	399	400	−1
	402	400	2
	400	400	0
3	402	400	2
	398	400	−2
	395	400	−5
4	400	400	0
	401	400	1
	400	400	0
5	408	400	8
	397	400	−3

3. Charts that plot all the individual subgroup values are useful when explaining the concept of variation within a subgroup.
4. Median and range charts, though less sensitive than variables control charts, can be used to study the variation of a process. Two different methods exist to create median and range charts.
5. Run charts can be constructed using either attribute or variables data. These charts show the performance of a particular characteristic over time.
6. Charts for variable subgroup size require a greater number of calculations because the values of A_2, D_4, and D_3 in the formulas for the control limits change according to the subgroup sample size.

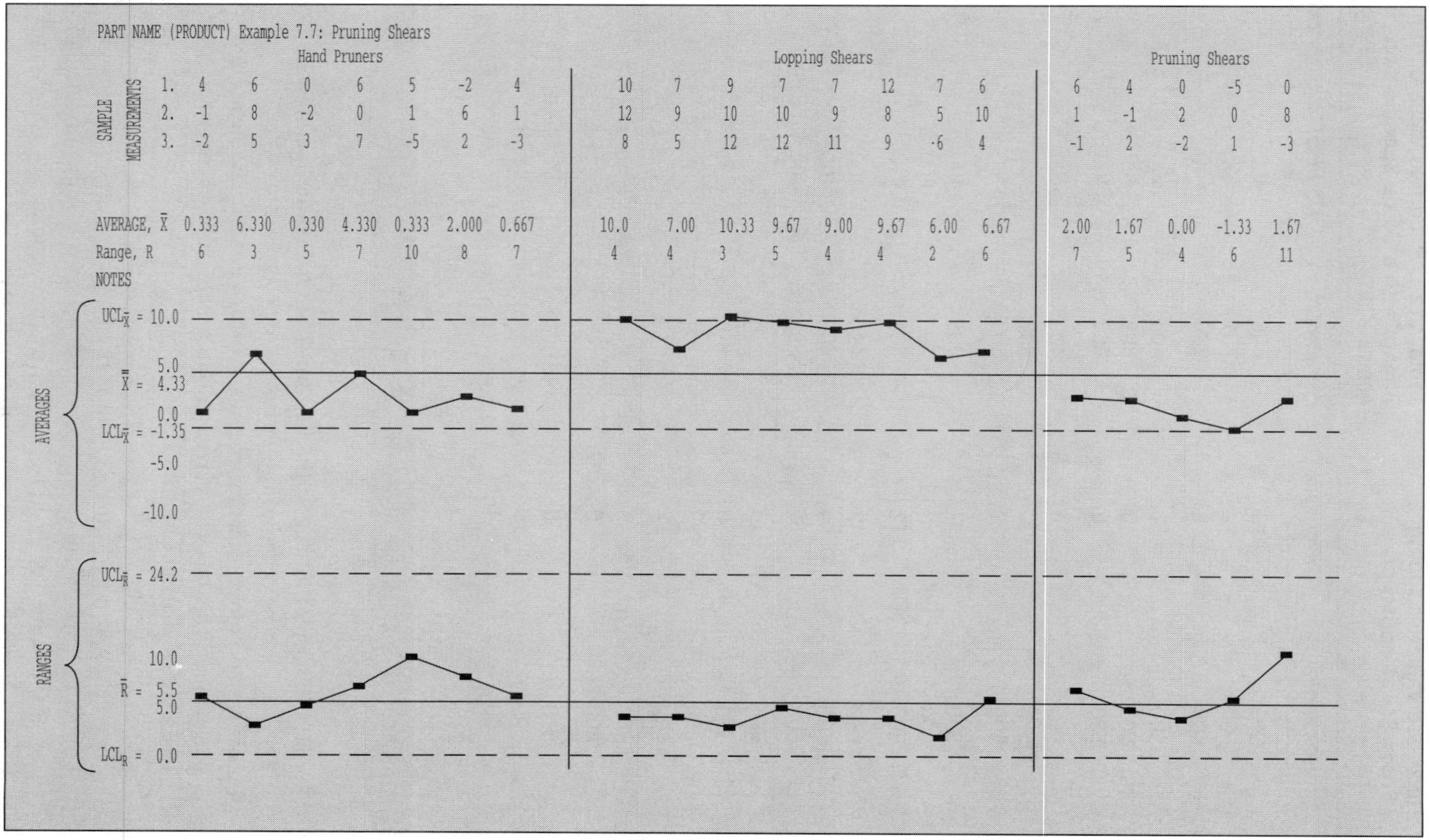

PART NAME (PRODUCT) Example 7.7: Pruning Shears

SAMPLE MEASUREMENTS	Hand Pruners							Lopping Shears								Pruning Shears				
1.	4	6	0	6	5	-2	4	10	7	9	7	7	12	7	6	6	4	0	-5	0
2.	-1	8	-2	0	1	6	1	12	9	10	10	9	8	5	10	1	-1	2	0	8
3.	-2	5	3	7	-5	2	-3	8	5	12	12	11	9	6	4	-1	2	-2	1	-3
AVERAGE, $\bar{X}$	0.333	6.330	0.330	4.330	0.333	2.000	0.667	10.0	7.00	10.33	9.67	9.00	9.67	6.00	6.67	2.00	1.67	0.00	-1.33	1.67
Range, R	6	3	5	7	10	8	7	4	4	3	5	4	4	2	6	7	5	4	6	11
NOTES																				

AVERAGES

$UCL_{\bar{X}} = 10.0$

5.0

$\bar{\bar{X}} = 4.33$

0.0

$LCL_{\bar{X}} = -1.35$

-5.0

-10.0

RANGES

$UCL_{\bar{R}} = 24.2$

10.0

$\bar{R} = 5.5$

5.0

$LCL_{R} = 0.0$

Figure 7.16 Completed Short-Run Chart, Example 7.7

7. Precontrol charts compare the item being produced with specification limits. Precontrol charts are useful during machine setups and short production runs to check process centering.
8. The nominal $\overline{X}$ and R chart uses coded measurements to monitor process centering and spread on short production runs. ■

■ Formulas

Chart for Individuals with Moving-Range Chart

$$\overline{X}_i = \frac{\Sigma X_i}{m}$$

$$\overline{R} = \frac{\Sigma R_i}{m - 1}$$

$$UCL_X = \overline{X}_i + 2.66\overline{R}$$
$$LCL_X = \overline{X}_i - 2.66\overline{R}$$

$$UCL_R = 3.27\overline{R}$$

Charts for Moving Average and Moving Range

$$\text{Centerline } \overline{\overline{X}} = \frac{\Sigma X_i}{m}$$

$$\overline{R} = \frac{\Sigma R_i}{m}$$

$$UCL_{\overline{X}} = \overline{\overline{X}} + A_2\overline{R}$$
$$LCL_{\overline{X}} = \overline{\overline{X}} - A_2\overline{R}$$

$$UCL_R = D_4\overline{R}$$
$$LCL_R = D_3\overline{R}$$

Median and Range Charts

$$UCL_{Md} = \overline{X}_{Md} + A_6\overline{R}_{Md}$$
$$LCL_{Md} = \overline{X}_{Md} - A_6\overline{R}_{Md}$$

$$UCL_R = D_4\overline{R}_{Md}$$
$$LCL_R = D_3\overline{R}_{Md}$$

or

$$UCL_{Md} = Md_{Md} + A_5R_{Md}$$
$$LCL_{Md} = Md_{Md} - A_5R_{Md}$$

$$UCL_R = D_6R_{Md}$$
$$LCL_R = D_5R_{Md}$$

Short-Run Charts: Nominal $\overline{X}$ and R

$$\text{Centerline} = \frac{\sum \text{coded } \overline{X}}{m}$$

$$UCL_X = \text{centerline} + A_2\overline{R}$$
$$LCL_X = \text{centerline} - A_2\overline{R}$$

$$\overline{R} = \frac{\sum R_i}{m}$$

$$UCL_R = D_4\overline{R}$$
$$LCL_R = D_3\overline{R}$$

Chapter Problems

Individuals Chart with Moving Range

1. Create a chart for individuals with a moving range for the measurements given below. (Values are coded 21 for 0.0021 mm.) After determining the limits, plotting the values, and interpreting the chart, calculate σ using $\overline{R}/d_2$. Is the process capable of meeting the specifications of 0.0025 ± 0.0005 mm?

   21 22 22 23 23 24 25 24 26 26 27 27 25
   26 23 23 25 25 26 23 24 24 22 23 25

2. Create a chart for individuals with a moving range from the Dow Jones Industrial Average data given below:

Sept.	3650	Sept.	3900
Oct.	3550	Oct.	3900
Nov.	3700	Nov.	3800
Dec.	3750	Dec.	3700
Jan.	3900	Jan.	3900
Feb.	3900	Feb.	3950
Mar.	3850	Mar.	4000
Apr.	3650		
May	3650		
June	3775		
July	3700		
Aug.	3775		

   How is the market performing?

Moving-Average with Moving-Range Charts

3. Use the following miles-per-gallon (mpg) data to create the centerline and control limits for a moving-average and a moving-range chart (n = 3). Comment on the chart. What is the process capability?

	mpg		*mpg*
1	36	9	28
2	35	10	27
3	37	11	35
4	38	12	36
5	32	13	35
6	35	14	34
7	36	15	37
8	35		

4. Eighteen successive heats of a steel alloy are tested for R_C hardness. The resulting data are shown below. Set up control limits for the moving-average and moving-range chart for a sample size of n = 3.

Heat	*Hardness*
1	0.806
2	0.814
3	0.810
4	0.820
5	0.819
6	0.815
7	0.817
8	0.810
9	0.811
10	0.809
11	0.808
12	0.810
13	0.812
14	0.810
15	0.809
16	0.807
17	0.807
18	0.800

5. Create a moving-average and moving-range chart (n = 3) using the following Dow Jones Industrial Average information. How is the market performing? How does the chart for individuals (Problem 2) compare with the chart for moving averages for the same data?

Sept.	3650	Sept.	3900
Oct.	3550	Oct.	3900
Nov.	3700	Nov.	3800
Dec.	3750	Dec.	3700
Jan.	3900	Jan.	3900
Feb.	3900	Feb.	3950
Mar.	3850	Mar.	4000
Apr.	3650		
May	3650		
June	3775		
July	3700		
Aug.	3775		

Individuals Chart

6. "Score depth" is the depth of a partial cut on a piece of metal. Pop-top lids with pull tabs are scored to make it easy for the customer to remove the lid but still keep the container closed during shipment and storage. Plot the following score-depth values on a chart plotting all individuals. The unit of measure is coded for 0.0001 inch. Create the limits using the traditional $\overline{X}$ and R chart formulas presented in Chapter 5. Discuss how the spread of the individuals is reflected in the R chart. (n = 3)

Sample No.	1	2	3	4	5	6	7	8	9	10	11	12	13	14	15
	27	28	28	27	31	28	26	26	25	26	29	27	28	26	25
	33	32	29	33	34	34	34	33	32	32	36	34	35	33	33
	32	31	33	32	30	33	33	33	31	33	32	32	33	32	31
	26	27	33	28	28	29	27	27	28	28	29	27	31	28	25

Median and Range Chart

7. Create a median chart and a range chart with the following information. Be sure to plot the points, limits, and centerlines. (n = 3)

Subgroup					
1	2	3	4	5	6
6	10	7	8	9	12
9	4	8	9	10	11
5	11	5	13	13	10

8. "Expand curl height" refers to the curled lip on lids and tops. Curled edges of lids are crimped down to seal the top of the can. Using the following information on curl height, create median and range charts to study the process. The data are coded 74 for 0.0074. How is this process performing? (n = 3)

Curl Height

Sample	1	2	3	4	5	6	7	8	9	10	11	12	13	14	15	16	17	18	19
1	74	75	72	73	72	74	73	73	73	74	74	74	73	73	71	75	74	72	75
2	72	73	72	72	73	73	76	74	72	74	74	73	74	72	73	72	74	74	75
3	75	74	74	74	73	73	74	75	73	73	73	73	72	75	75	73	74	74	72

Run Charts

9. Create a run chart using the following information about the bond market. How is the market performing?

Sept.	5650	Sept.	5000
Oct.	5800	Oct.	5050
Nov.	5500	Nov.	4900
Dec.	5580	Dec.	5100
Jan.	5600	Jan.	5200
Feb.	5500	Feb.	5300
Mar.	5300	Mar.	5400
Apr.	5200		
May	5000		
June	5200		
July	5000		
Aug.	5100		

10. Create a run chart for the following torque measurements. (Values recorded left to right.)

Car Strut Nut Torque									
74.6	73.1	73.1	73.7	74.5	72.5	72.9	72.8	72.7	74.7
75.4	73.2	74.2	73.4	73.8	73.8	73.1	74.5	73.0	73.2
74.4	74.8	73.9	74.8	72.7	72.2				

11. Create a run chart for the following safety statistics from the security office of a major apartment building. The chart should track the total number of violations from 1989 to 1999. What does the chart tell you about their safety record and their efforts to combat crime?

	1989	*1990*	*1991*	*1992*	*1993*	*1994*	*1995*	*1996*	*1997*	*1998*	*1999*
Homicide	0	0	0	0	0	0	0	0	0	1	0
Aggravated Assault	2	1	0	1	3	4	5	5	6	4	3
Burglary	80	19	18	25	40	26	23	29	30	21	20
Grand Theft Auto	2	2	6	15	3	0	2	10	5	2	3
Theft	67	110	86	125	142	91	120	79	83	78	63
Petty Theft	37	42	115	140	136	112	110	98	76	52	80
Grand Theft	31	19	16	5	4	8	4	6	2	4	3
Alcohol Violations	20	17	16	16	11	8	27	15	10	12	9
Drug Violations	3	4	2	0	0	0	0	2	3	2	1
Firearms Violations	0	0	0	0	0	1	0	0	1	0	0

Precontrol

12. Describe the concept of precontrol. When is it used? How are the zones established?

13. NB Manufacturing has ordered a new machine. During today's runoff the following data were gathered concerning the runout for the diameter of the shaft machined by this piece of equipment. A precontrol chart was used to set up the machines. Recreate the precontrol chart from the following data. The tolerance associated with this part is a maximum runout value of 0.002 (upper specification). The optimal value is 0.000 (no runout), the lower specification limit.

Runout			
0.0021	0.0004	0.0009	0.0025
0.0013	0.0003	0.0020	0.0010
0.0018	0.0010	0.0012	
0.0007	0.0015	0.0021	
0.0002	0.0011	0.0022	
0.0030	0.0023	0.0004	
0.0024	0.0025	0.0027	
0.0006	0.0022	0.0020	
0.0002	0.0025	0.0011	
0.0006	0.0004	0.0018	

14. RY Inc. is in the business of producing medical and hospital products. Some of the products that they produce include surgical tables, stretchers, sterilizers, and examination tables. Surgical tables require spacer holes, which are drilled into the tables. When a drill bit becomes dull, the size of the hole being drilled becomes larger than the specifications of 0.250 cm ± 0.010. In the past, the operator would decide when to change the drill bits. RY Inc.'s quality engineer recently approved the use of precontrol charts to monitor the process. When parts enter the yellow zone, the operator knows that drill-bit changing time is approaching. As soon as a part enters the red zone, the operator changes the bit. Using this information, create and interpret a precontrol chart for the following information.

0.250
0.250
0.251
0.250
0.252
0.253
0.252
0.255
0.259
0.261
0.249
0.250
0.250
0.250
0.252

0.251
0.253
02.54
0.254
0.256
0.259
0.259
0.260
0.261
0.248
0.248

Short-Run Charts

15. In an automatic transmission, a part called a parking pawl is used to hold the vehicle in park. The size of the pawl depends on the model of vehicle. Max Manufacturing Inc. makes two different thicknesses of pawls—a thin pawl, which has a nominal dimension of 0.2950 inch, and a thick pawl, which has a nominal dimension of 0.6850 inch. Both parts are stamped on the stamp machine. Max Manufacturing runs a just-in-time operation and uses short-run control charts to monitor critical part dimensions. Pawl thickness for both parts is measured and recorded on the same short-run control chart. Create and interpret a short-run control chart for the following data:

0.2946	0.2947	0.6850	0.6853
0.2951	0.2951	0.6851	0.6849
0.2957	0.2949	0.6852	0.6852
0.2951	0.2947	0.6847	0.6848
0.2945	0.2950	0.6851	
0.2951	0.2951	0.6852	
0.2950	0.2952	0.6853	
0.2952	0.2949	0.6850	
0.2952	0.2944	0.6848	
0.2950	0.2951	0.6849	
0.2947	0.2948	0.6847	
0.2945	0.2954	0.6849	

16. A series of pinion gears for a van seat recliner are fine-blanked on the same press. The nominal part diameters are small (50.8 mm), medium (60.2 mm), and large (70.0 mm). Create a short-run control chart for the following data:

60.1	70.0	50.8
60.2	70.1	50.9
60.4	70.1	50.8
60.2	70.2	51.0
60.3	70.0	51.0
60.2	69.9	50.9
60.1	69.8	50.9
60.2	70.0	50.7
60.1	69.9	51.0
60.4		50.9
60.2		50.9
60.2		50.8

CASE STUDY 7.1
Precontrol

TIKI Inc. produces and distributes natural all-fruit drinks. They sell several sizes of bottles: 10-, 16-, and 32-oz. Each bottle-filling machine is capable of filling any of the three sizes of bottles. Each machine can also fill any of the seven different flavors of drinks offered by TIKI Inc. The versatility of the machines helps keep costs low.

Machine setup is required whenever the product line is changed. A change may be related to the size of bottle being filled or to the type of drink being placed in the bottle. For instance, often production scheduling will call for back-to-back runs of grape drink in all three sizes. While the mechanisms providing the grape drink are not altered, the machines must be set up to accommodate the three different bottle sizes. In order to ensure that the appropriate amount of product is placed in each bottle, careful attention is given to the setup process. TIKI Inc. uses precontrol to monitor the setup process and to prevent the operator from "tweaking" the machine unnecessarily.

Today's changeover requires a switch from the larger 32-oz family-size bottle of apple juice product to the smaller 10-oz lunch-size bottle. Since bottles that contain less than 10 oz of juice would result in fines from the Bureau of Weights and Measures, the specification limits associated with the 10-oz bottle are 10.2 ± 0.2 oz. The operator uses the following steps to set up the charts:

1. She creates zones by dividing the width of the tolerance limits (upper specification limit minus lower specification limit) in half:

$$10.4 - 10.0 = 0.4 \text{ tolerance width}$$
$$0.4 \div 2 = 0.2$$

or 0.2 above the nominal dimension and 0.2 below the nominal. Additional zones are created by dividing the distance between each specification limit and the nominal in half:

$$10.4 - 10.2 = 0.2$$
$$0.2 \div 2 = 0.1$$
$$10.2 - 10.0 = 0.2$$
$$0.2 \div 2 = 0.1$$

As shown in Figure 1, the chart now has four sections: 10.4 to 10.3, 10.3 to 10.2, 10.2 to 10.1, and 10.1 to 10.0. The center two sections, one above and one below the centerline, are combined to create the green, or "go," section of the chart: 10.1 to 10.3. The two sections nearest the upper or lower specification limit, the yellow, or "caution," sections, include 10.3 to 10.4 and 10.0 to 10.1. The

PRECONTROL SHEET

MACHINE ______________ OPERATION ______________ PART NUMBER ________

DEPARTMENT ______________ SHIFT ______ DATE ________ SHEET NO. ________

Specification/Dimension ______________

DATE																																				
TIME																																				
MAX. TOL. 10.4																																				
1/2 MAX. TOL. 10.3																																				
NOMINAL 10.2																																				
1/2 MIN. TOL. 10.1																																				
MIN. TOL. 10.0																																				

Specification/Dimension ______________

DATE																																				
TIME																																				
MAX. TOL.																																				
1/2 MAX. TOL.																																				
NOMINAL																																				
1/2 MIN. TOL.																																				
MIN. TOL.																																				

Figure 1

areas above and below the upper and lower specification limits—10.4 and 10.0, respectively—are colored in red.

2. Having created her chart, the operator sets up and starts running the 10-oz apple juice job. During the setup process the machine can be operated by hand to allow the operator to produce one bottle, stop the machine, check the fill level, and then start the machine to produce another bottle.

3. Beginning with the first bottle filled, the operator measures the fill level of the product. She records the measurement on the chart. The first sample is 10.35, a value not outside the upper specification limit but in the yellow zone (10.3 to 10.4). According to the setup rules, if one piece is inside specification limits but outside the green limits, the operator must check the next piece. She fills another bottle and checks it. She finds it to be 10.41, which is outside the specification limits. Now she must apply the setup rule that specifies that if a piece is found to be outside the specification limits, the process must be reset.

After resetting the process, the operator fills the third bottle and checks the fill rate. This time the fill rate is 10.31. While this value is very close to the green zone, according to the setup rules the operator has to check another bottle before making any changes to the process. She fills another bottle and finds it to be 10.22, well inside the green zone (10.1 to 10.3). In this case, since this piece falls inside the green zone, the setup rules allow her to continue the process, which will be reset only when two pieces in a row fall in the yellow zone.

While it is not necessary to reset the machine at this point, she still must take further samples and check fill levels to meet the setup criteria. When five pieces in a row fall in the green zone, she can begin full operation of the machine.

The next five fill rates recorded by the operator are 10.2, 10.19, 10.2, 10.23, and 10.21 oz, all within the green zone. The operator may now continue to operate the process normally and begin sampling for the creation of other charts such as the $\overline{X}$ and R chart.

If at any time the run decision rules are violated by pieces in the yellow or red zones, she must begin the process again at step 2 and proceed with the sampling until five successive pieces fall inside the green zone.

Assignment

Following the method presented in the chapter and the format of this case, use the run rules to monitor the following output from the process.

Sample Data in Pairs
10.20, 10.25
10.18, 10.12
10.21, 10.22
10.25, 10.27
10.26, 10.28
10.31, 10.26
10.28, 10.26
10.25, 10.22
10.21, 10.20
10.17, 10.15
10.09, 10.11
10.08, 10.09

CASE STUDY 7.2
Run Charts

MYRY Inc. designs and manufactures equipment used for making books. Their equipment includes binderies, printers, control systems, and material-handling devices. Until recently, they have been having difficulty during the final assembly process. Far too often, during the final assembly of a machine, a needed part would not be present. There has been no correlation between the missing parts and the type of machine being assembled. If a part is not available, an assembly person writes up an unplanned issue request. Once the assembly person receives a signature from the line supervisor, he or she can go to the stockroom and receive the part immediately. Unplanned issue parts are undesirable for two reasons. First, when a part is needed, work stops on the line. This time lost during assembly idles workers and equipment, costs money, and could cause delays in shipping. Second, the unplanned issued part must be taken from inventory, which means it may not be available when it is needed for the next machine.

The project engineers in charge of isolating the root cause of this problem and correcting it have used cause-and-effect diagrams to isolate the problem. Their investigation has shown that the computerized parts scheduling and tracking system is not up to the task. The data from their investigation convinced management to purchase and install a new parts scheduling and tracking system. The new system is a highly disciplined system, where all parts needed for constructing a piece of equipment are tracked from the moment the order is placed until construction is complete. In order for a part to be on the assembly floor, a requirement must exist in the system. The computer system turned out to be much more than a means of installing a counting system. Not only does the system track parts availability more accurately, when implementing the new system, MYRY had to improve their ordering and manufacturing processes. The new system enables all areas of the company to work together and communicate more efficiently. For example, purchasing can relay information directly to their vendors and track incoming part quality, eliminating stockouts due to late deliveries or quality problems. Bills of material from the engineering department can be transferred immediately to the factory floor or purchasing, wherever the information is needed. Manufacturing process changes have enabled them to route the equipment under construction through the factory more efficiently, avoiding lost parts and decreasing assembly time.

Assignment

Create a run chart to track the performance of the parts scheduling and tracking system both before and after the change. Can you pick out where the change occurs? How do you know? Does the new system make a significant difference? How do you know? Can you see the period of time where MYRY was getting used to the new system? What was the average number of unplanned issues before the change? During? After? What do you think the new level of total unplanned issues per week will average out to?

Total Unplanned Issues per Week	
Week 1	164
Week 2	160
Week 3	152
Week 4	156
Week 5	153
Week 6	155
Week 7	150
Week 8	140
Week 9	145
Week 10	148
Week 11	138
Week 12	150
Week 13	147
Week 14	152
Week 15	168
Week 16	159
Week 17	151
Week 18	149
Week 19	154
Week 20	160
Week 21	108
Week 22	105
Week 24	92
Week 25	85
Week 26	60
Week 27	42
Week 28	31
Week 29	15
Week 30	6
Week 31	5
Week 32	6
Week 33	4
Week 34	5

III

Control Charts for Attributes

8

Probability

Learning Opportunities:

1. To become familiar with seven probability theorems
2. To become familiar with the discrete probability distributions: hypergeometric, binomial, and Poisson
3. To review the normal continuous probability distribution
4. To become familiar with the binomial approximation to the hypergeometric, the Poisson approximation to the binomial, and other approximations ■

Thirty People at a Party

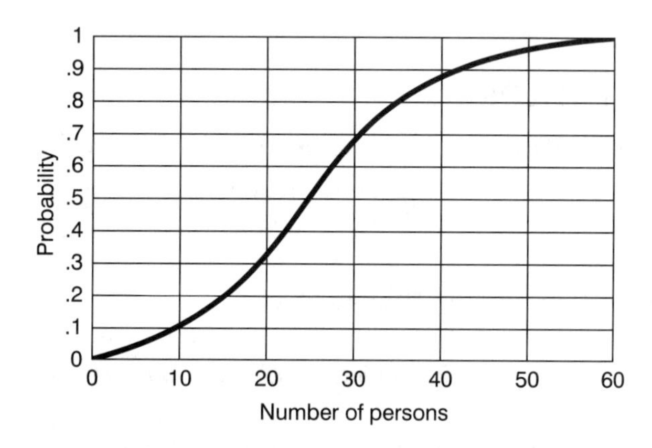

The Likelihood of Sharing a Birthday

Often we are surprised at a party or in a class to discover that someone has the same birthday as we do. It seems like such an unusual coincidence. In fact, as seen in the chart above, as the number of people gathered together increases, the chance of two people in the group having the same birthday increases dramatically. Try verifying this phenomenon in your class.

Probability is the chance that an event will occur. When an event has a certain probability, that doesn't mean the event will definitely *happen; probabilities specify the* chance *of an event happening. This chapter discusses probability and its application to quality.*

Probability affects all aspects of our lives. Intuitively we understand things like the chance it will rain, the probability of there being a traffic jam on the commute to school or to work, or the chance that a particular concert will sell out on the first day that tickets go on sale. ***Probability** is the chance that something will happen*. Probabilities *quantify* the chance that an event will occur. Having a probability attached to an event does not mean that the event will definitely happen.

A person's first exposure to probability theory usually occurs early in life in the form of the coin toss. A fair coin is tossed and the winner is the person who correctly predicts whether the coin will land face up or face down. After observing the way the coin lands a number of times, we see a pattern emerge: approximately half of the tosses land face up, half face down. In future coin tosses, whenever a pattern inconsistent with this 50-50 split emerges, we wonder if the coin is "fair."

Probability plays a role in the quality of products being produced or services being provided. There is the chance that a tool will break, that a line will clog, that a person will be late for an appointment, or that a service will not be performed on time.

PROBABILITY THEOREMS

The probability of an occurrence is written as P(A) *and is equal to*

$$P(A) = \frac{\text{number of occurrences}}{\text{total number of possibilities}} = \frac{s}{n}$$

EXAMPLE 8.1 Determining Probabilities

$$P(\text{opening a 309-page book to page 97}) = \frac{1}{309}$$

There is only one successful case possible, only one page 97, out of the total cases of 309.

Or

$$P(\text{heads}) = \frac{\text{coin lands head side up}}{\text{two sides of a coin}} = \frac{1}{2}$$

Or during a production run, two test parts are mixed in with a box of 25 good parts. During an inspection, the probability of randomly selecting one of the test parts is

$$P(\text{selecting a test part}) = \frac{2}{27}$$

Or a 104-key computer keyboard has been disassembled and the keys have been placed in a paper sack. If all of the keys, including the function keys, number keys, and command keys, are in the bag, what is the probability that one of the 12 function keys will be drawn randomly from the bag?

$$P(\text{selecting one of the 12 function keys}) = \frac{12}{104}$$

Theorem 1: Probability Is Expressed as a Number between 0 and 1:

$$0 \leq P(A) \leq 1$$

If an event has a probability value of 1, then it is a certainty that it will happen; in other words, there is a 100 percent chance the event will occur. At the other end of the spectrum, if an event will not occur, then it will have a probability value of 0. In between the certainty that an event will definitely occur or not occur exist probabilities defined by their ratios of desired occurrences to the total number of occurrences, as seen in Example 8.1.

Theorem 2: The Sum of the Probabilities of the Events in a Situation is Equal to 1.00:

$$\Sigma P_i = P(A) + P(B) + \cdots + P(N) = 1.00$$

EXAMPLE 8.2 The Sum of the Probabilities

A manufacturer of piston rings receives raw materials from three different suppliers. In the stockroom there are currently 20 steel rolls from supplier A, 30 rolls of steel from supplier B, and another 50 rolls from supplier C. From these 100 rolls of steel in the stockroom, a machinist will encounter the following probabilities in selecting steel for the next job:

$$P(\text{steel from supplier A}) = 20/100 = 0.20 \text{ or } 20\%$$
$$P(\text{steel from supplier B}) = 30/100 = 0.30 \text{ or } 30\%$$
$$P(\text{steel from supplier C}) = 50/100 = 0.50 \text{ or } 50\%$$

And from theorem 2:

$P(\text{steel from A}) + P(\text{steel from B}) + P(\text{steel from C}) = 0.20 + 0.30 + 0.50 = 1.00$

Theorem 3: If P(A) Is the Probability That an Event A Will Occur, Then the Probability That A Will Not Occur Is

$$P(A') = 1.00 - P(A)$$

EXAMPLE 8.3 Determining the Probability That an Event Will Not Occur

Two students have been sitting in the park enjoying the fine spring weather. For entertainment they have been watching a dog and its owner play Frisbee. The dog is a very good Frisbee player, catching nearly every Frisbee before it touches the ground. If the dog has caught 8 of the last 10 Frisbees, what is the probability that it will not catch the Frisbee?

$$P(A') = 1.00 - P(A)$$

If $P(A) = P(\text{dog catches Frisbee}) = 0.80$, then $P(\text{dog does not catch Frisbee}) = 1.00 - 0.80 = 0.20$

Events are considered ***mutually exclusive*** *if they cannot occur simultaneously*. Mutually exclusive events can happen only one at a time. When one event occurs it prevents the other from happening. Rolling a die and getting a 6 is an event mutually exclusive of getting any other value on that roll of the die.

Theorem 4: For Mutually Exclusive Events, the Probability That Either Event A or Event B Will Occur Is the Sum of Their Respective Probabilities:

$$P(A \text{ or } B) = P(A) + P(B)$$

Theorem 4 is called "the additive law of probability." Notice that the "or" in the probability statement is represented by a "+" sign.

EXAMPLE 8.4 Probability in Mutually Exclusive Events I

At a party you and another partygoer are tied in a competition for the door prize. Those hosting the party have decided that each of you should select two values on a single die. When the die is rolled, if one of your values comes up then you will be the winner. You select 2 and 4 as your lucky numbers. What is the probability that either a 2 or a 4 will appear on the die when it is rolled?

$$P(\text{rolling a 2 or a 4 on a die}) = P(2) + P(4)$$
$$P(2) = 1/6 \qquad P(4) = 1/6$$
$$P(\text{rolling a 2 or a 4 on a die}) = 1/6 + 1/6 = 1/3$$

EXAMPLE 8.5 Probability in Mutually Exclusive Events II

At the piston ring factory in Example 8.2, a machine operator visits the raw materials holding area. If 20 percent of the steel comes from supplier A, 30 percent from supplier B, and 50 percent from supplier C, what is the probability that the machinist will randomly select steel from either supplier A or supplier C?

$$P(\text{steel from supplier A}) = 0.20$$
$$P(\text{steel from supplier B}) = 0.30$$
$$P(\text{steel from supplier C}) = 0.50$$

Since the choice of steel from supplier A precludes choosing either supplier B or C, and vice versa, these events are mutually exclusive. Applying theorem 4, we have

$$\begin{aligned} P(\text{steel from A or steel from C}) &= P(A) + P(C) \\ &= 0.20 + 0.50 \\ &= 0.70 \end{aligned}$$

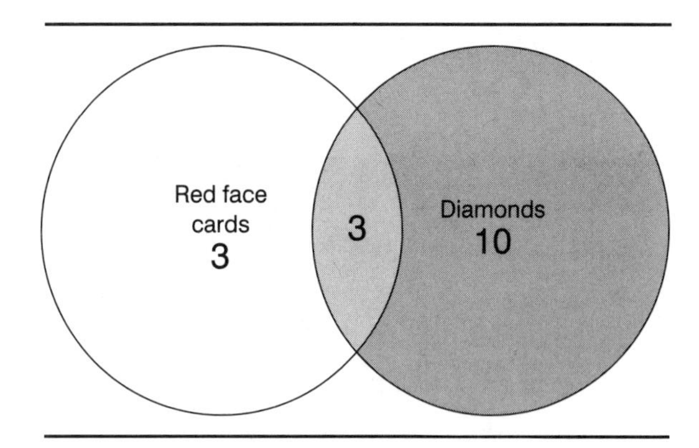

Figure 8.1 Venn Diagram

Events are considered ***nonmutually exclusive events*** *when they may occur simultaneously*. If both can happen, then theorem 4 must be modified to take into account the overlapping area where both events can occur simultaneously (Figure 8.1).

Theorem 5: When Events A and B Are Not Mutually Exclusive Events, the Probability That Either Event A or Event B or Both Will Occur Is

$$P(A \text{ or } B \text{ or both}) = P(A) + P(B) - P(\text{both})$$

EXAMPLE 8.6 Probability in Nonmutually Exclusive Events I

A quality-assurance class at the local university consists of 32 students, 12 females and 20 males. The professor recently asked, "How many of you are out-of-state students?" Eight of the women and five of the men identified themselves as out-of-state students (Figure 8.2). Apply theorem 5 and determine the probability that a student selected at random will be female, out-of-state, or both:

$$P(\text{female or out-of-state or both}) = P(F) + P(O) - P(\text{both})$$

Counting the number of occurrences, we find

$$\begin{aligned} P(\text{female}) &= 12/32 \\ P(\text{out-of-state}) &= 13/32 \\ P(\text{both}) &= 8/32 \\ P(\text{female or out-of-state or both}) &= 12/32 + 13/32 - 8/32 \\ &= 17/32 \end{aligned}$$

If the overlap of eight students who are both female and out-of-state had not been subtracted out, then, as shown by Figure 8.2, the total (25/32) would have been incorrect.

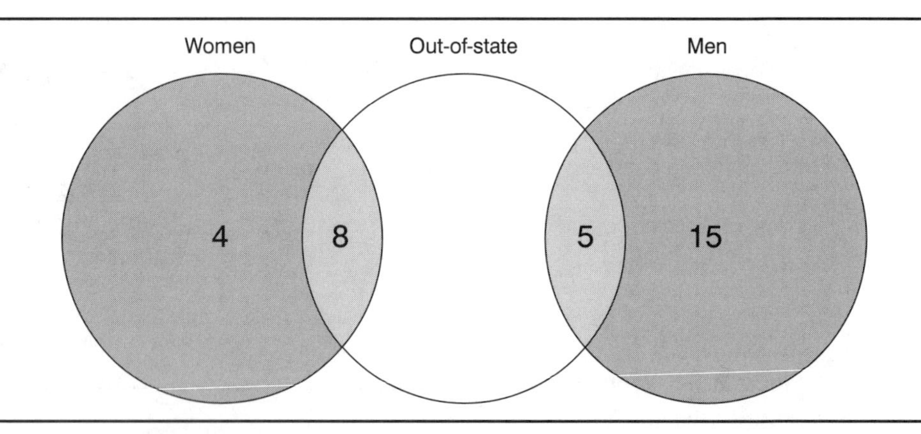

Figure 8.2 A Venn Diagram for Example 8.6

EXAMPLE 8.7 Probability in Nonmutually Exclusive Events II

Owners of a local gas station/convenience store are trying to determine what purchases their customers make most often. Knowing this information will help the managers of the station plan what to sell. The owners have uncovered the following information:

$$\text{P(customer purchasing a soft drink)} = 0.12$$
$$\text{P(custoner purchasing milk)} = 0.29$$

Customers who purchase soft drinks may also purchase milk or vice versa. Therefore, these events are not mutually exclusive, and

$$\text{P(customer purchases both a soft drink and milk)} = 0.07$$

The managers would like to know the probability that the customer will purchase either milk *or* a soft drink. Applying theorem 5, we find

$$\begin{aligned} \text{P(milk or soft drink)} &= \text{P(milk)} + \text{P(soft drink)} - \text{P(both)} \\ &= 0.29 + 0.12 - 0.07 \\ &= 0.34 \end{aligned}$$

It is not unusual for one outcome or event to affect the outcome of another event. *When the occurrence of one event alters the probabilities associated with another event, these events are considered* ***dependent.***

Theorem 6: If A and B Are Dependent Events, the Probability That Both A and B Will Occur Is

$$P(A \textit{ and } B) = P(A) \times P(B|A)$$

In this theorem, the occurrence of B is dependent on the outcome of A. This relationship between A and B is represented by $P(A|B)$. The vertical bar (|) is translated as "given that." The probability that both A and B will occur is the probability that A will occur multiplied by the probability that B will occur, given that A has already occurred.

EXAMPLE 8.8 Probability in Dependent Events I

Suppose that a student from Example 8.6 has been chosen at random from the 32 in the class. If the student selected is female, what is the probability that she will be from out of state?

This is a conditional probability: the answer is conditioned on the selection of a female student. The formula in theorem 6 will need to be rearranged to determine the answer:

$$P(B|A) = \frac{P(\text{A and B})}{P(A)}$$

$$P(\text{out-of-state}|\text{female}) = \frac{P(\text{female and out-of-state})}{P(\text{female})}$$

The probability of selecting one of the 12 females in the class is equal to $P(\text{female}) = 12/32$. Since being female and being from out of state are dependent events, the probability of being both female and out-of-state ($P(\text{female and out-of-state})) = 8/32$. These values can be verified by looking at Figure 8.2.

Completing the calculation, we find

$$P(\text{out-of-state}|\text{female}) = \frac{8/32}{12/32} = 2/3$$

Given that the student selected is female, the probability that she will be from out of state is 2/3.

EXAMPLE 8.9 Probability in Dependent Events II

In order to attend a business meeting in Los Angeles, a business traveler based in Cincinnati must first fly to Chicago and board a connecting flight. Because of the nature of the flights, the traveler has very little time to change planes in Chicago. On his way to the airport, he ponders the possibilities. On the basis of past traveling experience, he knows that there is a 70 percent chance that his plane will be on time in Chicago [$P(\text{Chicago on time}) = 0.70$]. This leaves a 30 percent chance that his Cincinnati-based plane will arrive late in Chicago. Knowing that if he arrives late, he may not be able to make the connecting flight to Los Angeles, he decides that the probability that he will be late in Chicago and on time in Los Angeles is 20 percent [$P(\text{Chicago late and Los Angeles on time}) = 0.20$]. Suppose that the traveler arrives late in Chicago. Given that he arrived late in Chicago, what is the probability that he will reach Los Angeles on time?

$$\text{P(L.A. on time} \mid \text{Chicago late)} = \frac{\text{P(Chicago late and L.A. on time)}}{\text{P(Chicago late)}}$$

From reading the problem, we know

$$\text{P(Chicago on time)} = 0.70$$
$$\text{P(Chicago late)} = 1 - 0.70 = 0.30$$
$$\text{P(Chicago late and L.A. on time)} = 0.20$$
$$\text{P(L.A. on time} \mid \text{Chicago late)} = \frac{0.20}{0.30} = 0.07$$

When events are ***independent,*** *one event does not influence the occurrence of another.* The result of one outcome or event is unaffected by the outcome of another event. Mathematically, events are considered independent if the following are true:

$$P(A \mid B) = P(A) \quad \text{and} \quad P(B \mid A) = P(B) \quad \text{and} \quad P(A \text{ and } B) = P(A) \times P(B)$$

EXAMPLE 8.10 Probability in Independent Events

A local mail-order catalog business employs 200 people in the packaging and shipping departments. The personnel department maintains the records shown in Table 8.1. Questions have arisen concerning whether there has been a tendency to place female workers in the packaging department instead of in the shipping department. Management feels that there is no relationship between being female and working in the packaging department. In other words, they feel that these two events are independent. A quick probability calculation can enable the firm to determine whether this is true.

If an employee is selected at random from the 200 total employees, what is the probability that the employee works for the packaging department?

$$\text{P(Packaging)} = 80/200 = 0.4$$

	Department		
Sex	***Packaging***	***Shipping***	***Total***
Female	32	48	80
Male	48	72	120
Total	80	120	200

Table 8.1 Employee Records

If the employee selected is female, what is the probability that the employee works in the packaging department?

$$P(\text{packaging} \mid \text{female}) = \frac{P(\text{female and packaging})}{P(\text{female})} = \frac{32/200}{80/200} = 0.4$$

Independence can be established if the following are true:

$$P(\text{packaging} \mid \text{female}) = P(\text{packaging})$$
$$0.4 = 0.4$$

and

$$P(\text{female} \mid \text{packaging}) = P(\text{female})$$
$$\frac{32/200}{80/200} = 0.4$$

and from Table 8.1, it can be seen that

$$P(\text{female and packaging}) = P(\text{female}) \times P(\text{packaging})$$
$$32/200 = 80/200 \times 80/200$$
$$0.16 = 0.16$$

Since these probabilities are equal, they are independent. 

Theorem 7: If A and B Are Independent Events, Then the Probability That Both A and B Will Occur Is

$$P(A \textit{ and } B) = P(A) \times P(B)$$

This is often referred to as a *joint probability*, meaning that both A and B can occur at the same time.

EXAMPLE 8.11 Joint Probability I

A company purchases fluorescent lightbulbs from two different suppliers. Sixty percent of the fluorescent bulbs come from WT Corporation, 40 percent from NB Corporation. Both suppliers are having quality problems. Ninety-five percent of the bulbs coming from WT Corporation perform as expected. Only 80 percent of the bulbs from NB Corporation perform.

What is the probability that a bulb selected at random will be from WT Corporation and that it will perform as expected?

Since the lightbulbs from supplier WT are separate and independent of the lightbulbs from supplier NB, theorem 7 may be used to answer this question:

$$P(\text{WT and perform}) = P(\text{WT}) \times P(\text{perform})$$
$$= (0.60) \times (0.95) = 0.57$$

Similarly, the probability that a bulb selected at random will be from NB Corporation and perform is

$$\begin{aligned} P(\text{NB and perform}) &= P(\text{NB}) \times P(\text{perform}) \\ &= (0.40) \times (0.80) = 0.32 \end{aligned}$$

The probabilities associated with the bulbs not performing can be calculated in a similar fashion:

$$\begin{aligned} P(\text{NB and not performing}) &= (0.40) \times (0.20) = 0.08 \\ P(\text{WT and not performing}) &= (0.60) \times (0.05) = 0.03 \end{aligned}$$

EXAMPLE 8.12 Joint Probability II

The company is interested in determining the probability that any one bulb selected at random, regardless of the supplier, will not perform. This value can be found by remembering theorem 2: The sum of the probabilities of the events of a situation is equal to 1.00:

$$P(A) + P(B) + \cdots + P(N) = 1.00$$

and theorem 3: If P(A) is the probability that event A will occur, then the probability that A will not occur is

$$P(A) = 1.00 - P(A)$$

In this instance, four events can take place:

The bulb tested is from WT Corporation and it performs.
The bulb tested is from WT Corporation and it does not perform.
The bulb tested is from NB Corporation and it performs.
The bulb tested is from NB Corporation and it does not perform.

If the probabilities from these four events are summed, they will equal 1.00 (theorem 2).

We already know the probabilities associated with the bulb performing:

$$\begin{aligned} P(\text{WT and perform}) &= (0.60) \times (0.95) = 0.57 \\ P(\text{NB and perform}) &= (0.40) \times (0.80) = 0.32 \end{aligned}$$

Therefore, using theorem 1, the probability that any one bulb selected at random, regardless of the supplier, will not perform is

$$\begin{aligned} P(\text{bulb not performing}) &= 1.00 - P(\text{WT and perform}) - P(\text{NB and perform}) \\ &= 1.00 - 0.57 - 0.32 \\ &= 0.11 \end{aligned}$$

PERMUTATIONS AND COMBINATIONS

As the number of ways that a particular outcome may occur increases, so does the complexity of determining all possible outcomes. For instance, if a part must go through four different machining operations and for each operation there are several machines available, then the scheduler will have a number of choices about how the part can be scheduled. Permutations and combinations are used to increase the efficiency of calculating the number of different outcomes possible. A ***permutation*** *is the number of arrangements that* n *objects can have when* r *of them are used:*

$$P_r^n = \frac{n!}{(n - r)!}$$

The order of the arrangement of a set of objects is important when calculating a permutation.

EXAMPLE 8.13 Calculating a Permutation I

While waiting in an airport, you strike up a conversation with a fellow traveler. She is trying to decide whether or not to phone a friend. Unfortunately she can remember only the first four digits (555-1???) of the friend's phone number and she knows that the last three digits are not the same. How many different permutations are there when determining the last three digits?

To solve this equation, it is important to remember that the order is important. For instance, if the three needed numbers are a 1, a 2, and a 3, are the last three digits in the phone number 123 or 321 or 312 or 213 or 132 or 231? For each missing number there are 10 choices (0–9). Since no one number is repeated, those 10 choices must be taken three at a time:

$$P_r^n = \frac{n!}{(n - r)!} = \frac{10!}{(10 - 3)!} = 720$$

Should she try to discover her friend's phone number?

EXAMPLE 8.14 Calculating a Permutation II

At a local manufacturing facility a scheduler is facing a dilemma. A part must go through the following three different machining operations in order and for each of these operations there are several machines available. The scheduler's boss has asked him to list all the different ways that the part can be scheduled. The scheduler is reluctant to begin the list because there are so many different permutations. Using the following information, he calculates the number of permutations possible:

Grinding (4 machines): The part must go to two different grinding machines.
Heat-treating (3 units): Heat-treating the part takes only one heat-treat unit.
Milling (5 machines): The part must go through three different milling machines.

The scheduler knows that order is important and so the number of permutations for each work center must be calculated. These values will then be multiplied together to determine the total number of schedules possible.

Grinding (four machines selected two at a time):

$$P_r^n = \frac{n!}{(n - r)!} = \frac{4!}{(4 - 2)!} = 12$$

This can also be found by writing down all the different ways that the two machines can be selected. Remember, order is important:

12 13 14 23 24 21 34 32 31 43 42 41

Heat-treating (three machines selected one at a time):

$$P_r^n = \frac{n!}{(n-r)!} = \frac{3!}{(3-1)!} = 3$$

Milling (five machines selected three at a time):

$$P_r^n = \frac{n!}{(n-r)!} = \frac{5!}{(5-3)!} = 60$$

Since there are 12 different permutations to schedule the grinding machines, 3 for the heat-treat units, and 20 for the milling machines, the total number of permutations that the scheduler will have to list is $12 \times 3 \times 60 = 2160$!

When the order in which the items are used is not important, the number of possibilities can be calculated by using the formula for a **combination.** The calculation for a combination uses only the number of elements, with no regard to any arrangement:

$$C_r^n = \frac{n!}{r!(n-r)!}$$

EXAMPLE 8.15 Calculating a Combination

A problem-solving task force is being created to deal with a situation at a local chemical company. Two different departments are involved, chemical process engineering and the laboratory. There are seven members of the chemical process engineering group and three of them must be on the committee. The number of different combinations of members from the chemical process engineering group is

$$C_r^n = \frac{n!}{r!(n-r)!} = \frac{7!}{3!(7-3)!} = 35$$

There are five members of the laboratory group and two of them must be on the committee. The number of different combinations of members from the laboratory group is

$$C_r^n = \frac{n!}{r!(n-r)!} = \frac{5!}{2!(5-2)!} = 10$$

For the total number of different arrangements on the committee, multiply the number of combinations from the chemical process engineering group by the number of combinations from the laboratory group:

$$\text{Total} = C_3^7 \times C_2^5 = 35 \times 10 = 350$$

DISCRETE PROBABILITY DISTRIBUTIONS

For a probability distribution to exist, a process must be defined by a random variable for which all the possible outcomes and their probabilities have been enumerated. Discrete probability distributions count attribute data, the occurrence of nonconforming activities or items or nonconformities on an item.

Hypergeometric Probability Distribution

When a random sample is taken from a small lot size, the hypergeometric probability distribution will determine the probability that a particular event will occur. The hypergeometric is most effective when the sample size is greater than 10 percent of the size of the population ($n/N > 0.1$) and the total number of nonconforming items is known. To use the hypergeometric probability distribution, the population must be finite and samples must be taken randomly, without replacement. The following formula is used to calculate the probability an event will occur:

$$P(d) = \frac{C_d^D \, C_{n-d}^{N-D}}{C_n^N}$$

where

D = number of nonconforming or defective units in lot
d = number of nonconforming or defective units in sample
N = lot size
n = sample size
$N - D$ = number of conforming units in lot
$n - d$ = number of conforming units in sample

The hypergeometric distribution is an exact distribution; the $P(d)$ translates to the probability of exactly d nonconformities. In the numerator, the first combination is the combination of all nonconforming items in the population and in the sample. The second combination is for all of the conforming items in the population and the sample. The denominator is the combination of the total population and the total number sampled.

EXAMPLE 8.16 Using the Hypergeometric Probability Distribution

To help in their forgery investigation, officials have hidden 4 counterfeit stock certificates with 11 authentic certificates. What is the probability that the culprits will select one counterfeit certificate in a random sample (without replacement) of 3?

D = number of counterfeits in lot = 4
d = seeking the probability that one counterfeit will be found = 1
N = lot size = 15
n = sample size = 3
$N - D$ = number of authentic certificates in lot = 11
$n - d$ = number of authentic certificates in sample = 2

$$P(1) = \frac{C_1^4 C_2^{11}}{C_3^{15}}$$

$$= \frac{\left[\frac{4!}{1!(4-1)!}\right]\left[\frac{11!}{2!(11-2)!}\right]}{\frac{15!}{3!(15-3)!}} = 0.48$$

If the officials were interested in determining the probability of finding two or fewer of the counterfeit certificates, the result would be

$$P(2 \text{ or fewer}) = P(0) + P(1) + P(2)$$

$$P(0) = \frac{C_0^4 C_{3-0}^{15-4}}{C_3^{15}}$$

$$= \frac{\left[\frac{4!}{0!(4-0)!}\right]\left[\frac{11!}{3!(11-3)!}\right]}{\frac{15!}{3!(15-3)!}} = 0.36$$

$$P(1) = 0.48$$

$$P(2) = \frac{C_2^4 C_{3-2}^{15-4}}{C_3^{15}}$$

$$= \frac{\left[\frac{4!}{2!(4-2)!}\right]\left[\frac{11!}{1!(11-1)!}\right]}{\frac{15!}{3!(15-3)!}} = 0.15$$

The probability of finding two or fewer counterfeit certificates is

$$P(2 \text{ or fewer}) = P(0) + P(1) + P(2) = 0.48 + 0.36 + 0.15 = 0.99$$

The above probability can also be found by calculating

$$P(2 \text{ or fewer}) = 1 - P(3)$$

What is the probability of finding all four counterfeits? This can't be calculated because the sample size is only three.

Binomial Probability Distribution

The ***binomial probability distribution*** was developed by Sir Issac Newton *to categorize the results of a number of repeated trials and the outcomes of those trials*. The "bi" in binomial refers to two conditions: The outcome is either a success or a failure. In terms of a product being produced, the outcome is either conforming or nonconforming. The distribution was developed to reduce the number of calculations associated with a large number of trials containing only two possible outcomes: success (s) or failure (f). Table 8.2 shows how complicated the calculations become as the number of trials held increases.

Number of Trials	Possible Outcomes*
1	s f
2	ss ff sf fs
3	sss fff ssf sff sfs fss ffs fsf
4	ssss ffff sfff ssff sffs sssf fsss ffss fffs fssf sfsf fsfs ssfs sfss ffsf fsff
5	sssss fffff sffff sfffs sffss sfsss fssss fsssf fssff fsfff sfsfs fsfsf ssfff ssffs ssfss ffsss ffssf ffsff ssssf sssff fffss ffffs sfsff fsfss sffsf fssfs sfssf fsffs ssfsf ffsfs fffsf sssfs

*Success = s; failure = f.

Table 8.2 Binomial Distribution: Outcomes Associated with Repeated Trials

The binomial probability distribution can be used if two conditions are met:

1. There is an infinite number of items or a steady stream of items being produced.
2. The outcome is seen as either a success or a failure. Or in terms of a product, it is either conforming or nonconforming. The binomial formula for calculating the probability an event will occur is

$$P(d) = \frac{n!}{d!(n-d)!}p^d q^{n-d}$$

where

d = number of nonconforming units, defectives, or failures sought
n = sample size
p = proportion of nonconforming units, defectives, or failures in population
$q = (1 - p)$ = proportion of good or conforming units or successes in population

This distribution is an exact distribution, meaning that the P(d) translates to the probability of exactly d nonconforming units or failures occurring. The mean of the binomial distribution is $\mu = np$. The standard deviation of the binomial distribution is $\sigma = \sqrt{np(1-p)}$. The binomial distribution tables are found in Appendix 3.

EXAMPLE 8.17 Using the Binomial Probability Distribution I

The billing department of a local department store sends monthly statements to the store's customers. In order for those statements to reach the customer in a timely fashion, the addresses on the envelopes must be correct. Occasionally errors are made with the addresses. The billing department estimates that errors are made 2 percent of the time. For this continuous process, in which an error in the address is considered a nonconforming unit, what is the probability that in a sample of size eight, one address will be incorrect?

$$P(d) = \frac{n!}{d!(n-d)!}p^d q^{n-d}$$

where

d = number of nonconforming units sought = 1
n = sample size = 8
p = proportion of population nonconforming = 0.02
q = (1 − p) = proportion of conforming units = 1 − 0.02

$$P(1) = \frac{8!}{1!(8-1)!}0.02^1 0.98^{8-1} = 0.14$$

Cumulative binomial distribution tables are included in Appendix 3. To utilize the appendix, match the sample size (n) with the number of nonconforming/defective units sought (d) and the proportion of nonconforming units/defectives in the population (p). Since this table is cumulative and we are seeking the probability of exactly 1 (P(1)), we must subtract the probability of exactly 0 (P(0)). For this example,

From Appendix 3:

$$\begin{aligned} P(1) &= \text{Cum } P(1) - \text{Cum } P(0) \\ &= 0.9897 - 0.8508 \\ &= 0.1389 \quad \text{or 0.14 when rounded} \end{aligned}$$

d = 1
p = 0.02
n = 8

n = 8

d \ p	*0.01*	*0.02* ↓	*0.03*	*0.04*	*0.05*	*0.06*	*0.07*
0	0.9227	0.8508	0.7837	0.7214	0.6634	0.6096	0.5596
→1	0.9973	0.9897	0.9777	0.9619	0.9428	0.9208	0.8965
2	0.9999	0.9996	0.9987	0.9969	0.9942	0.9904	0.9853

EXAMPLE 8.18 Using the Binomial Probability Distribution II

From experience, a manufacturer of ceramic floor tiles knows that 4 percent of the tiles made will be damaged during shipping. Any chipped, scratched, dented, or broken tile is considered a nonconforming unit. If a random sample of 14 tiles is taken from a current shipment, what is the probability that 2 or fewer tiles will be damaged?

The binomial distribution can be applied because a steady stream of tiles is being manufactured and a tile can be judged either conforming or nonconforming. In order to calculate the probability of finding two or fewer, theorems 2 and 3 must be applied:

$$P(\text{2 or fewer}) = P(0) + P(1) + P(2)$$

$$P(d) = \frac{n!}{d!(n-d)!}p^d q^{n-d}$$

where

d = number of nonconforming units sought
n = sample size
p = proportion of population that is nonconforming
q = (1 − p) = proportion of conforming units

$$P(0) = \frac{14!}{0!(14-0)!}0.04^0 0.96^{14-0} = 0.57$$

$$P(1) = \frac{14!}{1!(14-1)!}0.04^1 0.96^{14-1} = 0.33$$

$$P(2) = \frac{14!}{2!(14-2)!}0.04^2 0.96^{14-2} = 0.09$$

Summing these values,

$$\begin{aligned} P(2 \text{ or fewer}) &= P(0) + P(1) + P(2) \\ &= 0.57 + 0.33 + 0.09 = 0.99 \end{aligned}$$

The binomial distribution tables from Appendix 3 may also be used in this example. This time, however, the probability being sought, P(2 or fewer), is cumulative. To utilize the appendix, match the sample size (n) with the number of nonconforming/defective units sought (d) and the proportion of nonconforming units/defectives in the population (p). For this example,

From Appendix 3:

$$P(2 \text{ or fewer}) = 0.9833$$

The differences here are from rounding.

d = 2 or fewer

p = 0.04

n = 14

n = 14

d \ p	*0.01*	*0.02*	*0.03*	↓ *0.04*	*0.05*	*0.06*	*0.07*
0	0.8687	0.7536	0.6528	0.5647	0.4877	0.4205	0.3620
1	0.9916	0.9690	0.9355	0.8941	0.8470	0.7963	0.7436
→2	0.9997	0.9975	0.9923	0.9833	0.9699	0.9522	0.9302
3	1.0000	0.9999	0.9994	0.9981	0.9958	0.9920	0.9864
4	1.0000	1.0000	1.0000	0.9998	0.9996	0.9990	0.9980

Poisson Probability Distribution

First described by Simeon Poisson in 1837, the ***Poisson probability distribution*** *quantifies the count of discrete events*. To use the Poisson distribution successfully, it is important to identify a well-defined, finite region known as the *area of opportunity* in which the discrete, independent events may take place. This finite region, or area of opportunity, may be defined as a particular space, time, or product. The Poisson distribution is often used when calculating the probability that an event will occur when there is a large area of opportunity, such as in the case of rivets on an airplane wing.

The Poisson distribution is also used when studying arrival-rate probabilities. The formula for the Poisson distribution is

$$P(c) = \frac{(np)^c}{c!}e^{-np}$$

where

np = average count or number of events in sample
c = count or number of events in sample
$e \approx 2.718281$

EXAMPLE 8.19 Using the Poisson Probability Distribution

Each week the manager of a local bank determines the schedule of working hours for the tellers. On average the bank expects to have two customers arrive every minute. What is the probability that three customers will arrive at any given minute?

$$P(3) = \frac{2^3}{3!}e^{-2} = 0.18$$

where

np = average number of customers in sample = 2
c = number of customers in sample = 3
$e \approx 2.718281$

While calculating probabilities associated with the Poisson distribution is simpler mathematically than using either the hypergeometric or the binomial distribution, these calculations can be further simplified by using a Poisson table (Appendix 4). To establish the probability of finding an expected number of nonconformities (c) by using the table, np and c must be known.

EXAMPLE 8.20 Using a Poisson Table

The local branch office of a bank is interested in improving staff scheduling during peak hours. For this reason the manager would like to determine the probability that three or more customers will arrive at the bank in any given minute. The average number of customers arriving at the bank in any given minute is two.

P(3 or more customers arriving in any given minute)

$$= P(3) + P(4) + P(5) + P(6) + \cdots$$

Since it is impossible to solve the problem in this fashion, theorems 2 and 3 must be applied:

P(3 or more customers arriving in any given minute)

$$= 1 - P(2 \text{ or fewer}) = 1 - [P(2) + P(1) + P(0)]$$

From the table in Appendix 4, np = 2, c = (2 or less). Using the column for cumulative values,

$$1 - 0.677 = 0.323$$

or by calculation:

$$1 - \left[\frac{(2)^2}{2!}e^{-2} + \frac{(2)^1}{1!}e^{-2} + \frac{(2)^0}{0!}e^{-2}\right]$$

where

np = average number of customers in sample = 2
c = number of customers in sample = 0, 1, 2
$e \approx 2.718281$

The Poisson probability distribution is also an exact distribution. P(c) is the probability of exactly c nonconformities. The mean of the Poisson distribution is $\mu = np$. The standard deviation of the Poisson distribution is $\sigma = \sqrt{np}$.

CONTINUOUS PROBABILITY DISTRIBUTION

Normal Distribution

In situations where the data can take on a continuous range of values, a discrete distribution, such as the binomial, cannot be used to calculate the probability that an event will occur. For these situations, the normal distribution, a continuous probability distribution, should be used. Covered in Chapter 4, this distribution is solved by finding the value of Z and using the Z tables in Appendix 1 to determine the probability an event will occur:

$$Z = \frac{X_i - \overline{X}}{s}$$

where

Z = standard normal value
X_i = value of interest
$\overline{X}$ = average
s = standard deviation

The values in the Z table can be interpreted either as frequencies or as probability values. The mean of the normal is np. The standard deviation of the normal is $\sqrt{npq}$.

EXAMPLE 8.21 Using the Normal Distribution

A tool on a stamping press is expected to complete a large number of strokes before it is removed and reworked. As a tool wears, the dimensions of the stamped part become smaller. Eventually, they become small enough that they are unable to meet specifications. As a tool wears, it is removed from the press and reground to bring the parts back into specification. For this particular example, a combination of $\overline{X}$ and R charts is being used to monitor part dimensions. The $\overline{X}$ chart tracks the dimension of the part as it gets smaller. After it reaches a certain point, the tool is pulled and reground. Those tracking the process have determined that the average number of strokes or parts the tool can complete is 60,000, with a standard deviation of 3,000. Determine the percentage of tools that will require regrinding before 55,000 strokes.

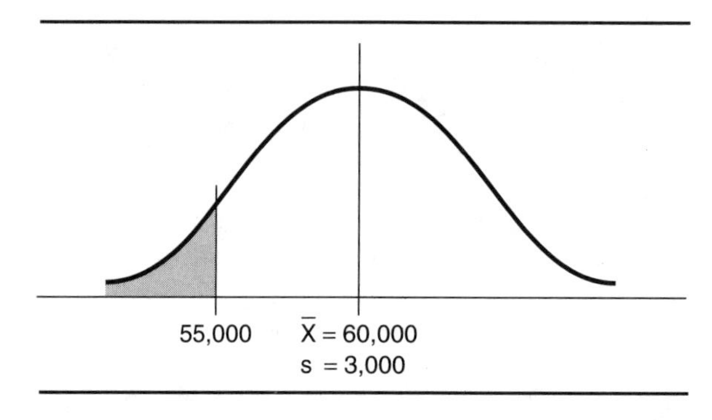

Figure 8.3 Normal Probability Distribution for Example 8.21

Figure 8.3 shows the normal probability distribution associated with this example. The area in question is shaded.

$$Z = \frac{55{,}000 - 60{,}000}{3{,}000} = -1.67$$

From the table in Appendix 1, the area to the left of 55,000 is equal to 0.0475. Only 4.75% of the tools will last for fewer than 55,000 strokes.

EXAMPLE 8.22 Using Normal Distribution

If the normal operating life of a car is considered to be 150,000 miles, with a standard deviation of 20,000 miles, what is the probability that a car will last 200,000 miles? Assume a normal distribution exists.

Figure 8.4 shows the normal probability distribution associated with this example. The area in question is shaded:

$$Z = \frac{200{,}000 - 150{,}000}{20{,}000} = 2.5$$

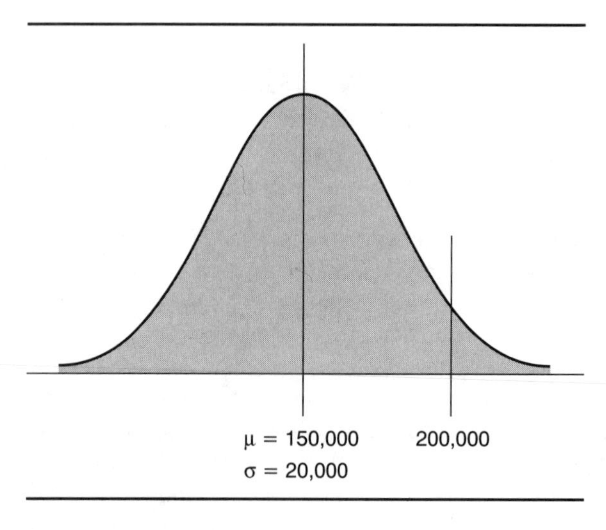

Figure 8.4 Normal Probability Distribution for Example 8.22

From the table in Appendix 1, the area to the left of 200,000 is equal to 0.9938. To find the area in question,

$$1 - 0.9938 = 0.0062$$

There is a probability of 0.62 percent that the car will last for more than 200,000 miles.

DISTRIBUTION INTERRELATIONSHIPS AND APPROXIMATIONS

Under certain situations the binomial, Poisson, and normal probability distributions can be used to approximate another distribution. Calculations for the hypergeometric distribution can be simplified under certain circumstances by using the binomial as an approximation. This substitution works best if the sample size n is less than 10 percent of the population lot size ($n/N \leq 0.10$). The hypergeometric can also be approximated by the Poisson, provided the following is true: $n/N \leq 0.10$, $p \leq 0.10$, and $np \leq 5$.

The Poisson distribution can be used to approximate the binomial distribution if three conditions can be met: The population of the lot can be assumed to be infinite, the average fraction nonconforming (p) is less than 0.10 ($p \leq 0.10$), and the value of $np \leq 5$.

EXAMPLE 8.23 Using the Poisson Approximation to the Binomial

In Example 8.17, the binomial distribution was used to determine the probability that in a sample of size eight one billing address will be incorrect. Use the Poisson approximation to the binomial to make the same calculation. Assume that the population of bills is nearly infinite.

Since the billing department estimates that errors are made 2 percent of the time, $p \leq 0.10$. With a sample size of eight, $np = (8 \times 0.02) = 0.16$; therefore, $np \leq 5$.

$$P(1) = \frac{0.16^1}{1!} e^{-0.16} = 0.14$$

where

$$np = \text{average number of customers in sample} = 0.16$$
$$c = \text{number of customers in sample} = 1$$
$$e \approx 2.718281$$

From Example 8.17,

$$P(1) = \frac{8!}{1!(8-1)!} 0.02^1 0.98^{8-1} = 0.14$$

In this case, the Poisson is an excellent approximation to the binomial.

When p nears 0.5 and $n \geq 10$, the normal distribution can be used to approximate the binomial. Since the binomial distribution is discrete and the normal distribution

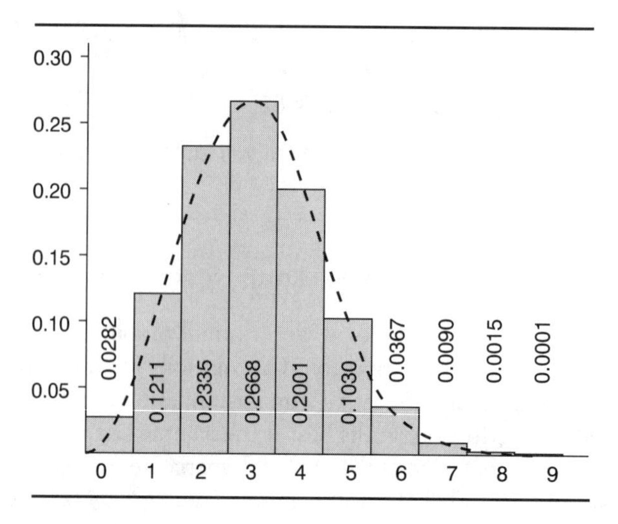

Figure 8.5 Binomial Distribution with Normal Distribution Overlay

is continuous, adjustments must be made to the normal distribution calculations when using the normal distribution to approximate the binomial. In Figure 8.5, a histogram has been constructed using binomial data. A normal curve has been overlaid on the histogram. To use the normal distribution to approximate the probability of two nonconforming [P(2)], the area of a rectangle centered at 2 must be determined. To capture this information using the normal distribution, the area between 1.5 and 2.5 must be calculated. Instead of determining P(2), we must find the $P(1.5 \leq 2 \leq 2.5)$. In summary, to use the normal distribution to approximate the binomial, 0.5 must be added to and/or subtracted from the desired value according to the situation.

EXAMPLE 8.24 Using the Normal Approximation to the Binomial

On average, 30 percent of the customers visiting a theme park take a ride on the train. Assume that visitors compose a steady stream and that they either ride on the train or they don't. Use the binomial distribution to determine the probability that 2 out of 10 randomly chosen visitors to the park will ride on the train:

$$P(2) = \frac{10!}{2!(10-2)!}0.30^2 0.70^{10-2} = 0.23$$

There is a 23 percent chance that 2 visitors from a sample of 10 will ride the train.

Refer to Figure 8.5 and use the normal approximation to the binomial to calculate the same probability. To determine P(2), $P(1.5 \leq 2 \leq 2.5)$ must be calculated. Using the normal distribution Z formula,

$$Z = \frac{X_i - \overline{X}}{s}$$

where

$$\overline{X} = np = 10(0.3) = 3$$
$$s = \sqrt{np(1 - p)} = \sqrt{10(0.3)(1 - 0.3)} = 1.45$$
$$P(1.5 \le 2 \le 2.5) = \frac{1.5 - 3}{1.45} \le Z \le \frac{2.5 - 3}{1.45}$$
$$-1.03 \le Z \le -0.34$$

Using the Z tables in Appendix 1,

$$P(1.5 \le 2 \le 2.5) = 0.22$$

This is a fairly close approximation to the binomial.

Use the normal approximation to the binomial to determine the probability that 2 or fewer visitors in a sample of 10 will ride on the train:

$$P(2 \text{ or fewer}) = P(0) + P(1) + P(2)$$
$$P(0 \le 2 \text{ or fewer} \le 2.5) = \frac{0 - 3}{1.45} \le Z \le \frac{2.5 - 3}{1.45}$$

Using the Z tables in Appendix 1,

$$-2.07 \le Z \le -0.34$$
$$0.3669 - 0.0192 = 0.3477, \text{ rounded to } 0.35$$

The same calculation performed using the binomial distribution will equal

$$P(2 \text{ or fewer}) = 0.38$$

In this situation, the normal approximation is a fairly close approximation to the binomial. Had the normal approximation not been adjusted, the approximation would not be close to the true probability value:

$$P(0 \le 2) = \frac{0 - 3}{1.45} \le Z \le \frac{2 - 3}{1.45}$$
$$-0.69 \le Z \le -2.07$$
$$0.2451 - 0.0192 = 0.23$$

SUMMARY

Probability, its theorems and distributions, plays an important role in understanding situations that arise in quality. Probability concepts also support the creation of control charts for attributes, which will be covered in the next chapter. The binomial distribution serves as the foundation for control charts for nonconforming units or activities. Control charts for nonconformities have the Poisson distribution as their basis.

Lessons Learned

1. Seven theorems exist to explain probability.
2. Discrete and continuous probability distributions exist to describe the probability that an event will occur.
3. The hypergeometric, binomial, and Poisson distributions are all discrete probability distributions.
4. The normal distribution is a continuous probability distribution.
5. The binomial and Poisson distributions can be used to approximate the hypergeometric distribution.
6. The Poisson and normal distributions can be used to approximate the binomial distribution. ■

Formulas

$$P(A) = \frac{\text{number of occurrences}}{\text{total number of possibilities}} = \frac{s}{n}$$

Theorem 1: Probability Is Expressed as a Number between 0 and 1:

$$0 \leq P(A) \leq 1$$

Theorem 2: The Sum of the Probabilities of the Events in a Situation Is Equal to 1.00:

$$\Sigma P_i = P(A) + P(B) + \cdots + P(N) = 1.00$$

Theorem 3: If P(A) Is the Probability That an Event A Will Occur, Then the Probability That A Will Not Occur Is

$$P(A') = 1.00 - P(A)$$

Theorem 4: For Mutually Exclusive Events, the Probability That Either Event A or Event B Will Occur Is the Sum of Their Respective Probabilities:

$$P(A \text{ or } B) \times P(A) + P(B)$$

Theorem 5: When Events A and B Are Not Mutually Exclusive Events, the Probability That Either Event A or Event B or Both Will Occur Is

$$P(A \text{ or } B \text{ or both}) = P(A) + P(B) - P(\text{both})$$

Theorem 6: If A and B Are Dependent Events, the Probability That Both A and B Will Occur Is

$$P(A \text{ and } B) = P(A) \times P(B|A)$$

Theorem 7: If A and B Are Independent Events, Then the Probability That Both A and B Will Occur Is

$$P(A \text{ and } B) = P(A) \times P(B)$$

Permutations

$$P_r^n = \frac{n!}{(n - r)!}$$

Combinations

$$C_r^n = \frac{n!}{r!(n - r)!}$$

Hypergeometric Probability Distribution

$$P(d) = \frac{C_d^D C_{n-d}^{N-D}}{C_n^N}$$

where

D = number of nonconforming or defective units in lot
d = number of nonconforming or defective units in sample
N = lot size
n = sample size
$N - D$ = number of conforming units in lot
$n - d$ = number of conforming units in sample

Binomial Probability Distribution

$$P(d) = \frac{n!}{d!(n - d)!} p^d q^{n-d}$$

where

d = number of nonconforming units, defectives, or failures sought
n = sample size
p = proportion of nonconforming units, defectives, or failures in population
$q = (1 - p)$ = proportion of good or conforming units, or successes, in population

Poisson Probability Distribution

$$P(c) = \frac{(np)^c}{c!} e^{-np}$$

where

np = average count or number of events in sample
c = count or number of events in sample
$e \approx 2.718281$

Normal Distribution

$$Z = \frac{X_i - \overline{X}}{s}$$

using the Z table in Appendix 1.

Approximations

Hypergeometric distribution can be approximated by the binomial when $n/N \leq 0.10$.

Hypergeometric distribution can be approximated by the Poisson when $n/N \leq 0.10$, $p \leq 0.10$, and $np \leq 5$.

Binomial distribution can be approximated by the Poisson when the population of the lot is assumed to be infinite, $p \leq 0.10$, and $np \leq 5$.

Binomial distribution can be approximated by the normal when p nears 0.5 and $n \geq 10$.

Chapter Problems

Probability Theorems

1. The probability of drawing a pink chip from a bowl of different-colored chips is 0.35, the probability of drawing a blue chip is 0.46, the probability of drawing a green chip is 0.15, and the probability of drawing a purple chip is 0.04. What is the probability that a blue or a purple chip will be drawn?
2. At a local county fair, the officials would like to a give a prize to 100 people selected at random from those attending the fair. As of the closing day, 12,500 people have attended the fair and completed the entry form for the prize. What is the probability that an individual who attended the fair and completed the entry form will win a prize?
3. At the county fair, the duck pond contains eight yellow ducks numbered 1 to 8, six orange ducks numbered 1 to 6, and ten gray ducks numbered 1 to 10. What is the probability of obtaining an orange duck numbered with a 5? Of obtaining an orange duck? Of obtaining a duck labeled with a 5?
4. If there are five different parts to be stocked but only three bins available, what is the number of permutations possible for five parts taken three at a time?
5. If a manufacturer is trying to put together a sample collection of her product and order is not important, how many combinations can be created with 15 items that will be placed in packages containing five items? If order is important, how many permutations can be created?
6. An assembly plant receives its voltage regulators from two different suppliers: 75 percent come from Hayes Voltage Co. and 25 percent come from Romig Voltage Co. The percentage of voltage regulators from Hayes that perform according to specifications is 95 percent. The voltage regulators from Romig perform according to specifications only 80 percent of the time. What is the probability that any one voltage regulator received by the plant performs according to spec?
7. If one of the voltage regulators from Problem 6 performed, what is the probability that it came from Hayes?

8. Suppose that the student selected in Example 8.8 is male. What is the probability that he is from out of state?
9. Suppose a firm makes couches in four different styles and three different fabrics. Use the table to calculate the probability that a couch picked at random will be made from fabric 1. If the couch is style 1, what is the probability that it will be made from fabric 2? What is the probability that a couch of style 4 will be selected at random? If a couch is made from fabric 3, what is the probability that it is a style 3 couch?

	Fabric			
Style	*F1*	*F2*	*F3*	*Total*
S1	150	55	100	305
S2	120	25	70	215
S3	80	60	85	225
S4	110	35	110	255
Total	460	175	365	1000

10. Use the information from Example 8.10 to calculate the following probabilities:
    a. If the employee selected at random is male, what is the probability that he works for the shipping department?
    b. If the employee selected at random is male, what is the probability that he works for the packing department?
    c. Is the probability of being male independent?

Hypergeometric Probability Distribution

11. The owner of a local office supply store has just received a shipment of copy machine paper. As the 15 cases are being unloaded off the truck, the owner is informed that one of the cases contains blue paper instead of white. Before the owner can isolate the case, it is mixed in with the other cases. There is no way to distinguish from the outside of the case which case contains blue paper. Since the cases sell for a different price than individual packages of paper, if the owner opens the case, it can not be replaced (sold as a case). What is the probability that the manager will find the case that contains blue paper in one of the first three randomly chosen cases?
12. A grocery store manager has just received a shipment of peanut butter. As the 12 cases are being unloaded, two of the cases are dropped. Before the manager can isolate the cases, they are mixed in with the other cases. The truck delivering the cases has departed, so there are no replacement cases. What is the probability that the manager will find the two broken cases in the first four randomly chosen cases?

13. A rather harried father is trying to find a very popular doll for his daughter for Christmas. He has a choice of ten stores to go to. From a radio announcement, he learned that the doll can definitely be found at four of the ten stores. Unfortunately, he did not hear which four stores! If the store he stops at does not have the doll, he will leave without buying anything; in other words, there is no replacement. He has time to go to only three stores. What is the probability that he will find the doll in two or fewer stops?

14. A lot of ten bottles of medicine has four nonconforming units. What is the probability of drawing two nonconforming units in a random sample of five? What is the probability that one or fewer nonconforming units will be chosen in a sample of five?

15. A group of 15 stock certificates contains 4 money-making stocks and 11 stocks whose performance is not good. If two of these stocks are selected at random and given to an investor (without replacement), what is the probability that the investor received one money-making stock?

16. Twenty water balloons are presented for inspection; four are suspected to have leaks. If two of these water balloons are selected at random without replacement, what is the probability that one balloon of the two selected has a hole in it?

17. A collection of 12 jewels contains 3 counterfeits. If two of these jewels are selected at random (without replacement) to be sold, what is the probability that neither jewel is counterfeit?

18. The local building inspector is planning to inspect ten of the most recently built houses for code violations. The inspector feels that four of the ten will have violations. Those that fail the inspection will be judged not suitable for occupation (without replacement). In a sample of three, what is the probability that two of the houses will fail inspection?

Binomial Probability Distribution

19. A steady stream of bolts is sampled at a rate of 6 per hour. The fraction nonconforming in the lot is 0.034. What is the probability that 1 or fewer of the 6 parts will be found nonconforming?

20. Plastic milk containers are produced by a machine that runs continuously. If 5 percent of the containers produced are nonconforming, determine the probability that out of four containers chosen at random, less than two are nonconforming.

21. A steady stream of product has a fraction defective of 0.03. What is the probability of obtaining 2 nonconforming in a sample of 20?

22. Returning to Example 8.18, what is the probability that more than four tiles will be damaged in a shipment?
23. A steady stream of newspapers is sampled at a rate of 6 per hour. The inspector checks the newspaper for printing legibility. If the first page of the paper is not clearly printed, the paper is recycled. Currently, the fraction nonconforming in the lot is 0.030. What is the probability that 2 of the 6 papers found will be nonconforming?
24. An assembly line runs and produces a large number of units. At the end of the line an inspector checks the product, labeling it as either conforming or nonconforming. The average fraction of nonconforming is 0.10. When a sample of size 10 is taken, what is the probability that 5 nonconforming units will occur?
25. An insurance company processes claim forms on a continuous basis. These forms, when checked by adjusters, are either filed as written or, if there is an error, returned to sender. Current error rates are running at 10 percent. In a sample of 6, what is the probability that more than two claims will be rejected because of errors?
26. A random sample of 3 bottles is selected from the running conveyor system. When the bottles are inspected, they are either properly labeled or they are not. Rejected bottles are scrapped. The proportion nonconforming is 0.25. What is the probability of there being 1 or fewer nonconforming units in the sample?
27. An environmental engineer places monitors in a large number of streams to measure the amount of pollutants in the water. If the amount of pollutants exceeds a certain level, the water is considered nonconforming. In the past, the proportion nonconforming has been 0.04. What is the probability that one of the fifteen samples taken per day will contain an excessive amount of pollutants?

Poisson Probability Distribution

28. A receptionist receives an average of 0.9 calls per minute. Find the probability that in any given minute there will be at least 1 incoming call.
29. If on the average 0.3 customers arrive per minute at a cafeteria, what is the probability that exactly 3 customers will arrive during a five-minute span?
30. A computer software company's emergency call service receives an average of 0.90 calls per minute. Find the probability that in any given minute there will be more than 1 incoming call.
31. A local bank is interested in the number of customers who will visit the bank in a particular time period. What is the probability that more than 2 customers will visit the bank in the next 10 minutes if the average arrival rate of customers is 3 every 10 minutes?

Normal Probability Distribution

32. The mean weight of a company's racing bicycles is 9.07 kg, with a standard deviation of 0.40 kg. If the distribution is approximately normal, determine (a) the percentage of bicycles weighing less than 8.30 kg and (b) the percentage of bicycles weighing between 8.00 and 10.10 kg.

Approximations

33. Use the Poisson approximation to the binomial to calculate the answer to Problem 25. Is this a good approximation?
34. Use the Poisson approximation to the binomial to calculate the answer to Problem 26. Is this a good approximation?
35. Use the Poisson approximation to the binomial to calculate the answer to Problem 27. Is this a good approximation?
36. Use the normal approximation to the binomial to calculate the answer to Problem 24. How good is the approximation?
37. Use the normal approximation to the binomial to calculate the answer to Problem 26. How good is the approximation?

CASE STUDY 8.1
Probability

PART 1

The registrar's office at a nearby university handles thousands of student class registrations every term. To take care of each student's needs efficiently and correctly, the registrar sees to it that a significant amount of time is spent in training the staff. Each staff member receives one of two types of training: A refresher course is given for current employees, and a general training course is given for new hires.

At the end of each of the training courses, a short exam tests the staff member's ability to locate errors on a registration form. A group of 15 registration forms contains 5 forms with errors. During the test, the staff member is asked to randomly select four forms and check them for errors. The sampling is to be done without replacement.

A major concern of those administering the test is that the forms with errors will not be selected by the staff member who is randomly selecting four forms.

Assignment

Find the probability that in a lot of 15 forms, a sample of 4 forms will have no errors.

PART 2

The registrar receives a steady stream of registration forms for the three weeks prior to each term. While the office tries to provide easy-to-comprehend forms with complete instructions, a few forms are completed that are incorrect. Forms are considered incorrectly filled out if any key information is missing. Key information includes the student's local or permanent address, a course number or title, program or area of study, etc. When incorrect forms are received by the registrar's office they are sent back to the student for correction. This process delays the student's registration. While methods, forms, and instructions have been improving, the registrar feels that 5 percent of the forms are filled out incorrectly. To help continue improving the forms and instructions, he has 10 forms sampled each day to determine the types of errors being made. This information is shared with a quality improvement team whose members are currently working on improving the existing forms.

Assignment

Given that the process has a fraction nonconforming (p) of 5 percent ($p = 0.05$), what is the probability that in a sample of 10 forms, 3 or fewer will be found with errors?

PART 3

For the three weeks prior to the beginning of every term and for the first two weeks of the term, the registrar's office is very busy. To help determine staffing needs, the registrar decides to study the number of students and/or parents utilizing the office's services during any given minute. From past experience he knows that on average three new customers arrive every minute.

Assignment

Find the probability that in any given minute, five new customers will arrive. What is the probability that more than seven new customers will arrive in a given minute?

CASE STUDY 8.2
Normal Probability Distribution

This case is the first of three related cases found in Chapters 8, 9, and 11. These cases seek to link information from the three chapters in order to resolve quality issues. Although they are related, it is not necessary to complete the case in this chapter in order to understand or complete the cases in the following chapters.

PART 1

Max's B-B-Q Inc. manufactures top-of-the-line barbeque tools. The tools include forks, spatulas, knives, spoons, and shish-kebab skewers. Max's fabricates both the metal parts of the tools and the resin handles. These are then riveted together to create the tools (Figure 1). Recently, Max's hired you as a process engineer. Your first assignment is to study routine tool wear on the company's stamping machine. In particular, you will be studying tool-wear patterns for the tools used to create knife blades.

In the stamping process, the tooling wears slightly during each stroke of the press as the punch shears through the material. As the tool wears, the part features become smaller. The knife has specifications of 10 mm ± 0.025 mm; undersized parts must be scrapped. The tool can be resharpened to bring the parts produced back into specification. To reduce manufacturing costs and simplify machine scheduling, it is critical to pull the tool and perform maintenance only when absolutely necessary. It is very important for scheduling, costing, and quality purposes that the average num-

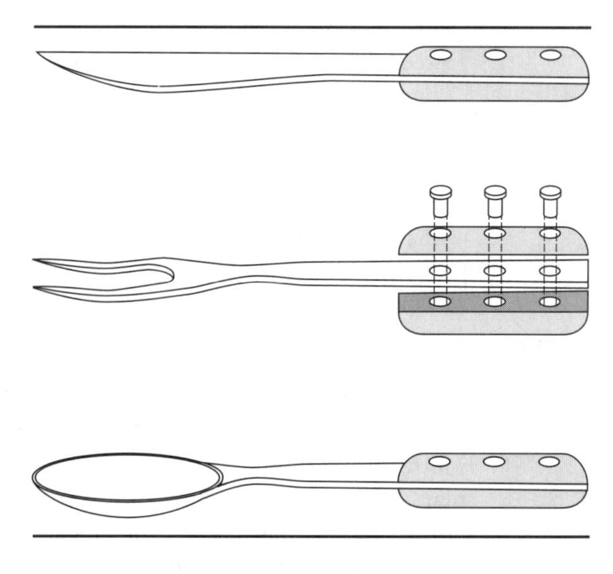

Figure 1 Barbeque Tools

ber of strokes, or tool run length, be determined. Knowing the average number of strokes that can be performed by a tool enables routine maintenance to be scheduled.

It is the plant manager's philosophy that tool maintenance be scheduled proactively. When a tool is pulled unexpectedly, the tool-maintenance area may not have time to work on it immediately. Presses without tools don't run, and if they are not running, they are not making money. As the process engineer studying tool wear, you must develop a prediction for when the tool should be pulled and resharpened.

The following information is available from the tool-maintenance department.

- The average number of strokes for a tool is 45,000.
- The standard deviation is 2,500 strokes.
- A punch has a total of 25 mm that can be ground off before it is no longer useful.
- Each regrind to sharpen a punch removes 1 mm of punch life.
- The cost to regrind is

  2 hours of press downtime to remove and reinsert tool, at $300 per hour
  5 hours of tool maintenance time, at $65 per hour
  5 hours of downtime while press is not being used, at $300 per hour

- The average wait time for unplanned tool regrind is 15 hours at $300 per hour.
- Because of the large number of strokes per tool regrind, this is considered to be a continuous distribution. The normal curve probability distribution is applicable.

Assignment

One percent of the tools wear out very early in their expected productive life. Early tool wearout—and, thus, an unplanned tool pull—can be caused by a variety of factors, including changes in the hardness of the material being punched, lack of lubrication, the hardness of the tool steel, and the width of the gap between the punch and the die. Key part dimensions are monitored using $\overline{X}$ and R charts. These charts reveal when the tool needs to be reground in order to preserve part quality. Use the normal probability distribution and the information provided to calculate the number of strokes that would result in an early wearout percentage of 1 percent or fewer. If the plant manager wants the tool to be pulled for a regrind at 40,000 strokes, what is the chance that there will be an early tool wearout failure before the tool reaches 40,000 strokes?

PART 2

Now that you have been at Max's B-B-Q Inc. for a while, the plant manager asks you to assist the production-scheduling department with pricing data on a high-volume job requiring knife blades for the company's best customer. As you know, it is the plant manager's philosophy to be proactive when scheduling tool maintenance (regrinds) rather than have to unexpectedly pull the tool. However, pricing will be a very important factor in selling this job to the customer. Essentially, the plant manager wants no unplanned tool pulls, but sales needs pricing cost reductions. The production scheduler would like a tool regrind schedule that results in minimal inventory.

Assignment

You will soon be meeting with the plant manager and the managers from sales and production scheduling. They are expecting you to have an answer to the question: given the need to balance maximizing tool use, minimizing inventory, minimizing production disruption, and minimizing cost, how many strokes should you recommend to run this tool before pulling for a regrind?

Create a graph that shows the number of unplanned pulls versus the number of strokes. The graph should comprise at least six data points. Next, complete the spreadsheet in Figure 2 showing the costs of each individual's plan. Using the graph and the spreadsheet, prepare a response for the question, How many strokes should the tool be run before pulling it for a regrind? Your analysis should include answers to the following questions: How will this number balance tool use, cost, inventory, and production disruption? What are the economics of this situation?

	Plant Manager	***Production Scheduler***	***Sales Manager***	***You***
Strokes Before Pull	40,000	42,000	43,000	
Number of Pulls	25	25	25	
Production Over Life of Tool	1,000,000	1,050,000	1,075,000	
Cost of Each Pull	$2,425	$2,425	$2,425	$2,425
Additional Cost of an Unplanned Pull	$4,500	$4,500	$4,500	$4,500
Chance of Unplanned Pulls				
Total Additional Cost Due to Unplanned Pulls				
Total Cost				

Figure 2

9

Quality Control Charts for Attributes

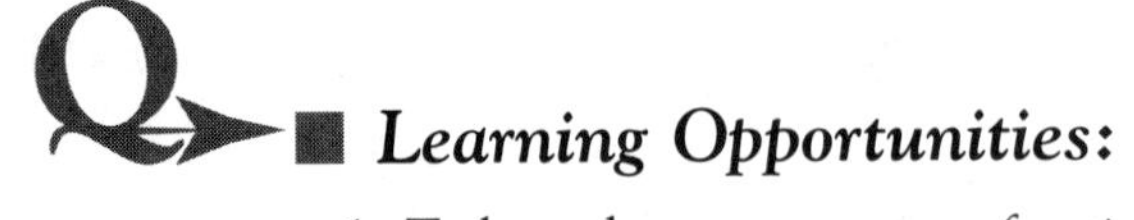

Learning Opportunities:

1. To learn how to construct fraction nonconforming (p) charts for both constant and variable sample sizes
2. To learn how to construct number nonconforming (np) charts
3. To learn how to construct percent nonconforming charts
4. To learn how to construct charts for counts of nonconformities (c charts)
5. To learn how to construct charts for nonconformities per unit (u charts) for both constant and variable sample sizes
6. To understand how to interpret p, np, c, and u charts

I Want an Ocean View!

Many service providers strive to give their customers what they want, even if it means moving heaven and earth—or in this case, the ocean to Lake Michigan. How does a hotel or other service industry keep track of the quality of the services they are providing? What is the best way for a company to study the number of complaints or returns? How does a firm assess the nonconforming product being produced? Attribute charts, discussed in this chapter, can assist in monitoring processes providing goods or services. These charts can be used whenever counts or percentages of nonconformities can be obtained. Attribute charts enable users to track performance, monitor process stability, and discover where improvements can be made.

ATTRIBUTES

Attributes *are characteristics associated with a product or service*. These characteristics either do or do not exist, and they can be counted. Examples of attributes include the number of leaking containers, of scratches on a surface, of on-time deliveries, or of errors on an invoice. Attribute charts are used to study the stability of processes over time, provided that a count of nonconformities can be made. Attribute charts are used when measurements may not be possible or when measurements are not made because of time or cost issues.

Attribute data are relatively easy and inexpensive to collect. The product, when studied, either conforms to specifications or it does not. The most difficult part about collecting attribute data lies in the need to develop precise operational definitions of what is conforming and what is not. For instance, at first glance, creating specifications to judge the surface finish of a television screen appears straightforward. The surface should be free from flaws and imperfections. Simple enough, until questions arise concerning what is a flaw? what is considered conformance? under what conditions?

There are some disadvantages to using attribute charts. Attribute charts do not give any indication about why the nonconformity occurred, nor do the charts provide much detail. The charts do not provide information to answer questions like: do several nonconformities exist on the same product? is the product still usable? can it be reworked? what is the severity or degree of nonconformance? The charts measure the increases and decreases in the level of quality of the process but provide little information about why the process changes.

Types of Charts

When nonconformities are investigated, two conditions may occur. The product or service may have a quality characteristic or characteristics that prevent it from being used. These conditions are called "nonconforming." In other situations, the nonconforming aspect may reduce the desirability of the product but not prevent its use; consider, for example, a dented washing machine that still functions as expected. These are called "nonconformities." *When the interest is in studying the proportion of products rendered unusable by their nonconformities, a fraction nonconforming* (p) *chart, a number nonconforming* (np) *chart, or a percent nonconforming chart should be used. When the situation calls for tracking the count of nonconformities, a number of nonconformities* (c) *chart or a number of nonconformities per unit* (u) *chart is appropriate.*

CHARTS FOR NONCONFORMING UNITS

Fraction Nonconforming (p) Charts: Constant Sample Size

The ***fraction nonconforming chart, or* p *chart,*** *based on the binomial distribution, is used to study the proportion of nonconforming products or services being provided.* A nonconforming product or service is considered unacceptable because of some deviation from an expected level of quality. For a p chart, nonconformities render the product or service unusable and therefore nonconforming. Formerly called "charts for defective or discrepant items," these charts are used to study situations where the product or service can be judged to be either good or bad, correct or incorrect, working or not working. For example, a container is either leaking or it is not, a light is either lit or it is not, and an order delivered to a restaurant patron is either correct or incorrect.

p charts, because of the structure of their formulas, can be constructed using either a constant or variable sample size. A p chart (Figure 9.1) for constant sample size is constructed using the following steps:

1. *Gather the data.* In constructing any attribute chart, careful consideration must be given to the process and what characteristics should be studied. The choice of the attributes to monitor should center on the customer's needs and expectations as well as on current and potential problem areas. Once the characteristics have been identified, time must be spent to define the acceptance criteria. When gathering the data concerning the attributes under study, identify nonconforming units by comparing the inspected product with the specifications. The number nonconforming (np) is tracked. Since a p chart studies the proportion of a process that is nonconforming, acceptance specifications should clearly state the expectations concerning conformance. In some cases, go/no-go gauges are used; in others, pictures of typical conforming and nonconforming products are helpful. In the service industry, details of incorrect bills, faulty customer service, or other performance criteria should be clearly established.

Once the characteristics have been designated for study and a clear understanding has been reached about what constitutes a conforming product, there are two aspects to gathering the data that must be dealt with: the sample size n and the frequency of sampling. The sample sizes for attribute charts tend to be quite large (for example, $n = 6,250$). Large sample sizes are required to maintain sensitivity to detect process performance changes. The sample size should be large enough to include nonconforming items in each subgroup. When process quality is very good, large sample sizes are needed to capture information about the process. When selected, samples must be random and representative of the process.

2. *Calculate* p, *the proportion nonconforming.* The proportion nonconforming (p) is plotted on a fraction nonconforming chart. As the products or services are inspected, each subgroup will yield a number nonconforming (np). The proportion nonconforming (p), plotted on the p chart, is calculated using n, the number of inspected items, and np, the number of nonconforming items found:

$$p = \frac{np}{n}$$

3. *Plot the proportion nonconforming* (p) *on the control chart.* Once calculated, the values of p for each subgroup are plotted on the chart. The scale for the p chart should reflect the magnitude of the data.

4. *Calculate the centerline and control limits.* The centerline of the control chart

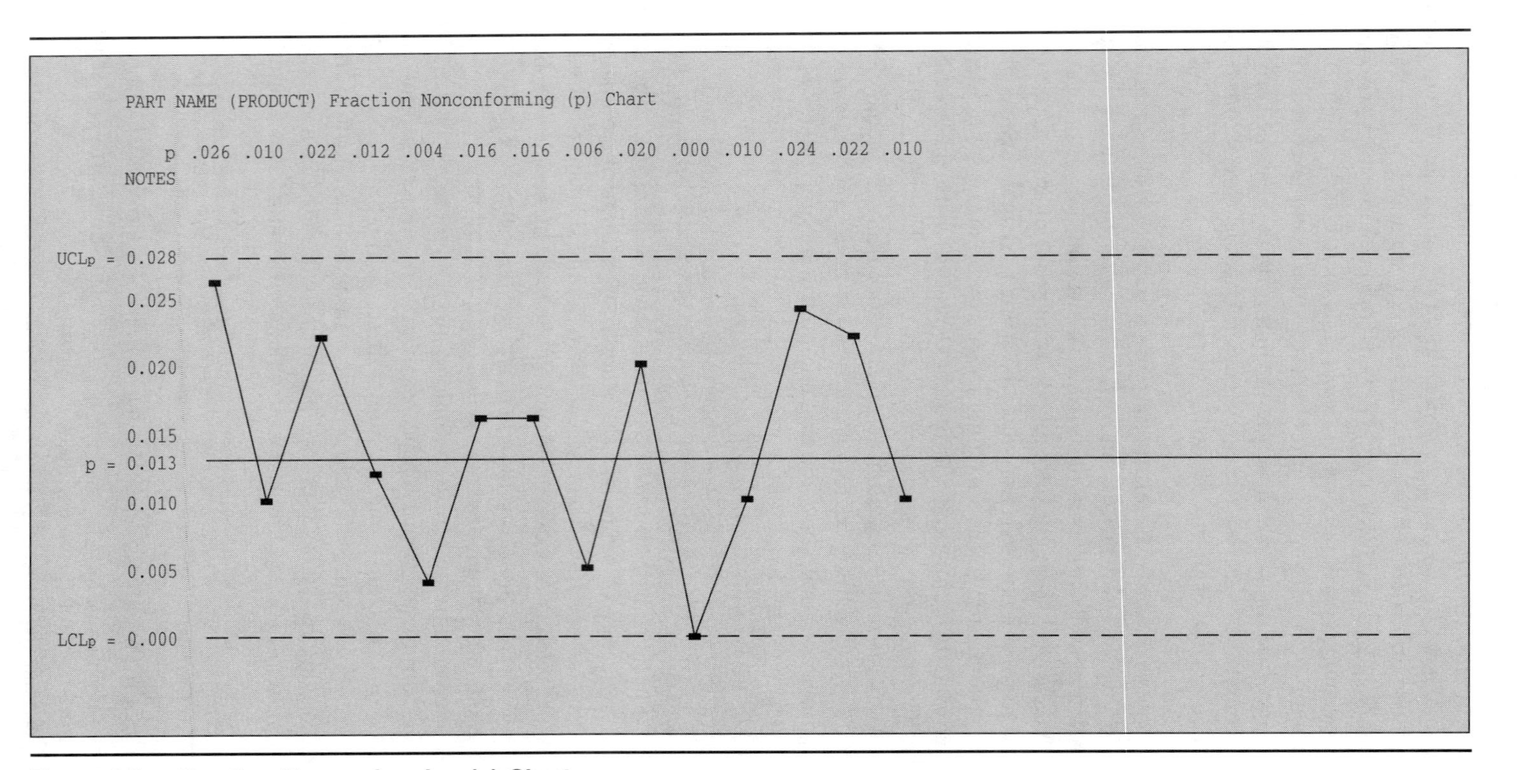

Figure 9.1 Fraction Nonconforming (p) Chart

is the average of the subgroup proportion nonconforming. The number nonconforming values are added up and then divided by the total number of samples:

$$\text{Centerline } \bar{p} = \frac{\sum_{i=1}^{n} np}{\sum_{i=1}^{n} n}$$

The control limits for a p chart are found using the following formulas:

$$UCL_p = \bar{p} + 3\frac{\sqrt{\bar{p}(1 - \bar{p})}}{\sqrt{n}}$$

$$LCL_p = \bar{p} - 3\frac{\sqrt{\bar{p}(1 - \bar{p})}}{\sqrt{n}}$$

On occasion, the lower control limit of a p chart may have a negative value. When this occurs, the result of the LCL_p calculation should be rounded up to zero.

5. *Draw the centerline and control limits on the chart.* Using a solid line to denote the centerline and dashed lines for the control limits, draw the centerline and control limits on the chart.

6. *Interpret the chart.* The interpretation of a fraction nonconforming chart is similar in many aspects to the interpretation of a variables control chart. As with variables charts, in interpreting attribute charts emphasis is placed on determining if the process is operating within its control limits and exhibiting random variation. As with a variables control chart, the data points on a p chart should flow smoothly back and forth across the centerline. The number of points on each side of the centerline should be balanced, with the majority of the points near the centerline. There should be no patterns in the data, such as trends, runs, cycles, or sudden shifts in level. All of the points should fall between the upper and lower control limits. Points beyond the control limits are immediately obvious and indicate an instability in the process. One difference between the interpretation of a variables control chart and a p chart is the desirability in the latter of having points that approach the lower control limits. This makes sense because quality-improvement efforts reflected on a fraction nonconforming chart should show that the number nonconforming is being reduced, the ultimate goal of improving a process. This favorable occurrence should be investigated to determine what was done right and whether or not there are changes or improvements that should be incorporated into the process on a permanent basis. Similarly, a trend toward zero nonconforming or shift in level that lowers the number of nonconforming should be investigated.

The process capability is the $\bar{p}$, the centerline of the control chart.

EXAMPLE 9.1 Making a p Chart with Constant Sample Size

Special Plastics, Inc., has been making the blanks for credit cards for a number of years. They use p charts to keep track of the number of nonconforming cards that are created each time a batch of blank cards is run. Use the data in Table 9.1 to create a fraction nonconforming (p) chart.

Step 1. Gather the Data. The characteristics that have been designated for study include blemishes on the card's front and back surfaces; color inconsistencies; white spots or bumps caused by dirt; scratches; chips; indentations; or other flaws. Several pictures are maintained at each operator's workstation to provide a clear understanding of what constitutes a nonconforming product.

Batches of 15,000 blank cards are run each day. Samples of size 500 are randomly selected and inspected. The number of nonconforming units (np) is recorded on data sheets (Table 9.1).

Step 2. Calculate p, the Proportion Nonconforming. After each sample is taken and the inspections are complete, the proportion nonconforming (p) is calculated using n = 500, and np, the number of nonconforming items, is found. (Here we work p to three decimal places.) For example, for the first value,

$$p = \frac{np}{n} = \frac{20}{500} = 0.040$$

The remaining calculated p values are shown in Table 9.1.

Sample Number	n	np	p
1	500	20	0.040
2	500	21	0.042
3	500	19	0.038
4	500	15	0.030
5	500	18	0.036
6	500	20	0.040
7	500	19	0.038
8	500	28	0.056
9	500	17	0.034
10	500	20	0.040
11	500	19	0.038
12	500	18	0.036
13	500	10	0.020
14	500	11	0.022
15	500	10	0.020
16	500	9	0.018
17	500	10	0.020
18	500	11	0.022
19	500	9	0.018
20	500	8	0.016
	10,000	312	

Table 9.1 Data Sheet: Credit Cards

Step 3. Plot the Proportion Nonconforming on the Control Chart. As they are calculated, the values of p for each subgroup are plotted on the chart. The p chart in Figure 9.2 has been scaled to reflect the magnitude of the data.

Step 4. Calculate the Centerline and Control Limits. The centerline of the control chart is the average of the subgroup proportion nonconforming. The number of nonconforming values from Table 9.1 are added up and then divided by the total number of samples:

$$\text{Centerline } \bar{p} = \frac{\sum_{i=1}^{n} np}{\sum_{i=1}^{n} n} = \frac{312}{20(500)} = 0.031$$

The control limits for a p chart are found using the following formulas:

$$\begin{aligned} UCL_p &= \bar{p} + 3\frac{\sqrt{\bar{p}(1-\bar{p})}}{\sqrt{n}} \\ &= 0.031 + 0.023 = 0.054 \end{aligned}$$

$$\begin{aligned} LCL_p &= \bar{p} - 3\frac{\sqrt{\bar{p}(1-\bar{p})}}{\sqrt{n}} \\ &= 0.031 - 0.023 = 0.008 \end{aligned}$$

Step 5. Draw the Centerline and Control Limits on the Chart. The centerline and control limits are then drawn on the chart in Figure 9.3, with a solid line denoting the centerline and dashed lines the control limits.

Step 6. Interpret the Chart. The process capability for this chart is $\bar{p}$, 0.031, the centerline of the control chart. Point 8 in Figure 9.3 is above the upper control limit and should be investigated to determine if an assignable cause exists. If one is found, then steps should be taken to prevent future occurrences. When the control chart is studied for any nonrandom conditions such as runs, trends, cycles, or points out of control, connecting the data points can help reveal any patterns. Of great interest is the significant decrease in the fraction nonconforming after point 13. This should be studied to determine what changes have been made to the process.

Revising the p Chart

Once an assignable cause has been isolated and the process has been modified to prevent its recurrence, then the centerline and control limits can be recalculated to reflect the changes. The points that have been isolated as due to an assignable cause will be removed from the calculations. To revise the centerline and control limits of a p chart,

$$\bar{p}_{new} = \frac{\sum_{i=1}^{n} np - np_d}{\sum_{i=1}^{n} n - n_d}$$

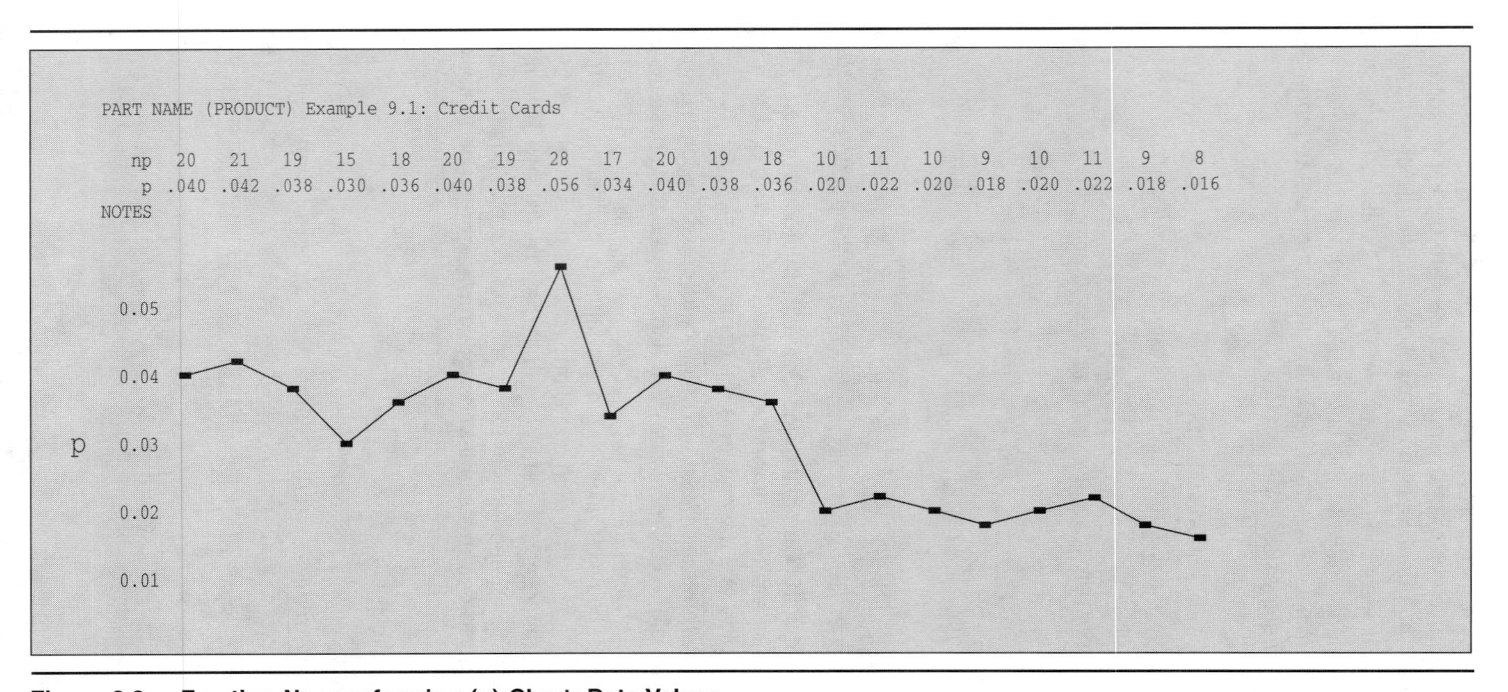

Figure 9.2 **Fraction Nonconforming (p) Chart: Data Values**

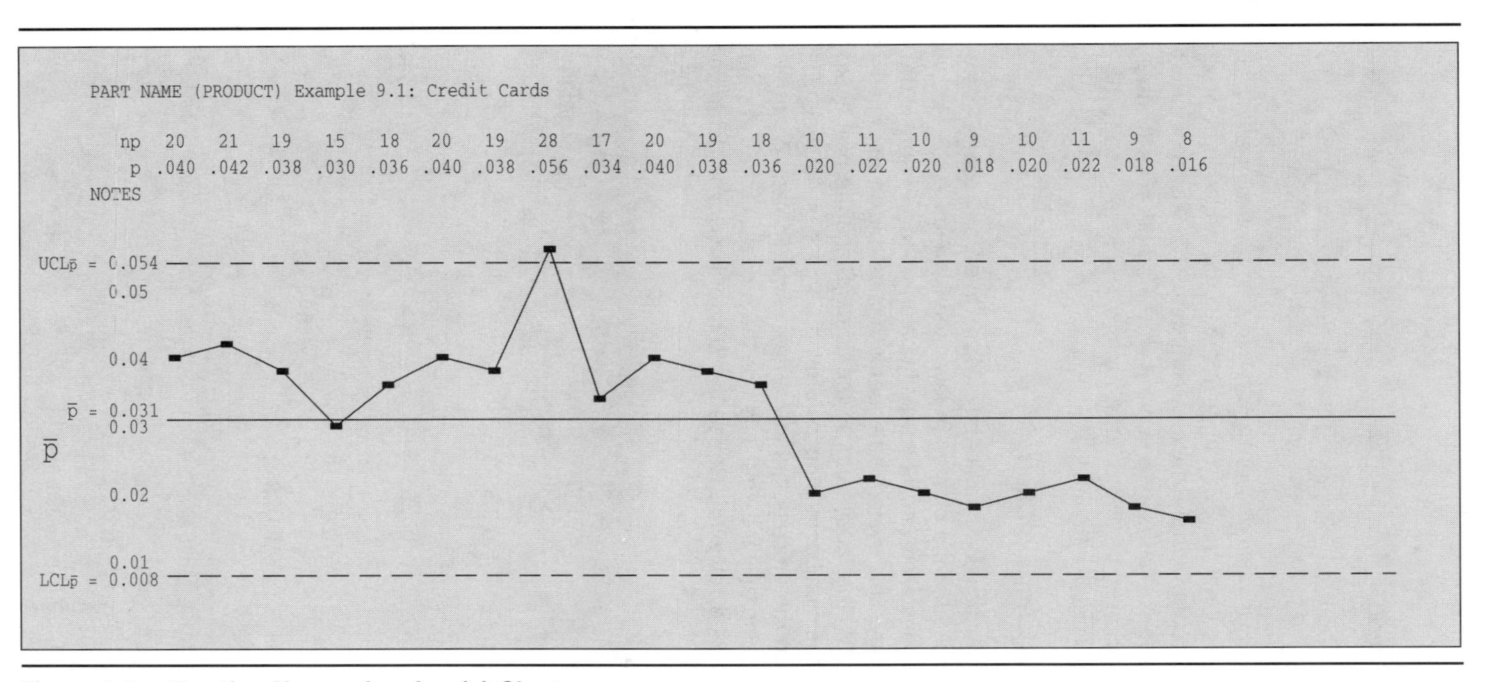

Figure 9.3 Fraction Nonconforming (p) Chart

$$UCL_{\bar{p}_{new}} = \bar{p}_{new} + 3\frac{\sqrt{\bar{p}_{new}(1 - \bar{p}_{new})}}{\sqrt{n}}$$

$$LCL_{\bar{p}_{new}} = \bar{p}_{new} - 3\frac{\sqrt{\bar{p}_{new}(1 - \bar{p}_{new})}}{\sqrt{n}}$$

The process capability of the revised chart is the newly calculated p_{new}.

There are cases in which improvements are made to the process that dramatically alter the process. Under these circumstances, it is appropriate to revise the centerline and control limits by removing all the points prior to the change and calculating a new centerline and control limits working only with the points that occurred after the changes were made.

EXAMPLE 9.2 Revising a p Chart

Special Plastics, Inc., has been involved in several quality improvement efforts that have resulted in the change in level seen in Figure 9.3. Recently a new machine has been installed and is currently being used to improve the printing and color consistency. This system safeguards against dirt in the printing ink and prevents white spots from appearing on the cards. Since the new equipment has been installed, the number of nonconforming cards has decreased (Figure 9.3). Those monitoring the process feel that a new centerline and control limits should be calculated using only the data following the process changes; that is, points 1 through 12 should be removed. Revise the control limits and determine the new process capability:

$$\bar{p}_{new} = \frac{\sum_{i=1}^{n} np - np_d}{\sum_{i=1}^{n} n - n_d} = \frac{312 - 20 - 21 - 19 - 15 - \cdots - 18}{10{,}000 - 12(500)}$$

$$= 0.020$$

$$UCL_{\bar{p}_{new}} = \bar{p}_{new} + 3\frac{\sqrt{\bar{p}_{new}(1 - \bar{p}_{new})}}{\sqrt{n}}$$

$$= 0.020 + 3\frac{\sqrt{0.020(1 - 0.020)}}{\sqrt{500}}$$

$$= 0.039$$

$$LCL_{\bar{p}_{new}} = \bar{p}_{new} - 3\frac{\sqrt{\bar{p}_{new}(1 - \bar{p}_{new})}}{\sqrt{n}}$$

$$= 0.020 - 3\frac{\sqrt{0.020(1 - 0.020)}}{\sqrt{500}}$$

$$= 0.001$$

The process capability of the revised chart is 0.020, the newly calculated $\overline{p}_{new}$. In the future, the process will be expected to conform to the new limits shown in Figure 9.4.

Fraction Nonconforming (p) Charts: Variable Sample Size

In a manufacturing or service industry it is not always possible to sample the same amount each time. When the amount sampled varies, fraction nonconforming charts can be easily adapted to varying sample sizes. Constructing a p chart in which the sample size varies requires that the control limits be calculated for each different sample size, changing the n in the control-limit formulas each time a different sample size is taken. Calculating the centerline and interpreting the chart will be the same.

EXAMPLE 9.3 Making a p Chart with Variable Sample Size

A local grocery has started to survey customers as they leave the store. The survey is designed to determine if the customer had a pleasant experience while shopping at the store. A nonconforming visit is one in which the customer has a complaint such as not being able to find a particular item; receiving unfriendly service; waiting too long to be served at the deli, meat, bakery, or seafood counter; or otherwise not having expectations met. Since the number of customers surveyed varies from day to day, a fraction nonconforming control chart for variable sample size is chosen.

Step 1. Gather the Data. Table 9.2 shows the results of four weeks of sampling.

Step 2. Calculate p, the Proportion Nonconforming, for Each of the Samples. For this example, the values are calculated to three decimal places. For example, for the second sample,

$$p = \frac{np}{n} = \frac{1}{100} = 0.010$$

Table 9.2 presents the calculated p values.

Step 3. Plot the Proportion Nonconforming (p) on the Control Chart (Figure 9.5).

Step 4. Calculate the Centerline and Control Limits.

$$\text{Centerline } \overline{p} = \frac{\sum_{i=1}^{n} np}{\sum_{i=1}^{n} n} = \frac{48}{2385} = 0.020$$

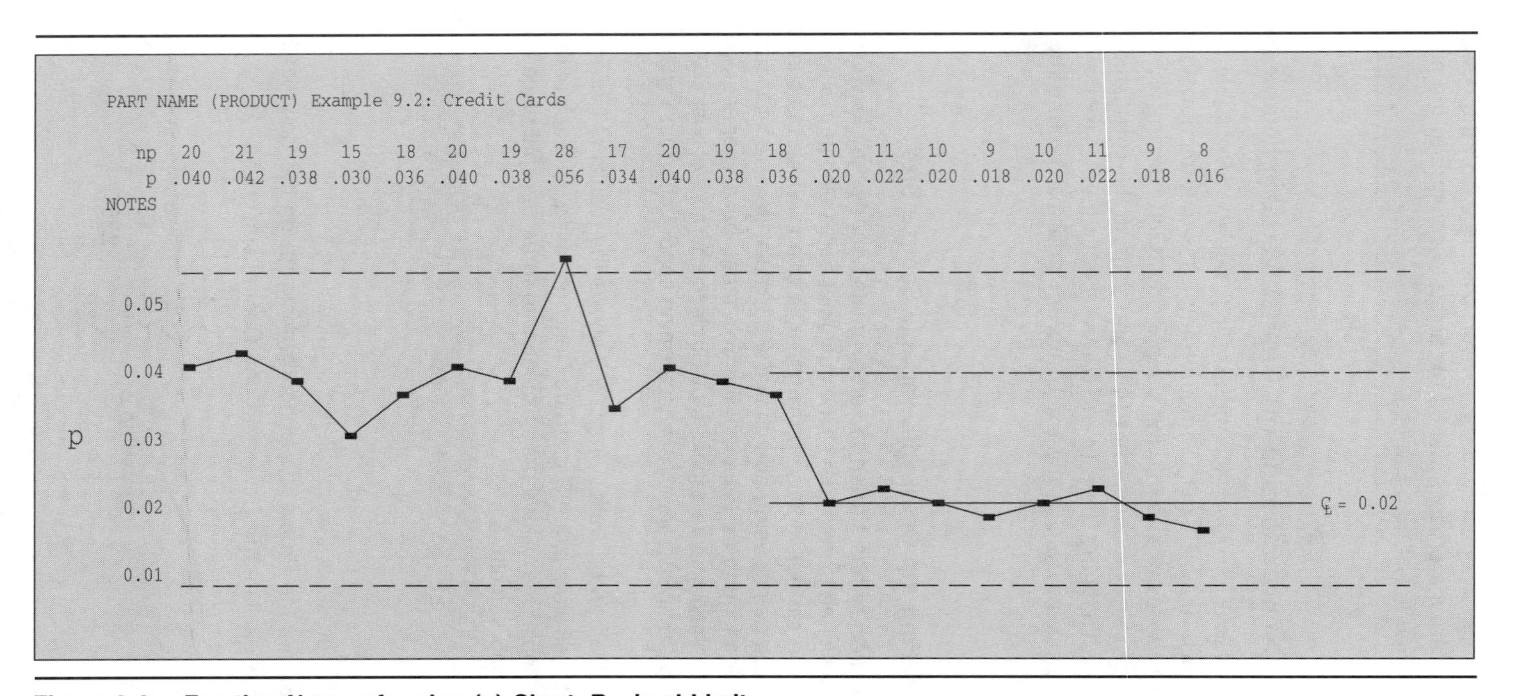

Figure 9.4 Fraction Nonconforming (p) Chart: Revised Limits

Sample Number	n	np	p
1	120	0	0.000
2	100	1	0.010
3	105	6	0.029
4	110	1	0.009
5	95	1	0.011
6	110	2	0.018
7	115	0	0.000
8	110	1	0.009
9	100	0	0.000
10	100	2	0.030
11	110	1	0.009
12	105	3	0.028
13	110	0	0.000
14	115	2	0.017
15	120	7	0.058
16	110	2	0.018
17	105	1	0.010
18	110	2	0.018
19	105	0	0.000
20	110	1	0.009
21	115	8	0.069
22	105	7	0.067

Table 9.2 Data Sheet: Grocery Store

At this point the control limits must be calculated for each different sample size. The control limits for a p chart are found using the following formulas and varying the sample size as needed. For example:

$$UCL_p = \bar{p} + 3\frac{\sqrt{\bar{p}(1-\bar{p})}}{\sqrt{n}}$$

$$UCL_p = 0.020 + 3\frac{\sqrt{0.020(1-0.020)}}{\sqrt{120}} = 0.058$$

$$UCL_p = 0.020 + 3\frac{\sqrt{0.020(1-0.020)}}{\sqrt{115}} = 0.059$$

$$UCL_p = 0.020 + 3\frac{\sqrt{0.020(1-0.020)}}{\sqrt{105}} = 0.061$$

$$UCL_p = 0.020 + 3\frac{\sqrt{0.020(1-0.020)}}{\sqrt{95}} = 0.063$$

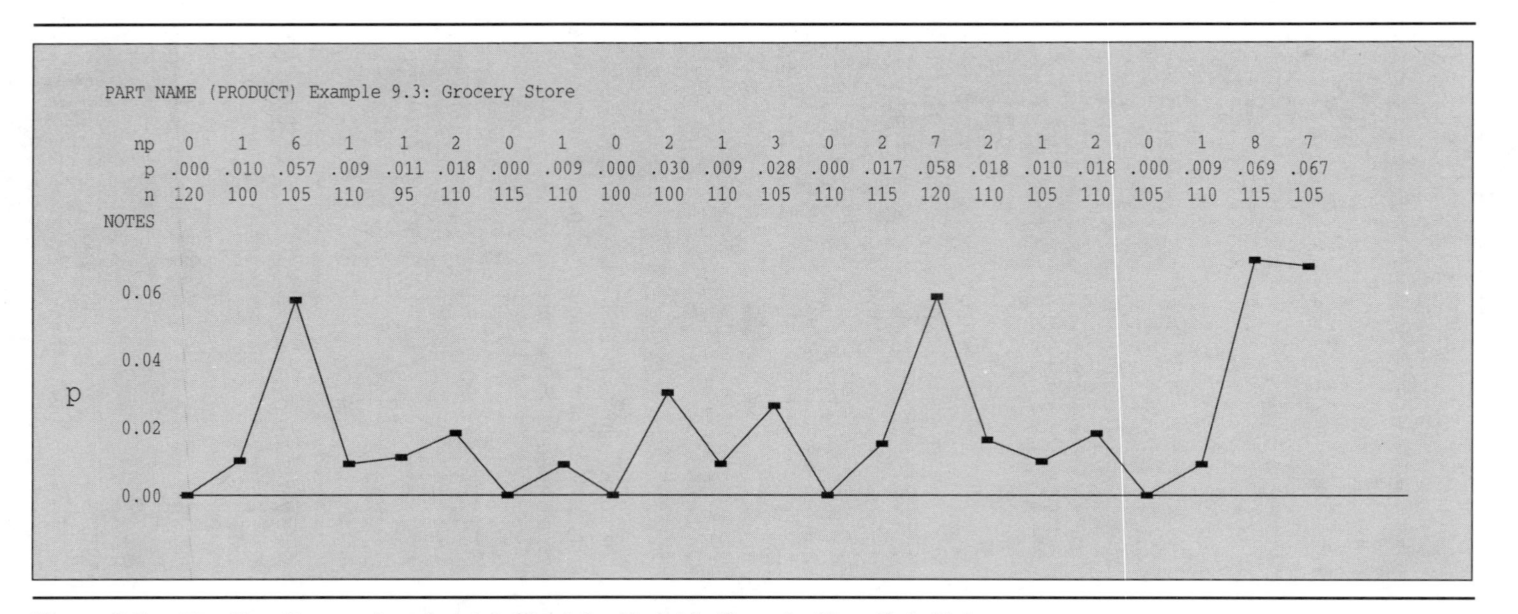

Figure 9.5 Fraction Nonconforming (p) Chart for Variable Sample Size: Data Values

Notice that the values for the upper control limits change slightly for each different sample size. As the sample size *decreases,* the control limits get wider. As the sample size *increases,* the control limits get tighter. Larger sample sizes enable us to learn more about the process and what it produces. The tighter control limits are a reflection of this increased knowledge.

$$LCL_p = \bar{p} - 3\frac{\sqrt{\bar{p}(1-\bar{p})}}{\sqrt{n}}$$

$$= 0.020 - 3\frac{\sqrt{0.020(1-0.020)}}{\sqrt{100}} = -0.022 = 0$$

Since the lower control limits for each different sample size yield values less than zero, they will all be rounded to zero. Figure 9.6 presents calculated upper and lower control limits.

Step 5. Draw the Centerline and Control Limits on the Chart. Draw the centerline on the chart using a solid line, just as before. The dashed lines for the control limits will increase and decrease according to the sample size.

Step 6. Interpret the Chart. Control charts with variable subgroup sizes are interpreted on the basis of where the point falls in relation to the centerline and each point's respective control limits. Unusual patterns are interpreted as before. In this chart (Figure 9.6), there are two points out of control at the end of the chart (points 21 and 22) but no unusual patterns.

Calculating p *Chart Control Limits Using* n_{ave}

Calculating control limits for each different sample size is time-consuming, especially if it must be done without the aid of a computer. To simplify the construction of a fraction nonconforming chart, n_{ave} can be used. The value n_{ave} can be found by summing the individual sample sizes and dividing by the total number of times samples were taken:

$$n_{ave} = \frac{\sum_{i=1}^{n} n}{m}$$

The value n_{ave} can be used whenever the individual sample sizes vary no more than 25 percent from the calculated n_{ave}. If, in a group of samples, several of the sample sizes vary more than 25 percent from n_{ave}, then individual limits should be calculated and used to study the process.

Larger sample sizes provide more information about process quality than smaller sample sizes. With a larger number of samples, more of the production is being studied, thus providing more information. Because of this, you may have already noticed that when sample sizes are large, the control limits are tighter. This is because you know more about what is being produced. The standard deviation will also be smaller,

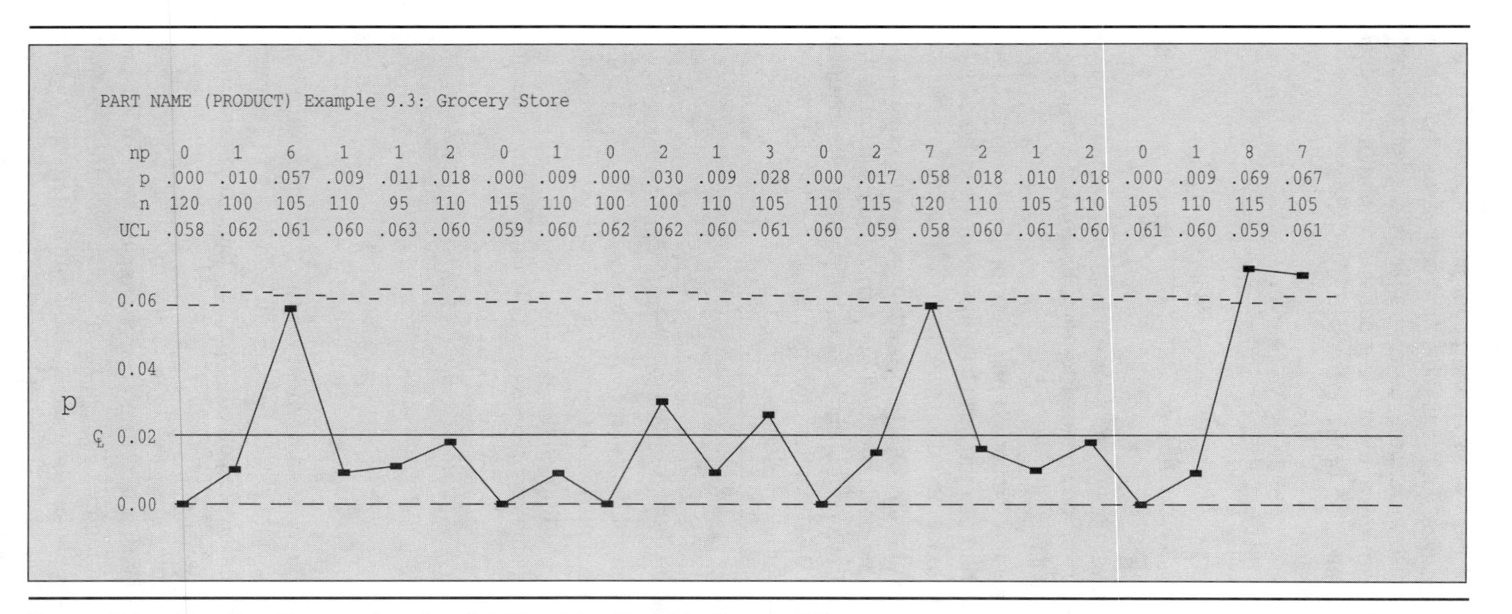

Figure 9.6 Fraction Nonconforming (p) Chart for Variable Sample Size

reflecting greater confidence in the information. When sample sizes are small, we really don't get a very complete picture of the process. For this reason, the control limits are wider, allowing for a larger margin for error, a wider estimate of where the product produced in the future will fall. A small sample size also results in a larger standard deviation. This relationship holds true for both variables and attributes data.

Control limits calculated using n_{ave} are based on the average value of all of the individual sample sizes. Points lying above or near the control limits must be scrutinized to determine whether or not their individual sample size will affect the interpretation of the point's location. Those values near the upper control limits indicate a situation where the quality of the product has deteriorated; it is those points that should be studied in an effort to determine the causes. Points near the centerline receive less emphasis because of their proximity to the average fraction nonconforming. The interpretation of a fraction nonconforming chart created using n_{ave} can be simplified by the use of four cases.

Case I A Case I situation occurs if the point falls inside the limits, but relatively close to the upper control limit, and $n_{ind} < n_{ave}$. Here the individual limits, created using a smaller sample size, will be wider than the limits calculated for n_{ave}. Therefore, there is no need to check the specific value of the individual control limits; the point, which is under control for the average limit, will definitely be under control for the individual limits.

Case II In Case II, the point falls inside the control limits calculated using n_{ave}, and in this case $n_{ind} > n_{ave}$. Any time a greater number of items is investigated, the limits will contract toward the centerline. A larger number of samples taken tells the user more about the process, thus enabling the process to be more discerning between good and bad quality. So in this case, the control limits associated with the individual value will be narrower than the control limits created using n_{ave}. Under these circumstances, the individual control limits should be calculated and the point checked to see if it is in control.

Case III In Case III, the point falls outside the control limits for n_{ave}, and $n_{ind} > n_{ave}$. Because n_{ave} is less than the sample size for the individual data point, the n_{ave} limits are more forgiving than are those for the individual sample size. If the point is out of control for the n_{ave} limits, then it will also be out of control for the narrower n_{ind} limits. There is no need to calculate the specific individual limits because the point will be out of control for both.

Case IV In Case IV, the point is also out of control. Here $n_{ind} < n_{ave}$. Since the individual sample size is less than that of the average sample size, the control limits for the individual sample will be wider than those for the average sample. The point should be tested to determine if it falls out of the control limits for its individual sample size.

EXAMPLE 9.4 Calculating Control Limits for a p Chart I

To simplify calculations, the grocer of Example 9.3 has decided to create a control chart using n_{ave}. Using the data provided in Table 9.2, he creates a fraction nonconforming control chart using n_{ave}. He will study the chart and interpret it using the four cases.

Begin by calculating n_{ave}:

$$n_{ave} = \frac{\sum_{i=1}^{n} n}{m} = \frac{120 + 100 + 105 + 110 + \cdots + 110 + 115 + 105}{22}$$

$$= 108.4, \text{ which is rounded to } 108.$$

Verify that the values for n in Table 9.2 do not exceed 108 ± 25 percent. Calculate the centerline and control limits:

$$\text{Centerline } \overline{p} = \frac{\sum_{i=1}^{n} np}{\sum_{i=1}^{n} n} = \frac{48}{2385} = 0.020$$

$$UCL_p = 0.020 + 3\frac{\sqrt{0.020(1 - 0.020)}}{\sqrt{108}} = 0.060$$

$$LCL_p = 0.020 - 3\frac{\sqrt{0.020(1 - 0.020)}}{\sqrt{108}} = -0.020 = 0$$

After calculating the limits and placing them on the chart (Figure 9.7), check the points nearest to the upper control limit, using the four cases, to determine if they are under control. From Figure 9.7, check points 3, 15, 21, and 22 against the cases.

Case I A Case I situation occurs if the point falls inside the limits, but relatively close to the upper control limit, and $n_{ind} < n_{ave}$. Point 3 (0.057) on the chart is close to the upper control limit of 0.060. Because the individual limits are created using a smaller sample size, the individual limits will be wider than the limits calculated for n_{ave}. There is no need to check the specific value of the individual control limits for point 3. It is under control for the average limit and will be under control for the individual limits.

Case II In Case II, the point falls inside the control limits calculated using n_{ave}, and $n_{ind} > n_{ave}$. Point 15 (0.058) fits this situation. Since a greater number of parts has been investigated (120 versus 108), the limits for point 15 will contract toward the centerline. In this case, the control limits associated with the individual value will be narrower than the control limits created using n_{ave}, and the individual control limits will need to be calculated and the point checked to see if it is in control:

$$UCL_p = 0.020 + 3\frac{\sqrt{0.020(1 - 0.020)}}{\sqrt{120}} = 0.058$$

Point 15 is on the upper control limit and should be investigated to determine the cause behind such a high level of nonconforming product.

Case III In Case III, the point falls outside the control limits for n_{ave}, and $n_{ind} > n_{ave}$. Point 21, with a value of 0.069 and a sample size of 115, fits this situation. The n_{ave} control limits will not be as tight as those for the individual sample size. Since the

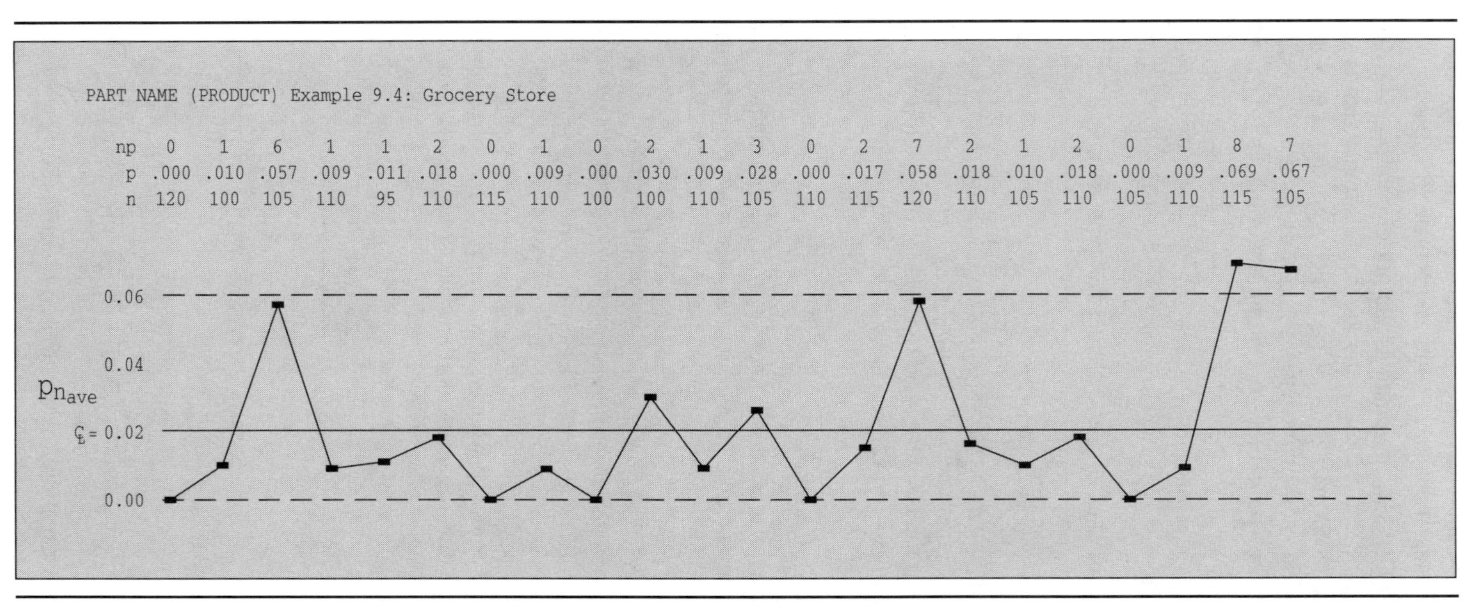

Figure 9.7 Fraction Nonconforming (p) Chart: Control Limits Calculated Using n_{ave}

point is out of control for the n_{ave} limits, then it will also be out of control for the narrower n_{ind} limits. There is no need to calculate the specific individual limits.

Case IV In Case IV, the point is also out of control and $n_{ind} < n_{ave}$. Point 22 has a value of 0.067. In this case the individual sample size is less than the average sample size (105 versus 108). When calculated, the control limits for the individual sample will be wider than those of the average sample. Point 22 will need to be tested to determine if it falls out of the control limits for its individual sample size:

$$UCL_p = 0.020 + 3\frac{\sqrt{0.020(1 - 0.020)}}{\sqrt{105}} = 0.061$$

Point 22 is out of control and should be investigated.

When comparing Examples 9.3 and 9.4 it should be noted that significantly fewer calculations are necessary when using n_{ave}. Care must be taken to properly interpret the chart using the four cases and to check that a particular sample size does not vary more than 25 percent from n_{ave}.

Revising the p Chart

p charts with variable sample sizes are revised in a manner similar to those revised with a constant sample size. Undesired values for p are removed using the formula

$$\bar{p}_{new} = \frac{\sum_{i=1}^{n} np - np_d}{\sum_{i=1}^{n} n - n_d}$$

If a point is removed when calculating $\bar{p}_{new}$, its corresponding n_{ave} should also be removed. The new n_{ave} is used when calculating the new control limits:

$$n_{ave} = \frac{\sum_{i=1}^{n} n - n_d}{m - m_d}$$

$$UCL_{p_{new}} = \bar{p}_{new} + 3\frac{\sqrt{\bar{p}_{new}(1 - \bar{p}_{new})}}{\sqrt{n}}$$

$$LCL_{p_{new}} = \bar{p}_{new} - 3\frac{\sqrt{\bar{p}_{new}(1 - \bar{p}_{new})}}{\sqrt{n}}$$

The process capability of the revised chart is the newly calculated $\bar{p}_{new}$.

EXAMPLE 9.5 Revising a p Chart II

Assume that assignable causes have been determined and process modifications are in place to prevent future recurrences. Compute the revised centerline and control limits for Example 9.4.

Recalculating n_{ave},

$$n_{ave} = \frac{\sum_{i=1}^{n} n - n_d}{m - m_d} = \frac{2385 - 115 - 105}{22 - 1 - 1} = \frac{2165}{20} = 108$$

The new centerline and control limits become

$$\bar{p}_{new} = \frac{\sum_{i=1}^{n} np - np_d}{\sum_{i=1}^{n} n - n_d} = \frac{48 - 8 - 7}{2385 - 115 - 105} = \frac{33}{2165} = 0.015$$

$$UCL_{p_{new}} = 0.015 + 3\frac{\sqrt{0.015(1 - 0.015)}}{\sqrt{108}} = 0.050$$

$$LCL_{p_{new}} = 0.015 - 3\frac{\sqrt{0.015(1 - 0.015)}}{\sqrt{108}} = -0.02 = 0$$

The process capability of the revised chart is 0.015, the newly calculated fraction nonconforming.

Percent Nonconforming Chart

Percent nonconforming charts are another variation on the fraction nonconforming (p) chart. Under certain circumstances they are easier to understand than the traditional p chart. Constructing a ***percent nonconforming chart*** is very similar to the construction of a fraction nonconforming chart, but here *the* p *values are changed to a percentage by multiplying by a factor of 100.*

The centerline for a percent nonconforming chart is $100\bar{p}$. The control limits are

$$UCL_{100p} = 100\left[\bar{p} + \frac{3\sqrt{\bar{p}(1 - \bar{p})}}{\sqrt{n}}\right]$$

$$LCL_{100p} = 100\left[\bar{p} - \frac{3\sqrt{\bar{p}(1 - \bar{p})}}{\sqrt{n}}\right]$$

A percent nonconforming chart is interpreted in the same manner as is a fraction nonconforming chart. The process capability of a percent nonconforming chart is the centerline or average percent nonconforming.

EXAMPLE 9.6 Making a Percent Nonconforming Chart

Create a percent nonconforming chart using the information presented in Example 9.1. The values to be plotted have been converted to percent and are given in Table 9.3. The centerline for a percent nonconforming chart is

$$100\bar{p} = 100(0.031) = 3.1\%$$

where:

$$\bar{p} = \frac{312}{10{,}000}$$

The control limits for a percent nonconforming chart are

$$UCL_{100p} = 100\left[0.031 + \frac{3\sqrt{0.031(1 - 0.031)}}{\sqrt{500}}\right] = 5.4\%$$

$$LCL_{100p} = 100\left[0.031 - \frac{3\sqrt{0.031(1 - 0.031)}}{\sqrt{500}}\right] = 0.8\%$$

The interpretation of this chart (Figure 9.8) will be the same as that for Example 9.1.

Sample Number	n	np	p	*100*p
1	500	20	0.040	4.0
2	500	21	0.042	4.2
3	500	19	0.038	3.8
4	500	15	0.030	3.0
5	500	18	0.036	3.6
6	500	20	0.040	4.0
7	500	19	0.038	3.8
8	500	28	0.056	5.6
9	500	17	0.034	3.4
10	500	20	0.040	4.0
11	500	19	0.038	3.8
12	500	18	0.036	3.6
13	500	10	0.020	3.0
14	500	11	0.022	2.2
15	500	10	0.020	2.0
16	500	9	0.018	1.8
17	500	10	0.020	2.0
18	500	11	0.022	2.2
19	500	9	0.018	1.8
20	500	8	0.016	1.6
	10,000	312		

Table 9.3 Data Sheet: Grocery Store

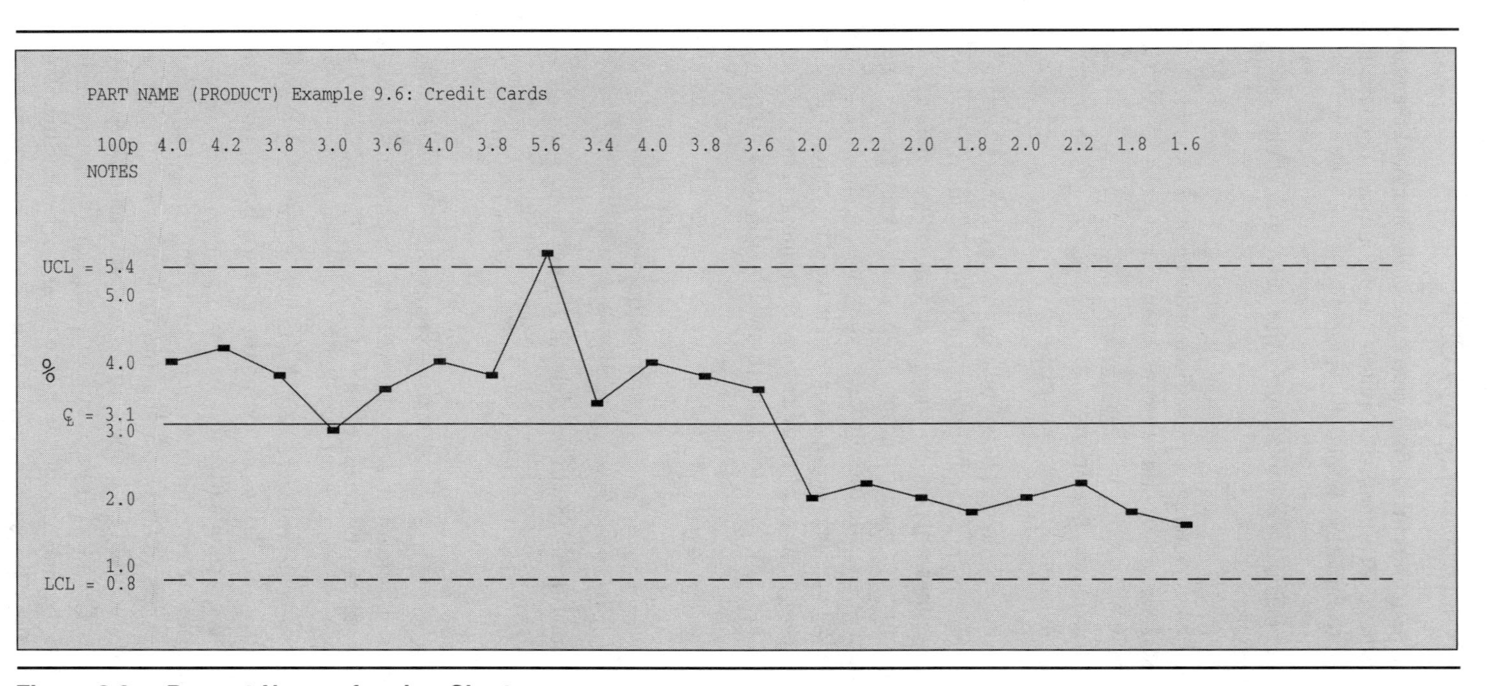

Figure 9.8 Percent Nonconforming Chart

Number Nonconforming (np) Chart

A ***number nonconforming*** **(np)** ***chart*** *tracks the number of nonconforming products or services produced by a process.* A number nonconforming chart eliminates the calculation of p, the fraction nonconforming.

1. *Gather the data.* We must apply the same critical elements we used in drafting the p chart:

- Specific characteristics or attributes designated for study
- A clear definition of a nonconforming product
- The sample size n
- The frequency of sampling
- Random, representative samples

Those studying the process keep track of the number of nonconforming units (np) by comparing the inspected product with the specifications.

2. *Plot the number of nonconforming units* (np) *on the control chart.* Once counted, the values of np for each subgroup are plotted on the chart. The scale for the np chart should reflect the magnitude of the data.

3. *Calculate the centerline and control limits.* The centerline of the control chart is the average of the total number nonconforming. The number nonconforming (np) values are added up and then divided by the total number of samples:

$$\text{Centerline } n\bar{p} = \frac{\sum_{i=1}^{n} np}{m}$$

The control limits for an np chart are found using the following formulas:

$$UCL_{np} = n\bar{p} + 3\sqrt{n\bar{p}(1 - \bar{p})}$$

$$LCL_{np} = n\bar{p} - 3\sqrt{n\bar{p}(1 - \bar{p})}$$

On occasion, the lower control limit of an np chart may have a negative value. When this occurs, the result of the LCL_{np} calculation should be rounded up to zero.

4. *Draw the centerline and control limits on the chart.* Using a solid line to denote the centerline and dashed lines for the control limits, draw the centerline and control limits on the chart.

5. *Interpret the chart.* Number nonconforming (np) charts are interpreted in the same manner as are fraction nonconforming charts. The points on an np chart should flow smoothly back and forth across the centerline with a random pattern of variation. The number of points on each side of the centerline should be balanced, with the majority of the points near the centerline. There should be no patterns in the data, such as trends, runs, cycles, points above the upper control

limit, or sudden shifts in level. Points approaching or going beyond the lower control limits are desirable. They mark an improvement in quality and should be studied to determine their cause and to ensure that it is repeated in the future. The process capability is $n\bar{p}$, the centerline of the control chart.

EXAMPLE 9.7 Making an np Chart

Plant Paradise, a nursery, receives shipments of plants from its suppliers by the truckload. The owners of Plant Paradise keep track of the number of damaged, weak, or dead plants found when the truck is unloaded. This information helps them make decisions about which suppliers to use in the future.

Step 1. Gather the Data. Those unloading the truck have a clear understanding of what constitutes a nonconforming plant. Unloaders have been trained to identify damaged, bug-infested, sick, weak, or dead plants by comparing the inspected plants with standards and setting the nonconforming plants aside to be counted. Because of the difference in the amount they deliver, trucks containing trees are tracked separately from trucks containing plants. For the purposes of this example, each plant shipment contains the same number of plants.

Plants are randomly sampled from each truckload with a sample size of n = 50.

Step 2. Plot the Number of Nonconforming Units (np) on the Control Chart. The results for the 20 most recently arrived trucks are shown in Table 9.4. The values

Sample Number	n	np
1	50	4
2	50	6
3	50	5
4	50	2
5	50	3
6	50	5
7	50	4
8	50	7
9	50	2
10	50	3
11	50	1
12	50	4
13	50	3
14	50	5
15	50	2
16	50	5
17	50	6
18	50	3
19	50	1
20	50	2

Table 9.4 Data Sheet: Plant Paradise

of np for each subgroup are plotted on the chart in Figure 9.9, which has been scaled to reflect the magnitude of the data.

Step 3. Calculate the Centerline and Control Limits. The average of the total number nonconforming is found by adding up the number of nonconforming values and dividing by the total number of samples. The example is worked to two decimal places:

$$\text{Centerline } n\bar{p} = \frac{\sum_{i=1}^{n} np}{m} = \frac{73}{20} = 3.65$$

The control limits for an np chart are found using the following formulas:

$$\bar{p} = \frac{\sum_{i=1}^{n} np}{\sum_{i=1}^{n} n} = \frac{73}{1000} = 0.073$$

$$\begin{aligned} UCL_{np} &= n\bar{p} + 3\sqrt{n\bar{p}(1 - \bar{p})} \\ &= 3.65 + 3\sqrt{3.65(1 - 0.073)} = 9.17 \end{aligned}$$

$$\begin{aligned} LCL_{np} &= n\bar{p} - 3\sqrt{n\bar{p}(1 - \bar{p})} \\ &= 3.65 - 3\sqrt{3.65(1 - 0.073)} = -1.87 = 0 \end{aligned}$$

Step 4. Draw the Centerline and Control Limits on the Chart. Figure 9.10 uses a solid line to denote the centerline and dashed lines for the control limits.

Step 5. Interpret the Chart. A study of Figure 9.10 reveals that the chart is under control. The points flow smoothly back and forth across the centerline. The number of points on each side of the centerline are balanced, with the majority of the points near the centerline. There are no trends, runs, cycles, or sudden shifts in level apparent in the data. All of the points fall between the upper and lower control limits.

CHARTS FOR COUNTS OF NONCONFORMITIES

Charts for counts of nonconformities monitor the number of nonconformities found in a sample. Two types of charts recording nonconformities may be used: a count of nonconformities (c) chart or a count of nonconformities per unit (u) chart. As the names suggest, one chart is for total nonconformities in the sample, and the other chart is nonconformities per unit.

Number of Nonconformities (c) Chart

The ***number of nonconformities chart, or c chart,*** *is used to track the count of nonconformities observed in a single unit of product of constant size.* p and np charts investigate the number of units that are nonconforming. Charts counting nonconformities are used when nonconformities are scattered through a continuous flow of product

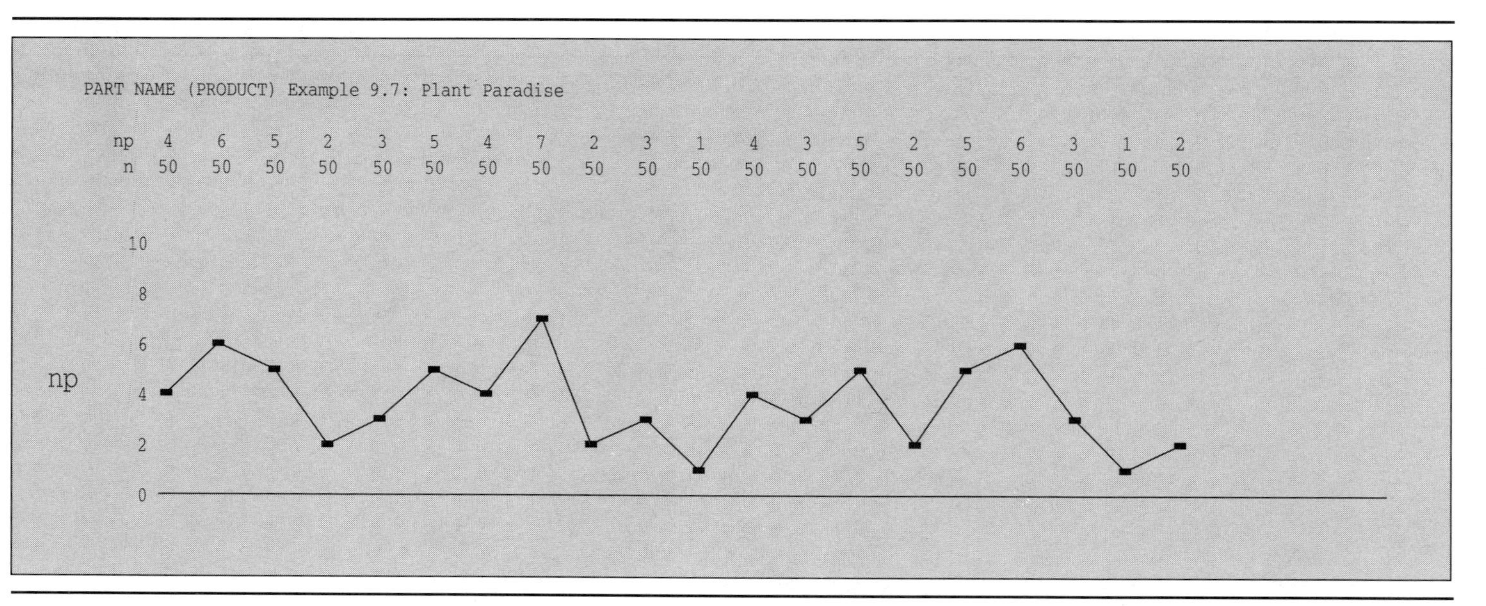

Figure 9.9 Number Nonconforming (np) Chart: Data Values

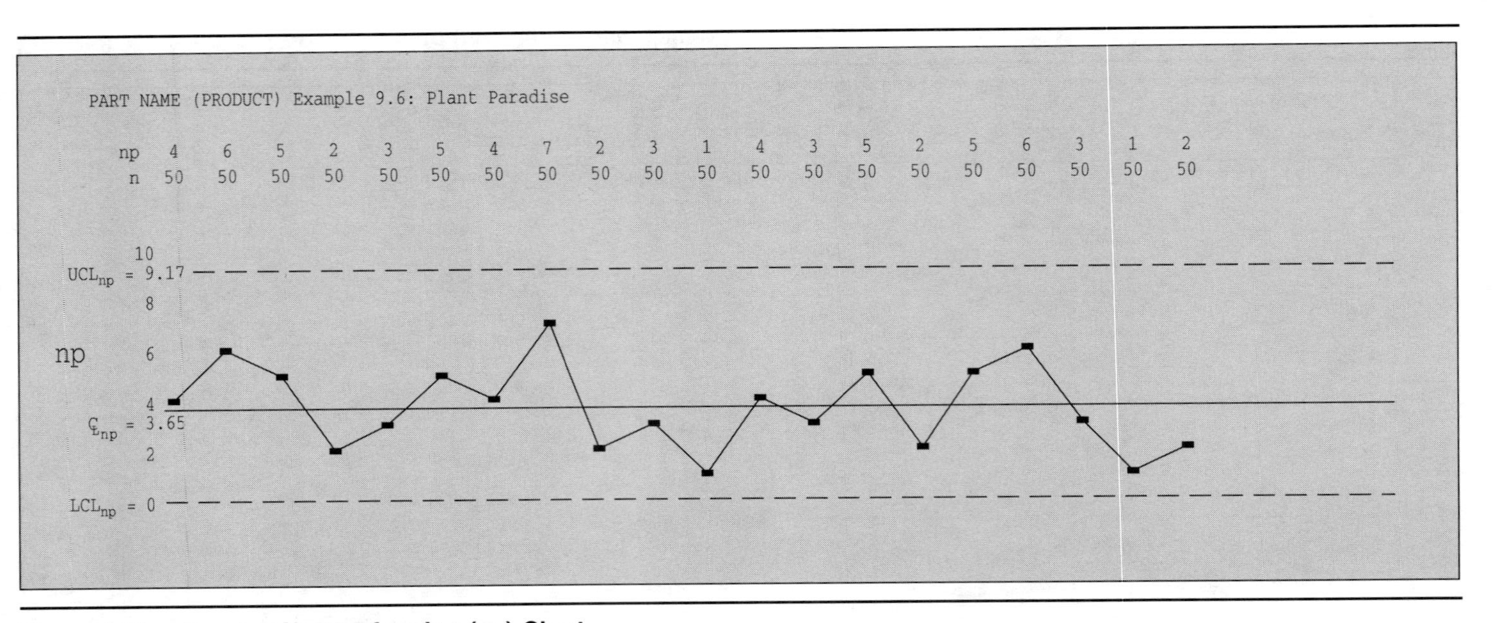

Figure 9.10 Number Nonconforming (np) Chart

such as bubbles in a sheet of glass, flaws in a bolt of fabric, discolorations in a ream of paper. To be charted, the nonconformities must be expressed as an average rate of nonconformities, such as four bubbles in a 3-square-foot pane of glass or two specks of dirt on an 8 1/2 × 11-inch sheet of paper. Count-of-nonconformities charts can combine the counts of a variety of nonconformities, such as the number of mishandled, dented, missing, or unidentified suitcases to reach a particular airport carousel. A constant sample size must be used when creating a count of nonconformities (c) chart.

1. *Gather the data.* A clear understanding of the nonconformities to be tracked is essential for successfully applying a count-of-nonconformities chart. Gathering the data requires that the area of opportunity of occurrence for each sample taken be equal. The rate of occurrences of nonconformities within a sample or area of opportunity (area of exposure) is plotted on the c chart. For this reason, the size of the piece of paper or fabric, the length of the steel, or the number of units must be equal for each sample taken. The number of nonconformities will be determined by comparing the inspected product with a standard and counting the deviations from the standard. On a c chart, all of the nonconformities have the same weight, regardless of the type of nonconformity that they are. The area of opportunity for these nonconformities should be large, with a very small chance of a particular nonconformity occurring at any one location.

2. *Count and plot* c, *the count of the number of nonconformities, on the control chart.* As it is inspected, each subgroup will yield a count of the nonconformities (c). It is this value that is plotted on the control chart. The scale for the c chart should reflect the number of nonconformities discovered.

3. *Calculate the centerline and control limits.* The centerline of the control chart is the average of the subgroup nonconformities. The number of nonconformities are added up and then divided by the total number of subgroups:

$$\text{Centerline } \bar{c} = \frac{\sum_{i=1}^{n} c}{m}$$

The control limits for a c chart are found using the following formulas:

$$UCL_c = \bar{c} + 3\sqrt{\bar{c}}$$

$$LCL_c = \bar{c} - 3\sqrt{\bar{c}}$$

On occasion, the lower control limit of a c chart may have a negative value. When this occurs, the result of the LCL_c calculation should be rounded up to zero.

4. *Draw the centerline and control limits on the chart.* Using a solid line to denote the centerline and dashed lines for the control limits, draw the centerline and control limits on the chart.

5. *Interpret the chart.* Charts for counts of nonconformities are interpreted similarly to interpreting charts for the number of nonconforming occurrences. The

charts are studied for changes in random patterns of variation. There should be no patterns in the data, such as trends, runs, cycles, or sudden shifts in level. All of the points should fall between the upper and lower control limits. The data points on a c chart should flow smoothly back and forth across the centerline and be balanced on either side of the centerline. The majority of the points should be near the centerline. Once again it is desirable to have the points approach the lower control limits or zero, showing a reduction in the count of nonconformities. The process capability is $\bar{c}$, the average count of nonconformities in a sample and the centerline of the control chart.

EXAMPLE 9.8 Making a c Chart

Pure and White, a manufacturer of paper used in copy machines, monitors their production using a c chart. Paper is produced in large rolls, 12 ft long and 6 ft in diameter. A sample is taken from each completed roll and checked in the lab for nonconformities. Nonconformities have been identified as discolorations, inconsistent paper thickness, flecks of dirt in the paper, moisture content, and ability to take ink. All of these nonconformities have the same weight on the c chart. A sample may be taken from anywhere in the roll so the area of opportunity for these nonconformities is large, while the overall quality of the paper creates only a very small chance of a particular nonconformity occurring at any one location.

Step 1. Gather the Data. (Table 9.5).

Step 2. Count and Plot c, the Count of the Number of Nonconformities, on the Control Chart. As each roll is inspected, it yields a count of the nonconformities (c) that is recorded in Table 9.5. This value is then plotted on the control chart shown in Figure 9.11.

Step 3. Calculate the Centerline and Control Limits. The centerline of the control chart is the average of the subgroup nonconformities. The number of nonconformities is added up and then divided by the total number of rolls of paper inspected:

$$\text{Centerline } \bar{c} = \frac{\sum_{i=1}^{n} c}{m} = \frac{210}{20} = 10.5$$

The control limits for a c chart are found using the following formulas:

$$\begin{aligned} UCL_c &= \bar{c} + 3\sqrt{\bar{c}} \\ &= 10.5 + 3\sqrt{10.5} = 20.2 \end{aligned}$$

$$\begin{aligned} LCL_c &= \bar{c} - 3\sqrt{\bar{c}} \\ &= 10.5 - 3\sqrt{10.5} = 0.8 \end{aligned}$$

Step 4. Draw the Centerline and Control Limits on the Chart. Using a solid line to denote the centerline and dashed lines for the control limits, draw the centerline and control limits on the chart (Figure 9.12).

Table 9.5 Data Sheet: Pure and White Paper

Sample Number	c
1	10
2	11
3	12
4	10
5	9
6	22
7	8
8	10
9	11
10	9
11	12
12	7
13	10
14	11
15	10
16	12
17	9
18	10
19	8
20	9
	210

Step 5. Interpret the Chart. Point 6 is out of control and should be investigated to determine the cause of so many nonconformities. There are no other patterns present on the chart. Except for point 6 the chart is performing in a very steady manner.

Revising the c Chart

If improvements have been made to the process or the reasons behind special-cause situations have been identified and corrected, c charts can be revised using the following formulas:

$$\text{Centerline } \bar{c}_{new} = \frac{\sum_{i=1}^{n} c - c_d}{m - m_d}$$

$$UCL_{c_{new}} = \bar{c}_{new} + 3\sqrt{\bar{c}_{new}}$$

$$LCL_{c_{new}} = \bar{c}_{new} - 3\sqrt{\bar{c}_{new}}$$

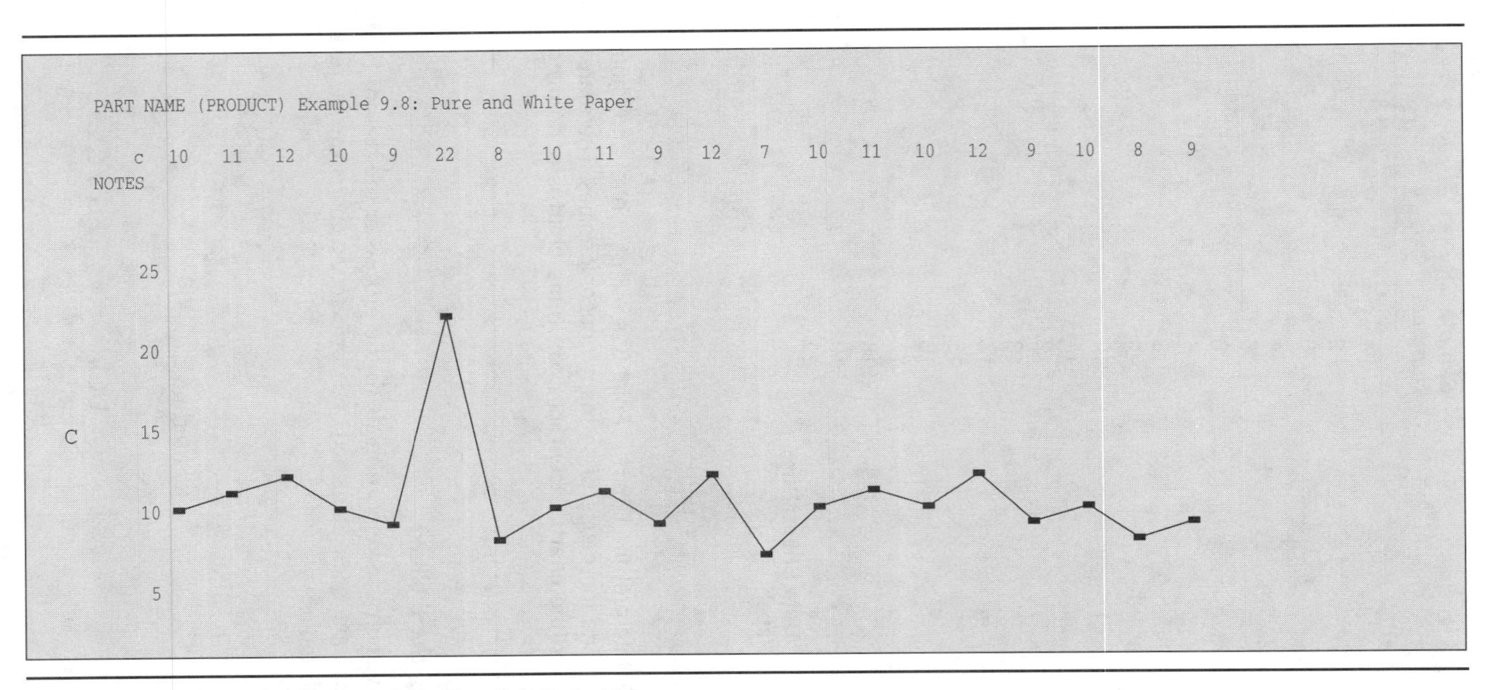

Figure 9.11 **Chart for Nonconformities (c): Data Values**

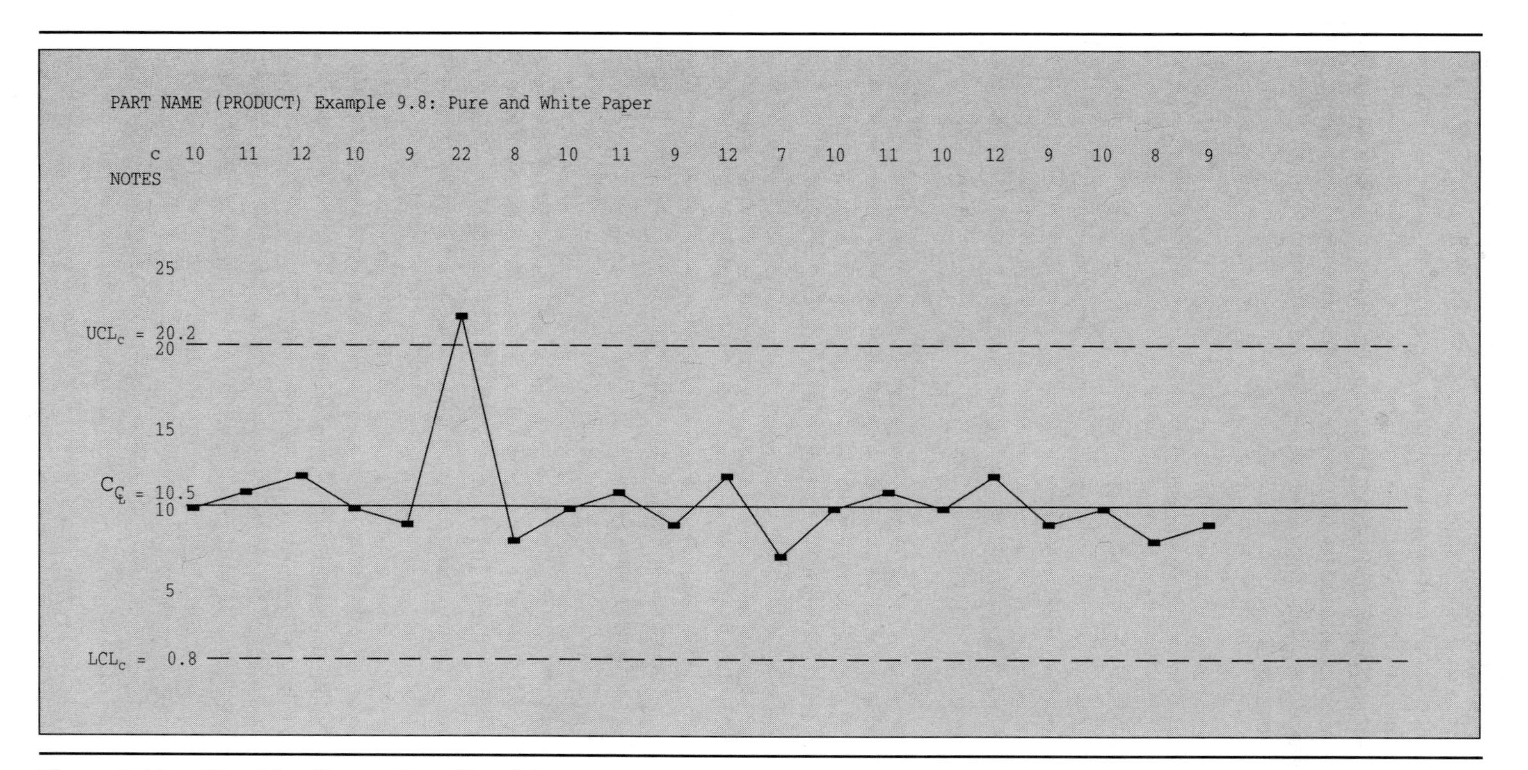

Figure 9.12 Chart for Nonconformities (c)

EXAMPLE 9.9 Revising a c Chart

Pure and White paper manufacturer investigated the out-of-control point on the count of nonconformities (c) chart in Figure 9.12 and determined that an operator was not adding sufficient bleach to the pulp-mixing tank. The operator was retrained in how to mix the correct combination of chemicals to make White and Pure paper. To revise the chart and determine the process capability, White and Pure followed this procedure:

$$\text{Centerline } \bar{c}_{new} = \frac{\sum_{i=1}^{n} c - c_d}{m - m_d} = \frac{210 - 22}{20 - 1} = 10$$

$$\begin{aligned} UCL_{\bar{c}_{new}} &= \bar{c}_{new} + 3\sqrt{\bar{c}_{new}} \\ &= 10 + 3\sqrt{10} = 19.5 \end{aligned}$$

$$\begin{aligned} LCL_{\bar{c}_{new}} &= \bar{c}_{new} - 3\sqrt{\bar{c}_{new}} \\ &= 10 - 3\sqrt{10} = 0.5 \end{aligned}$$

Number of Nonconformities per Unit (u) Charts: Constant Sample Size

A ***number of nonconformities per unit, or u, chart*** *is a chart that studies the number of nonconformities in a unit*. The u chart is very similar to the c chart; however, unlike c charts, u charts can also be used with variable sample sizes. To create a number of nonconformities per unit chart:

1. *Gather the data.*

2. *Calculate* u, *the number of nonconformities per unit*. As it is inspected, each sample of size n will yield a count of nonconformities (c). The number of nonconformities per unit (u), used on the u chart, is calculated using n, the number of inspected items, and c, the count of nonconformities found:

$$u = \frac{c}{n}$$

This value is plotted on the control chart.

3. *Calculate the centerline and control limits*. The centerline of the u chart is the average of the subgroup nonconformities per unit. The number of nonconformities are added up and then divided by the total number of samples:

$$\text{Centerline } \bar{u} = \frac{\sum_{i=1}^{n} c}{\sum_{i=1}^{n} n}$$

The control limits for the u chart are found using the following formulas:

$$UCL_u = \bar{u} + 3\frac{\sqrt{\bar{u}}}{\sqrt{n}}$$

$$LCL_u = \bar{u} - 3\frac{\sqrt{\bar{u}}}{\sqrt{n}}$$

On occasion, the lower control limit of a u chart may have a negative value. When this occurs, the result of the LCL_u calculation should be rounded to zero.

4. *Draw the centerline and control limits on the chart.* Using a solid line to denote the centerline and dashed lines for the control limits, draw the centerline and control limits on the chart.

5. *Interpret the chart.* A u chart is interpreted in the same manner as is a p, np, or c chart. When using variable sample sizes, follow the same procedure as used in creating and interpreting a p chart with variable sample sizes. Once again, when n_{ave} is used, the sample sizes must not vary more than 25 percent from n_{ave}. The chart should be studied for any nonrandom conditions such as runs, trends, cycles, or points out of control. The data points on a u chart should flow smoothly back and forth across the centerline. The number of points on each side of the centerline should be balanced, with the majority of the points near the centerline. Quality-improvement efforts are reflected on a u chart when the count of nonconformities per unit is reduced, as shown by a trend toward the lower control limit and therefore toward zero nonconformities. The process capability is $\bar{u}$, the centerline of the control chart, the average number of nonconformities per unit.

EXAMPLE 9.10 Making a u Chart with Constant Sample Size

At Special Plastics, Inc., on a separate production line from the credit card blanks, small plastic parts used to connect hoses are created. Special Plastics, Inc., uses u charts to collect data concerning the nonconformities per unit in the process.

Step 1. Gather the Data. During inspection, a random sample of size 400 is taken once an hour. The hose connectors are visually inspected for a variety of nonconformities, including flashing on inner diameters, burrs on the part's exterior, incomplete threads, flashing on the ends of the connectors, incorrect plastic compound, and discolorations.

Step 2. Calculate u, the Number of Nonconformities per Unit. Table 9.6 shows the number of nonconformities (c) that each subgroup of sample size $n = 400$ yielded. The number of nonconformities per unit (u) is calculated by dividing c, the number of nonconformities found, by n, the number of inspected items. Working the example to three decimal places, for the first sample,

$$u_1 = \frac{c}{n} = \frac{10}{400} = 0.025$$

Sample Number	n	c	u
1	400	10	0.025
2	400	11	0.028
3	400	12	0.030
4	400	10	0.025
5	400	9	0.023
6	400	22	0.055
7	400	8	0.020
8	400	10	0.025
9	400	11	0.028
10	400	9	0.023
11	400	12	0.030
12	400	7	0.018
13	400	10	0.025
14	400	11	0.028
15	400	10	0.025
16	400	12	0.030
17	400	9	0.023
18	400	10	0.025
19	400	8	0.020
20	400	9	0.023

Table 9.6 Data Sheet: Hose Connectors

Step 3. Calculate the Centerline and Control Limits. The centerline of the u chart is the average of the subgroup nonconformities per unit. The number of nonconformities are added up and then divided by the total number of samples:

$$\text{Centerline } \bar{u} = \frac{\sum_{i=1}^{n} c}{\sum_{i=1}^{n} n} = \frac{210}{20(400)} = 0.026$$

Find the control limits for the u chart by using the following formulas:

$$\begin{aligned} UCL_u &= \bar{u} + 3\frac{\sqrt{\bar{u}}}{\sqrt{n}} \\ &= 0.026 + 3\frac{\sqrt{0.026}}{\sqrt{400}} = 0.05 \end{aligned}$$

$$\begin{aligned} LCL_u &= \bar{u} - 3\frac{\sqrt{\bar{u}}}{\sqrt{n}} \\ &= 0.026 - 3\frac{\sqrt{0.026}}{\sqrt{400}} = 0.001 \end{aligned}$$

Step 4. Create and Draw the Centerline and Control Limits on the Chart. Using a solid line to denote the centerline and dashed lines for the control limits, draw the centerline and control limits on the chart. Values for u are plotted on the control chart (Figure 9.13).

Step 5. Interpret the Chart. Except for point 6, the chart appears to be under statistical control. There are no runs or unusual patterns. Point 6 should be investigated to determine the cause of such a large number of nonconformities per unit.

Number of Nonconformities per Unit (u) Charts: Variable Sample Size

When the sample size varies, n_{ave} is used in the formulas just as it was when creating fraction nonconforming charts for variable sample size. The sample sizes must not vary more than 25 percent from n_{ave}. If there is greater variance, then the individual control limits for the samples must be calculated.

EXAMPLE 9.11 Making a u Chart with Variable Sample Size

Occasionally the sample size on the hose connector line varies. Individual control limits are calculated, n_{ave} is used in the same manner as described for fraction nonconformities (p) charts. If n_{ave} is used, care must be taken in the interpretation of the nonconformities per unit chart.

Step 1. Gather the Data. Table 9.7 shows the most recent data collected concerning hose connectors. Note that the sample size varies.

Step 2. Calculate u, the Number of Nonconformities per Unit. Table 9.7 also shows the values of u calculated to three decimal places. Plot these values on the control chart (Figure 9.14).

Step 3. Calculate the Centerline and Control Limits.

$$\text{Centerline } \overline{u} = \frac{\sum_{i=1}^{n} c}{\sum_{i=1}^{n} n} = \frac{99}{3995} = 0.025$$

For each different sample size calculate the control limits. (In this example the control limits have been calculated to four decimal places.)

$$UCL_u = \overline{u} + 3\frac{\sqrt{\overline{u}}}{\sqrt{n}}$$

$$UCL_u = 0.025 + 3\frac{\sqrt{0.025}}{\sqrt{390}} = 0.0490$$

$$UCL_u = 0.025 + 3\frac{\sqrt{0.025}}{\sqrt{400}} = 0.0487$$

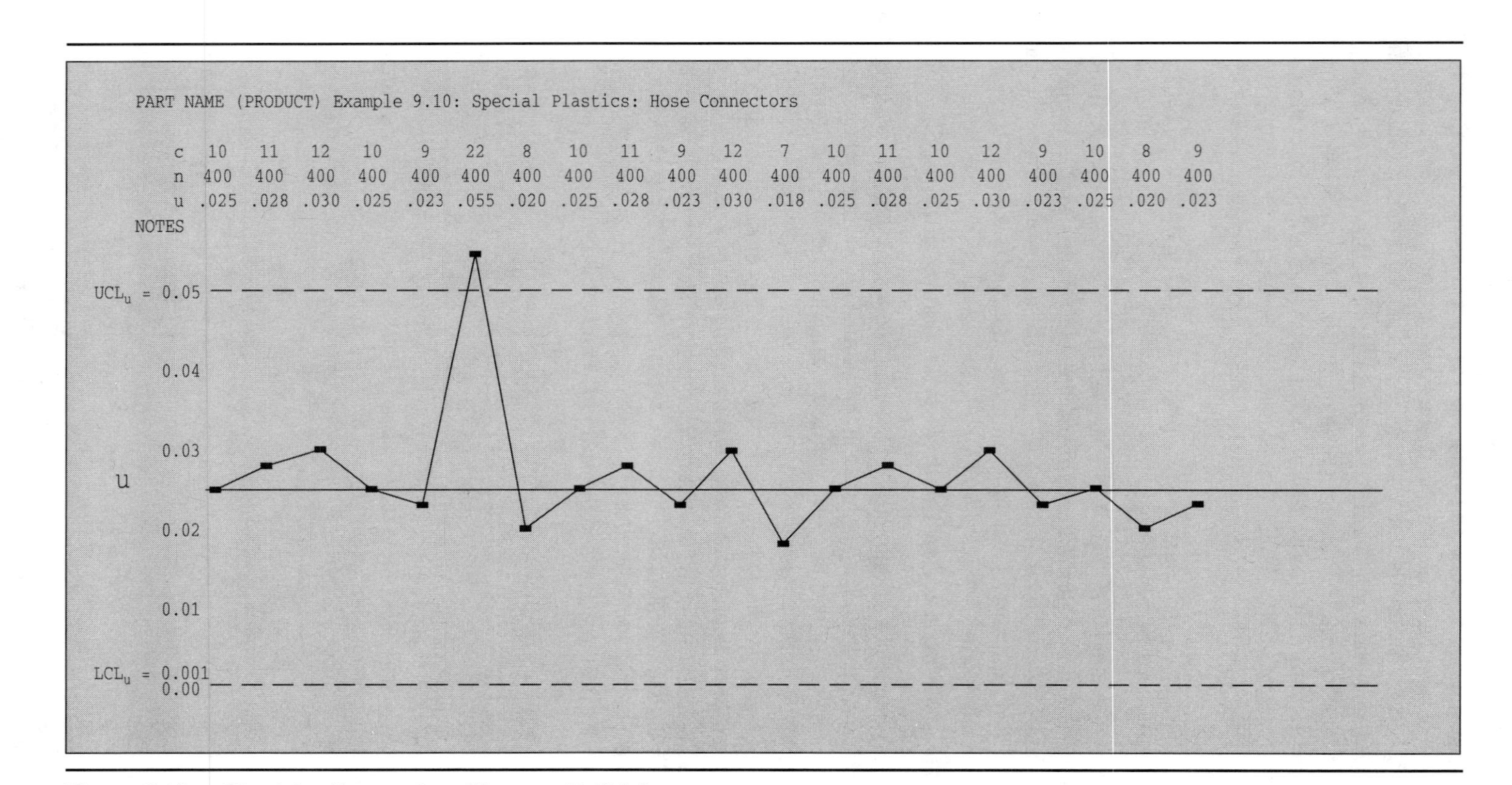

Figure 9.13 Chart for Nonconformities per Unit (u)

Table 9.7 Data Sheet: Hose Connectors Variable Sample Size

Sample Number	n	c	u
21	405	10	0.025
22	390	11	0.028
23	410	12	0.029
24	400	10	0.025
25	405	9	0.022
26	400	9	0.023
27	390	8	0.021
28	400	10	0.025
29	390	11	0.028
30	405	9	0.022

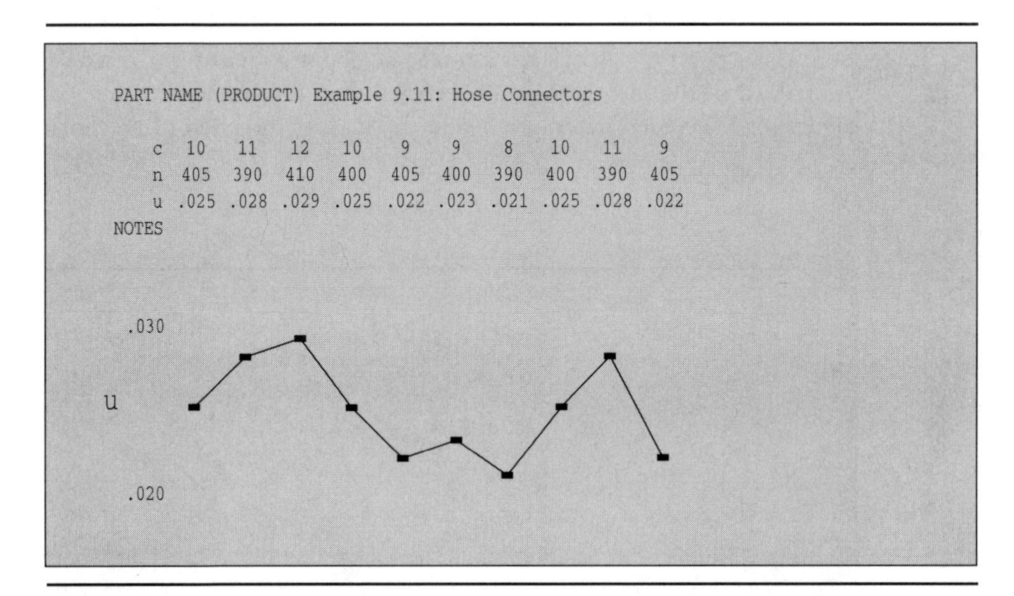

Figure 9.14 Nonconformities per Unit (u) Chart for Variable Sample Size: Data Values

$$\text{UCL}_u = 0.025 + 3\frac{\sqrt{0.025}}{\sqrt{405}} = 0.0485$$

$$\text{UCL}_u = 0.025 + 3\frac{\sqrt{0.025}}{\sqrt{410}} = 0.0484$$

$$\text{LCL}_u = \bar{u} - 3\frac{\sqrt{\bar{u}}}{\sqrt{n}}$$

$$LCL_u = 0.025 - 3\frac{\sqrt{0.025}}{\sqrt{400}} = 0.0013$$

$$LCL_u = 0.025 - 3\frac{\sqrt{0.025}}{\sqrt{405}} = 0.0014$$

The lower control limit may be rounded to zero.

Step 4. Draw the Centerline and Control Limits on the Chart. The centerline and control limits have been drawn in Figure 9.15.

Step 5. Interpret the Chart. The process shown in Figure 9.15 is a stable process.

Revising the u *Chart*

Both the u chart for a constant sample size and a u chart for a variable sample size are revised in the same fashion as are the other attribute charts. Once the assignable causes have been determined, and the process modified to prevent their recur-

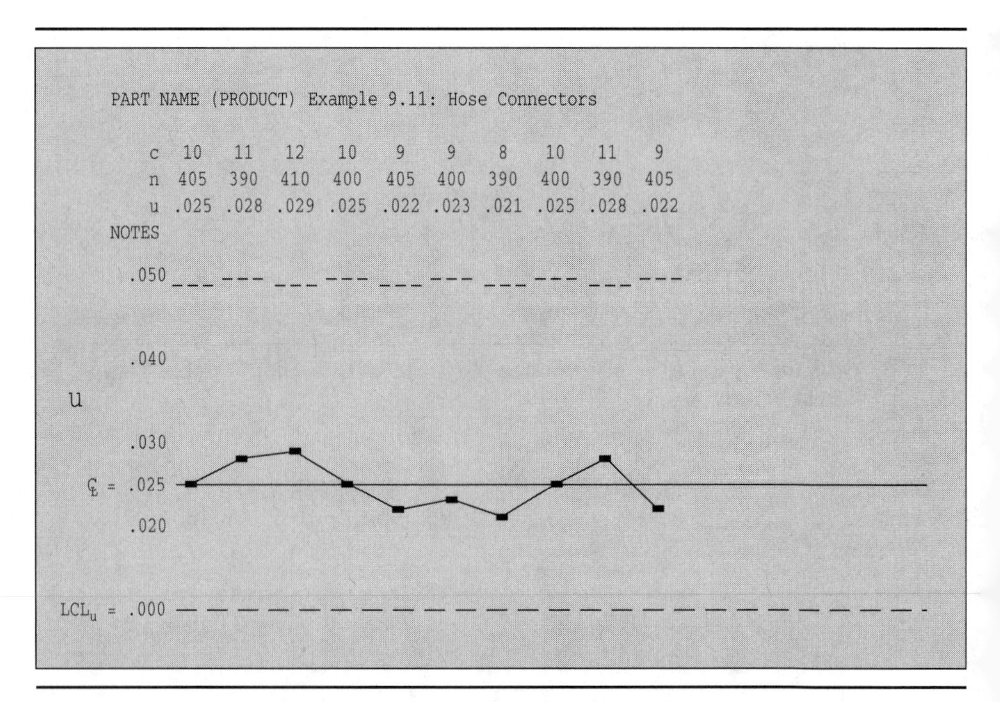

Figure 9.15 Nonconformities per Unit (u) Chart for Variable Sample Size

rence, the centerline and control limits can be recalculated using the following formulas. For the u chart with variable sample size, n_{ave} must be recalculated:

$$\text{Centerline } \bar{u} = \frac{\sum_{i=1}^{n} c - c_d}{\sum_{i=1}^{n} n - n_d}$$

The control limits for the u chart are found using the following formulas:

$$UCL_u = \bar{u}_{new} + 3\frac{\sqrt{\bar{u}_{new}}}{\sqrt{n}}$$

$$LCL_u = \bar{u}_{new} - 3\frac{\sqrt{\bar{u}_{new}}}{\sqrt{n}}$$

$$n_{ave} = \frac{\sum_{i=1}^{n} n - n_d}{m - m_d}$$

SUMMARY

You have now seen a variety of control charts. The choice of which chart to implement under what circumstances is less confusing than it first appears. To begin the selection process, care must be taken to identify the type of data to be gathered. What is the nature of the process under study? Are they variables data and therefore measurable? Or are they attribute data and therefore countable? Is there a sample size or is just one item being measured? Is there a sample size or are just the nonconformities in an area of opportunity being counted? The flowchart in Figure 9.16 can assist you in choosing a chart.

Choosing a Control Chart

I. Variables Data
  A. Use an $\bar{X}$ chart combined with an R or s chart
    1. The characteristic can be measured.
    2. The process is unable to hold tolerances.
    3. The process must be monitored for adjustments.
    4. Changes are being made to the process and those changes need to be monitored.
    5. Process stability and process capability must be monitored and demonstrated to a customer or regulating body.
    6. The process average and the process variation must be measured.

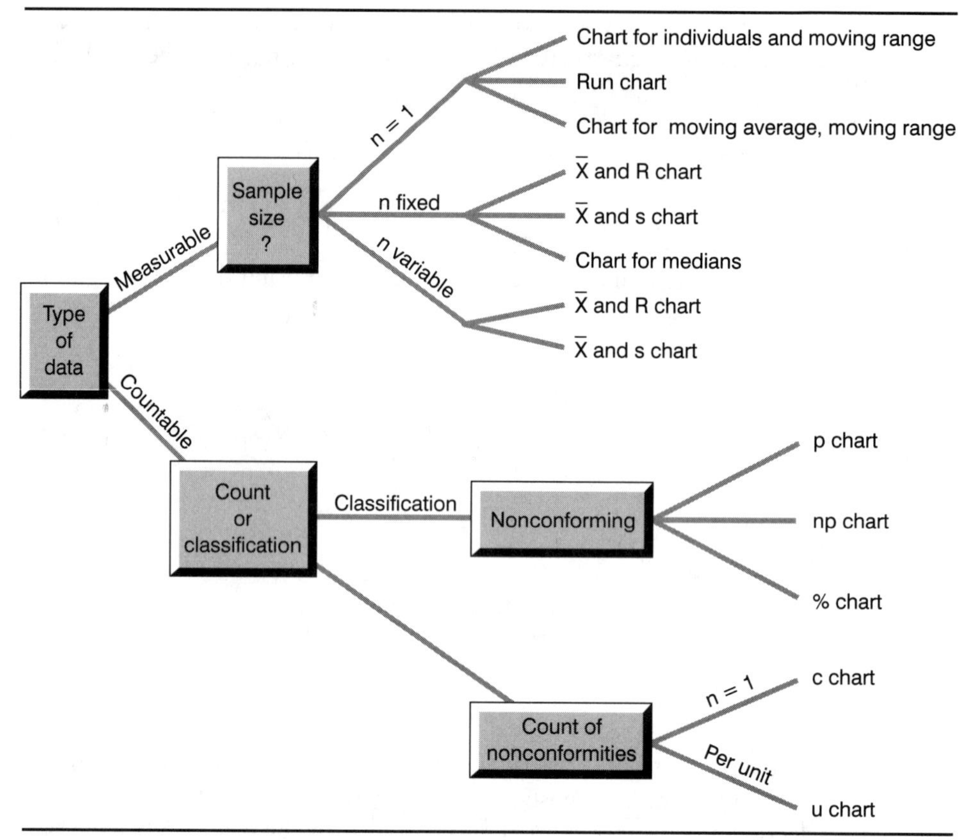

Figure 9.16 Flowchart

B. Use control charts for individuals (X_i and moving-range charts or moving-average and moving-range charts) when
   1. Destructive testing or other expensive testing procedures limit the number of products sampled.
   2. It is inconvenient or impossible to obtain more than one sample.
   3. Waiting for a large sample would not provide timely data.

II. Attribute Data

A. Use charts for nonconforming units or fraction nonconforming charts (p, np, 100p) when
   1. There is a need to monitor the portion of the lot that is nonconforming.
   2. The characteristics under study in the process can be judged either conforming or nonconforming.
   3. The sample size varies (p chart).
   4. Process monitoring is desired but measurement data cannot be obtained because of the nature of the product or the expense.

B. Use charts for nonconformities (c, u) when
   1. There is a need to monitor the number of nonconformities in a process (c charts for counts of nonconformities, u charts for nonconformities per unit).
   2. The characteristics under study in the process can be judged as having one or more nonconformities.
   3. Process monitoring is desired but measurement data cannot be obtained because of the nature of the product or the expense.
   4. The sample size varies (u chart).

Lessons Learned

1. Fraction nonconforming (p) charts can be constructed for both constant and variable sample sizes because the sample size can be isolated in the formula.
2. For the fraction nonconforming chart, the product or service being provided must be inspected and classed as either conforming or not conforming.
3. Number nonconforming (np) and percent nonconforming charts may be easier to interpret than fraction nonconforming (p) charts.
4. It is possible to construct either a p or a u chart for variable sample size. Individual control limits for the different sample sizes may be calculated or n_{ave} may be used. When n_{ave} is used, care must be taken to correctly interpret the points approaching or exceeding the upper control limit.
5. Charts for counts of nonconformities (c) and nonconformities per unit (u) are used when the nonconformities on product or service being inspected can be counted.
6. Charts for nonconformities per unit (u charts) can be constructed for both constant and variable sample sizes.
7. When interpreting p, np, c, or u charts, it is important to remember that values closer to zero are desirable. ■

Formulas

Fraction Nonconforming (p) Chart

$$p = \frac{np}{n}$$

$$\text{Centerline } \bar{p} = \frac{\sum_{i=1}^{n} np}{\sum_{i=1}^{n} n}$$

$$\mathrm{UCL_p} = \bar{p} + 3\frac{\sqrt{\bar{p}(1-\bar{p})}}{\sqrt{n}}$$

$$\mathrm{LCL_p} = \bar{p} - 3\frac{\sqrt{\bar{p}(1-\bar{p})}}{\sqrt{n}}$$

Revising:

$$\bar{p}_{new} = \frac{\sum_{i=1}^{n} np - np_d}{\sum_{i=1}^{n} n - n_d}$$

$$\mathrm{UCL}_{p_{new}} = \bar{p}_{new} + 3\frac{\sqrt{\bar{p}_{new}(1-\bar{p}_{new})}}{\sqrt{n}}$$

$$\mathrm{LCL}_{p_{new}} = \bar{p}_{new} - 3\frac{\sqrt{\bar{p}_{new}(1-\bar{p}_{new})}}{\sqrt{n}}$$

The process capability of the revised chart is the newly calculated p_{new}:

$$n_{ave} = \frac{\sum_{i=1}^{n} n}{m}$$

Number Nonconforming (np) Chart

$$\text{Centerline } n\bar{p} = \frac{\sum_{i=1}^{n} np}{m}$$

$$\mathrm{UCL_{np}} = n\bar{p} + 3\sqrt{n\bar{p}(1-\bar{p})}$$

$$\mathrm{LCL_{np}} = n\bar{p} - 3\sqrt{n\bar{p}(1-\bar{p})}$$

Percent Nonconforming (100p) Chart

The centerline for a percent nonconforming chart is $100\bar{p}$.

$$\mathrm{UCL_{100p}} = 100\left[\bar{p} + \frac{3\sqrt{\bar{p}(1-\bar{p})}}{\sqrt{n}}\right]$$

$$\mathrm{LCL_{100p}} = 100\left[\bar{p} - \frac{3\sqrt{\bar{p}(1-\bar{p})}}{\sqrt{n}}\right]$$

Count of Nonconformities (c) Chart

$$\text{Centerline } \bar{c} = \frac{\sum_{i=1}^{n} c}{m}$$

$$UCL_c = \bar{c} + 3\sqrt{\bar{c}}$$
$$LCL_c = \bar{c} - 3\sqrt{\bar{c}}$$

Revising:

$$\text{Centerline } \bar{c}_{new} = \frac{\sum_{i=1}^{n} c - c_d}{m - m_d}$$

$$UCL_{c_{new}} = \bar{c}_{new} + 3\sqrt{\bar{c}_{new}}$$
$$LCL_{c_{new}} = \bar{c}_{new} - 3\sqrt{\bar{c}_{new}}$$

Nonconformities per Unit (u) Chart

$$u = \frac{c}{n}$$

$$\text{Centerline } \bar{u} = \frac{\sum_{i=1}^{n} c}{\sum_{i=1}^{n} n}$$

$$UCL_u = \bar{u} + 3\frac{\sqrt{\bar{u}}}{\sqrt{n}}$$
$$LCL_u = \bar{u} - 3\frac{\sqrt{\bar{u}}}{\sqrt{n}}$$

Chapter Problems

1. What is the difference between a common cause and a special cause? Is this true for all charts—p, c, u, $\overline{X}$, and R?

2. How would you determine that a process is under control? How is the interpretation of a p, u, or c chart different from that of an $\overline{X}$ and R chart?

Charts for Fraction Nonconforming

3. Given the following information about mistakes made on tax forms, make and interpret a fraction nonconforming chart.

Sample Size	*Nonconforming*	*Sample Size*	*Nonconforming*
20	0	20	1
20	0	20	6
20	0	20	0
20	2	20	0
20	0	20	2
20	3	20	0
20	1	20	0
20	1	20	1
20	0	20	1
20	2	20	0
20	10	20	2
20	2	20	0
20	1	20	4
20	0	20	1
20	1	20	0

4. The following table gives the number of nonconforming product found while inspecting a series of 12 consecutive lots of galvanized washers for finish defects such as exposed steel, rough galvanizing, and discoloration. A sample size of n = 200 was used for each lot. Find the centerline and control limits for a fraction nonconforming chart. If the manufacturer wishes to have a process capability of $\overline{p} = 0.015$, is the process capable?

Sample Size	*Nonconforming*	*Sample Size*	*Nonconforming*
200	0	200	0
200	1	200	0
200	2	200	1
200	0	200	0
200	1	200	3
200	1	200	1

5. Thirst-Quench, Inc., has been in business for more than 50 years. Recently Thirst-Quench updated their machinery and processes, acknowledging their out-of-date style. They have decided to evaluate these changes. The engineer is to record data, evaluate those data, and implement strategy to keep quality at a maximum. The plant operates 8 hours a day, 5 days a week, and produces 25,000 bottles of Thirst-Quench each day. Problems that have arisen in the past include partially filled bottles, crooked labels, upside-down labels, and no labels. Samples of size 150 are taken each hour. Create a p chart.

Subgroup Number	*Number Inspected* n	*Number Nonconforming* np	*Proportion Nonconforming* p
1	150	6	0.040
2	150	3	0.020
3	150	9	0.060
4	150	7	0.047
5	150	9	0.060
6	150	2	0.013
7	150	3	0.020
8	150	5	0.033
9	150	6	0.040
10	150	8	0.053
11	150	9	0.060
12	150	7	0.047
13	150	7	0.047
14	150	2	0.013
15	150	5	0.033
16	150	7	0.047
17	150	4	0.027
18	150	3	0.020
19	150	9	0.060
20	150	8	0.053
21	150	8	0.053
22	150	6	0.040
23	150	2	0.013
24	150	9	0.060
25	150	7	0.047
26	150	3	0.020
27	150	4	0.027
28	150	6	0.040
29	150	5	0.033
30	150	4	0.027
Total	4500	173	

6. Nearly everyone who visits a doctor's office is covered by some form of insurance. For a doctor, the processing of forms in order to receive payment from an insurance company is a necessary part of doing business. If a form is filled out incorrectly, the form can not be processed and is considered nonconforming (defective). Within each office, an individual is responsible for inspecting and correcting the forms before filing them with the appropriate insurance company. A local doctor's office is interested in determining whether or not errors on insurance forms are a major problem. Every week they take a sample of 20 forms to use in creating a p chart. Use the following information to create a p chart. How are they doing?

Nonconforming	
1	2
2	5
3	8
4	10
5	4
6	7
7	6
8	3
9	7
10	5
11	2
12	3
13	17
14	5
15	2
16	4
17	5
18	2
19	3
20	2
21	6
22	4
23	5
24	1
25	2

7. For 15 years, a county has been keeping track of traffic fatalities. From the total number of fatal traffic accidents each year, a random sample is taken

and the driver is tested for blood alcohol level. For the sake of creating control charts, a blood alcohol level higher than 80 mg is considered nonconforming. Given the following information, create a p chart with a sample size that varies.

Consider the following information when interpreting the chart. In 1991, the county began a three-pronged attack against drunk driving. Mandatory seat-belt legislation was enacted, advertisements on radio and TV were used to increase public awareness of the dangers of drinking and driving, and harsher sentences were enacted for drunk driving. Based on your knowledge of chart interpretations, were these programs successful? How do you know?

Year	*No. of driver fatalities tested*	*No. of driver fatalities tested with a blood alcohol level over 80 mg*
1983	1,354	709
1984	1,501	674
1985	1,479	687
1986	1,265	648
1987	1,261	583
1988	1,209	668
1989	1,333	665
1990	1,233	641
1991	1,400	550
1992	1,181	545
1993	1,237	539
1994	1,186	535
1995	1,291	557
1996	1,356	537
1997	1,341	523
	19,627	9,061

8. Use the data from Problem 7 and create a p chart using n average. Use n average to interpret the chart.

9. The House of Bolts is one of the largest producers of industrial (Class III) bolts. The bolts are sampled daily with a sample size of 500. The bolts are inspected for thread management and overall design. A nonconforming bolt would display an irregular bolt size or poor thread engagement. The control chart shows the fraction of nonconforming bolts per sample size. Create a p chart with the following data. Assume assignable causes and revise the chart.

Subgroup Number	*Number Inspected* n	*Number Nonconforming* np	*Fraction Nonconforming* p
1	500	13	0.026
2	500	5	0.010
3	500	11	0.022
4	500	6	0.012
5	500	2	0.004
6	500	8	0.016
7	500	8	0.016
8	500	3	0.006
9	500	10	0.020
10	500	0	0.000
11	500	5	0.010
12	500	12	0.024
13	500	11	0.022
14	500	5	0.010
15	500	0	0.000
16	500	0	0.000
17	500	7	0.014
18	500	9	0.018
19	500	10	0.020
20	500	6	0.012
Total	10,000	131	

10. The House of Bolts has changed their data-collecting techniques. The bolts are inspected daily, but not at a constant sample size. Guidelines for nonconforming bolts have not changed. The bolts are inspected for thread management and overall design. As before, a nonconforming bolt would be an irregular bolt size or poor thread engagement. Create a p chart for variable subgroup size. Assume assignable causes and revise the chart. Does the control chart change significantly from the chart in Problem 9?

Subgroup Number	*Number Inspected* n	*Number Nonconforming* np	*Fraction Nonconforming* p
1	500	13	0.026
2	500	5	0.010
3	500	11	0.022
4	450	6	0.013
5	450	2	0.004

Subgroup Number	*Number Inspected* n	*Number Nonconforming* np	*Fraction Nonconforming* p
6	450	8	0.018
7	600	8	0.013
8	600	3	0.005
9	600	10	0.017
10	400	0	0.000
11	400	5	0.013
12	400	12	0.030
13	550	11	0.020
14	550	5	0.009
15	550	0	0.000
16	450	0	0.000
17	450	7	0.016
18	450	9	0.020
19	600	10	0.017
20	600	6	0.010
Total	10,050	131	

11. The following are data on 5-gal containers of paint. If the color mixture of the paint does not match the control color, then the entire container is considered nonconforming and is disposed of. Since the amount produced during each production run varies, use n_{ave} to calculate the centerline and control limits for this set of data. Carry calculations to four decimal places. Remember to round n_{ave} to a whole number; you can't sample part of a 5-gal pail.

Production Run	*Number Inspected*	*Number Defective*
1	2,524	30
2	2,056	84
3	2,750	76
4	3,069	108
5	3,365	54
6	3,763	29
7	2,675	20
8	2,255	25
9	2,060	48
10	2,835	10
11	2,620	86
12	2,250	25

Which points should be tested? Use the cases presented in the chapter to discuss why or why not. Show calculations if the points need to be tested.

12. Given the chart in Figure 1, discuss the state of control the charts are exhibiting. What percent nonconforming should the company expect in the future for process 1?

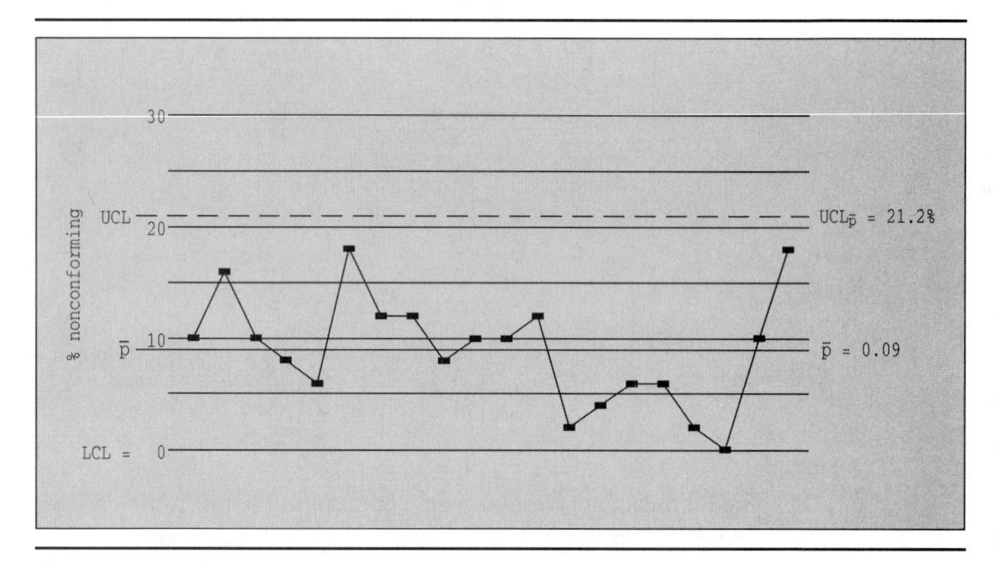

Figure 1 Problem 12

13. The board of directors is meeting to discuss quality on a particular product. The quality supervisor wants to create a chart that is easy to understand. What chart would you choose? What are the control limits and centerline for this chart? ($n = 50$, $p = 0.09$)

Sample No.	1	2	3	4	5	6	7	8	9	10
Nonconforming	5	4	3	2	6	18	5	8	2	2

14. Given Figure 2, complete the calculations. Assume the assignable causes for the circled points have been determined and corrected. Remove those points and recalculate the limits.

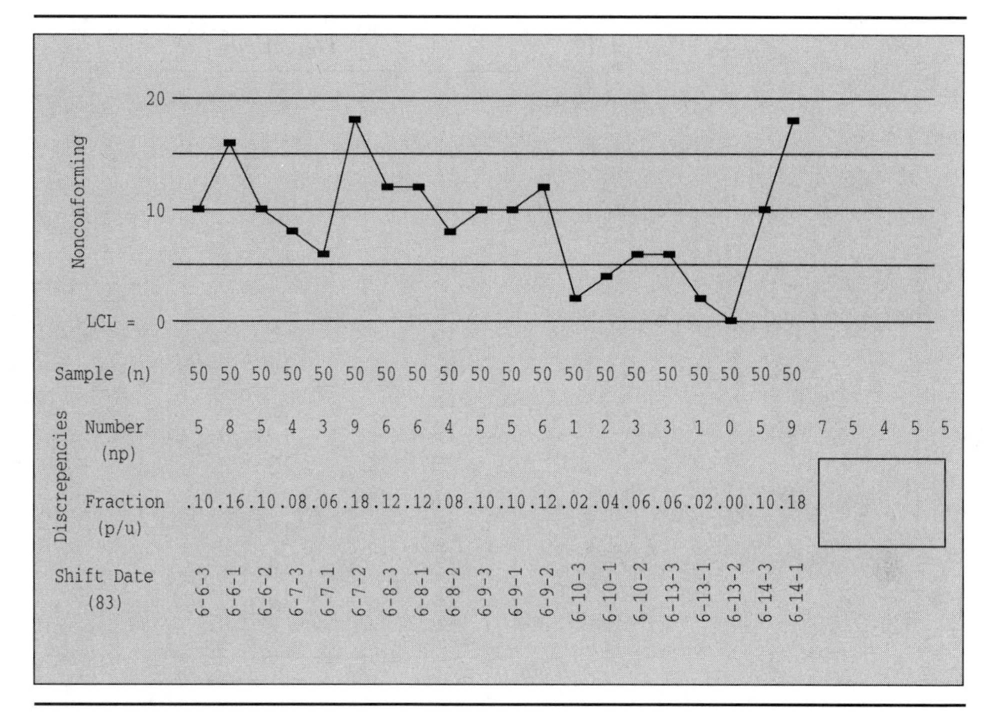

Figure 2 Problem 14

15. At Fruits and Such, a local frozen fruit concentrate manufacturer, frozen grape juice is packed in 12-ounce cardboard cans. These cans are formed on a machine by spinning them from preprinted cardboard stock and attaching two metal lids, one for the top and one for the bottom. After the filling operation, samples (n = 100) of the production are taken hourly. Inspectors look for several things that would make the can defective: how well the strip for opening the can operates, whether there are proper crimps holding the lids to the cardboard tube, if the lids are poorly sealed, etc. Create and interpret an np chart for the following data.

	Defectives
1	12
2	15
3	8
4	10
5	4
6	7
7	16
8	9

	Defectives
9	14
10	10
11	5
12	6
13	17
14	12
15	22
16	8
17	10
18	5
19	13
20	12
21	20
22	18
23	24
24	15
25	8

16. From a lot of 1,000 soup cans, a sample size 160 (n = 160) is taken. The following data show the results of 15 such subgroups. Use the data to create a number nonconforming chart.

Lot	n	np
1	160	12
2	160	15
3	160	13
4	160	15
5	160	11
6	160	11
7	160	10
8	160	15
9	160	13
10	160	12
11	160	16
12	160	17
13	160	13
14	160	14
15	160	11

17. A manufacturer of lightbulbs is keeping track of the number of nonconforming bulbs (the number that don't light). Create the centerline and control limits for a number nonconforming chart. What is the capability of the process? Assume assignable causes and revise the chart.

	n	np		n	np
1	300	10	11	300	31
2	300	8	12	300	32
3	300	9	13	300	10
4	300	12	14	300	8
5	300	10	15	300	12
6	300	11	16	300	10
7	300	9	17	300	11
8	300	10	18	300	9
9	300	12	19	300	10
10	300	11	20	300	8

18. The Tri-State Foundry Company is a large-volume producer of ball joints for various automotive producers. The ball joints are sampled four times daily. The sample size is a constant of 400 parts. The chart shows how many nonconforming parts are found in a sample size. Nonconforming quality comes from an outer diameter that is larger than 1.5 inches. The 1.5 inches is a strict guideline because any parts that are larger will not work with the female coupling assembly. Create an np chart with the following data. Assume assignable causes and revise the chart.

Subgroup Number	*Number Inspected* n	*Number Nonconforming* np
1	400	3
2	400	4
3	400	1
4	400	7
5	400	6
6	400	3
7	400	11
8	400	13
9	400	4
10	400	10
11	400	5
12	400	0

Subgroup Number	*Number Inspected* n	*Number Nonconforming* np
13	400	2
14	400	2
15	400	4
16	400	8
17	400	5
18	400	3
19	400	1
20	400	12

Charts for Nonconformities

19. A line produces leather jackets. Once each hour a jacket is selected at random from those made during that hour. Small imperfections (missed stitches, zipper catch, buttons missing, loose threads) are counted. Recent data show an average of eight imperfections (nonconformities) in each sample. Using the formulas for a c chart, calculate the control limits and centerline for this situation. If management wants a process capability of four nonconformities per sample, will this line be able to meet that requirement?

20. A production line manufactures compact discs. Twice an hour, one CD is selected at random from those made during the hour. Each disc is inspected separately for imperfections (defects), such as scratches, nicks, discolorations, and dents. The resulting count of imperfections from each disc is recorded on a control chart. Use the following data to create a c chart. If the customer wants a process capability of 1, will this line be able to meet that requirement?

	Imperfections
1	2
2	0
3	2
4	1
5	0
6	1
7	0
8	2
9	0
10	2
11	1
12	2

	Imperfections
13	0
14	1
15	1
16	0
17	0
18	2
19	4
20	1
21	2
22	0
23	2
24	1
25	0

21. A production line produces large plastic Santas. Once each hour a Santa is selected at random from those made during that hour. Small imperfections in the plastic Santa (bubbles, discolorations, and holes) are counted. The count of nonconformities is recorded on control charts. Recent data have averaged five imperfections (nonconformities) in each sample.
    a. Calculate the control limits and centerline for this c chart.
    b. If management wants a process capability of two nonconformities per sample, will this line be able to meet that requirement? Why or why not?
    c. Given the figure of several hours of production, how does this process look? In other words, interpret the chart in Figure 3.

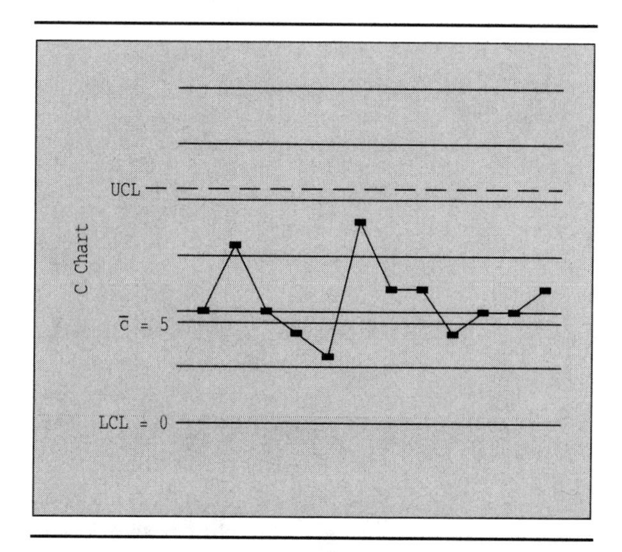

Figure 3 Problem 21

22. The Par Fore Golf Company is a producer of plastic divot fixers. They produce the fixers in bags of 30. They consider these bags a sample size of 1. The nonconformities range from overall size to individual flaws in the production of the plastic tool. These tools are inspected four times daily. Create a chart with the following data. Assume assignable causes and revise the chart.

	Serial Number	*Count of Nonconformities*	*Comment*
1	JG100	7	
2	JG101	3	
3	JG102	5	
4	JG103	4	
5	JG104	6	
6	JG105	6	
7	JG106	17	Wrong mold used
8	JG107	7	
9	JG108	2	
10	JG109	0	
11	JG110	3	
12	JG111	4	
13	JG112	2	
14	JG113	0	
15	JG114	21	Not enouth time in mold
16	JG115	3	
17	JG116	2	
18	JG117	7	
19	JG118	5	
20	JG119	9	
		113	

23. When printing full-color glossy magazine advertising pages, editors do not like to see blemishes or pics (very small places where the color did not transfer to the paper). One person at the printers has the job of looking at samples of magazine pages ($n = 20$) and counting blemishes. Below are her results. The editors are interested in plotting the nonconformities per unit on a u chart. Determine the centerline and control limits. Create a chart of this process. If the editors want only two nonconformities per unit on average, how does this compare with the actual process average?

Sample Number	*Count of Nonconformities*
1	65
2	80
3	60
4	50
5	52
6	42
7	35
8	30
9	30
10	25

24. A manufacturer of holiday light strings tests 400 of the light strings each day. The light strings are plugged in and the number of unlit bulbs are counted and recorded. Unlit bulbs are considered nonconformities, and they are replaced before the light strings are shipped to customers. This data are then used to aid process engineers in their problem-solving activities. Given the following information, create and graph the chart. How is their performance? What is the process capability?

	Sample Size	*Nonconformities*
1	400	50
2	400	23
3	400	27
4	400	32
5	400	26
6	400	38
7	400	57
8	400	31
9	400	48
10	400	34
11	400	37
12	400	44
13	400	34
14	400	32
15	400	50
16	400	49
17	400	54
18	400	38
19	400	29
20	400	47
	8,000	780

25. The Davis Plastic Company produces plastic Wiffle-ball bats for a major sporting goods company. These bats are inspected daily but are not inspected at a constant number each day. The control chart shows how many bats were inspected and how many nonconformities were found per unit. Nonconformities include wrong taper on bat, incomplete handle, bad knurl on grip, or deformed bat ends. Create a u chart with the following data. Assume assignable causes and revise the chart.

Date	*Number Inspected* n	*Count of Nonconformities* c	*Nonconformities per Unit* u
April			
17	126	149	1.18
18	84	91	1.08
19	96	89	0.93
20	101	97	0.96
21	100	140	1.40
24	112	112	1.00
25	79	89	1.13
26	93	119	1.28
27	88	128	1.46
28	105	117	1.11
May			
1	99	135	1.36
2	128	155	1.21
3	113	145	1.28
4	120	143	1.19
5	81	94	1.16
8	85	104	1.22
9	130	157	1.21
10	97	121	1.25
11	78	83	1.06
12	110	120	1.09
	2,025	2,388	

CASE STUDY 9.1
Attribute Control Charts: np Charts

PART 1

At Fruits and Such, a local frozen fruit concentrate manufacturer, frozen orange juice concentrate is packed in 12-ounce cardboard cans. These cans are formed on a machine by spinning them from preprinted cardboard stock and attaching two metal lids, one for the top and one for the bottom. After filling, the container is carefully inspected. The inspectors test how well the strip to open the can operates. If a strip breaks, the can is considered nonconforming and unusable. Other packaging nonconformities include improper crimps holding the lids to the cardboard tube or poorly sealed lids. Fill rates are carefully checked to ensure that the containers truly contain 12 ounces. Inspectors also check the cosmetics of the can. Incorrectly wrapped cans or crooked labels are considered nonconforming and unusable.

An np chart for tracking unusable nonconformities needs to be established to track the level of nonconformities produced on a daily basis. The first 30 subgroups of size n = 50 are found in Figure 1.

Assignment

Use the total number of defectives to create the np chart. How is the process currently performing? What is the process capability?

While the np chart keeps track only of the number of nonconformities and not the type, Figure 1 reveals that the inspectors have also been keeping track of the types of nonconformities.

Assignment

Perform a Pareto analysis and discuss which problems are most prevalent and which should be investigated first.

PART 2

An investigation of the Pareto analysis reveals that the two largest problems are the fill level of the orange juice cans and the breakage of the plastic opening strip. Since the fill level appears to be the largest problem, the investigators decide to start their problem-solving efforts there.

DEPT.	Forming	**PART NAME**	OJ can
PART NO.	OJ-1	**MACHINE**	Spinner
GROUP	1	**VARIABLE**	Seal

Sample	1	2	3	4	5
Time	07:00:00	07:16:00	07:32:00	07:48:00	08:04:00
Date	03/03/XX	03/03/XX	03/03/XX	03/03/XX	03/03/XX
Sample Size	50	50	50	50	50
Defective	12	15	8	10	4
STRIP BRK	4	5	2	4	2
CRIMP	0	2	0	1	0
WRAP	2	0	0	1	0
BASE	0	0	0	1	0
LABEL	2	2	2	1	0
FILL	4	6	4	2	2

Sample	6	7	8	9	10
Time	08:20:00	08:36:00	08:52:00	09:08:00	09:24:00
Date	03/03/XX	03/03/XX	03/03/XX	03/03/XX	03/03/XX
Sample Size	50	50	50	50	50
Defective	7	16	9	14	10
STRIP BRK	2	4	4	5	2
CRIMP	1	2	0	1	2
WRAP	0	2	0	1	0
BASE	0	0	0	1	1
LABEL	0	2	0	2	1
FILL	4	6	5	4	4

Sample	11	12	13	14	15
Time	09:30:00	09:46:00	10:02:00	10:18:00	10:34:00
Date	03/03/XX	03/03/XX	03/03/XX	03/03/XX	03/03/XX
Sample Size	50	50	50	50	50
Defective	5	6	17	12	22
STRIP BRK	2	3	8	4	8
CRIMP	0	0	2	2	0
WRAP	0	0	1	0	0
BASE	0	0	0	1	1
LABEL	1	0	0	2	2
FILL	2	3	6	3	11

Figure 1

DEPT.	Forming		**PART NAME**	OJ can
PART NO.	OJ-1		**MACHINE**	Spinner
GROUP	1		**VARIABLE**	Seal

Sample	16	17	18	19	20
Time	10:50:00	11:06:00	11:22:00	11:38:00	11:54:00
Date	03/03/XX	03/03/XX	03/03/XX	03/03/XX	03/03/XX
Sample Size	50	50	50	50	50
Defective	8	10	5	13	12
STRIP BRK	3	2	1	4	4
CRIMP	1	2	1	1	1
WRAP	0	0	0	1	2
BASE	0	0	0	1	0
LABEL	0	0	0	1	0
FILL	4	6	3	5	5

Sample	21	22	23	24	25
Time	12:10:00	12:26:00	12:42:00	12:58:00	13:14:00
Date	03/03/XX	03/03/XX	03/03/XX	03/03/XX	03/03/XX
Sample Size	50	50	50	50	50
Defective	20	18	24	15	9
STRIP BRK	7	7	10	6	4
CRIMP	1	1	2	1	0
WRAP	2	2	0	2	0
BASE	1	1	2	2	0
LABEL	1	1	3	0	0
FILL	8	6	7	4	5

Sample	26	27	28	29	30
Time	13:30:00	13:46:00	14:02:00	14:18:00	14:34:00
Date	03/03/XX	03/03/XX	03/03/XX	03/03/XX	03/03/XX
Sample Size	50	50	50	50	50
Defective	12	7	13	9	6
STRIP BRK	4	3	4	4	3
CRIMP	1	1	1	0	0
WRAP	1	0	1	0	0
BASE	1	0	2	0	0
LABEL	1	0	0	0	0
FILL	4	3	5	5	3

Figure 1 Continued

Upon checking the operation, they determine that there is a serious lack of precision in filling the orange juice cans. The current method fills the can by timing the flow of the product into the can. This type of filling method does not account for the change in pressure in the level of the supply tank. With a full storage tank of orange juice there is more pressure in the lines, resulting in the overfilling of cans. As the volume in the orange juice storage tank decreases, the cans are underfilled.

Replacing the existing tanks and delivery system, retrofitting pressure meters onto the system, or installing a weight-sensitive pad and fill monitor to the existing line are all suggestions for improving this process. An economic analysis reveals that the weight-sensitive pad and fill-monitoring equipment would be the most economical change. The can rests on the pad during the filling operation. When the can reaches a preset weight, the can is indexed to the next station. Previously, the can was indexed on the basis of time. This new method ensures that the volume in each can changes very little from can to can. The weighing device has been calibrated so that the variance in empty can weight and density will produce only a slight variation in the can fill volume.

Assignment

The decision has been made to add the weight-sensitive pad and fill-monitoring equipment for a trial period. Create an np chart and perform a Pareto analysis with the information in Figure 2. How does the chart look compared with the previous 30 subgroups of size n = 50? What is the process capability? Were there any changes found by the Pareto analysis? What should be investigated next?

DEPT.	Forming		**PART NAME**		OJ can
PART NO.	OJ-1		**MACHINE**		Spinner
GROUP	1		**VARIABLE**		Seal
Sample	1	2	3	4	5
Time	07:00:00	07:16:00	07:32:00	07:48:00	08:04:00
Date	03/16/XX	03/16/XX	03/16/XX	03/16/XX	03/16/XX
Sample Size	50	50	50	50	50
Defective	6	4	6	5	7
STRIP BRK	4	2	3	2	2
CRIMP	0	0	1	0	1
WRAP	1	0	0	1	1
BASE	0	0	0	1	1
LABEL	0	1	1	1	2
FILL	1	1	1	0	0

Figure 2

DEPT.	Forming		**PART NAME**	OJ can
PART NO.	OJ-1		**MACHINE**	Spinner
GROUP	1		**VARIABLE**	Seal

Sample	6	7	8	9	10
Time	08:20:00	08:36:00	08:52:00	09:08:00	09:24:00
Date	03/16/XX	03/16/XX	03/16/XX	03/16/XX	03/16/XX
Sample Size	50	50	50	50	50
Defective	6	5	3	7	6
STRIP BRK	4	1	1	5	0
CRIMP	1	0	1	1	1
WRAP	0	1	0	0	1
BASE	0	1	0	0	1
LABEL	1	1	1	1	2
FILL	0	1	0	0	1

Sample	11	12	13	14	15
Time	09:40:00	09:56:00	10:12:00	10:28:00	10:42:00
Date	03/16/XX	03/16/XX	03/16/XX	03/16/XX	03/16/XX
Sample Size	50	50	50	50	50
Defective	2	4	3	6	5
STRIP BRK	2	3	1	3	4
CRIMP	0	0	0	1	0
WRAP	0	0	1	1	0
BASE	0	0	0	0	0
LABEL	0	1	0	1	1
FILL	0	0	1	0	0

Sample	16	17	18	19	20
Time	11:00:00	11:16:00	11:32:00	11:48:00	12:30:00
Date	03/16/XX	03/16/XX	03/16/XX	03/16/XX	03/16/XX
Sample Size	50	50	50	50	50
Defective	4	5	6	7	5
STRIP BRK	2	4	2	4	1
CRIMP	0	0	0	0	1
WRAP	0	0	1	0	1
BASE	1	0	1	1	0
LABEL	1	1	1	1	1
FILL	0	0	1	1	1

Figure 2 Continued

PART 3

Correction of the strip-breakage problem involves some investigation into the type of plastic currently being used to make the strips. The investigators have determined that the purchasing department recently switched to a less expensive type of plastic from a new supplier. The purchasing department never checked into the impact that this switch would have in the manufacturing department.

The investigators have shown the purchasing agent the results of the Pareto analysis and control charts and the old supplier has been reinstated, effective immediately. The unused plastic from the new supplier has been taken out of use and returned.

In a separate incident, the investigators have observed that the operators are not well trained in changing the cardboard-roll stock material. A direct correlation exists between the changeover times and the situations when the label is off-center on the can. The operators involved in this operation are currently undergoing retraining.

Assignment

To verify that the changes discussed above have had an impact on process quality, create an np chart and a Pareto diagram on the newest data in Figure 3. Discuss the results. What is the process capability now compared with before the changes? What should be done now?

DEPT.	Forming		**PART NAME**		OJ can
PART NO.	OJ-1		**MACHINE**		Spinner
GROUP	1		**VARIABLE**		Seal
Sample	1	2	3	4	5
Time	07:00:00	07:18:00	07:36:00	07:54:00	08:12:00
Date	03/18/XX	03/18/XX	03/18/XX	03/18/XX	03/18/XX
Sample Size	50	50	50	50	50
Defective	2	1	2	0	1
STRIP BRK	0	0	0	0	1
CRIMP	0	0	1	0	0
WRAP	0	1	0	0	0
BASE	0	0	0	0	0
LABEL	2	0	1	0	0
FILL	0	0	0	0	0

Figure 3

DEPT.	Forming	**PART NAME**	OJ can
PART NO.	OJ-1	**MACHINE**	Spinner
GROUP	1	**VARIABLE**	Seal

Sample	6	7	8	9	10
Time	08:30:00	08:48:00	09:06:00	09:24:00	09:42:00
Date	03/18/XX	03/18/XX	03/18/XX	03/18/XX	03/18/XX
Sample Size	50	50	50	50	50
Defective	0	1	0	1	2
STRIP BRK	0	0	0	0	0
CRIMP	0	0	0	0	1
WRAP	0	0	0	0	0
BASE	0	1	0	1	0
LABEL	0	0	0	0	1
FILL	0	0	0	0	0

Sample	11	12	13	14	15
Time	10:00:00	10:18:00	10:36:00	10:54:00	11:12:00
Date	03/18/XX	03/18/XX	03/18/XX	03/18/XX	03/18/XX
Sample Size	50	50	50	50	50
Defective	1	0	2	1	2
STRIP BRK	0	0	1	0	0
CRIMP	0	0	0	1	1
WRAP	0	0	1	0	0
BASE	1	0	0	0	1
LABEL	0	0	0	0	0
FILL	0	0	0	0	0

Sample	16	17	18	19	20
Time	11:30:00	11:48:00	12:06:00	12:24:00	12:42:00
Date	03/18/XX	03/18/XX	03/18/XX	03/18/XX	03/18/XX
Sample Size	50	50	50	50	50
Defective	2	0	1	1	0
STRIP BRK	0	0	0	1	0
CRIMP	0	0	0	0	0
WRAP	0	0	0	0	0
BASE	0	0	1	0	0
LABEL	2	0	0	0	0
FILL	0	0	0	0	0

Figure 3 Continued

CASE STUDY 9.2
Attribute Control Charts: c Charts

This case is the second of three related cases found in Chapters 8, 9, and 11. These cases seek to link information from the three chapters in order to resolve quality issues. Although they are related, it is not necessary to complete the case in this chapter in order to understand or complete the cases in the other chapters.

PART 1

Max's B-B-Q Inc. manufactures top-of-the-line barbeque tools. The tools include forks, spatulas, knives, spoons, and shish-kebab skewers. Max's fabricates both the metal parts of the tools and the resin handles. These are then riveted together to create the tools (Figure 1). Recently, Max's hired a process engineer. His first assignment is to study routine tool wear on the company's stamping machine. In particular, he will be studying tool wear patterns for the tools used to create knife blades (Figure 2).

Each quarter, Max's makes more than 10,000,000 tools. Of these, approximately 240,000 are inspected for a variety of problems, including handle rivets, fork tines, handle cracks, dents, pits, butts, scratches, handle color, blade grinds, crooked blades, and others. Any tool that has a defect is either returned to the line to be reworked or it is scrapped. Figure 2 provides information for their first-quarter inspection results.

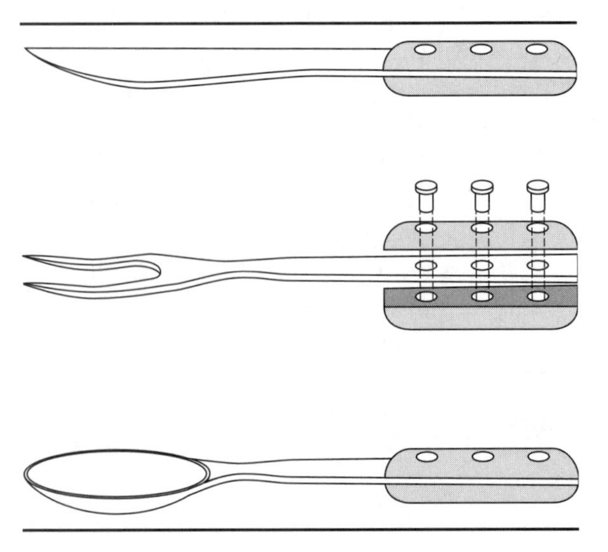

Figure 1 Barbeque Tools

Assignment

Create a Pareto chart with the information provided in Figure 2. Are any of the categories related? Can they be grouped in any way? What problems should be tackled first?

Assignment

Create a chart for nonconformities using the data in Figure 3. How is the process performing? What is the process capability?

Subgroup #	*n*	*c*
1	1250	43
2	1250	40
3	1250	45
4	1250	39
5	1250	44
6	1250	42
7	1250	41
8	1250	40
9	1250	46
10	1250	45
11	1250	42
12	1250	43
13	1250	40
14	1250	38
15	1250	43
16	1250	42
17	1250	45
18	1250	43
19	1250	37
20	1250	39
21	1250	43
22	1250	44
23	1250	45
24	1250	41
25	1250	46
	32000	1056

Figure 3 February Summary of Daily Final Inspection Data for Table Knife Production. For Nonconformities by Type, see Figure 2.

FINAL INSPECTION DEPARTMENT 8 FEBRUARY 1996	Table Knife	Steak Knife	Paring Knife	Filet Knife	Basting Spoon	Ladle Spoon	Perforated Spoon	Skewer
DEFECT								
Bad handle rivets							1	
Bad steel	37				1			
Bad tines								
Bad/bent points	16							
Bent					1			
Buff concave								
Burn	146	57						
Burned handles	1							
Burrs	24	14						
Cloud on blades								
Cracked handle		32						
Cracked steel	1							
Crooked blades								
DD edge	41	2						
Dented								
Edge								
Etch	31							
Finish								
Grind in blade							1	
Haft at rivets								
Hafting marks handles	64				4		3	
Hafting marks steel								
Handle color	27	182				5		3
Heat induct	110	37					2	
High handle rivets								
High-speed buff								
Hit blades								
Hit handle	239							
Holder marks								

Figure 2

FINAL INSPECTION DEPARTMENT 8 FEBRUARY 1996	Table Knife	Steak Knife	Paring Knife	Filet Knife	Basting Spoon	Ladle Spoon	Perforated Spoon	Skewer
DEFECT	**1759**	**1759W**	**1706**	**1707**	**1712**	**1712W**	**1713**	**1713W**
Honing								
Narrow blades	4							
Nicked and scratched	105	10			9	2	7	
Open at rivet								
Open handles							1	
Open steel								
Pitted blades		1					1	
Pitted bolsters								
Raw backs								
Raw fronts								
Rebend								
Recolor concave	2	43						
Rehone								
Reruns								
Rivet								
Scratched blades	105	53		1	12	2	4	
Scratched rivets								
Seams/holes	11	50						
Seconds				2				
Stained blades	84	18			5	1		
Stained rivets								
Vendor rejects								
Water lines					15			
Wrap	8							
Total Inspected	32,000	28,811	5	277	3,145	1,504	2,008	1,805
Total Accepted	30,944	28,312	5	274	3,098	1,494	1,988	1,802
Total Rejected	1,056	499	0	3	47	10	20	3

Figure 2 Continued

IV

Expanding the Scope of Quality

10

Reliability

■ *Learning Opportunities:*

1. To understand the importance of system reliability
2. To understand how performance during the life of a product, process, or system is affected by its design and configuration
3. To be able to compute the reliability of systems, including systems in series, parallel, and hybrid combinations
4. To know what to look for in a comprehensive reliability program ■

Cars in a Snowstorm

Product designers and manufacturers must pay careful attention to the customer's expectations concerning a product's intended function, life, and environmental conditions for use. Customers judge a product's performance on much more than what can be seen at the moment of purchase. Consider the figure above. What are the intended function, life, and conditions for use of a passenger car? Most readers would agree that the intended function of a passenger car is to carry passengers from one location to another. Expectations concerning the life of such a vehicle may vary; a reasonable estimate may be 100,000 to 150,000 miles of use. The prescribed environmental conditions may at first appear to be easily defined: transport passengers on paved surface roads. But what about the vehicle's performance in snow, ice, or floods? How often does the car need to be started? Should the car start easily in weather significantly below zero? What temperature should it be able to endure before overheating? How deep a flood can it drive through and still run? These are just a few of the factors that define reliability. Reliability concepts play a key role in the consumer's perception of quality.

RELIABILITY

A car, the Internet, an automated teller machine, pagers, the telephone—are these everyday items and things that you take for granted? Have you ever contemplated what happens if they fail? What happens if your car won't start? Do you get a ride from someone? Stay home? Take the bus or a cab? What happens if the Internet fails? Loss of access to information? Millions of missing e-mail messages? Lost Web pages? And if the telephone doesn't work? Can't call home? Can't call the hospital? What happens if the data-processing centers, which tell ATMs whether or not to hand out money, have a system failure? No cash for you?

Our society depends on the high-tech devices that make our lives easier. We have become so used to their presence in our lives that their sudden failure can bring about chaos. Yet, how much attention do we pay to the reliability of the devices we take for granted? Perhaps we become aware of reliability issues only when the device fails at a particularly critical time.

Reliability, *or quality over the long term, is the ability of a product to perform its intended function over a period of time and under prescribed environmental conditions.* The reliability of a system, subsystem, component, or part is dependent on many factors, including the quality of research performed at its conception, the original design and any subsequent design changes, the complexity of the design, the manufacturing processes, the handling received during shipping, the environment surrounding its use, the end user, and numerous other factors. The causes of unreliability are many. Improper design, less-than-specified construction materials, faulty manufacturing or assembly, inappropriate testing leading to unrealistic conclusions, and damage during shipment are all factors that may contribute to a lack of product reliability. Once the product reaches the user, improper start-ups, abuse, lack of maintenance, or misapplication can seriously affect the reliability of an item. Reliability studies are looking for the answers to such questions as Will it work? how long will it last?

EXAMPLE 10.1 Sudden Failure

Consider the chaos caused in 1998, when a digital beeper communication satellite failed, rendering digital pagers across the United States silent. Imagine the effect on a personal level. Romances on the rocks when couples unaware of the problem try unsuccessfully to page each other. (Why didn't you return my page ?) Parents unable to check in with their kids and vice versa. (Where were you?) Imagine not being able to contact doctors, nurses, or police to respond to emergencies. (My wife is having a baby . . . where is the doctor?) The satellite relayed signals for nearly 90 percent of the 45 million pagers in the United States as well as signals for dozens of broadcasters and data networks. Besides the silent pagers, gas stations couldn't take credit cards; broadcasting stations went off the air; some data information networks were inoperable; in some buildings, elevator and office Musak was silenced; and many flights were delayed as controllers and pilots sought information on high-altitude weather.

Eight of the nation's 10 biggest paging companies used this satellite because of its large area of coverage, the entire United States and the Caribbean. To regain communications, users had to switch to other satellites. For some, this involved climbing up on roofs and manually reorienting thousands of satellite dishes. Others were able to get up and running more quickly by temporarily shifting to other satellites.

All this chaos occurred because of the failure of the on-board computer that kept a five-year-old communication satellite pointed at the earth in a synchronized orbit 22,300 miles high. The $250 million, nine-foot cube with two 50-foot solar-panel wings had a projected failure rate of less than one percent. No backup systems were in place or considered necessary. Is a failure rate of less than one percent acceptable? How can system reliability be determined? What can be done to improve system reliability? This chapter provides an introduction to reliability concepts.

Not worried about pagers? Remember, automatic teller machines share data-processing centers, air-traffic controllers share regional radar centers, and the Internet depends on a relatively small number of server computers for routing communications . . .

"The Day the Beepers Died," *Newsweek,* June 1, 1998, p. 48.
"One Satellite Fails, and the World Goes Awry," *The Wall Street Journal,* May 21, 1998.

PRODUCT LIFE CYCLE CURVE

The life cycle of a product is commonly broken down into three phases: early failure, chance failure, and wear-out (Figure 10.1). The early failure, or infant mortality, phase is characterized by failures occurring very quickly after the product has been produced or put into use by the consumer. The curve during this phase is exponential, with the number of failures decreasing the longer the product is in use. Failures at this stage have a variety of causes. Some early failures are due to inappropriate or inadequate materials, marginal components, incorrect installation, or poor manufacturing techniques. Incomplete testing may not have revealed design weaknesses which become apparent as the consumer uses the system. Inadequate quality checks could have allowed substandard products to leave the manufacturing area. The manufacturing processes or tooling may not have been capable of producing to the specifications needed by the designer. The consumer also plays a role in product reliability. Once the product reaches the consumer, steps must be taken to ensure that the

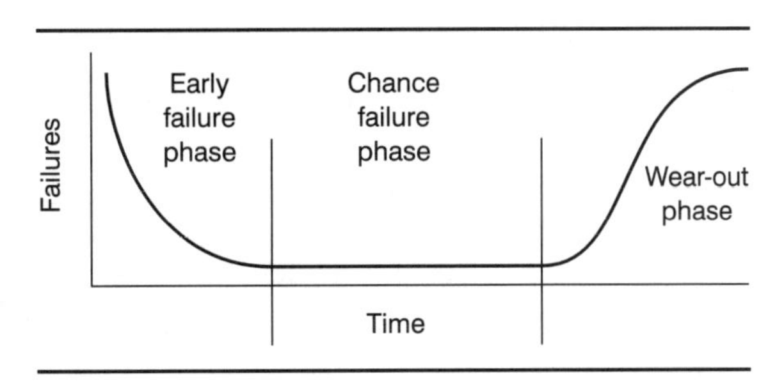

Figure 10.1 Life Cycle Curve

consumer understands the appropriate environment for product use. Consumers need to receive training on how to operate and handle the product. Using improper procedures will affect the reliability of a product.

During the chance failure portion of a product's useful life, failures occur randomly. This may be due to inadequate or insufficient design margins. Manufacturing or material problems have the potential to cause intermittent failures. At this stage the consumer can also affect product reliability. Misapplication or misuse of the product by the consumer can lead to product failure. Overstressing the product is a common cause of random failures. Because of their unfamiliarity with product capabilities, consumers may inadvertently overstress the product.

As the product ages, it approaches the final stage of its life cycle, the wear-out phase. During this phase, failures increase in number until few, if any, of the product are left. Wear-out failures are due to a variety of causes, some related to actual product function, some cosmetic. A system's reliability, useful operation, or desirability may decrease if it becomes scratched, dented, chipped, or otherwise damaged. Age and the associated wear, discolorations and brittleness may lead to material failure. Normal wear could decrease reliability through misalignments, loose fittings, and interference between components. Combined stresses placed on the product (like the zipper or latches on a suitcase) during its lifetime of use are a source of decreased reliability. Neglect or inadequate preventive maintenance lessens product reliability and shortens product life.

MEASURES OF RELIABILITY

Systems are orderly arrangements or combinations of parts, components, and subassemblies. These elements interact with each other and with external factors, such as humans and other systems, to perform their intended functions. Overall system reliability depends on the individual reliabilities associated with the parts, components, and subassemblies. Reliability values are sought to determine the performance of a product. Reliability studies also reveal any recurring patterns of failure and the underlying causes of those failures.

Reliability tests exist to aid in determining if distinct patterns of failure exist during the product's or system's life cycle. Reliability tests determine what failed, how it failed, and the number of hours, cycles, actuations, or stresses it was able to bear before failure. Once these data are known, decisions can be made concerning product reliability expectations, corrective action steps, maintenance procedures, and costs of repair or replacement. Several different types of tests exist to judge the reliability of a product, including failure-terminated, time-terminated, and sequential tests. The name of each of these tests says a good deal about the type of the test. *Failure-terminated tests* are ended when a predetermined number of failures occur within the sample being tested. The decision concerning whether or not the product is acceptable is based on the number of products that have failed during the test. A *time-terminated test* is concluded when an established number of hours is reached.

For this test, product is accepted on the basis of how many products failed before reaching the time limit. A *sequential test* relies on the accumulated results of the tests.

Failure Rate, Mean Life, and Availability

When system performance is time-dependent, such as the length of time a system is expected to operate, then reliability is measured in terms of mean life, failure rates, availability, mean time between failures, and specific mission reliability. As a system is used, data concerning failures become available. This information can be utilized to estimate the mean life and failure rate of the system. Failure rate, λ, the probability of a failure during a stated period of time, cycle, or number of impacts, can be calculated as

$$\lambda_{\text{estimated}} = \frac{\text{number of failures observed}}{\text{sum of test times or cycles}}$$

From this, θ, the average life, can be calculated:

$$\theta_{\text{estimated}} = \frac{1}{\lambda}$$

or

$$\theta_{\text{estimated}} = \frac{\text{sum of test times or cycles}}{\text{number of failures observed}}$$

The average life θ is also known as the mean time between failure or the mean time to failure. Mean time between failure (MTBF), how much time has elapsed between failures, is used when speaking of repairable systems. Mean time to failure is used for nonrepairable systems.

EXAMPLE 10.2 Calculating Failure Rate and Average Life

Twenty windshield-wiper motors are being tested using a time-terminated test. The test is concluded when a total of 200 hours of continuous operation have been completed. During this test, the number of windshield wipers that fail before reaching the time limit of 200 hours is counted. If three wipers failed after 125, 152, and 189 hours, calculate the failure rate λ and the average life θ:

$$\lambda_{\text{estimated}} = \frac{\text{number of failures observed}}{\text{sum of test times or cycles}}$$

$$= \frac{3}{125 + 152 + 189 + (17)200} = 0.0008$$

From this, θ, the average life, can be calculated:

$$\theta_{\text{estimated}} = \frac{1}{\lambda} = \frac{1}{0.0008} = 1250 \text{ hours}$$

or

$$\theta_{\text{estimated}} = \frac{\text{sum of test times or cycles}}{\text{number of failures observed}}$$

$$= \frac{125 + 152 + 189 + (17)200}{3} = 1289 \text{ hours}$$

Here, the difference between λ estimated and θ estimated is due to rounding. 

Mean times between failures (MTBF) and mean times to failure (MTTF) describe reliability as a function of time. Here the amount of time that the system is actually operating is of great concern. For example, without their radar screen, air traffic controllers are sightless and therefore out of operation. To be considered reliable, the radar must be functional for a significant amount of the expected operating time. Since many systems need preventive or corrective maintenance, a system's reliability can be judged in terms of the amount of time it is available for use:

$$\text{Availability} = \frac{\text{mean time to failure (MTTF)}}{\text{MTTF} + \text{mean time to repair}}$$

MTBF values can be used in place of MTTF.

EXAMPLE 10.3 Determining Availability

Windshield-wiper motors are readily available and easy to install. Calculate the availability of the windshield wipers on a bus driven eight hours a day, if the mean time between failure or average life θ is 1250 hours. When the windshield-wiper motor must be replaced, the bus is out of service for a total of 24 hours.

$$\text{Availability} = \frac{\text{mean time between failure (MTBF)}}{\text{MTBF} + \text{mean time to repair}}$$

$$= \frac{1250}{1250 + 24} = 0.98$$

The bus is available 98 percent of the time. 

Calculating System Reliability

When system performance is dependent on the number of cycles completed successfully, such as the number of times a coin-operated washing machine accepts the coins and begins operation, then reliability is measured in terms of the probability of successful operation. This type of reliability calculation has probability theory as its basis. *Reliability is the probability that a product will not fail during a particular time period.* Like probability, reliability takes on numerical values between 0.0 and 1.0. A reliability value of 0.78 is interpreted as 78 out of 100 parts will function as expected during a particular time period and 22 will not. If n is the total number of units being tested and s represents those units performing satisfactorily, then reliability R is given by

$$R = \frac{s}{n}$$

System reliability is determined by considering the reliability of the components and parts of the system. To calculate system reliabilities, the joint probabilities of their independent failure rates must be utilized. As systems become more complex, these calculations become increasingly more difficult. Interdependence of components in a system cannot be overlooked. In a system, the failure of one component could influence the failure of another component through load transferal, changes to the immediate environment, and many other factors.

Reliability in Series

A ***system in series*** *exists if proper system functioning depends on whether all the components in the system are functioning.* Failure of any one component will cause system failure. Figure 10.2 portrays an example of a system in series. In the diagram, the system will function only if all four components are functioning. If one component fails, the entire system will cease functioning. The reliability of a system in series is dependent on the individual component reliabilities. To calculate series system reliability the individual independent component reliabilities are multiplied together:

$$R_s = r_1 \cdot r_2 \cdot r_3 \cdot \cdots \cdot r_n$$

where

$$R_s = \text{reliability of series system}$$
$$r_i = \text{reliability of component}$$
$$n = \text{number of components in system}$$

EXAMPLE 10.4 Calculating Reliability in a Series System I

The reliability values for the flashlight components pictured in Figure 10.3 are listed below. If the components work in series, what is the reliability of the system?

Battery	0.75	Screw cap	0.97
Lightbulb	0.85	Body	0.99
Switch	0.98	Lens	0.99
Spring	0.99		

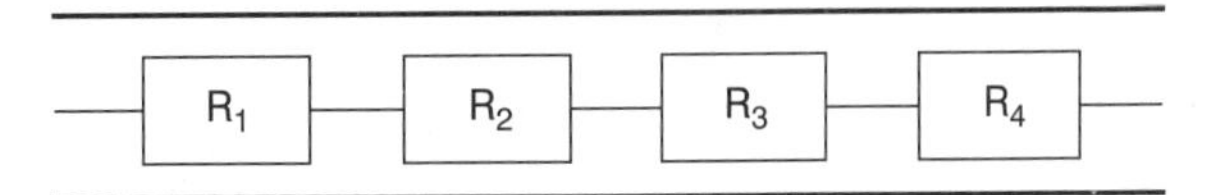

Figure 10.2 System in Series

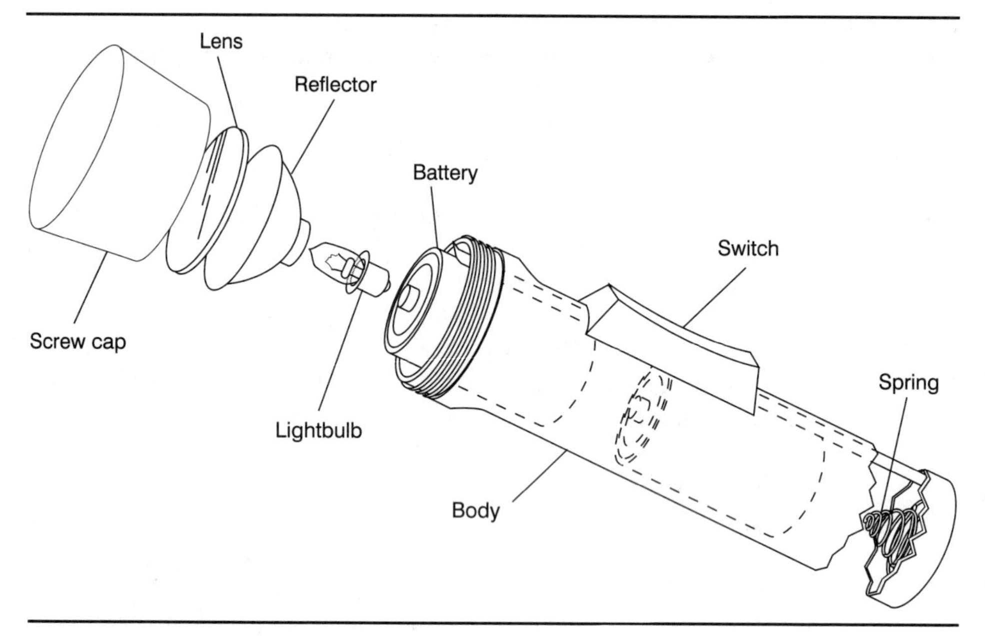

Figure 10.3 Flashlight

$$
\begin{aligned}
R_s &= r_1 \cdot r_2 \cdot r_3 \cdot \cdots \cdot r_n \\
&= 0.75 \cdot 0.98 \cdot 0.85 \cdot 0.99 \cdot 0.97 \cdot 0.99 \cdot 0.99 \\
&= 0.59
\end{aligned}
$$

This flashlight has a reliability of only 0.59. Perhaps the owner of this flashlight ought to keep two on hand!

EXAMPLE 10.5 Calculating Reliability in a Series System II

Figure 10.4 displays a block diagram representing three clean-water supply systems, each system increasing in complexity. The first filter filters out leaves and dirt that may clog the pump. Following the pump, a charcoal filter cleans microorganisms out of the water before it is chlorinated. Calculate the reliability of each of the systems.

$$
\begin{aligned}
R_{s1} &= r_1 \cdot r_2 \cdot r_3 \cdot \cdots \cdot r_n \\
&= 0.95 \cdot 0.98 \cdot 0.90 \cdot 0.99 = 0.83
\end{aligned}
$$

$$
\begin{aligned}
R_{s2} &= r_1 \cdot r_2 \cdot r_3 \cdot \cdots \cdot r_n \\
&= 0.95 \cdot 0.98 \cdot 0.98 \cdot 0.90 \cdot 0.99 = 0.81
\end{aligned}
$$

$$
\begin{aligned}
R_{s3} &= r_1 \cdot r_2 \cdot r_3 \cdot \cdots \cdot r_n \\
&= 0.95 \cdot 0.98 \cdot 0.98 \cdot 0.90 \cdot 0.99 \cdot 0.98 = 0.80
\end{aligned}
$$

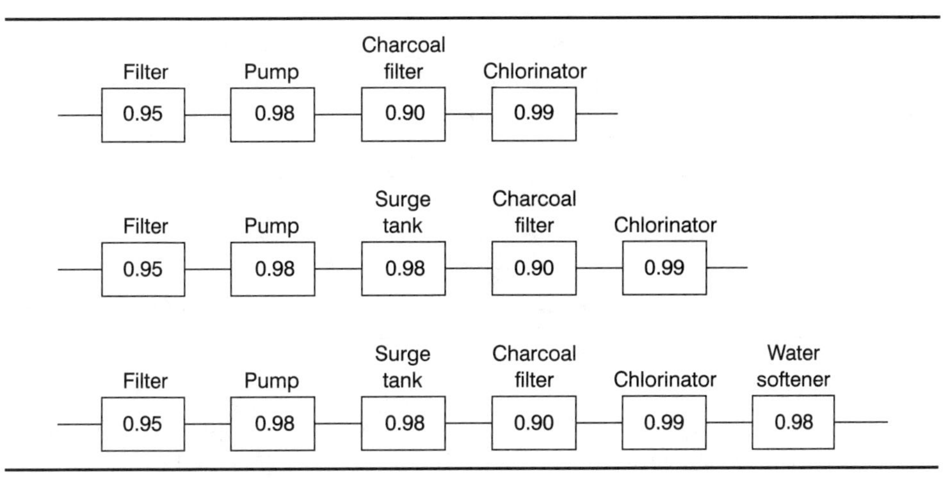

Figure 10.4 Clean-Water Supply Systems

Even though the reliability of the additional components is high, as the system increases in complexity, the overall system reliability decreases.

Notice from this example that the reliability of a system decreases as more components are added in series. This means that the series-system reliability will never be greater than the reliability of the least reliable component. Sustained performance over the life of a product or process can be enhanced in several ways. Parallel systems and redundant or standby component configurations can be used to increase the overall system reliability.

Reliability in Parallel

A ***parallel system*** *is a system that is able to function if at least one of its components is functioning*. Figure 10.5 displays a system in parallel. The system will function provided at least one component has not failed. In a parallel system, all of the components must fail in order to have a system failure. This is the opposite of a series system in which, if one component fails, the entire system fails. Since the duplicated or paralleled component takes over functioning for the failed part, the reliability of

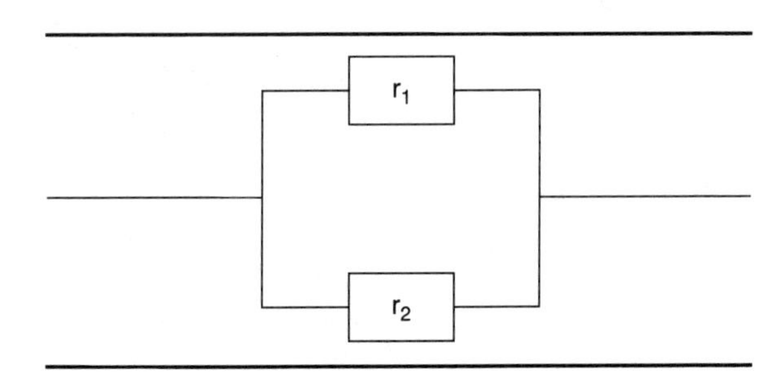

Figure 10.5 Parallel System

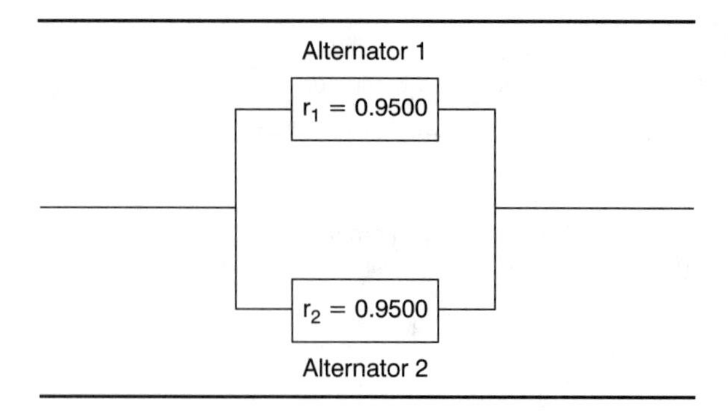

Figure 10.6 Systems in Parallel: Alternators

this type of system is calculated on the basis of the sum of the probabilities of the favorable outcomes: the probability that no components fail and the combinations of the successful operation of one component but not the other(s). The reliability of a parallel system is given by

$$R_p = 1 - (1 - r_1)(1 - r_2)(1 - r_3) \cdots (1 - r_n)$$

where

$$R_p = \text{reliability of parallel system}$$
$$r_i = \text{reliability of component}$$
$$n = \text{number of components in system}$$

EXAMPLE 10.6 Determining Reliability in a Parallel System I

On a twin-engine aircraft, two alternators support a single electric system. As shown in Figure 10.6, these alternators each have a reliability of 0.9500. Calculate their parallel reliability:

$$R_p = 1 - (1 - r_1)(1 - r_2)$$
$$= 1 - (1 - 0.9500)(1 - 0.9500) = 0.9975$$

Even though the alternators' reliability values individually are 0.9500, when combined they have a system reliability of 0.9975.

EXAMPLE 10.7 Determining Reliability in a Parallel System II

In the 1920s, an airplane called a Ford Tri-motor was built (Figure 10.7). As the name suggests, this plane had three motors to increase its reliability. Today, triple-engine-type aircraft include DC-10s and L-1011s. Since the duplicated or paralleled motor takes over functioning for the failed motor, the reliability of this type of system is calculated on the basis of the sum of the probabilities of the favorable outcomes: the probability that no components fail and the combinations of the successful operation of one component but not the other(s). If each of the motors on this aircraft has a reliability of 0.9500, what is the overall reliability?

$$R_p = 1 - (1 - r_1)(1 - r_2)(1 - r_3)$$
$$= 1 - (1 - 0.9500)(1 - 0.9500)(1 - 0.9500) = 0.9999$$

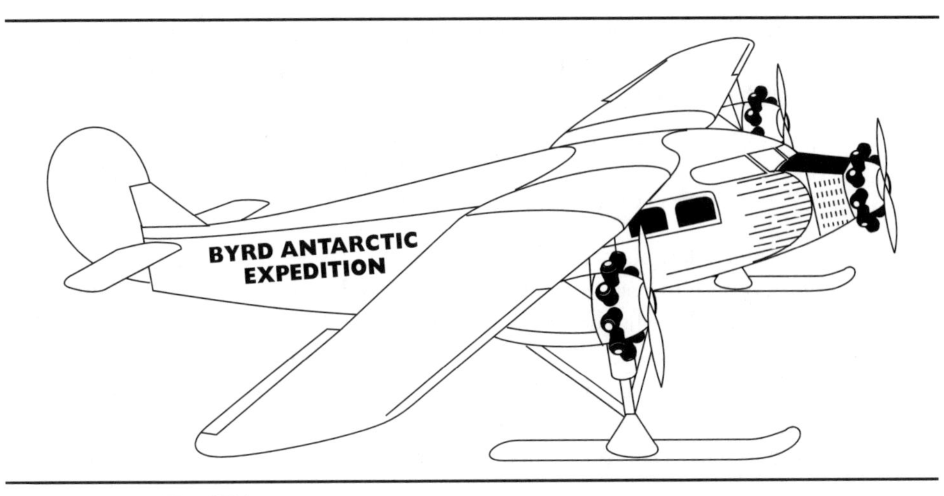

Figure 10.7 Ford Tri-motor

System reliability increases as components are added to a parallel system. Reliability in a parallel system will be no less than the reliability of the most reliable component. Parallel redundancy is often used to increase the reliability of a critical system; however, at some point there is a diminishing rate of return where the added costs outweigh the increased reliability. Parallel components are just one type of redundancy used to increase system reliability.

Reliability in Redundant Systems and Backup Components

Backup or spare components, used only if a primary component fails, increase overall system reliability. The likelihood of needing to access the spare or backup component in relation to the reliability of the primary component is shown mathematically as

$$R_b = r_1 + r_b(1 - r_1)$$

where

$$R_b = \text{reliability of backup system}$$
$$r_1 = \text{reliability of primary component}$$
$$r_b = \text{reliability of backup component}$$
$$1 - r_1 = \text{chance of having to use backup}$$

EXAMPLE 10.8 Calculating Reliability in a Redundant System

A local hospital uses a generator to provide backup power in case of a complete electrical power failure. This generator is used to fuel the equipment in key areas of the hospital such as the operating rooms and intensive care units. In the past few years, the power supply system has been very reliable (0.9990). There has been only one incident where the hospital lost power. The generator is tested frequently to ensure that it is capable of operating at a moment's notice. It too is very reliable (0.9900). Calculate the reliability of this system:

$$R_b = r_1 + r_b(1 - r_1)$$
$$= 0.9990 + 0.9900(1 - 0.9990) = 0.9999$$

From a design point of view, redundancy is achieved through the use of design margins. These factors of safety are added to design calculations to provide coverage for the variability that exists in manufacturing, material strengths, potential misuse, and other stresses.

Invariably, redundancy increases the costs associated with a system. Whether these are initial costs or costs incurred in maintaining the system, there are trade-offs to be made between reliability, costs, and safety.

Reliability in Systems

From a cost point of view, utilizing parallel systems for every situation is ineffective. Imagine equipping every car with two sets of headlights, two sets of taillights, two or four more doors, two engines, two batteries, and so on. Designers often use the advantages of parallel systems on critical components of an overall system. These systems, containing parallel components and components in series, are called combination systems. As the next example shows, calculating the reliability of such a system involves breaking down the overall system into groups of series and parallel components.

EXAMPLE 10.9 Calculating the Reliability of a Combination System

In an endeavor to discourage thievery, a local firm has installed the alarm system shown in Figure 10.8. This system contains series, parallel, and backup components. Calculate the reliability of the system.

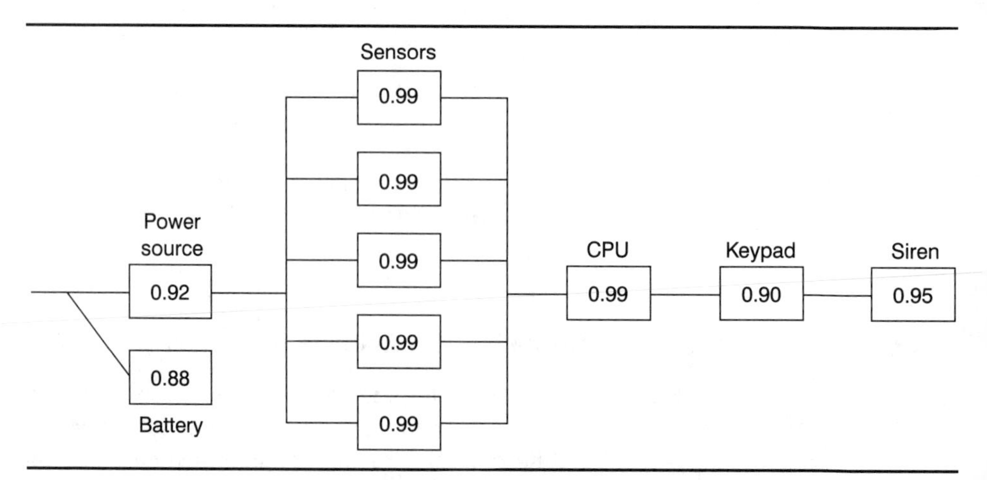

Figure 10.8 Reliability Diagram: Alarm System

To calculate the overall system reliability, begin by determining the overall reliability of the components in parallel. The combined reliability of the five sensors is

$$R_p = 1 - (1 - r_1)(1 - r_2)(1 - r_3)\cdots(1 - r_n)$$
$$= 1 - (1 - 0.99)(1 - 0.99)(1 - 0.99)(1 - 0.99)(1 - 0.99)$$
$$= 1.0$$

The combined reliability of the power source and battery is

$$R_b = r_1 + r_b(1 - r_1)$$
$$= 0.92 + 0.88(1 - 0.92)$$
$$= 0.99$$

The overall reliability of the system is

$$R_s = r_p \cdot r_{cpu} \cdot r_{keypad} \cdot r_{siren} \cdot r_b$$
$$= 1.0 \cdot 0.99 \cdot 0.90 \cdot 0.95 \cdot 0.99$$
$$= 0.84$$

RELIABILITY PROGRAMS

Reliability programs endeavor to incorporate reliability concepts into system design. Reliability issues surface in nearly every facet of system design, development, creation, use, and service. A well-thought-out reliability program will include the areas of design, testing, manufacture, raw material and component purchases, production, packaging, shipping, marketing, field service, and maintenance. A sound reliability program developed and implemented to support an entire system will consider the following aspects.

1. The entire system. What composes the system? What goals does the system meet? What are the components of the system? How are they interrelated? How do the interrelationships influence system reliability? What are the system reliability requirements set by the customer?

Critical to ensuring system reliability is a complete understanding of the purpose of the product. Everyone involved in providing the system must have a complete understanding of the product. The consumer's reliability needs and expectations must be a part of product reliability. Also important is an understanding of the life of a product. How is it shipped? How is it transported and stored? What type of environment will it face during usage? Comprehensive knowledge of a product and its life is the foundation of a strong reliability program.

2. The humans in the system. What are their limitations? What are their capabilities? What knowledge do they have of the product? How will this knowledge affect their use of the product? How might they misuse the product?

Product reliability can be increased if greater emphasis is placed on proper training and education of the users of the product. Appropriate steps must also be taken to ensure that sales does not promise more than the product can deliver. Setting the expectations of the users is as important as showing them how to use a product.

3. Maintenance of the system. Can the components or subassemblies be maintained separately from the system? Are the system components accessible? Is the

system designed for replacement components or subsystems? Are those components or subsystems available? Can those components or subsystems be misapplied or misused? Will maintenance be performed under difficult circumstances?

If and when a component or a system fails, those using the system or component will be interested in the length of time it takes to return the failed system, subsystem, or component to full operational status. Changes to product, process, or system design based on maintainability considerations are sure to include an investigation of factors such as ease of repair or replacement of components, costs, and time. Well-designed policies, practices, and procedures for effective preventive maintenance before failures occur and timely corrective maintenance after failure increase overall system reliability.

4. Simplicity. Will a simple design be more effective than a complicated one? Will a reduction in the number of elements increase the system reliability? Will the addition of elements increase the system reliability?

Simple, straightforward designs will increase system reliability. Designs of this sort are less likely to break down and are more easily manufactured. In some instances, the effort to impress with the current technological wizardry decreases the product's reliability. The greater the number of components, the greater the complexity and the easier it is for some aspect of the product, process, or system to fail.

5. Redundant and fail-safe features. Can the addition of redundant components or fail-safe features prevent overall system failure? At what cost?

Incremental increases in reliability should be balanced against costs. Products can be overdesigned for their purposes. There is a diminishing return on investment for this approach to achieving a reliable product. Is it preferable to avoid specially designed parts and components for the sake of easy availability? Will interchangeability decrease maintenance complexity?

6. Manufacturing methods and purchasing requirements. Have the chosen manufacturing methods enhanced system quality and reliability?

Purchasing and manufacturing must be made aware of how critical it is to purchase and use the materials deemed appropriate for this product.

7. Maintenance of complete product or system performance records. Can this information be used to increase future system reliability?

System reliability information from tests or from actual experience with the product or system should be gathered in a manner that will be meaningful when used for decisions concerning product or system design changes, development of new products or systems, product or system manufacture, operation, maintenance, or support.

8. Communication. Have clear channels of communication been established among all those involved in the design, manufacture, shipment, maintenance, and use of the product or system?

While perhaps not a complete list, the above areas for consideration form the basis of a well-structured reliability program. When the program provides answers to the questions, system reliability is enhanced.

EXAMPLE 10.10 Disasters That Could Have Been Averted

Sometimes it is difficult to comprehend that a complete reliability program improves overall system reliability. Consider some examples from real-life events where the critical aspects we discussed above were ignored:

Aspect 1. The entire system

In 1981, a walkway at the Kansas City Hyatt Regency Hotel collapsed, resulting in the loss of 100 lives. The failure was caused by a design change that substituted two short rods in place of the original long rod. An understanding of the entire system could have averted this disaster. The use of two rods doubled the load on the washer and nut holding the assembly together, but when the design was changed, the washer and nut were not strengthened. The undersized nut and washer allowed the rod to be pulled through the box beam. The entire system, including the nut and washer, should have been considered in the new design.

Aspect 1. The entire system

In 1989, United Airlines Flight 232 crashed in Sioux City, Iowa, despite heroic efforts on the part of the flight crew. The DC-10's tail engine exploded in flight and destroyed the plane's hydraulic lines. Although the plane could fly on its two wing-mounted engines, the loss of hydraulic power rendered the plane's main control systems (flaps and gear) inoperable. A study of the plane's design revealed that, although backup hydraulic systems existed, all three hydraulic lines, including backups, went through one channel, which rested on top of the third engine in the tail. When the engine failed, debris from the engine severed all the hydraulic lines at the same time. In his March 28, 1997 *USA Today* article, "Safety Hearings End Today on Cracks in Jet Engines," Robert Davis stated, "The nation's airlines could be forced to spend more time and money using new techniques for inspecting engines." Those familiar with reliability issues know that the problems are more complicated than an airline's maintenance practices.

Aspect 1. The entire system

During the summer of 1996, power outages plagued Western states, leaving offices, businesses, and homes without lights, air-conditioning, elevators, computers, cash registers, faxes, electronic keys, ventilation, and a host of other services. Traffic chaos resulted when traffic signals failed to function. During one outage, 15 states were without power. This particular power failure occurred when circuit breakers at a transmission grid were tripped. The interrupted flow of power caused generators further down the line to overload. Automatic systems, activated in the event of an overload, cut power to millions of customers. Six weeks later, a second power-grid failure resulted when sagging transmission lines triggered a domino effect similar to the previous outage. This outage affected ten states and at least 5.6 million people for nearly 24 hours. When millions of people rely upon highly complex and integrated systems such as power transmission and satellites, these systems need careful consideration from a reliability point of view.

Aspect 3. Maintenance of the system

Sinks, showers, and bathtubs are all common items found in homes. These systems are so reliable that we take them for granted—that is, until they break. In many homes, it is these items' very reliability that causes designers and builders to overlook creating access to the plumbing. When no access has been provided, maintenance of the system may require cutting holes in walls. With some forethought about system repair needs, maintenance access panels can be installed.

Aspect 4. Simplicity

The design of the rocket joints on the space shuttle Challenger was based on the highly successful Titan III rocket. The joints of the Titan contained a single O-ring. When adapting the design for the space shuttle, designers felt that adding a second O-ring would make the design even more reliable. Investigation of the design after the explosion revealed that the additional O-ring had contributed to the disaster. Designers have redesigned the rocket to include a *third* O-ring, but will it increase reliability?

Aspect 5. Redundant and fail-safe features

The interchangeability of air filters from the lunar module and the command ship became an issue for the astronauts on Apollo 13. The air filters for the command ship could not be used on the lunar module and vice versa. There were no redundant or backup systems available when the main systems failed. Only through the ingenuity of the scientists and engineers at NASA, who managed to use available materials on both ships to create a substitute filter, was disaster prevented.

Aspect 5. Redundant and fail-safe features

In September of 1991, American Telephone and Telegraph Co. experienced its third major system breakdown. Besides disrupting long-distance service for 1 million people, the system failure abruptly cut off contact between air-traffic controllers and airline pilots in the Northeast. While no serious accidents occurred—fast-thinking pilots communicated their positions to each other—a massive tie-up in air traffic occurred across the country. In the months prior to this disruption, the Federal Aviation Administration had lobbied unsuccessfully to gain permission from the General Services Administration to switch to a phone system with greater reliability and redundancy.

Aspect 5. Redundant and fail-safe features

On December 20, 1997, air traffic over the central United States was disrupted by a multiple-level power failure. Air-traffic-control systems maintain three backup systems. The combined reliability of such systems is considered to be greater than 99.5 percent. On December 20, however, one system was shut down for scheduled maintenance, the commercial power failed, and a technician mistakenly pulled a circuit card from the remaining power system. With no other backups, the air-traffic-control information system, including computers, software, and displays, was inoperable for about five hours. The unexpected failure of the redundant and backup systems needs to be carefully considered as part of overall reliability.

RELIABILITY ENGINEERS

Reliability engineers are challenged to meet customer demands through enhanced product and process performance. Reliability engineers work to design reliability into products by defining product and process configurations, specifying materials and applications, and determining optimal operating methods and environments. During production, reliability engineers test and demonstrate the reliability of systems and their components. In circumstances where limited information is available, they attempt to predict the reliability of components and systems on the basis of information from available tests. Reliability engineers attempt to quantify the optimal maintenance schedule to reduce time between failures and increase the safe use of a product.

During the course of work, a reliability engineer may participate in design reviews; plan and conduct reliability tests; analyze the test data; and use the information gained from the tests to assist production, design engineering, quality assurance, sales, and purchasing. The information from the tests may be used to identify causes of reliability degradation. Reliability engineers may estimate the reliability of a system and try to determine the life cycle of a product. They may write specifications for purchased items or for the items being manufactured.

SUMMARY

The objective of the study of product or system reliability is to attempt to predict its useful life. Reliability tests are conducted to determine if the predicted reliability has been achieved. Reliability programs are designed to improve product or system reliability through improved product design, manufacturing processes, maintenance, and servicing.

■ *Lessons Learned*

1. Reliability refers to quality over the long term. The system's intended function, expected life, and environmental conditions all play a role in determining system reliability.
2. The three phases of a product's life cycle are early failure, chance failure, and wear-out.
3. Reliability tests aid in determining if distinct patterns of failure exist.
4. Failure rates can be determined by dividing the number of failures observed by the sum of their test times.
5. θ, or the average life, is the inverse of the failure rate.
6. A system's availability can be calculated by determining the mean time to failure and dividing that value by the total of the mean time to failure plus the mean time to repair.
7. Reliability is the probability that failure will not occur during a particular time period.
8. For a system in series, failure of any one component will cause system failure. The reliability of a series system will never be greater than that of its least reliable component.
9. For a system in parallel, all of the components must fail to have system failure. The reliability in a parallel system will be no less than the reliability of the most reliable component.

10. Overall system reliability can be increased through the use of parallel, backup, or redundant components.
11. Reliability programs are enacted to incorporate reliability concepts into system designs. A well-thought-out reliability program will include the eight considerations presented in this chapter. ■

Formulas

$$\lambda_{\text{estimated}} = \frac{\text{number of failures observed}}{\text{sum of test times or cycles}}$$

$$\theta_{\text{estimated}} = \frac{1}{\lambda}$$

or

$$\theta_{\text{estimated}} = \frac{\text{sum of test times or cycles}}{\text{number of failures observed}}$$

$$\text{Availability} = \frac{\text{mean time to failure (MTTF)}}{\text{MTTF + mean time to repair}}$$

(MTBF values can be used in place of MTTF.)

$$R = \frac{s}{n}$$

$$R_s = r_1 \cdot r_2 \cdot r_3 \cdot \cdots \cdot r_n$$

$$R_p = 1 - (1 - r_1)(1 - r_2)(1 - r_3)\cdots(1 - r_n)$$

$$R_b = r_1 + r_b(1 - r_1)$$

Chapter Problems

1. Define reliability in your own words. Describe the key elements and why they are important.
2. Study your house. What do you consider key convenience items to be? How reliable do you consider them to be? What would you do if they failed? How can you make them more reliable?
3. Describe the three phases of the life history curve. Draw the curve and label it in detail (the axes, phases, type of product failure, etc.).

Failure Rate, Mean Life, and Availability

4. Determine the failure rate λ for the following: You have tested circuit boards for failures during a 500-hour continuous-use test. Four of the 25 boards

failed. The first board failed in 80 hours, the second failed in 150 hours, the third failed in 350 hours, the fourth in 465 hours. The other boards completed the 500-hour test satisfactorily. What is the mean life of the product?

5. Determine the failure rate for a 90-hour test of 12 items where 2 items fail at 45 and 72 hours, respectively. What is the mean life of the product?

6. A power station has installed ten new generators to provide electricity for a local metropolitan area. In the past year (8,776 hours), two of those generators have failed, one at 2,460 hours and one at 5,962 hours. It took five days, working 24 hours a day, to repair *each* generator. Using one year as the test period, what is the mean time between failure for these generators? Given the repair information, what is the availability of all ten generators?

System Reliability

7. Why is a parallel system more reliable than a system in series?

8. Given the system in Figure 1, what is the system reliability?

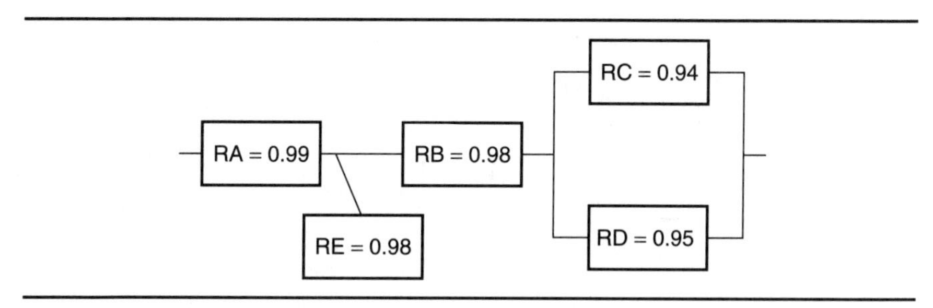

Figure 1 Problem 8

9. Given the diagram of the system in Figure 2, what is the system reliability?

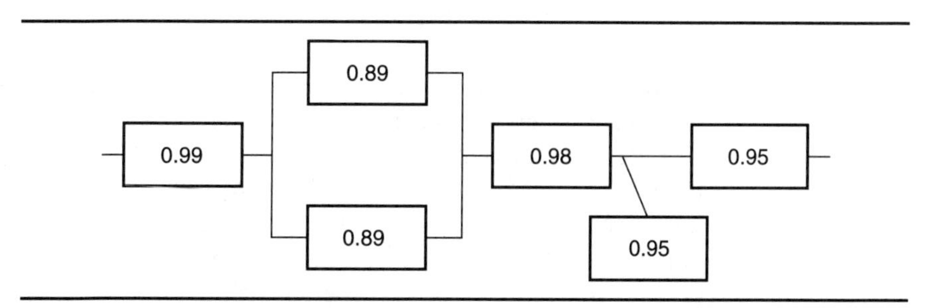

Figure 2 Problem 9

10. What is the reliability of the system in Figure 3?

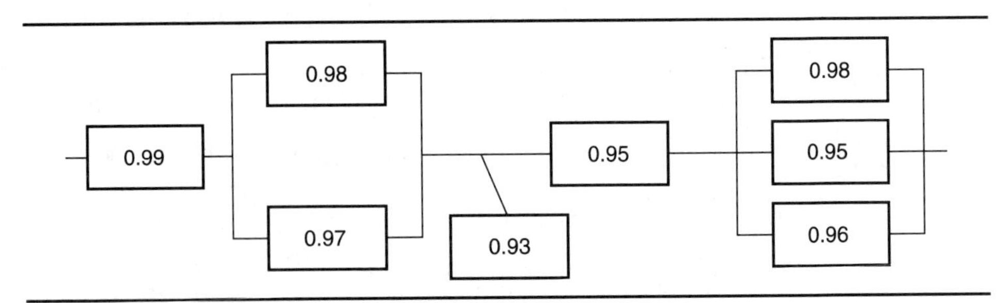

Figure 3 Problem 10

11. A paranoid citizen has installed the home-alert system shown in Figure 4. What is the overall system reliability?

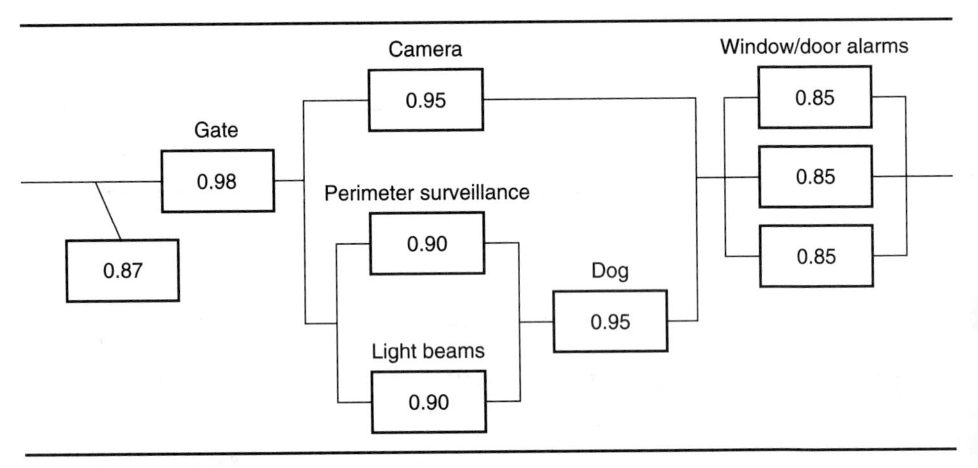

Figure 4 Problem 11

12. Pilots prefer to land airplanes with the wheels down and locked into position. A "wheels-down" landing is significantly more comfortable than a "belly slide"; it also gives the passenger a much more positive impression of the pilot's capabilities. In order to ensure a wheels-down landing, several systems work in series and parallel. A landing-gear-down light signals when the gear are in a down-and-locked position. Often pilots carry backup lightbulbs because the reliability of each bulb is only 0.75. The hydraulic system contains two parallel hydraulic pumps with a reliability of 0.98 each. The hydraulic system operates when signaled by an electric impulse from the gear-down switch. This switch is in series with the rest of the system and has a reliability of 0.96. What is the reliability of the entire system?

13. Pilots also have the option of bypassing the entire system described in Problem 12 with a hand crank. This hand crank serves as the final backup when the entire electric and hydraulic system fails. The hand crank operates with a reliability of 0.95. Recalculate the reliability of the system, including the hand-crank backup.
14. The next-door neighbors are currently installing a new cistern that will provide water for their household. The different components for the system and their reliability are described in the diagram in Figure 5. Determine the reliability of the system.

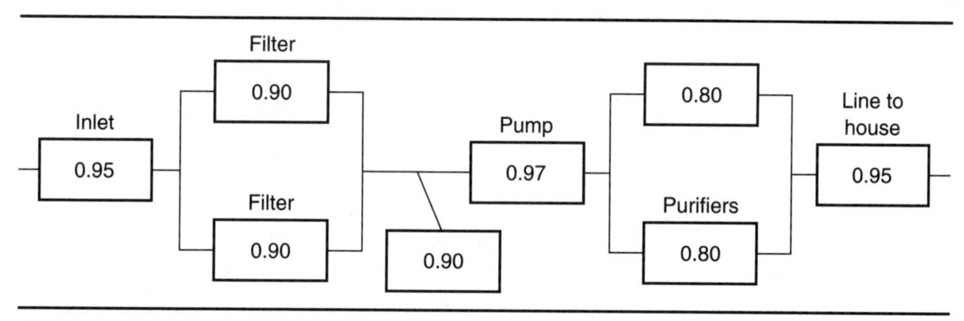

Figure 5 Problem 14

CASE STUDY 10.1
Reliability: The Entire System

In February of 1965, Boeing began to produce the 737 twin-engine jet. This jet, a shorter version of the 707/727 with a more radically swept-back wing, filled a market need for a more fuel-efficient, no-frills aircraft. The smaller size of the 737 enables the jet to fly cost-effectively to smaller airports and more remote locations. Three decades and a few design and fuel-efficiency modifications later, the 737 continues to be manufactured. Over the years, it has become the best-selling aircraft ever built.

Of the three thousand 737s created, most are still flying, which means that many have lasted over a quarter of a century. When the 737 was originally conceived, engineers at Boeing predicted a life expectancy of 75,000 flights, or takeoff-and-landing cycles. Based on the behavior of the current planes still flying long after the predicted 75,000 flights, a Boeing 737 may last as long as 195,000 cycles (flights).

Besides longevity, the 737 jets have the advantage of cockpit design and systems commonality with other Boeing jets. The marked similarity in cockpit design has simplified pilot training. Cockpit similarities allow pilots to transfer easily from cockpit to cockpit, and cockpit familiarity increases flight safety since in emergency situations pilots are not confused by control-surface differences.

Cockpit and systems commonality provides the benefits of interchangeable parts, components, subassemblies, and subsystems. Maintenance is easier since repair people familiar with one plane can transfer that knowledge to another. Parts availability, and therefore aircraft availability, increase with interchangeable parts.

Redundant and fail-safe systems are also a part of the Boeing 737. The plane has two engines but is able to fly with only one. The two engines support two electrical systems. Two fuel systems exist on the 737. If a loss of pressurization occurs on an aircraft, an emergency backup oxygen system exists.

All aircraft, whether private or commercial, maintain aircraft logs. In these logs, pilots record all flights and information pertaining to the length of flight, weather conditions, system failures, or repairs needed. The logs provide a record of routine maintenance, as well as information about loss of pressurization, turbine failure, engine overheat, false alarms with the warning lights, and other such failures. Extensive tests at Boeing and field-tested information from the 3,000 aircrafts' actual logs of flight experiences have been used to make system design changes, improve manufacturing methods, modify quality checks, and select the most reliable components.

In an industry where 10 to 15 percent of the operating costs are devoted to maintenance, these aircraft are maintained in such a manner that they will last indefinitely. Planes are systematically inspected, cleaned, repaired, reinforced, rebolted, and resealed. This type of maintenance exceeds original factory standards and protects the planes from the deterioration due to wear and tear and corrosion. However, as the planes age, the maintenance cost per flight hour increases significantly. The airline industry rule of thumb says that at the 25-year mark, a plane's maintenance will cost approximately double the maintenance costs associated with a new plane.

The mean life of a jet engine is approximately 15,000 hours. Jet engines can be rebuilt, overhauled, or replaced to prolong the life of the entire jet. With the Boeing 737, pilots and mechanics have discovered that engine wear can be reduced by decreasing the amount of engine thrust by 4 to 5 percent, enabling the engine to run 30 to 40 degrees cooler. Communication of this sort has resulted in significantly reduced engine wear and prolonged engine life.

Assignment

Reliability programs have been used to prolong the lives of Boeing 737 airplanes. Using this chapter as a guide, describe the elements of Boeing's reliability program. Then discuss the 737 in terms of the definition of reliability. Be sure to discuss the plane's intended function and environmental conditions.

CASE STUDY 10.2
Reliability

In mid-1998, the Federal Aviation Administration (FAA) ordered the immediate inspection of Boeing's 737 aircraft. This order grounded 15 percent (297 planes) of the 737s flown in the United States. The planes in question were 737-100s and 737-200s with more than 50,000 hours, as well as later models (737-300, 400, and 500) with between 40,000 and 50,000 hours.

A Boeing 737 has two fuel tanks, one in each wing. Two fuel pumps move fuel out of each tank. Even though the wires are encased in conduit (Figure 1), vibrations prevalent during flights cause the wires to chafe against each other and wear away the protective insulation.

Affected planes were not to be returned to service until a complete inspection had been made of their wiring and, when needed, repairs made. Mechanics searched for wear on the insulation of the fuel-pump wires and the wires themselves. Worn wires were to be replaced and an additional Teflon cover was to be put in place to add protection. On average, this inspection and repair took 15 hours to complete.

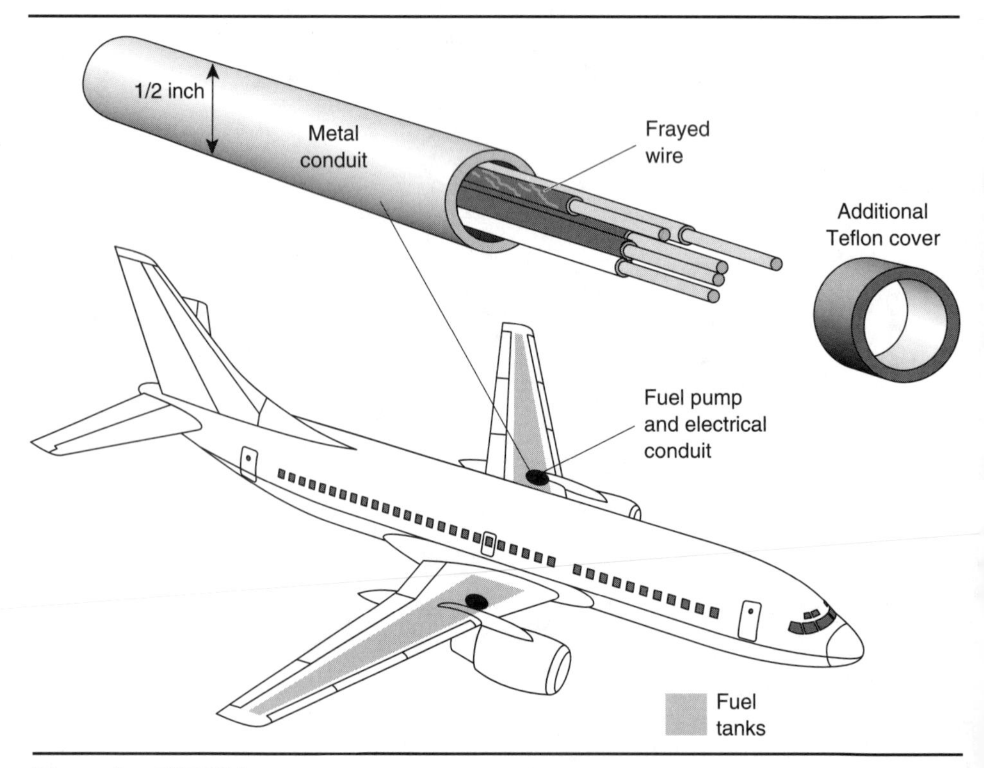

Figure 1 737 Wiring

This FAA action was prompted when a routine inspection found a wiring bundle that showed clear signs of arcing in one area and bare wires in another. These problems were found on a 737 with 60,000 hours of flying time. A similar situation was found on another 737 with 74,000 hours of flying time. In this instance, mechanics found fuel leaking out of two pin-sized holes in a tube that houses wires running through one of the plane's wing fuel tanks. The combination of fuel, electricity, and air could ignite a fuel tank. In the total sample of 26 planes, 13 of the 26 main pump lines inspected showed signs of wear on the wire insulation.

These wiring problems are related to the aging of the aircraft. Nearly all the more than 3,000 737s built are still flying. Many of them are more than a quarter of a century old. The 737, a highly respected workhorse of a plane, was expected to have a life expectancy of 75,000 flights (takeoff and landing cycles). Based on its current performance, this relatively trouble-free aircraft may last as long as 195,000 cycles. Longevity has its complications though, as this wiring example shows.

Assignment

Investigate this 1998 incident with the Boeing 737 aircraft. If we consider the long life of the 737 aircraft, what reliability considerations might have been overlooked during the design, construction, or use of the 737s? In other words, what could have been done differently to improve the 737 aircraft reliability?

Assignment

Calculate the mean life and availability associated with the 26 airplanes sampled. Use the information provided as well as the following assumptions:

- A failure is considered to be any sign of worn wiring or insulation.
- Of the 13 failures, one occurred at 60,000 hours, the second at 74,000 hours, and the remainder at 50,000 hours.
- The remaining planes showed no signs of failure when the test was terminated at 75,000 hours.
- The average time to inspect and/or repair any of the planes was 15 hours.

CASE STUDY BIBLIOGRAPHY

Adcock, S. "FAA Orders Fix on Older 737s." *Newsday*, May 8, 1998.

Field, D. "FAA Grounds Oldest 737's." *USA Today*, May 11, 1998.

Field, D. "Order Causes Little Turbulence." *USA Today*, May 11, 1998.

McMarthy, M. "No. 19603 Still Flies After 27 Years' Service to a Number of Airlines." *The Wall Street Journal*, August 9, 1995.

"U.S. FAA: FAA Orders Immediate Inspection for High-time Boeing 737s, Extends Inspection Order, M2." *PressWIRE*, May 11, 1998.

11

Quality Costs

■ *Learning Opportunities:*

1. To familiarize the reader with the concept of quality costs
2. To create an understanding of the interrelationships between and among the different types of quality costs
3. To gain an understanding of how quality costs can be used in decision making. ■

Ship Meets Iceberg

On an April night in 1912 the Titanic *struck an iceberg. The ship thought to be unsinkable sank in a little over four hours. In today's business environment, a provider of goods or services must carefully navigate around icebergs that threaten to sink a company. Quality costs are similar to icebergs in that the majority of the costs associated with poor-quality products or services are hidden from view. This chapter defines the cost of failing to provide a quality product or service.*

WHAT ARE QUALITY COSTS?

Two factions often find themselves at odds with each other when discussing quality. There are those who believe that no "economics" of quality exists, that it is never economical to ignore quality. There are others who feel it is uneconomical to have 100 percent perfect quality all of the time. Cries of "Good enough is not good enough" will be heard from some, while others will say that achieving perfect quality will bankrupt a company. Should decisions about the level of quality of a product or service be weighed against other factors such as meeting schedules and cost? To answer this question, an informed manager needs to understand the concepts surrounding the costs of quality. Investigating the costs associated with quality provides managers with an effective method to judge the economics and viability of a quality-improvement system. Quality costs serve as a baseline and a benchmark for selecting quality-improvement projects and for later evaluating their success.

DEFINING QUALITY COSTS

A variety of terms are used to describe the costs associated with providing a quality product or service, including cost of quality, poor-quality cost, cost of poor quality, and costs related to quality. A ***quality cost*** *is considered to be any cost that the company would not have incurred if the quality of the product or service were perfect. Quality costs are the portion of the operating costs brought about by providing a product or service that does not conform to performance standards. Quality costs are also the costs associated with the prevention of poor quality.* The most commonly listed costs of quality include scrap, rework, and nonconformities. As Figure 11.1 shows, these easily identified quality costs are merely the tip of the iceberg.

Quality costs can originate from nearly anywhere within a company. No single department has the corner of the market on making mistakes that might affect the quality of a product. Even departments far removed from the day-to-day operations of a firm can affect the quality of a product or service. The receptionist, often the first person the customer has contact with, can affect a customer's perceptions of the firm. The cleaning people provide an atmosphere conducive to work. Mistakes, oversights, and errors can and will affect the quality of an item. Salespeople must clearly define the customer's needs as well as the capabilities of the company. Figure 11.2 provides a few more examples of quality costs. Every department within a company should identify, collect, and monitor quality costs within their control.

Quality, in both the service and manufacturing industries, is a significant factor in allowing a company to maintain and increase its customer base. As a faulty product or service finds its way to the customer, the costs associated with the error increase. Preventing the nonconformity before it is manufactured or prepared to serve the customer is the least costly approach to providing a quality product to service. Potential problems should be identified and dealt with during the design and planning stage. An error at this phase of product development will require effort to solve, but in most cases the changes can be made before costly investments in equipment or customer service are made.

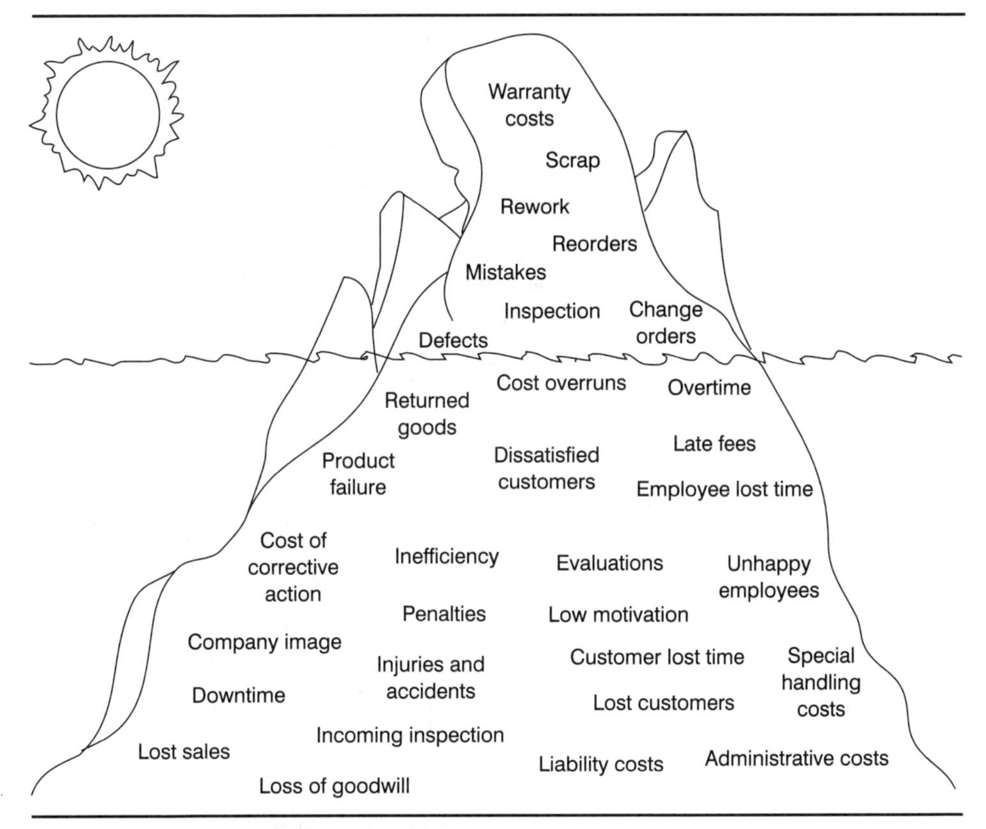

Figure 11.1 The Iceberg of Quality Costs
SOURCE: Adapted from *Principles of Quality Costs,* ed. J. Campanella. ASQC Quality Press.

Rewriting an insurance policy to match customer expectations
Redesigning a faulty component that never worked right
Reworking a shock absorber after it was completely manufactured
Retesting a computer chip that was tested incorrectly
Rebuilding a tool that was not built to specifications
Repurchasing because of nonconforming materials
Responding to a customer's complaint
Refiguring a customer's bill when an error was found
Replacing a shirt the dry cleaner lost
Returning a meal to the kitchen because the meat was overcooked
Retrieving lost baggage

Figure 11.2 Examples of Quality Costs

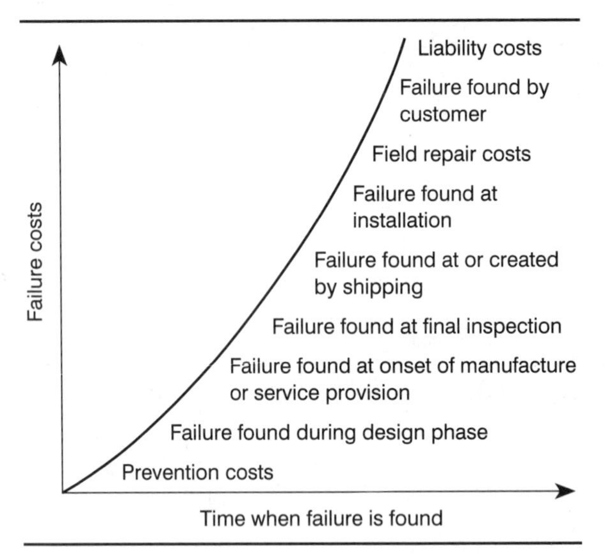

Figure 11.3 Increases in Quality Costs as Faulty Product Reaches Consumer

If the nonconformity with the product or service is located during the manufacturing cycle, or behind the scenes at a service industry, then the nonconformity can be corrected internally. The cost is greater here than at the design phase because the product is in some stage of completion. Product may need to be scrapped or, at a minimum, reworked to meet the customer's quality expectations.

If the nonconforming product or service reaches the customer, the company providing it incurs the greatest costs. Customers who locate this nonconformity will experience a variety of feelings, the least of which will be dissatisfaction. Even if the customers are later satisfied by providing them with a new product, repairing their existing product, or repeating the service, the damage has already been done to the company's reputation. Disgruntled customers rarely fail to tell others about their experience. Mistakes that reach the customer result in the loss of goodwill between the present customer, future customers, and the provider. Failure cost increases as a function of the failure's detection point in the procss (Figure 11.3).

EXAMPLE 11.1 The Cost of Poor Quality

Jim and Sue Homeowners recently purchased a home on a large lot. Using their small push mower to cut the lawn proved to be a tedious, time-consuming job, so over the winter, they saved to purchase a riding lawn mower. Early in spring, they checked a variety of lawn mower sales and finally settled on Brand X. Unknown to them, this particular mower had a design flaw that had gone unnoticed by the manufacturer. The mower deck, which covers the blades, is held onto the mower through the use of bolts. Because there is too little clearance between the mower-deck fastener bolts and the blade, under certain circumstances the blades come in contact with the bolts, shearing them off and releasing the mower deck. The damage to the blades is unrepairable.

This design error was not apparent in the store lot where the Homeowners tried the mower. The lot was smooth and the problem only becomes obvious when the mower is used on uneven ground. The Homeowners went home to try out their new lawn mower and within minutes the mower was moving along uneven terrain. A loud shearing sound was heard, the mower deck fell off, mangling the blades in the process. The Homeowners immediately called the salesperson, who, understanding the importance of happy customers, quickly sent a repair person to investigate.

The repair person identified the problem, returned to the store with the mower, and replaced the first mower with a second. Even though the repair person knew what had caused the failure, he had no way of informing the manufacturer. Previous attempts to contact the manufacturer to give them information about other problems had not been successful. The manufacturer thus lost a valuable bit of information that would have improved the design.

Although they had used most of a Saturday getting the problem resolved, the Homeowners felt satisfied with the service they had received . . . until they tried the new mower. The new mower experienced the same problem on a different patch of grass. This time, the store was unable to replace the mower until the next weekend. The Homeowners resorted to their push mower.

The next Saturday was not a good day. Another mower was promptly delivered and tried out in the presence of the salesperson and the repair person. Same problem. The Homeowners returned the mower and resolved to purchase no other yard-care products—no weed whips, wagons, push mowers, or edging tools—from this company. At a barbecue that weekend, they told their neighbors about their troubles with Brand X products. Over the next few weeks, the Homeowners investigated the types of lawn mowers used by yard-service firms and golf courses. Even though it was a bit more expensive, they bought Brand Y because of its proven track record and quality reputation.

In this situation, an unsuitable product reached a consumer. Quality service from the store could not compensate for the original design flaw. As Figure 11.4 shows, failure to spot the design flaw during the design review, or during production and assembly, or during testing dramatically increased costs. The quality costs associated with providing a poorly designed lawn mower included here not only the three damaged lawn mowers but also sales of related equipment to the original customers and future sales to their friends.

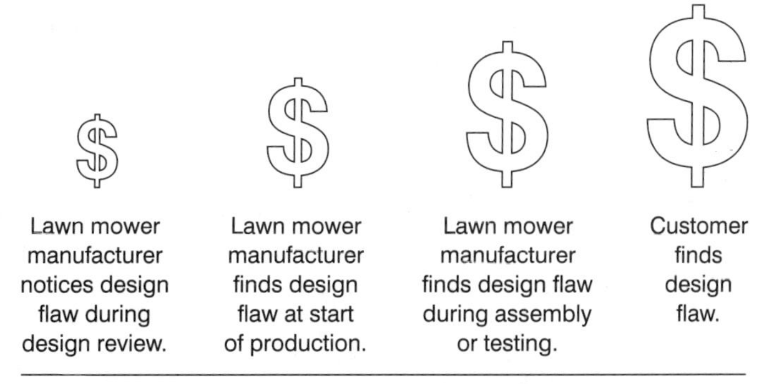

Figure 11.4 Increasing Costs Associated with Faulty Lawn Mower

TYPES OF QUALITY COSTS

It is possible to categorize the costs that a manufacturer of goods or a provider of services incurs when creating a quality product or service. The costs associated with preventing nonconformities from occurring, with appraising products or services as they are being produced or provided, and with failure can be defined. With an understanding of these costs, a manager can make decisions concerning the implementation of quality improvement projects. Using quality costs, a manager can determine the usefulness of investing in a process, changing a standard operating procedure, or revising a product or service design.

Prevention Costs

Prevention costs *are those costs that occur when a company is performing activities designed to prevent poor quality in products or services*. Prevention costs are often seen as front-end costs designed to ensure that the product or service is created to meet the customer requirements. Examples of such costs are design reviews, education and training, supplier selection and capability reviews, process capability reviews and quality improvement projects. Prevention activities must be reviewed to determine if they truly bring about improvement in the most cost-effective manner.

Prevention efforts try to determine the root causes of problems and eliminate them at the source so reoccurrences do not happen. Preventing poor quality stops companies from incurring the cost of doing it over again. Essentially, if they had done it right the first time, they would not have to repeat their efforts. In his book *Quality Is Free*, Philip Crosby emphasizes the need to invest in preventing quality problems in order to reap the benefits of reduced appraisal and failure costs.* The initial investment in improving processes is more than compensated by the resulting cost savings, so, in essence, quality is free.

Appraisal Costs

Appraisal costs *are the costs associated with measuring, evaluating, or auditing products or services to make sure that they conform to specifications or requirements*. Appraisal costs can be considered to be the cost of evaluating the product or service during the production of the product or the providing of the service to determine if, in its unfinished or finished state, it is capable of meeting the requirements set by the customer. Appraisal activities are necessary in an environment where poor quality is found. Appraisal costs can be found anywhere the raw materials, work-in-process (activities-in-process for the service industries), or finished product is reviewed, inspected, or otherwise checked to ensure that it meets requirements. Examples of appraisal costs include incoming inspection, work-in-process inspection, final inspection or testing, material reviews, and calibration of measuring or testing equipment. When the quality of the product or service reaches high levels, then appraisal costs can be reduced.

*Philip Crosby, *Quality Is Free*. New York: McGraw-Hill, 1979.

Failure Costs

Failure costs *occur when the completed product or service does not conform to customer requirements*. Two types exist: internal and external. ***Internal failure costs*** *are those costs associated with product nonconformities or service failures found before the product is shipped or the service is provided to the customer.* Internal failure costs are the costs of correcting the situation. This failure cost may take the form of scrap, rework, remaking, reinspection, or retesting. ***External failure costs*** *are the costs that occur when a nonconforming product or service reaches the customer.* External failure costs include the costs associated with customer returns and complaints, warranty claims, or product recalls. Since external failure costs have the greatest impact on the corporate pocketbook, they must be reduced to zero. Because they are highly visible, external costs often receive the most attention. Unfortunately, internal failure costs may be seen as necessary evils in the process of providing good-quality products to the consumer. Nothing could be more false. Doing the work twice, through rework or scrap, is not a successful strategy for operating in today's economic environment.

Intangible Costs

How a customer views a company and its performance will have a definite impact on long-term profitability. ***Intangible costs,*** *the hidden costs associated with providing a nonconforming product or service to a customer, involve the company's image*. Intangible costs of poor quality, because they are difficult to identify and quantify, are often left out of quality-cost determinations. They must not be overlooked, or disregarded. Is it possible to quantify the cost of missing an important deadline? What will be the impact of quality problems or schedule delays on the company's image? Intangible costs of quality can be three or four times as great as the tangible costs of quality. Even if these costs can only be named, and no quantifiable value can be placed on them, it is important for decision makers to be aware of their existence.

The four types of quality costs are interrelated. In summary, ***total quality costs*** *are considered to be the sum of prevention costs, appraisal costs, failure costs, and intangible costs*. Figure 11.5 shows some of the quality costs from Figure 11.1 in their respective categories. Investments made to prevent poor quality will reduce internal and external failure costs. Consistently high quality reduces the need for many appraisal activities. Suppliers with strong quality systems in place can reduce incoming inspection costs. High appraisal costs combined with high internal failure costs signal that poor-quality products or services are being produced. Efforts made to reduce external failure costs will involve changes to efforts being made to prevent poor quality. Internal failure costs are a portion of the total production costs, just as external failure costs reduce overall profitability. A trade-off to be aware of when dealing with quality costs is the need to ensure that appraisal costs are well spent. Companies with a strong appraisal system need to balance two points of view: Is the company spending too much on appraisal for its given level of quality performance or is the company risking excessive failure costs by underfunding an appraisal program? In all three areas—prevention, appraisal, and failure costs—the activities undertaken must be evaluated to

Prevention Costs

- Quality planning
- Quality program administration
- Supplier-rating program administration
- Customer requirements/expectations market research
- Product design/development reviews
- Quality education programs
- Equipment and preventive maintenance

Appraisal Costs

- In-process inspection
- Incoming inspection
- Testing/inspection equipment
- Audits
- Product evaluations

Internal Failure Costs

- Rework
- Scrap
- Repair
- Material-failure reviews
- Design changes to meet customer expectations
- Corrective actions

External Failure Costs

- Returned goods
- Corrective actions
- Warranty costs
- Customer complaints
- Liability costs
- Penalties

Intangible Costs

- Customer dissatisfaction
- Company image
- Lost sales
- Loss of customer goodwill

Figure 11.5 Categories of Quality Costs

ensure that the efforts are gaining further improvement in a cost-effective manner. Figure 11.6 reveals that as quality costs are reduced or—in the case of prevention quality costs—invested wisely, overall company profits will increase. The savings associated with doing it right the first time show up in the company's bottom line.

EXAMPLE 11.2 Uncovering Quality Costs

A sheet metal stamping firm recently bid $75,000 as the price of providing an order of aluminum trays. After stamping a large number of these trays, the company investigated their costs and determined that if production continued, the cost of providing the order would be $130,000! This was nearly double what they had bid.

They immediately set out to figure out what was causing such a dramatic increase in their costs. They investigated their product design and found it to be satisfactory. Prevention costs associated with product design and development were negligible. The customer had not returned any of the completed trays. There were no customer complaints and therefore no corrective-action costs. Their investigation showed that external failure costs were zero. Appraisal costs were normal for a product of this type. Incoming material and in-process inspection costs were similar to the amount projected. The product had not required any testing upon completion.

It was only when they investigated the internal failure costs that the problem became apparent. At first, they felt that internal failure costs were also negligible since there was almost no scrap. It wasn't until a bright individual observed the operators

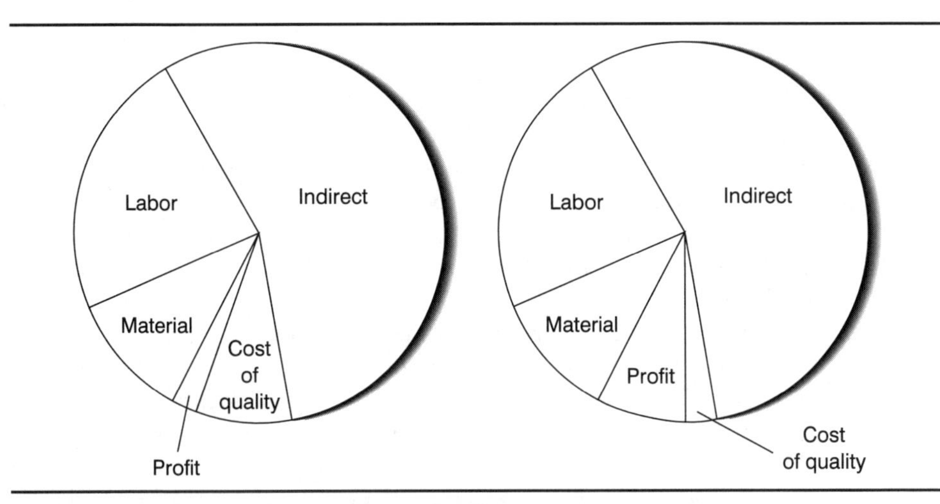

Figure 11.6 Cost of Quality versus Profit

feeding the trays through the stamping operation twice that they were able to locate the problem. When questioned, operators told the investigators that they knew that the product was not stamped to the correct depth on the first pass. Since the operators knew that the trays would not pass inspection, they decided among themselves to rework the product immediately, before sending it to inspection. The near doubling of the manufacturing costs turned out to be the internal failure costs associated with reworking the product.

Essentially, Company X incurred a cost of quality equal to the standard labor and processing costs of the operation. To meet specifications, a significant number of trays were being handled and processed twice. Only by preventing the need for double processing can the costs of quality be reduced.

This type of internal failure cost actually began as a prevention cost. If the manufacturer had investigated the capability of the stamping operation more closely, they would have determined that maintenance work was necessary to improve the machines' performance and therefore improve product quality.

QUALITY-COST MEASUREMENT SYSTEM

In the struggle to meet the three conflicting goals of quality, cost, and delivery, identifying and quantifying quality costs helps ensure that quality does not suffer. When quality costs are discussed as a vague entity, their importance in relation to cost and delivery is not understood. By quantifying quality costs, all individuals producing a product or providing a service understand what it will cost if quality suffers.

A quality-cost measurement system should be designed to keep track of the different types of quality costs. Being able to define and quantify quality costs enables quality to be managed more effectively. Once quantified, these costs can be used to determine which projects will allow for the greatest return on investment and which projects have been most effective at lowering failure and appraisal quality costs. A quality-cost measurement system should use quality costs as a tool to help justify improvement actions.

Quality costs will help to zero in on which problems, if solved, will provide the greatest return on investment. A quality-cost measurement system should try to capture and utilize significant quality costs. Cost-improvement efforts should be focused on locating the largest cost contributors to the improvement. Try not to be tempted to spend time chasing insignificant, incremental measures of quality costs. Use a Pareto analysis to identify the projects with the greatest return on investment; those should be tackled first.

Those establishing a quality-cost measurement system need to keep in mind that the success of such a system is based on using quality-cost information to guide improvement. Each cost has a root cause that can be identified and prevented in the future. Remember that preventing quality problems in the first place is always cheaper. Those implementing a quality-cost measurement system need to establish a basic method of identifying correctable problems, correcting the problems, and achieving a new level of performance.

UTILIZING QUALITY COSTS FOR DECISION MAKING

Quality costs can be used as a justification for actions taken to improve the product or service. Typically, investments in new equipment, materials, or facilities require the project sponsor to determine which projects will provide the greatest return on investment. These calculations traditionally include information on labor savings, production time savings, and ability to produce a greater variety of products with better quality. The "better quality" aspect of these calculations can be quantified by investigating the costs of quality, particularly the failure costs. It is important to determine the costs of in-process and incoming material inspection, sorting, repair, and scrap as well as the intangible costs associated with having a nonconforming product or service reach the customer. Making a decision with more complete quality information, such as product appraisal costs, can help determine the true profitability of a product or service.

Once quality costs are identified, those reviewing the project can determine if the money for the project is being well spent to prolong the growth of the company. Identifying and quantifying quality costs has a twofold benefit. Cost savings are identified and quality is improved. By improving the quality performance of a company, the company also improves its quality costs.

SUMMARY

With any quality-measurement or improvement system, the information and improvements realized need to be applied to any future activities. Deming's plan-do-study-act cycle plays a significant role. Once improvements are planned and made, they must be acted upon to ensure that future activities maintain the new level of quality and the decreased level of quality costs. Identifying quality costs will allow management to judge quality improvement investments and profit contributions. A quality program is only as valuable as its ability to contribute to customer satisfaction and, ultimately, to company profits.

Lessons Learned

1. Collecting quality costs provides a method to evaluate system effectiveness.
2. Quality costs can serve as a basis or benchmark for measuring the success of internal improvements.
3. Prevention costs are the costs associated with preventing the creation of errors or nonconformities in the first place.
4. Appraisal costs are the costs incurred when a product or service is studied to determine if it conforms to specifications.
5. Internal failure costs are the costs associated with errors found before the product or services reach the customer.
6. External failure costs are the costs associated with errors found by the customer.
7. Intangible failure costs are the costs associated with the loss of customer goodwill and respect.
8. Once quantified, quality costs enhance decision making if they are used to determine which projects will allow for the greatest return on investment and which projects are most effective at lowering failure and appraisal quality costs.
9. Quality-cost information should be used to guide improvement. ■

Chapter Problems

1. Which of the following statements about quality costs is true? Why, why not?
   a. Quality cost reduction is the responsibility of the quality control department.
   b. Quality costs provide a method of determining problem areas and action priorities.
   c. The efficiency of any business is measured in terms of the number of defective products.
2. There are four specific types of quality costs. Clearly define two of them. Give an example of each kind of quality cost you have defined.
3. What is a prevention cost? How can it be recognized? Describe where prevention costs can be found.
4. Describe the two types of failure costs. Where do they come from? How will a person recognize either type of failure cost?
5. How do the four types of quality costs vary in relation to each other?
6. Describe the relationship between prevention costs, appraisal costs, and failure costs. Where should a company's efforts be focused? Why?

7. A lamp manufacturing company has incurred the following quality costs. Analyze how this company is doing on the basis of these costs. Creating a graph of the values may help you see trends. Values given are in percentage of total cost of lamp.

	Costs			
Year	*Prevention*	*Appraisal*	*Internal Failure*	*External Failure*
1	1.2	3.6	4.7	5.7
2	1.6	3.5	4.3	4.6
3	2.2	3.8	4.0	3.8
4	2.5	2.7	3.5	2.2
5	3.0	2.0	2.8	0.9

8. How can quality costs be used for decision making?
9. Where should dollars spent on quality issues be invested in order to provide the greatest return on investment? Why?
10. Choose two industries and list specific examples of their costs of quality.
11. Why is the following statement true? The further along the process that a failure is discovered, the more expensive it is to correct.
12. What should a quality-cost program emphasize?
13. What are the benefits of having, finding, or determining quality costs?
14. On what premise is the strategy for using quality costs based?
15. For each of the following activities, identify whether the costs incurred should be allocated to the prevention, appraisal, failure (internal, external), or intangible costs category:
    a. Customer inquires about incorrect billing statement.
    b. Maintenance technician is unable to gain access to customer's premises during a service call because the service is not being performed when scheduled and no one notified the customer.
    c. Workers check/test their work before allowing it to leave their work area.
    d. Management carries out a quality systems audit.
    e. The accounts department is having difficulty obtaining monthly figures from the production department.
    f. A customer's purchase breaks because of inadequate training in the use of the item and must be replaced.
    g. Management must deal with written complaints from a customer about continuous poor service.

h. A technician keeps an appointment after driving across town to pick up equipment from an outside supplier upon finding it out of stock in his own warehouse.
i. An instruction manual and set of procedures are developed for maintaining a new piece of equipment.
j. The company conducts a survey of customer requirements.

CASE STUDY 11.1
Costs of Quality

PART 1

The medicinal supplies distribution department of a large hospital provides the nursing floors and medical departments with all of the supplies needed to service the patients requiring medical attention. These items include medicines, bandages, I.V. tubing, catheters, syringes, tourniquets, strainers, scalpels, swabs, I.D. bracelets, and other items. Each floor and department has different medicinal needs, and carts carry supplies to the floor or area.

Over time, the medicinal supplies distribution department's workload has increased, while staff size has remained constant. Concerns over staffing problems, budget constraints, and maintaining the quality of service have prompted the formation of a continuous improvement task force to investigate the problems. Their mission is to provide quick, accurate service to the medical staff while continuously improving the provision process to make it more efficient, more compatible with medical staff needs, and more cost-effective.

To begin the problem-solving process, a flowchart of the process of supplying medical items is created (Figure 1). This chart details the activities involved in stocking the carts, checking inventory on the carts, reordering necessary items, and taking carts to their respective floors. Records of the items in inventory are kept by a computerized inventory-control system. When a patient uses items from the cart, nurses and other medical staff record the chargeable items on the patient's chart. These items are billed to the patient when accounts payable figures the bill. Computer inventory records are compared with a complete list of items charged to all patients' bills. When inventory records and items taken from the cart and billed don't match, a lost charge report is generated. This report is used by the medical staff to determine which patients should be charged for what items.

While patient care is first and foremost in importance, significant costs from incorrect charges are being incurred by the hospital. Other problems with medicinal supplies distribution involve items not found on carts, items out of stock on carts, incorrect inventory counts leading to incorrect reordering or product spoilage, and lengthy waits for items to be brought to the floor from inventory storage (Figure 2).

Task force members have decided to investigate this situation from a cost-of-quality point of view. At first, some of the group members are concerned about ensuring that quality patient care is not overlooked. Others feel that the emphasis on quality patient care has caused costs to be ignored. Fortunately, the task force leader understands the use of quality costs in decision making. Using quality costs, a manager can determine the usefulness of investing in new equipment, changing a standard

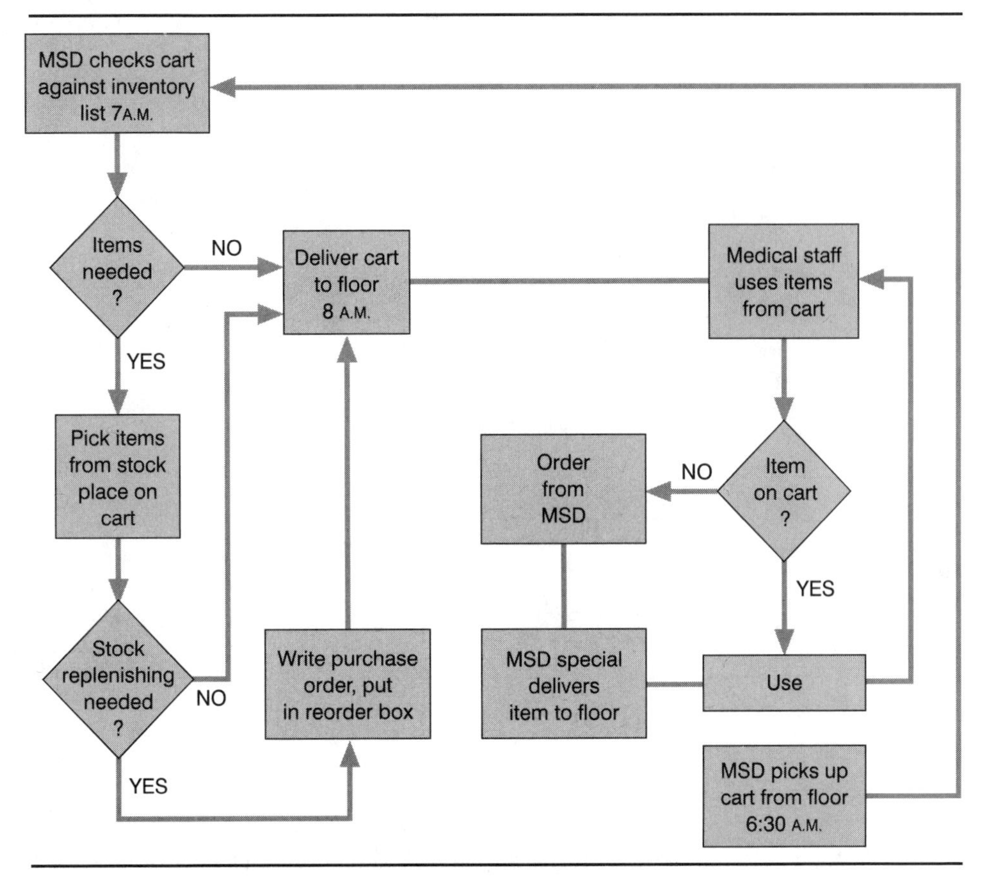

Figure 1 Flowchart: Process of Supplying Medical Items

operating procedure, or revising a service design. The task force intends to use quality costs to guide the changes they will make to the inventory system.

The first step is to convince group members to study this situation from a cost-of-quality point of view. The leader reminds group members that quality costs are any cost that the hospital would not have incurred if the quality of the product or service were perfect. Quality costs are the portion of the operating costs brought about by providing a medical service that does not conform to performance standards. Quality costs are also the costs associated with the prevention of poor quality. To identify quality costs, the leader encourages a brainstorming session.

Assignment

From reading the case, and from your knowledge of hospitals, identify prevention costs, appraisal costs, failure costs, and intangible costs in this situation.

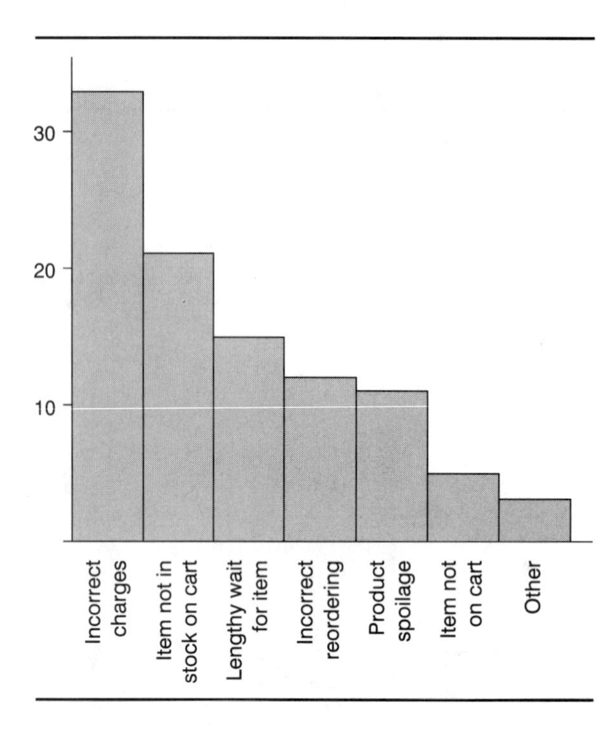

Figure 2 Pareto Analysis of One Week's Calls to the Medicinal Supplies Distribution Department

PART 2

By investigating hospital records, the task force captures the dollars associated with the quality costs identified in part 1. Their relative values are shown in Figure 3. This figure clearly shows that the appraisal and failure costs associated with this problem necessitate investigating the situation.

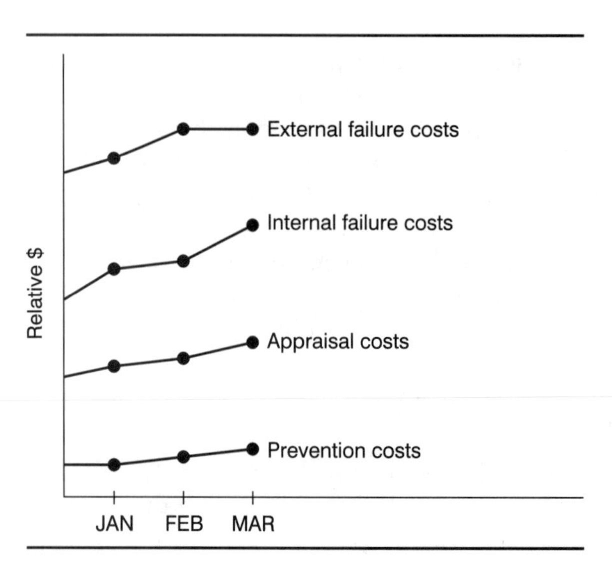

Figure 3 Quality Control Chart

Timely patient care delivery, quality of service, and the costs of providing those services are all interrelated. By studying the process of delivering medicinal items to the patients, the task force seeks to uncover methods of meeting patient demands, increasing the quality of the services to patients, doctors, nurses, and other medical staff and reducing the inventory cost problems currently existing at the hospital.

Inventory usage now becomes the subject of a three-month study. The study is limited to the cart supplying the second floor, west wing in the hospital. Time studies are made of the times associated with stocking the carts, checking the carts, and obtaining items when the items are not on the cart. The number of calls for needed items is recorded with a check sheet (Figure 4).

After all the data are gathered, the times are investigated. The time to record the information on the computer is similar from person to person and dependent on computer processing time. Picking of needed items and restocking the carts involves the number of items on the carts and the number used. It is fairly obvious that the more items on the cart and the greater usage of those items, the more time it takes to restock the cart. This points to a need to understand the supplies usage. Since supplies usage is the driving force behind the stock and restock times, this becomes the area to be most closely studied.

When the material usage summaries of the three months are studied, the investigators find that there are quite a few items that are not used at all or are used very rarely. A study of the usage percentage from the usage report determines four categories of supplies usage: no use, infrequent use, some use, and frequent use. Organized in this manner, the results of the study are shared with the medical staff of the unit. Based on the study, they remove the items they feel they do not need at all and identify others that can be reduced in quantity.

Inventory Item Check Sheet Showing Number of Calls for Each Item over a Three-Month Period

Item No.	*Usage*	*On Hand*	*Avg/Day*	*Std Dev*	*% Usage*
1	3	0	0.033	0.181	
2	157	1350	1.744	2.256	11.6
7	7	90	0.078	0.308	7.8
12	74	180	0.822	0.894	41.1
13	70	180	0.778	0.871	38.9
22	16	450	0.178	0.646	3.6
25	5	90	0.056	0.230	5.6
31	110	0	1.222	10.582	
32	145	360	1.611	6.475	40.3
36	10	0	0.111	0.608	
38	6	0	0.067	0.292	
40	28	360	0.311	0.612	7.8
47	2	90	0.022	0.148	2.2

Figure 4 Quality Control Chart

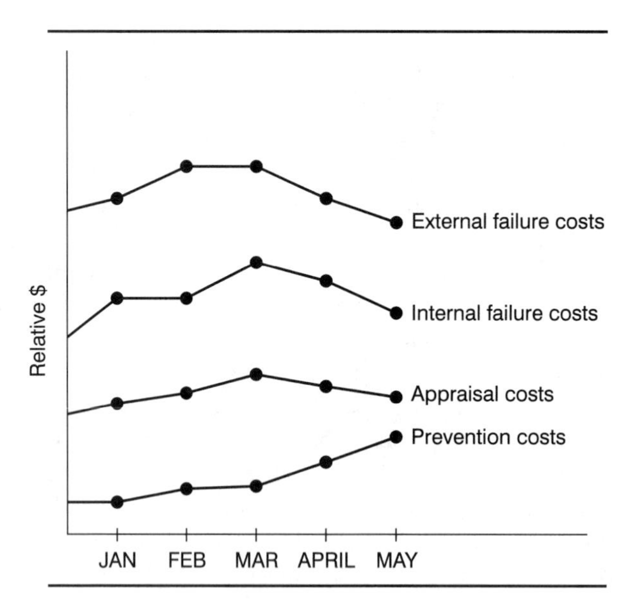

Figure 5 Quality Control Chart

Following this, the overstocking situation is investigated. In some instances, overstocking of the cart leads to product spoilage or breach of sterilization. Overstockage rates are discussed with the staff and eventually the item amounts are decreased. Both of these decreases in inventory on the carts reduce the stocking time as well as the cost of product spoilage. Figure 5 shows the result of investing in prevention costs and allowing process changes based on continuous improvement investigations. Failure costs decrease and a small reduction in appraisal costs is noticed.

Another major change is suggested by the task force. Inventorying the carts every other day instead of every day will save on appraisal costs. As long as failure costs associated with stockouts can be prevented or limited, this change will reduce the burden on the current staff without increasing costs. A less-harried stocking staff member will be less likely to make a mistake in overlooking an expiration date or an understocked item.

Switching to an every-other-day schedule requires studying the usage summary reports for the past three months. Calculations based on these reports determine an average usage per day of each item and the standard deviation associated with the usage. From this, usage levels for a two-day period are determined. Using the standard deviation to understand the average usage reduces the number of stockouts and midday replenishing. To fail to incorporate the standard deviation in the calculations would have increased the internal failure costs (running out of stock) instead of limiting them.

After this change is approved by the staff, the amount of supplies on the cart is changed to reflect usage rates and the carts are restocked every other day. This new process is monitored to determine if costs are really reduced and if new costs of quality have not developed. The amount of usage is noted as well as the number of calls

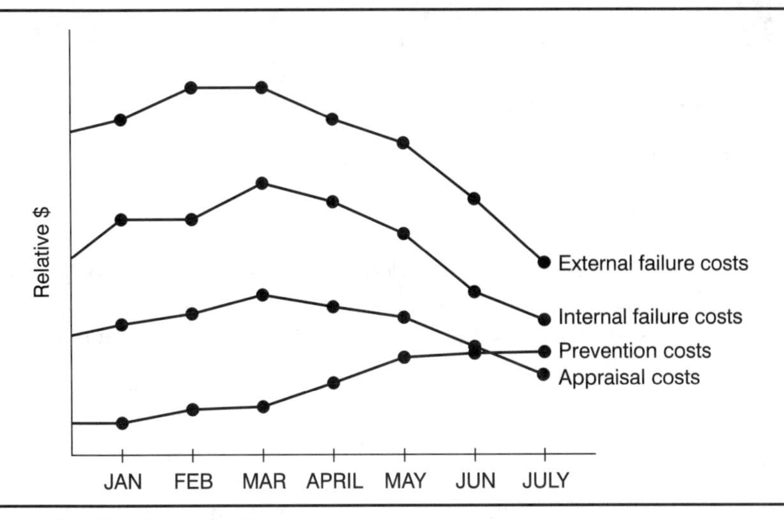

Figure 6 Quality Control Chart

to replenish the item between cart restocking. If too many calls are made, the amount is increased on the cart. Figure 6 shows the changes in appraisal costs in relation to the other costs.

Other cost-of-quality issues are identified during this investigation. For example, the hospital does not bill patients for very low cost items (swabs, alcohol, etc.). Following the study the limit is raised so that more items are not charged. While this means that fewer charges for minor items reach the patient, the hospital feels that these costs are recouped by eliminating the appraisal costs associated with checking for these items on patient bills and lost charges lists. Another result of the study is a noticeable reduction in the materials being wasted because of overstocking on the carts and in inventory. The preventive quality costs of studying the carts have been balanced by the savings incurred by not wasting products (a failure cost). These changes have also resulted in reductions of failure and intangible quality costs that arise when outdated materials jeopardize the quality of patient care.

Assignment

On the basis of this reading, discuss how quality costs were used in decision making.

CASE STUDY 11.2
Quality Costs

This case is the third of three related cases found in Chapters 8, 9, and 11. These cases seek to link information from the three chapters in order to resolve quality issues. It is not necessary to have completed the cases in the other chapters in order to understand or complete this case.

PART 1

Max's B-B-Q Inc. manufactures top-of-the-line barbeque tools. The tools include forks, spatulas, knives, spoons, and shish-kebab skewers. Max's fabricates both the metal parts of the tools and the resin handles. These are then riveted together to create the tools (Figure 1).

Max's performs a final inspection before the tools are placed into kits for selling. Each quarter, Max's makes more than 1,000,000 tools. Of these, approximately 240,000 are inspected for a variety of problems, including handle rivets, fork tines, handle cranks, dents, pits, burrs, scratches, handle color, blade grinds, crooked blades, and others. Any tool that has a defect is either returned to the line to be reworked or is scrapped. Figure 2 provides information for their first-quarter inspection results.

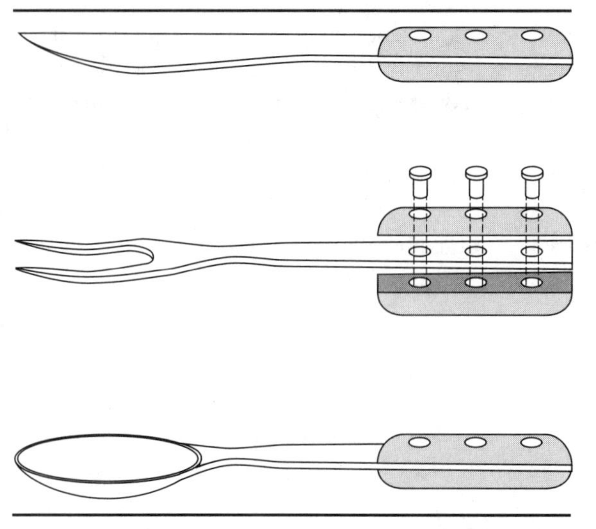

Figure 1 Barbeque Tools

Assignment

Create a Pareto chart with the information provided in the Figure 2. Are any of the categories related? Can they be grouped in any way? What problems should be tackled first?

Assignment

Given the present situation, what types of quality costs will Max's incur? Do you agree with the way they are operating? Why or why not? Where should they be making their investment? In what types of prevention costs do you feel they should be investing in order to avoid other quality costs?

PART 2

Recently, Max's hired you as a process engineer. Your first assignment is to study routine tool wear on the company's stamping machine. In particular, you will be studying tool-wear patterns for the tools used to create knife blades.

In the stamping process, the tooling wears slightly during each stroke of the press as the punch shears through the material. As the tool wears, the part features become smaller. Although the knife has specifications of 10 mm ± 0.025 mm, undersized parts must be scrapped. During their use, tools can be resharpened to enable them to produce parts within specification. To reduce manufacturing costs and simplify machine scheduling, it is critical to pull the tool and perform maintenance only when absolutely necessary. It is very important for scheduling, costing, and quality purposes that the average number of strokes, or tool run length, be determined. Knowing the average number of strokes that can be performed by a tool enables routine maintenance to be scheduled.

It is the plant manager's philosophy that tool maintenance be scheduled proactively. When a tool is pulled unexpectedly, the tool maintenance area may not have time to work on it immediately. Presses without tools don't run, and if they are not running they are not making money. As the process engineer studying tool wear, you must develop a prediction for when the tool should be pulled and resharpened.

The following information is available from the tool maintenance department.

- The average number of strokes for a tool is 45,000.
- The standard deviation is 2,500 strokes.
- A total of 25 mm can be ground off a punch before it is no longer useful.
- Each regrind to sharpen a punch removes 1 mm of punch life (25 total regrinds per tool).

FINAL INSPECTION DEPARTMENT 8 FEBRUARY 1996	Table Knife	Steak Knife	Paring Knife	Filet Knife	Basting Spoon	Ladle Spoon	Perforated Spoon	Skewer
DEFECT								
Bad handle rivets							1	
Bad steel	37				1			
Bad tines								
Bad/bent points	16							
Bent					1			
Buff concave								
Burn	146	57						
Burned handles	1							
Burrs	24	14						
Cloud on blades								
Cracked handle		32						
Cracked steel	1							
Crooked blades								
DD edge	41	2						
Dented								
Edge								
Etch	31							
Finish								
Grind in blade							1	
Haft at rivets								
Hafting marks handles	64				4		3	
Hafting marks steel								
Handle color	27	182				5		3
Heat induct	110	37					2	
High handle rivets								
High-speed buff								
Hit blades								
Hit handle	239							
Holder marks								

Figure 2 Inspection Results

FINAL INSPECTION DEPARTMENT 8 FEBRUARY 1996	Table Knife	Steak Knife	Paring Knife	Filet Knife	Basting Spoon	Ladle Spoon	Perforated Spoon	Skewer
DEFECT								
Honing								
Narrow blades	4							
Nicked and scratched	105	10			9	2	7	
Open at rivet								
Open handles							1	
Open steel								
Pitted blades		1					1	
Pitted bolsters								
Raw backs								
Raw fronts								
Rebend								
Recolor concave	2	43						
Rehone								
Reruns								
Rivet								
Scratched blades	105	53		1	12	2	4	
Scratched rivets								
Seams/holes	11	50						
Seconds				2				
Stained blades	84	18			5	1		
Stained rivets								
Vendor rejects								
Water lines					15			
Wrap	8							
Total Inspected	32,000	28,811	5	277	3,145	1,504	2,008	1,805
Total Accepted	30,944	28,312	5	274	3,098	1,494	1,988	1,802
Total Rejected	1,056	499	0	3	47	10	20	3

Figure 2 **Continued**

- The cost to regrind is

  2 hours of press downtime to remove and reinsert tool, at $300/hour
  5 hours of tool maintenance time, at $65/hour
  5 hours of downtime while press is not being used, at $300/hour

- Average wait time for unplanned tool regrind is 15 hours at $300 per hour.
- Because of the large number of strokes per tool regrind, this is considered to be a continuous distribution. The normal curve probability distribution is applicable.

Assignment

Your first assignment as process engineer is to develop a prediction routine for tool wear. As mentioned previously, the plant manager's chief concern is ensuring planned tool regrinds. Early wearout—and thus an unplanned tool pull—can be caused by a variety of factors, including changes in the hardness of the material being punched, lack of lubrication, the hardness of the tool steel, and the width of the gap between the punch and the die. Key part dimensions are monitored using $\overline{X}$ and R charts. These charts reveal when the tool needs to be reground in order to preserve part quality.

When a tool wears out earlier than expected, the tool room may not have time to work on the tool immediately. While waiting for the tool to be reground, the press and its operators will be idle. The plant manager would prefer that tools be pulled early in their wearout phase in order to avoid the chance of an unexpected tool pull. He would like to pull the tool for regrind at 40,000 strokes. What is the chance that the tool will be pulled for regrind before 40,000 strokes?

PART 3

Now that you have been at Max's B-B-Q Inc. for a while, the plant manager asks you to assist the production scheduling department with pricing data on a high-volume job requiring knife blades for the company's best customer. As you know, it is the plant manager's philosophy to be proactive when scheduling tool maintenance (regrinds) rather than to have to unexpectedly pull the tool. However, pricing will be a very important factor in selling this job to the customer. Essentially, the plant manager wants no unplanned tool pulls but sales needs pricing cost reductions. The production scheduler would like a tool-regrind schedule that results in minimal inventory.

Assignment

You will soon be meeting with the plant manager and the managers from sales and production scheduling. They are expecting you to have an answer to the question, Given the need to balance maximizing tool use, minimizing inventory, minimizing production disruption, and minimizing cost, how many strokes should you recommend to run this tool before pulling for a regrind?

Create a graph that shows the number of unplanned pulls versus the number of strokes. The graph should comprise at least six data points. Next complete the following spreadsheet, showing the costs of each individual's plan. Using the graph and the spreadsheet, prepare a response for the question, How many strokes should the tool be run before pulling it for a regrind? Your analysis should include answers to the following questions: How will this number balance tool use, cost, inventory, and production disruption? What are the economics of this situation? What quality costs exist?

	Plant Manager	*Production Scheduler*	*Sales Manager*	*You*
Strokes Before Pull	40,000	42,000	43,000	
Number of Pulls	25	25	25	
Production Over Life of Tool	1,000,000	1,050,000	1,075,000	
Cost of Each Pull	$2,425	$2,425	$2,425	$2,425
Additional Cost of an Unplanned Pull	$4,500	$4,500	$4,500	$4,500
Chance of Unplanned Pulls				
Total Additional Cost due to Unplanned Pulls				
Total Cost				

12

Product Liability

■ *Learning Opportunities:*

1. To introduce the concept of product liability
2. To broaden the reader's understanding of how designs are affected by product liability issues
3. To show how product design and development programs can help protect companies from product liability issues
4. To show how quality-assurance programs can help protect companies from product liability issues ■

- Only two of the original eighteen American firms making football helmets exist, due in part to multiple product liability suits. (*Business Week,* February 7, 1994)
- The only safe and effective antinausea drug, Bendectin, has been taken off the market because the cost of product liability insurance reached $10 million a year, over 80 percent of the annual sales. (*Nation's Business,* February 1986)
- Keen Corporation is the 18th asbestos manufacturer to declare bankruptcy due to product liability claims. (*Business Week,* February 7, 1994)
- A nationwide shortage of vaccine to protect children from diphtheria and whooping cough developed when only one company chose to continue to manufacture it, given the high liability insurance costs. (*Nation's Business,* February 1986)
- Twenty percent of a stepladder's cost goes to pay for liability insurance. (*Nation's Business,* February 1986)
- In the Washington, D.C. area, the Girl Scouts must use the profits from 87,000 boxes of cookies to cover the organization's liability insurance even though they have never been sued. (*Newsweek,* March 20, 1995)
- Due to product liability costs, production of private aircraft at Piper Aircraft Corporation fell from 17,000 planes in 1979 to only 1,535 by 1989. (*Forbes,* November 8, 1993)

Examples of the Results of Product Liability Losses

A perfectly safe product is one that can do no objectionable harm at any time, under any circumstances. Relatively few such perfect products exist. Since this is the case, much must be done in the design and manufacture of a product or the provision of a service to give the consumer the safest product or service possible. In a society so highly dependent on products and services produced by others, are there ways of providing for society's needs without exposing the provider to excessive product liability costs? Product liability costs can add 15 to 30 percent to the cost of a product. This chapter provides some background in product liability as well as ideas for designing and creating safer products and services.

THE EVOLUTION OF PRODUCT LIABILITY

The ability of an individual to recover damages after being injured by a product or service has changed considerably over time. Product liability law is case law, meaning that there are few, if any, federal mandates or legislative actions. Because it is based on common law—the decisions of judges and juries in individual cases—product liability law reflects changing social and economic conditions. Product liability laws and the outcomes of the trials vary from state to state.

A ***lawsuit*** *is a civil suit seeking money damages for injuries to either a person or property, or both.* Product liability lawsuits involve two parties: plantiffs and defendants. ***Plaintiffs*** *are the injured parties who file suit in a court of law. The* ***defendant*** *is the person or company against whom a claim or suit is filed.*

Negligence and Privity

Negligence *occurs when the person manufacturing the product or providing the service has been careless or unreasonable.* Negligence cases focus on the conduct of the manufacturer. In negligence cases, the injured party must prove that the defendant failed to exercise due care in the creation of the product or provision of the service. An example of failure to exercise due care would be the case where a manufacturer or service provider fails to keep current with technology. Negligence cases place the burden on the plaintiff, because it is the plaintiff's responsibility to show that the defendant's conduct was not what it should have been.

One of the earliest product liability cases based on negligence, *Winterbottom* v. *Wright* (1842), involved a defective mail coach. The case was settled on the basis of the concept of privity. Under ***privity,*** *a suit cannot be filed unless a contract (an agreement) exists between the parties of the suit.* In today's world, the concept of privity means that you would not be able to sue the maker of a toaster oven that caused a fire, unless you had bought that toaster oven directly from the manufacturer. It wasn't until 1916 that privity was no longer considered a suitable defense for negligence cases.

In the 1916 landmark case, *MacPherson* v. *Buick Motor Company*, the court found that an auto manufacturer had a product liability obligation to a car buyer whose car wheels were defective, even though the sales contract was between the buyer and the dealer. Presiding Judge Kellogg held that

> under the circumstances the defendant owed a duty to all purchasers of its automobiles to make a reasonable inspection and test to ascertain whether the wheels purchased and put in use by it were reasonably fit for the purposes for which it used them, and, if it fails to exercise care in that respect, that it is responsible for any defect which would have been discovered by any such reasonable inspection and test.*

Without the concept of privity, though the injured party must still prove negligence, a manufacturer's duties extend to the ultimate user of the product.

**MacPherson* v. *Buick Motor Company*, 1916.

Strict Liability

In 1962, *Greenman* v. *Yuba* brought another change to product liability cases by introducing the concept of strict liability. Under ***strict liability,*** *a manufacturer who makes and sells a defective product has committed a fault*. A *product is considered defectively made if it has a defect that causes it to be unreasonably dangerous and the defect is the reason for an injury*. Even though there is no evidence of negligence, the manufacturer can be held strictly liable in tort. By placing the item on the market, the manufacturer has said the product is fit for use. The concept of strict liability is basically sound although it ensures that the costs of injuries are carried by the manufacturer. It assumes that *the person was injured through no fault of his or her own; in other words, there was no* ***contributory negligence.*** The court's mission has changed from finding fault to determining the amount of compensation due the injured party. Some people feel that the explosion of product liability cases can be traced to the strict liability in tort doctrine.

WARRANTIES

Two kinds of warranties exist: expressed and implied. An ***expressed warranty*** *is part of the conditions of sale*. Expressed warranties can be written or oral, found in advertising or contracts. An ***implied warranty*** *is implied by law rather than by the seller*. Merchantability, the act of putting the product on the market or providing a service, implies that the product or service is fit for normal foreseeable use. The buyer purchases the product on the reasonable assumption that the product will be reliable. Merchantability also implies that the manufacturer contemplated the known risks associated with the product or service, who the intended users of the product or service were, what the necessary warnings and instructions were, and the foreseeable misuse. In 1960, the *Henninsen* v. *Bloomfield* breach of implied warranty case removed the concept of privity from contract cases. The judge determined that merchantability and fitness for use extends to the reasonably expected end user of the product.

PRODUCT LIABILITY SUITS

The Plaintiff's Actions

The plaintiff, the injured party, may file a lawsuit that alleges that the provider of a product or service is guilty of breach of warranty and negligence as well as being strictly liable. Plaintiffs are required to show that the product was defective or unreasonably dangerous, that the defect was the proximate cause of harm, and that the defect was in the product when it left the manufacturer. In strict liability cases, plaintiffs do not need to prove manufacturer's negligence. In strict liability cases it does not matter if the injured party acted in a careless and unreasonable manner (contributory negligence). Strict liability focuses on the product and its quality. The behavior of the manufacturer and the reasonableness in which it acted is not a factor. A plaintiff who chooses to sue under warranty law needs to prove that the product was not fit for the purposes for which it was advertised.

The Defendant's Actions

In a product liability suit, the defendant must establish that the product or service provided is not defective or unreasonably dangerous. The defendant must prove that no defect existed in the product at the time of manufacture or when the service was provided. It is also important that the defendant show that the product or service was not the actual cause of the accident or injury. The defendant may do this by showing that the plaintiff exhibited bad judgment, improperly used the product, conducted improper maintenance, or changed the product after its manufacture. A manufacturer's best defense against warranty problems involves developing advertising and warranty claims that proclaim completely accurate, truthful information concerning the product's life, safety, and quality. Advertising claims must be truthful about a product's life cycle, safety, fabric durability, environmental side effects, and usage. Any advertisement is seen to support a valid company product application and usage.

The Court's Actions

When deciding a product liability case, the courts will look at the defective aspect of a product or service. They will also consider the designs of similar and perhaps safer products or services. Some products or services are inherently dangerous (for example, knives and bungees for bungee jumping), so the court will expect that the public had some common knowledge of the product's dangerous aspects. The court also questions the probability of an injury occurring, given reasonable care on the part of the user. The adequacy of the instructions and warnings, as well as the environment where the product is used, also plays a role. Courts have expanded their view of warranties and no longer look solely at the character of the product; they now also consider the consumer and public interest factors.

A court judges whether or not a product or service is unreasonably dangerous by considering several factors. Would a reasonable company market the product or service if they knew of this defect? Products and services are also judged on the basis of consumer expectations. Is the product or service more dangerous than a consumer could contemplate? Given the public's ordinary knowledge of the product, is the danger open and obvious? This approach raises the issue of who is a reasonable consumer with ordinary knowledge of the product or service. Can a four-year-old who uses a swimming pool understand its dangers? Can an average adult understand the consequences of using chemicals in the yard? Risk of injury versus the utility of the product is also judged. When an inherently dangerous product, such as a knife, is judged, its usefulness to society is considered.

The Expert Witness

Both defendants and plaintiffs may choose to support their arguments with the testimony of expert witnesses. Expert witnesses are chosen to inform the court about aspects of the accident, injury, or product. They must be technically competent in their field as evidenced by their degrees, years of experience, and professional activities. The behavior and credentials of an expert witness must be above reproach. Good expert witnesses are able to communicate in common language with the judge and the jury.

LIABILITY LOSS CONTROL PROGRAMS

Liability exposure exists at every phase of product development or service provision, from conception to the first preliminary design; through all design stages, development, and testing of prototypes; into the manufacturing phases, including negotiation of contracts with vendors; through actual manufacture, testing, and assembly; into packaging, labeling, and use instructions; product service; and ultimately to the methods of promotion, marketing, and distribution, including warranty periods and service arrangements. In a product liability suit, a plaintiff must show that a defect was the cause of the injury or loss. Product or service defects can come from several sources: a flaw in the manufacturing process, a design defect, a packaging defect, or failure to warn of the dangers. Manufacturing defects may be evident in a weld that breaks under normal usage or a cracked part that causes an equipment malfunction. Design defects such as those that caused the collapse of the Kansas City hotel walkway may be due to material or weld failures, sharp edges or corners, failure to provide guarding and safety devices, a concealed danger, or just poor design, such as locating a chain saw muffler below the handle. Defective packaging may cause an accident or injury. The Tylenol poisonings are an example where proper packaging could have alerted consumers to tampering. After these poisonings, many packages have added shrink-wrapped safety seals to make tampering more visible.

Service and information industries are also vulnerable to product liability action. A city may be sued if it fails to trim roadside shrubbery and an accident occurs because of limited sight distances. A creator of software may be sued because a software glitch dropped an item from a quotation or bid. The U.S. government may be sued because a National Weather Service forecast predicted fair weather when a hurricane was approaching.

In the case of a design defect, an inadvertent error may have occurred, from misplacement of a decimal or an error in reading a table. A design defect may have resulted from a conscious design choice that was considered acceptable at the time. When a manufacturing defect causes the injury or loss, the defendants have little with which to defend themselves. To protect themselves, many companies have developed liability loss control programs. These programs have two major aspects: product design and development and quality assurance. In some companies, reliability testing may also be a component of a liability loss control program.

Product Design and Development

To avoid creating a dangerous product, designers should consider the usefulness and desirability of their product. Once it has been determined that the product has value for the consumer, companies should follow several steps to minimize risks.

1. *Design to remove unsafe aspects*. First and foremost, the product or service should be designed without a possible danger or hazard. This is the most important step to making a product or service safe for the consumer. All avenues of design modifications should be studied before allowing a product to be manufactured or a service to be provided if that service or product is inherently dangerous. Other, perhaps safer, products or services on the market should be studied.

Designers should consider the user and determine the likelihood of injury and the seriousness of that injury. What level of understanding can a reasonable and prudent user of the product or service be expected to have? How obvious is the danger? Does the public have common knowledge of the danger? And finally, if a danger exists in a product or service, can the danger be eliminated without seriously affecting the design, form, or function of the product? If the answer to this is yes, then it is the company's responsibility to do so.

Manufacturers and service providers should also investigate the area of product or service misuse. For every product or service on the market, there exists the potential for misuse by the consumer. Designers should strive to determine every foreseeable use for their product or service and design out potential hazards. Creative brainstorming sessions with future customers are often fertile grounds for discovering future misuses of the product. Designing with misuse in mind encourages designers to recheck their designs, guards, and product warnings.

2. *Guard against unsafe use*. If the danger cannot be designed out of the product, or if the product is inherently dangerous, the second step a designer should take involves guarding the user against the danger. Guards prevent users from gaining access to the area that would cause them injury. When the potential exists for misuse of the product or service, guards should be in place to prevent the misuse. Care must be taken to ensure that the guards are not made so inconvenient that people circumvent them, thus defeating the purpose.

3. *Provide product warnings*. If the product hazards cannot be designed or guarded out of a pocket, then a designer's final recourse is to warn the user. Warnings must be specific and must describe the consequences of their being ignored. Warnings should be highly visible and easily comprehensible. In today's world, warnings should be pictorial in nature or multilingual to be effective. If the warning is written, it should be clearly worded in simple language. Warnings should discuss safe installation, operation, cleaning, and repair practices. The consequences of misusing the product should be clearly stated. Warnings are a last resort, to be used when designing out or guarding against an inherent danger has proven infeasible. People tend to become accustomed to warnings and ignore them, even where there is a high probability of injury.

4. *Design to standards*. A product should be designed to comply with industry standards. While designing to standards does not ensure a safe product, standards do tend to create safer products. Standards also encourage those providing products and services to remain current in their field. State-of-the-art knowledge enhances the design and production of a product. Companies must keep up with new products and state-of-the-art advances that apply to their field.

5. *Advertise and market wisely*. Occasionally, a company creates potential product misuse situations through its advertisements, marketing materials, and sales personnel. Advertisements and marketing materials that show the product performing tasks it is not designed to do set the stage for misuse of the product by the consumer. Product liability loss prevention is not the sole responsibility of the product designer or manufacturing; misrepresentation and exaggeration in advertisements and marketing materials may also be involved. Proper training of

distributors, dealers, and sales personnel can lower product liability costs by enabling them to give guidance and training to purchasers of the product. Increasing a consumer's knowledge of the product and its uses can greatly reduce the potential for harm.

Quality-Assurance Involvement

A program to reduce the risks associated with product liability must embrace design, marketing, manufacturing, service provision, packaging, shipping, and service. There should be a thorough, systematic review of new products and services for all basic designs. Such a review should consider any hazards that might occur. Strict quality control should be maintained during the production of a product. Production, quality, and reliability test data, as well as data from incoming material inspections and product or service audits, should be examined to determine if improvements can be made to the product or service that would enhance its safety. Information from consumer complaints, warranty claims, and product recalls should be carefully investigated and the results of those investigations used to improve the product or service in the future.

New products and services and their components should be tested to ensure their reliability. Subjecting a product or procedure to tests simulating actual use will reveal the conditions that might lead to a breakdown in the system. For instance, if a hospital creates a new procedure to dispense medication to patients, this procedure should be tested and monitored to determine its reliability. Reliability testing will pinpoint aspects of the procedure that could lead to an error in dispensing the medication. Information from reliability tests should be returned to the product or process designers so that revisions can be made.

In order to ensure that product liability issues are tackled before they become a major expense for a company, some companies designate a product liability expert. This individual develops a thorough technical knowledge of all of the products or services provided by the company. Normally this person also has an understanding of the legal issues surrounding product liability cases. Individuals best suited for this type of work display a high degree of professionalism and diplomacy as well as technical knowledge.

EXAMPLE 12.1 Company Consciousness of Product Liability

In the past decade, the design of lawn mowers and their attachments has changed dramatically. An inherently dangerous product, lawn mowers are commonly used pieces of equipment in the life of most suburban householders. The necessity of mowing a lawn with something larger than an old-style human-powered lawn mower has created a product that consumers see as useful, desirable, and valuable. In the past decade, lawn mower manufacturers have taken the steps described in this chapter to minimize risks.

Step 1. Design to Remove Unsafe Aspects. Since lawn mowers are inherently dangerous, it is not possible to completely design out every possible danger or hazard. But within the lawn mower there are many localized changes that can be made to the design to improve the safety of the equipment. One such modification is to place the muffler in a location that makes it difficult for users to burn themselves on the hot cover. There have been lawn mowers that have been designed so that the muffler or a hot engine component is placed near the gas tank opening. Designs such as these reflect little thought to designing out hazards. The user has not been considered nor has the likelihood of burn injuries or the seriousness of those injuries. Good designers investigate all avenues of design modifications before allowing a product to be manufactured or a service to be provided that is inherently dangerous.

Step 2. Guard Against Unsafe Use. It is in this area that lawn mowers have undergone the greatest amount of change. Since all the dangerous aspects cannot be designed out of lawn mowers, the addition of guards has changed their look. Guards on the platforms of riding lawn mowers and on the backs of push lawn mowers prevent users from gaining access to the blades. One major safety issue surrounds cleaning the duct that spews out cut grass. Here a high potential exists for misuse of the mower. People reach in to clear the duct, placing their hands or feet in close proximity to the blades. Push lawn mowers currently on the market have handles that the operator must hold in order to keep the blades spinning. It is extremely difficult for a person to reach the duct while holding the handle; thus this type of guard prevents the person from cleaning the duct while the blades are still turning.

Step 3. Provide Product Warnings. Lawn mowers currently on the market are accompanied by a long list of warnings. Since the product hazards can not be completely designed or guarded out, the user must be warned. Figure 12.1 shows warnings that are specific and that describe the consequences of ignoring them. The warnings are pictorial and simply and clearly worded. In the instruction booklet, the warnings discuss safe installation, operation, cleaning, and repair practices.

Step 4. Design to Standards. Many of the design, guarding, and warning changes made to lawn mowers are a direct result of industry standards, which were created to help ensure a safer product.

Many mower manufacturers have also instituted quality-assurance programs. At the plant level, these programs link the activities of design, marketing, manufacturing, service provision, packaging, shipping, and service. New product design reviews study the basic designs of lawn mowers being offered. These reviews consider hazards, quality information, and consumer feedback to improve mower design. Information from consumer complaints, warranty claims, and product recalls is carefully investigated and the results of those investigations are used to improve the product.

Product liability costs manifest themselves in many ways, including recall, replacement, and repair costs. Prevention and appraisal quality costs will also increase when a company is confronted with a product liability issue. While most people recognize the

PROTECT CHILDREN

Keep children and others away when you operate the machine.

Do not let children operate or ride on the mower.

BEFORE YOU BACK UP:

CAUTION

• Look behind the tractor for children, pets, and objects.

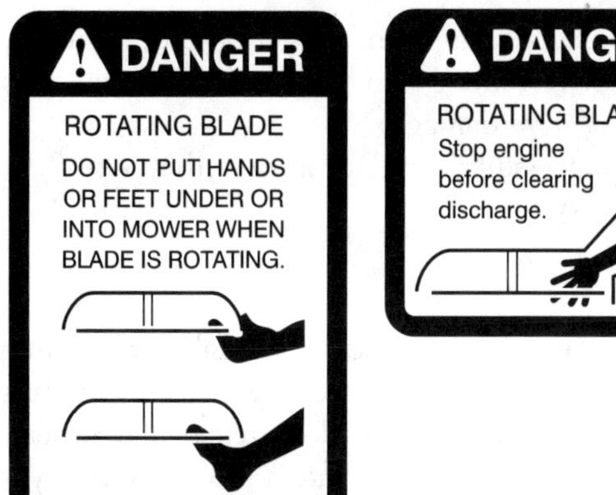

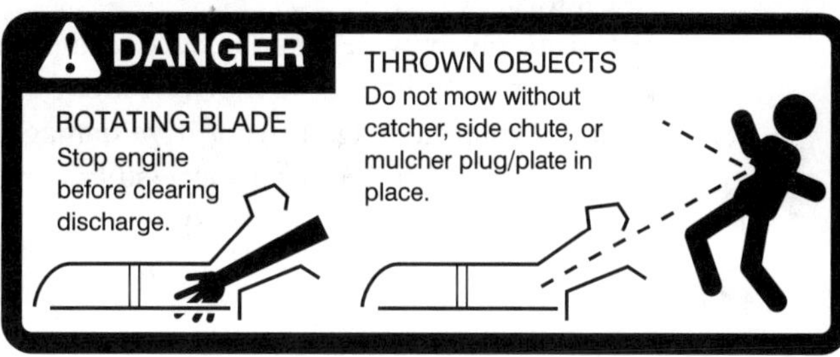

Figure 12.1 Product Warnings

costs associated with legal defenses, awards, and liability insurance, few realize that product liability has an adverse impact in other areas. Faulty products or services create dissatisfied customers; coupled with product liability disputes, they may damage a company's reputation on a broad scale. Product liability costs may cause a company to decide against introducing new products or to discontinue existing product lines entirely. If new product introductions have decreased, then there is little encouragement to continue product research. Other companies report lost market share as an impact of product liability issues. Lost market share can result in laid-off workers, closed production plants, or the decision to move production offshore. To remain competitive, companies must emphasize product design and quality assurance to minimize the impact of product liability.

FUTURE CONSIDERATIONS

Many different interest groups have concerns about the product liability situation in the future. The following are some suggested areas for further investigation:

- Enact a statute of limitations for filing liability claims.
- Provide a release from liability for the original manufacturer if the item is modified or the safety devices removed or altered.
- Judge products according to state of the art at the time of manufacture or provision of service.
- Create a standard code of awards.
- Reduce or eliminate punitive damages.
- Regulate attorney's fees.
- Institute comparative negligence judgments.
- Establish standards describing types of behavior that are grounds for punitive damages.
- Establish enhanced procedures for including scientific evidence of causes of injuries.
- Create nationwide standards for awards and litigation procedures.
- Accept compliance with government standards as a valid defense in lawsuits and a guarantee against punitive damages.
- Assess the legal fees of a successful defendant against the plantiff.

SUMMARY

Regardless of how the consumer uses the product, responsibility for the product or service goes beyond the provider's doors. Immediate responsiveness to unsatisfactory quality through product service or replacement can do a great deal to lessen the chance of a product liability suit. Companies should institute a product liability loss control program that involves designing and providing a high-quality product or service rather than simply responding as necessary to periodic crises. An inherent, built-in approach to product safety will reduce product liability risks. An old saying stresses the importance of taking care of the little things: For want of a nail, the shoe was lost; for want of a shoe, the horse was lost; for want of the horse, the goods were lost. Companies need to recognize that product liability problems can occasionally be traced to routine and relatively minor decisions made daily in nearly all areas of a company's operation and at all levels of responsibility. Just as with the nail and the horseshoe, the size and source of the error may have little relationship to the magnitude of potential loss. From this perspective, a simple decision made during the manufacturing of a product may be the cause of a product liability suit. Strong total quality assurance and loss prevention programs can prevent product liability problems from occurring.

Lessons Learned

1. Civil suits seek money damages for injuries to either a person or property or both. They involve two parties—the plaintiff, the injured party, and the defendant, the person or company against whom the suit is filed.
2. When a product or service has been provided in a careless or unreasonable manner, negligence occurs.
3. If a manufacturer makes and sells a defective product or provides a defective service, they can be held strictly liable if the defect renders the product or service unreasonably dangerous and the defect causes an injury.
4. Under contributory negligence, the plaintiff is found to have acted in such a way as to contribute to the injury.
5. An expressed warranty can be either written or oral, and it is part of the conditions of sale.
6. An implied warranty is implied by law.
7. An expert witness informs the court about the technical aspects of the accident, injury, or product.
8. Product liability exposure can be reduced through thoughtful design and development of products or services. Designers should design to remove unsafe aspects, guard against unsafe use, provide product warnings, design to standards, and advertise and market the product or service appropriately.
9. A program to reduce the risks associated with product liability must embrace design, marketing, manufacturing, service provision, packaging, shipping, and service. ■

Chapter Problems

Product Liability

1. Airline A uses a traditional navigation system that has been successful in the past. A new, more sophisticated and accurate system is available at a higher cost, and most other airlines have switched to the new system. During a storm, a flight of airline A hits a mountain. The follow-up investigation finds that a defective navigation system was the cause of the accident.
   a. Who is the defendant in this case?
   b. Who is the plaintiff?
   c. What could the company/individual have done to avoid liability?

2. Define negligence.
3. Describe strict liability.
4. Describe contributory negligence.
5. What is the role of the defendant in a strict liability case?
6. What is the role of the plaintiff in a strict liability case?
7. Describe the role of an expert witness during a product liability court case.
8. Discuss the types of changes that may be seen in future product liability laws. Why are these changes necessary? What abuses of the system might they prevent?
9. Investigate recent developments in product liability. Are there any changes or additions to the laws? Have the judges or juries adjusted their verdicts by taking into account contributory negligence? Have any statutes of limitations been enacted? If yes, for what products?

Product Liability Suits

Investigate each product liability case in Problems 10–14. Answer the following questions.

- What event or series of events occurred to cause the need for a product liability suit?
- Who are the plaintiffs?
- What are they seeking?
- What motivated them to file a product liability suit?
- What is their point of view or defense?
- What was the verdict?
- Were there punitive damages?
- Was the case appealed?
- Did the judge change the amount of the award?
- Who were the expert witnesses involved?
- What kind of loss control program would have helped the defendant?
- How could quality assurance have been involved?

10. A young couple in their twenties driving an on/off-road vehicle picked up two friends to go out for a drive. The young man went to a local off-road course. There he proceeded to drive down a very steep hill, even though course safety rules indicated that this hill was an "uphill drive only." For some reason, perhaps braking, the vehicle went down end over end, crashing upside down. Three of the people were killed instantly—two when the roll bar snapped forward, crushing them. The third individual died from head trauma. The fourth individual, a young woman, was paralyzed, having apparently been struck by the back of the vehicle.

11. After being injured in an accident in her 1991 Toyota Tercel, the plaintiff filed suit against Toyota Motor Sales U.S.A. alleging that the Tercel was poorly designed and defective because it did not have an airbag. Toyota said that the belt-restraint system provided in the Tercel met the 1966 National Traffic and Motor Vehicle Safety Act. This act, applicable to the 1991 model, does not require airbags.
12. In 1994 an elderly woman burned herself when she spilled the hot coffee she had just purchased at a McDonald's drive-through window.
13. During the 1990s, Dow Corning Inc. was forced into bankruptcy over their silicon breast implants.
14. Throughout the 1990s a variety of controversial suits were filed against cigarette manufacturers by smokers.

CASE STUDY 12.1
Product Liability*

XYZ Aircraft Co. prides themselves on their total quality assurance efforts and their excellent safety record. They have instituted preventive maintenance and training programs as part of their efforts to comply with and exceed Federal Aviation Administration (FAA) requirements.

As part of their pilot training program, XYZ Aircraft has been instructing its pilots on the use of autopilots. The training has taught the pilots to check the autopilot for malfunctions and the proper way to handle in-flight malfunctions. Although XYZ had not had an in-flight incident, it was during training and subsequent preventive maintenance checks that the company realized that uncommanded actions might occur when the plane was on autopilot. Since this discovery, XYZ has been working with the autopilot suppliers to develop corrective actions.

XYZ Aircraft is sincerely interested in the safety of their passengers and crew. This interest is the driving force behind their total quality assurance efforts. XYZ Aircraft is also very aware that from a product liability point of view, they could be held strictly liable for any incident involving their planes. In a strict liability case, they could be found to have committed a fault if their service (air transportation) is considered defective, if their aircraft had a defect that caused it to be unreasonably dangerous and the defect was the reason for an injury. They would be strictly liable even if they were not negligent. For this reason, they are diligently attempting to correct the autopilot anomalies.

Most of XYZ's aircraft are equipped with the same model autopilot. This facilitates training and ensures that pilots flying different planes from day to day are familiar with the operation of the autopilot. Nearly all autopilot systems work in the same fashion. The autopilot receives an instruction from the pilot, senses the present aircraft situation, determines the adjustments to make to align the pilot's instructions with the aircraft's attitude, and moves the controls accordingly. Essentially there are three areas that could malfunction: the sensing mechanisms that monitor the plane's position, the computation system that calculates the difference between the command and the aircraft position, and the control area that moves the aircraft controls into the appropriate position. Each of these areas has been investigated.

To avoid incurring product liability costs, XYZ Aircraft is prepared to study the design and manufacture of the autopilots and use quality-assurance information to isolate the potential cause of the problems. Their most immediate step during the

*This case is fictitious and is not meant to resemble any particular company or incident.

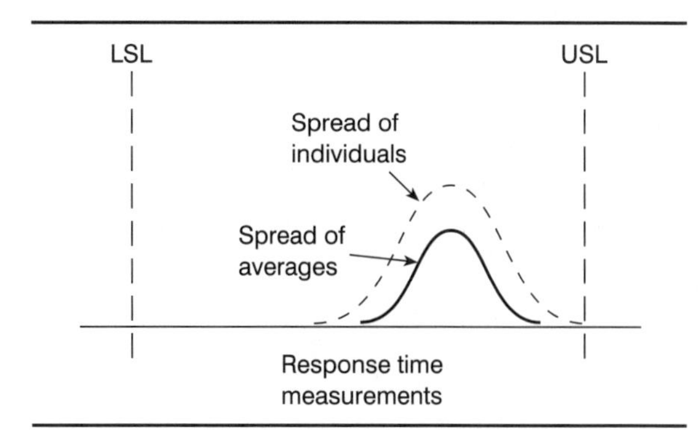

Figure 1 Histogram of Response Times

process is to effectively warn XYZ pilots of the incidents of uncommanded movements and errors being made by autopilots. They also continue to train pilots on recognizing and handling in-flight malfunctions.

After preliminary investigations into the three areas, attention begins to focus on the computation system of the autopilot. A manufacturer of a signaling component in the system has noted high response times in the sensor switches. Although the response times were high, the switches were within specification, and the manufacturing engineer had signed off on the product as acceptable to use. The switches were inserted into subassemblies and consequently into the autopilots. Several of these autopilots were later installed on XYZ planes.

Now that an investigation has pinpointed the switches as a potential problem source, the quality-assurance records for the switches are scrutinized. It is determined that while the parts were within response-time specifications, the times were close to the upper specification limit (Figure 1). Process capability calculations (C_p, C_{pk}) reveal that the process is capable but not centered. Causes for this are investigated and a Pareto diagram created (Figure 2). Selecting the most frequently occurring problems to study, a production-line team has suggested improvements, which have been incorporated into the process.

Although the switch could not be pinpointed as the total cause of the problem, the supplier decided to install new switches on the existing autopilots. Retrofitting is very costly: The company will have to supply the new switch and send out trained technicians, engineers, and marketing people all over the world to aircraft service centers to replace the switch. One hundred percent inspection is instituted on the parts. Reliability tests are run. During these tests, parts are put through 2,000 cycles under electrical and mechanical load to detect infant mortality.

When deciding upon a course of action, the costs of retrofitting, inspecting, and reliability testing are weighed against the costs of lawsuits and catastrophic-loss suits. Management feels that product liability suits will inevitably damage the reputation and future of the company, greatly outweighing the cost of retrofitting.

The engineers who had signed off on the original switches are now instructed in the importance of determining the causes of deviations and the potential consequences

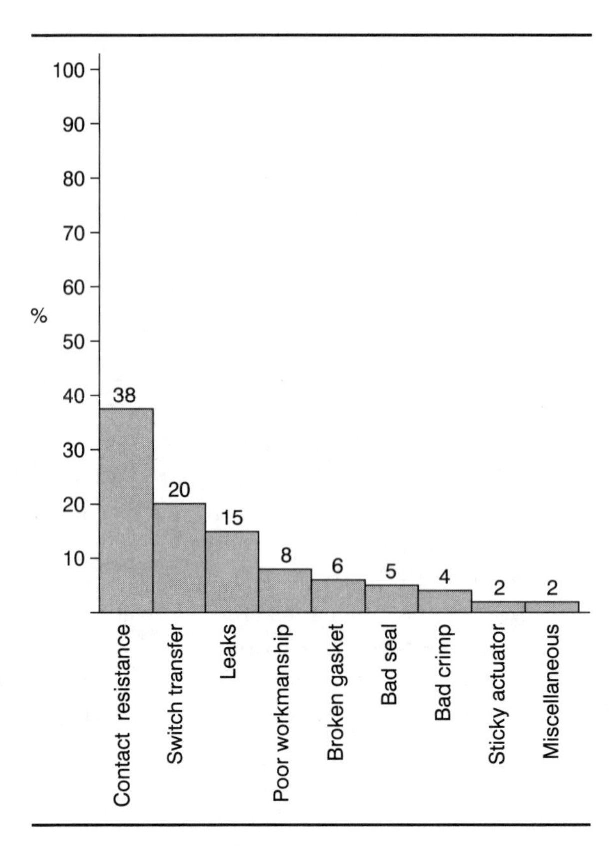

Figure 2 Pareto Diagram of Causes of Increased Response Times

of such deviations before signing off on a change. The engineers were unaware that quality and reliability issues were of significant importance in decisions related to the creation of a part or provision of a service. They were also unaware that any paper they sign their names to can become evidence in a liability suit. Therefore care must be taken to determine the causes of the deviations and their consequences.

For instance, if a manufacturing engineer, advised by an operator that an assembly tool is not functioning properly, has the tool fixed and then signs off on the tool as okay to use, that engineer can be held responsible if the tool produces parts that are used in an assembly that proves to have a quality problem. On items that may be discrepant but are found to be acceptable, engineers should be sure to look into the issue before signing off. What is the discrepancy? What are the consequences? If a problem is encountered and an investigation ensues, the engineer's name will be on the documentation and the engineer will be ultimately responsible.

Product liability issues can be reduced when proper attention is paid to the details of creating a product or providing a service.

Assignment

1. Why would XYZ be held strictly liable if an autopilot malfunctioned during flight?
2. What activities would constitute negligent behavior on the part of XYZ Aircraft?
3. How did this particular company use quality-assurance data to help avoid a product liability problem?
4. What type of questions should an engineer ask before signing off on a product change or deviation?

CASE STUDY BIBLIOGRAPHY

"NTSB Recommends FAA Action on Faulty Autopilots." *Avionics*, December 1993, p. 46.

G. Picou. "What Autopilots Go Nuts." *IFR*, September 1994, pp. 16–19, 26.

13

Quality Systems: ISO 9000, Supplier Certification Requirements, and the Malcolm Baldrige Award

Learning Opportunities:

1. To become familiar with the requirements of ISO 9000
2. To become familiar with the requirements of QS 9000
3. To become familiar with the requirements of the Malcolm Baldrige National Quality Award

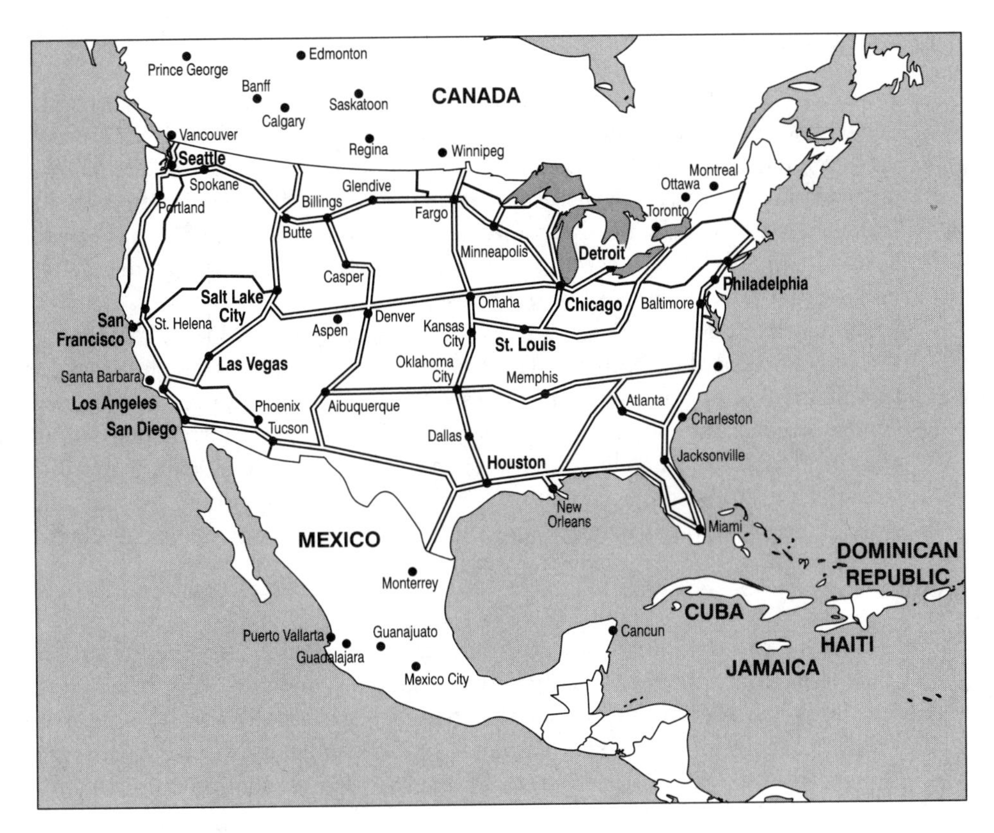

Road Map to Success

Throughout this text, different ideas, concepts, and techniques have been presented. Individually, each idea, concept, and technique has merit and together they make up the supporting pieces of a total quality system. ISO 9000, supplier certification requirements, and the Malcolm Baldrige Award present more unified approaches to quality, encouraging companies to create quality systems. They provide a road map for companies to follow in their quest for quality.

QUALITY SYSTEMS

In order to best fulfill customer needs, requirements, and expectations, organizations create quality systems. *Within a* ***quality system,*** *the necessary ingredients exist to enable the organization's employees to identify, design, develop, produce, deliver, and support products or services that the customer wants.* A quality system is dynamic. It is able to adapt and change to meet the needs, requirements, and expectations of its customers. Managers of a quality system coordinate the efforts of those in the system to effectively provide products or services for their customers.

Management begins to develop a system by creating a vision or mission that sets the direction for the company. The vision is supported by strategies, which are in turn supported by goals and objectives. Developing and shaping visions, missions, strategies, goals, and objectives is a complicated process. Standards such as ISO 9000 and QS 9000 and awards like the Malcolm Baldrige National Quality Award provide guidance for establishing a quality system's structure, maintaining records, and integrating the use of quality techniques.

ISO 9000

Continued growth in international trade revealed the need for a set of quality standards to facilitate the relationship between suppliers and purchasers. The creation of the ISO 9000 series of international standards began in 1979 with the formation of a technical committee with participants from 20 countries. Named the International Organization for Standardization, this Geneva-based association continues to revise and update the standards. The name "ISO 9000" has its origin in the Greek word "isos," meaning equal. The intent of the standards is to make comparisons between companies equal.

The purpose of the ISO standards is to facilitate the multinational exchange of products and services by providing a clear set of quality system requirements. Companies competing on a global basis are finding it necessary to adopt and adhere to these standards. The standards provide a baseline against which an organization's quality system can be judged. This baseline has as its foundation the achievement of customer satisfaction through multidisciplinary participation in quality-improvement efforts, documentation of systems and procedures, and the basic structural elements necessary to quality systems. The generic nature of the standards allows the interested company to determine how the standards apply to its organization. Many companies use ISO 9000 as the foundation for their continuous improvement efforts.

ISO 9000 is applicable to nearly all organizations, including manufacturers of pieces, parts, assemblies, and finished goods; developers of software; producers of processed materials, including liquids, gases, solids, or combinations; and service providers. ISO 9000 is actually composed of five basic standards: ISO 9000, 9001, 9002, 9003, and 9004. ISO 9000 provides definitions and concepts, quality-management and quality-assurance standards, guidelines for auditing quality systems, and quality-assurance requirements for measuring equipment. ISO 9000 also explains how to use the other standards in the series. The remaining standards in the series—ISO

9001, 9002, 9003, 9004—all contain standards and practices specific to quality assurance. Compliance to the standards is monitored through registration and audits. Figure 13.1 shows the flow of a typical registration process.

ISO 9000 is the standard for design, development, production, installation, and service. This standand applies to manufacturing companies that design and build products. There are 20 quality-system requirements that must be met in ISO 9000 (Figure 13.2). ISO 9002 is for companies who produce, install, and/or service products. When compared with the requirements for ISO 9000, 19 of the 20 are the same. Only the section on design control is not applicable. ISO 9003 deals with final inspection and test. Testing laboratories, farms, and equipment distributors are users of this standard. Sixteen of the 20 ISO 9000 requirements are used in ISO 9003; design control, purchasing, process control, and servicing are not utilized. Table 13.1 summarizes which requirements apply to each of these sections. Companies use ISO 9004 when they are interested in applying the above standards for their inherent benefits but are not seeking registration. Guidelines for quality management and quality systems are also found in ISO 9004. Table 13.2 outlines the information found in ISO 9004.

A quick comparison of Figure 13.2 and the table of contents of this text will reveal how the different quality topics discussed in this text are integrated into the ISO 9000 requirements. Chapter 2, Quality Advocates and Quality Management, provides background for the requirements found in Section 4.1 of ISO 9000. Chapter 11, Quality Costs, provides information to be analyzed and used in Section 4.1.5. ISO 9000 Sections 4.2, Quality Systems, and 4.14, Corrective and Preventive Action, contain elements of teamwork and problem solving discussed in Chapter 3 of this text. The chapters on statistics and probability support Section 4.20, Statistical Techniques. Chapters 5, 6, 7, 9, and 10 are used in conjunction with Sections 4.9, Process Control, and 4.10, Inspection and Testing. In the next chapter, the topic of Auditing, ISO Section 4.17, will be covered. ISO 9000 provides the framework to integrate a wide variety of quality information.

Documentation and record keeping are important aspects of ISO 9000. The documentation includes three distinct areas:

1. The companywide quality manual containing policies, commitment statement, and objectives
2. The procedures section containing operating procedures and the who, what, where, when, why, and how
3. The work instructions section containing operating instructions for processes, equipment, customer requirements, process control, and statistical analysis

ISO 9000 requires records of many plant activities, including employee training records: procedures, policies, and instructions; process control charts and capability records; purchasing records; test and reliability data; audit records; incoming and final inspection records; and equipment calibration records. Companies following the ISO 9000 criteria need to keep records of any information that is useful in the operation of the organization. Evidence that procedures, policies, and instructions are being followed must also accompany these records.

Decide to implement ISO 9000 Form management committee Develop strategic plan Begin ISO 9000 training Educate workforce Do cost assessment Do self-assesment	Establish steering committee Communicate intentions to entire corporation Select and train audit teams Continue ISO 9000 training Continue education Begin implementing standards	Conduct internal audits Organize quality system Select registrar Continue to train audit teams Continue ISO 9000 training Continue education Define areas for improvement	Continue internal audits Begin documentation: Analyze processes Write procedures Continue to train audit teams Continue ISO 9000 training Continue education Create quality manual Implement new procedures	Implement and document procedures Initial visit made	Revise, improve, update, review, take corrective action based on initial visit Revise quality manual Conduct management review	Begin preassessment procedures Correct deficiencies Document and implement practices	Conduct registration assessment	Complete registration Strive for continuous improvement Conduct surveillance audits Continue management reviews

Figure 13.1 ISO 9000 Registration Cycle

4.1 Management Responsibility
- .1 Quality Policy
- .2 Organization
  - Responsibility and Authority
  - Resources
  - Management
  - Organizational Interfaces
- .3 Management Review
- .4 Business Plan
- .5 Analysis and Use of Company-Level Data
- .6 Customer Satisfaction

4.2 Quality System
- .1 General
- .2 Quality System Procedures
- .3 Quality Planning
  - Quality Planning
  - Special Characteristics
  - Use of Cross-Functional Teams
  - Feasibility Reviews
  - Process Failure Mode and Effects Analysis
  - The Control Plan

4.3 Contract Review
- .1 General
- .2 Review
- .3 Amendment to a Contract
- .4 Records

4.4 Design Control
- .1 General
- .2 Design and Development Planning
- .3 Organizational and Technical Interfaces
- .4 Design Input
- .5 Design Output
- .6 Design Review
- .7 Design Verification
- .8 Design Validation
- .9 Design Changes

4.5 Document and Data Control
- .1 General
- .2 Document Approval and Issue
- .3 Document and Data Changes

4.6 Purchasing
- .1 General
- .2 Evaluation of Subcontractors
  - Subcontractor Development
  - Scheduling Subcontractors
- .3 Purchasing Data

Figure 13.2 ISO 9000 Quality System Requirements

.4 Verification of Purchased Product
Supplier Verification at Subcontractors
Customer Verification of Subcontracted Product
4.7 Control of Customer-Supplied Product
4.8 Product Identification and Traceability
4.9 Process Control
Government Safety and Environmental Regulations
Designation of Special Characteristics
Preventive Maintenance
.1 Process Monitoring and Operator Instructions
.2 Preliminary Process Capability Requirements
.3 Ongoing Process Performance Requirements
.4 Modified Preliminary or Ongoing Capability Requirements
.5 Verification of Job Setups
.6 Process Changes
.7 Appearance Items
4.10 Inspection and Testing
.1 General
Acceptance Criteria
Accredited Laboratories
.2 Receiving Inspection and Testing
Incoming Product Quality
.3 In-Process Inspection and Testing
.4 Final Inspection and Testing
.5 Inspection and Test Records
4.11 Control of Inspection, Measuring and Test Equipment
.1 General
.2 Control Procedure
.3 Inspection, Measuring, and Test Equipment Records
.4 Measurement System Analysis
4.12 Inspection and Test Status
4.13 Control of Nonconforming Product
.1 General
.2 Review and Disposition of Non-conforming Product
.3 Control of Reworked Product
.4 Engineering Approved Product Authorization
4.14 Corrective and Preventive Action
.1 General
.2 Corrective Action
.3 Preventive Action
4.15 Handling, Storage, Packing, Preservation and Delivery
.1 General
.2 Handling
.3 Storage
.4 Packaging
.5 Preservation
.6 Delivery

Figure 13.2 Continued

4.16 Control of Quality Records
 .1 Record Retention
 .2 Superseded Parts
4.17 Internal Quality Audits
 Inclusion of Working Environment
4.18 Training
4.19 Servicing
4.20 Statistical Techniques
 .1 Identification of Need
 .2 Procedures
  Selection of Statistical Tools
  Knowledge of Basic Statistical Control

Figure 13.2 Continued

SOURCE: This material is reprinted from ISO 9000 with permission of the American National Standards Institute (ANSI) under an exclusive licensing agreement with the International Organization for Standardization. *Not for resale.* No part of ISO 9000 may be copied or reproduced in any form, electronic retrieval system or otherwise, without the prior written consent of the American National Standards Institute, 11 West 42nd Street, New York, NY 10036.

Table 13.1 Summary of ISO 9000 Requirements

Category	***9001***	***9002***	***9003***
4.1 Management responsibility	X	X	*
4.2 Quality system	X	X	*
4.3 Contract review	X	X	X
4.4 Design control	X		
4.5 Document and data control	X	X	X
4.6 Purchasing	X	X	
4.7 Control of customer-supplied product	X	X	X
4.8 Product identification and traceability	X	X	*
4.9 Process control	X	X	
4.10 Inspection and testing	X	X	*
4.11 Control of inspection, measuring, and test equipment	X	X	X
4.12 Inspection and test status	X	X	X
4.13 Control of Nonconforming product	X	X	*
4.14 Corrective and preventive action	X	X	*
4.15 Handling, Storage, Packaging, Preservation, and Delivery	X	X	X
4.16 Control of quality records	X	X	*
4.17 Internal quality audits	X	X	*
4.18 Training	X	X	*
4.19 Servicing	X	X	
4.20 Statistical techniques	X	X	*

*Partial applicability.

Table 13.2 Outline of ISO 9004

Goods Producers	***Service Providers***
Definitions	Definitions
Management responsibility	Characteristics of services
Quality system elements	Service and service-delivery characteristics
Financial considerations of quality systems	Control of service and service-delivery characteristics
Quality in marketing	Quality system principles
Quality in specification and design	Key aspects of a quality system
Quality in purchasing	Management responsibility
Quality of processes	Personnel and material resources
Control of processes	Quality system structure
Product verification	Interface with customers
Control of inspection, measuring, and test equipment	Quality system operational elements
Control of nonconforming product	Marketing process
Corrective action	Design process
Postproduction activities	Service-delivery process
Quality records	Service performance analysis and improvement
Personnel	
Product safety	
Use of statistical methods	

A great deal of emphasis is placed on the need for excellent record keeping. In most cases, since the product has left the manufacturing facility or the service has been performed, only clearly kept records can serve as evidence of product or service quality. Sloppy or poorly maintained records give the impression of poor quality. High-quality records are easy to retrieve, legible, appropriate, accurate, and complete. Necessary records may originate internally or be produced externally. Customer or technical specifications and regulatory requirements are considered external records. Internally produced records include forms, reports, drawings, meeting minutes, problem-solving documentation, and process control charts. A high-quality documentation control system will contain records that are easily identified and used in the decision-making process.

Companies seeking registration must have their compliance with the appropriate ISO 9000 standard judged by an independent ISO 9000–certified registrar. Before the ISO 9000 governing body grants certification, a registrar conducts a thorough audit to verify that the company does indeed meet the requirements as set forth in the applicable ISO 9000 standard. When the company desiring accreditation feels it is ready, it invites an auditor to observe the company's operations and determine its level of compliance with the standards. Many firms find that conducting an internal audit before the actual registrar visit is more effective than a single audit. During this preaudit, deficiencies in the company's methods can be identified and corrected prior to the registrar's official visit. Once registration has been achieved, surveillance audits are conducted, often unannounced, approximately every six months. These audits are intended to ensure continued compliance.

Obtaining ISO 9000 certification is a time-consuming and costly process. Depending on the current state of an organization's quality system, preparation for certification may take several thousand employee-hours and cost thousands of dollars. Costs depend on the company size, the strength of the organization's existing quality system, and the number of plants within the company requesting certification.

As with any major process improvement, the opportunity to fail exists. Attempts to incorporate ISO 9000 into the way a company does business may be hindered by a variety of forces, including insufficient management involvement in the process, inadequate resources, lack of an implementation plan, or lack of understanding about ISO 9000 and its benefits. This last force, a lack of understanding about ISO 9000, is particularly crucial. ISO 9000 requires significant documentation. The additional burden of paperwork, without an understanding of how this newfound information can be used in decision-making, leads to problems. It is important to realize that standardized procedures and organized information go a long way toward preventing errors that lead to poor quality products and services. Individuals who are unaware of how to access the power that procedures and information provide may miss out on improvement opportunities. It is up to management to encourage the use of this information, thus gaining the maximum benefits from, rather than merely giving lip service to, ISO 9000.

The high cost of certification is counterbalanced by the benefits an organization will receive by using the requirements as a guide to improve their quality system. Quality becomes more consistent, and the percentage of "done right the first time" jobs increases. Improved procedures and removal of redundant operations also dramatically improve a company's effectiveness. The strength of the ISO 9000 standards lies in their ability to facilitate international trade and dramatically improve record keeping.

SUPPLIER CERTIFICATION REQUIREMENTS: QS 9000

Major corporations often purchase raw materials, parts, subassemblies, and assemblies from an outside source. To ensure quality products, the suppliers of these parts and materials are subjected to rigorous requirements. Purchasers establish these requirements and judge conformance to them by visiting the supplier's plant site and reviewing the supplier's quality systems.

Among motor vehicle manufacturers, though purchasers developed their own requirements, strong similarities existed in quality system and documentation requirements. Redundant requirements and multiple plant visits from purchasers placed a significant burden on suppliers. Conforming to several different, yet similar, sets of requirements meant unnecessarily expended time, effort, and money. Recognizing the overlap in requirements, the major automotive manufacturers—General Motors, Ford, and Chrysler—as well as truck manufacturers created a task force to develop a quality system that has as its foundation ISO 9000. Named "Quality System Requirements QS 9000," this comprehensive quality system was intended to develop fundamental quality systems that provide for continuous improvement. QS 9000 emphasizes defect prevention as well as the reduction of variation and waste. QS 9000 eliminates redundant requirements while maintaining customer-specific, division-specific, and commodity-specific requirements. Internal and external suppliers of production and service parts, subassemblies, materials, components, or other items to the major motor vehicle manufacturers must conform to the requirements set forth by QS 9000.

QS 9000 has three separate sections. The first section parallels information discussed in ISO 9001:1994, Section 4. This section is followed by information concerning sector-specific requirements (Figure 13.3). The third section covers company-specific, division-specific, and commodity-specific requirements. Since QS 9000 contains information supplemental to ISO 9000, ISO 9000 registration alone is not sufficient.

ISO 14000

The ISO 14000 series of standards covers the area of environmental management. Divided into two main classifications, Organization/Process-Oriented Standards and Product-Oriented Standards, these voluntary standards were developed as an approach to protecting the environment. A company complying with these standards is monitoring its processes and products to determine their effect on the environment. Within the two classifications, six topic areas are covered: Environmental Management Systems, Environmental Performance Evaluation, Environmental Auditing, Life-Cycle Assessment, Environmental Labeling, and Environmental Aspects in Product Standards. The ISO 14000 series of standards enables a company to improve environmental management voluntarily. The standards do not establish product or performance stan-

Section I: ISO 9000

Section II: Sector Specific Requirements

1. Production Part Approval Process
   - .1 General
   - .2 Engineering Change Validation
2. Continuous Improvements
   - .1 General
   - .2 Quality and Productivity Improvement
   - .3 Techniques for Continuous Improvement
3. Manufacturing Capabilities
   - .1 Facilities, Equipment, and Process Planning and Effectiveness
   - .2 Mistake Proofing
   - .3 Tool Design and Fabrication
   - .4 Tooling Management

Section III: Customer Specific Requirements

Figure 13.3 QS 9000 Quality System Requirements

- ISO 14001
  Environmental Management Systems—Specification with Guidance for Use
- ISO 14004
  Environmental Management Systems—General Guidelines on Principles, Systems, and Supporting Techniques
- ISO 14010
  Guidelines for Environmental Auditing—General Principles on Environmental Auditing
- ISO 14011/1
  Guidelines for Environmental Auditing—Audit Procedures—Auditing of Environmental Management Systems
- ISO 14012
  Guidelines for Environmental Auditing—Qualification Criteria for Environmental Auditors
- ISO 14014
  Initial Reviews
- ISO 14015
  Environmental Site Assessments
- ISO 14031
  Evaluation of Environmental Performance
- ISO 14020
  Goals and Principles of All Environmental Labeling
- ISO 14021
  Environmental Labels and Declarations—Self-Declaration Environmental Claims—Terms and Definitions

Figure 13.4 ISO 14000 Process-Oriented Document Requirements

dards, establish mandates for emissions or pollutant levels, or specify test methods. The standards do not expand upon existing government regulations. Figures 13.4 and 13.5 provide the general structure of the ISO 14000 series of standards.

MALCOLM BALDRIGE NATIONAL QUALITY AWARD

The Malcolm Baldrige National Quality Award was established in 1987 by the U.S. Congress. Similar to Japan's Deming Prize, it sets a national standard for quality excellence. The award has three eligibility categories: manufacturing companies, service companies, and small businesses. It is managed by the U.S. Department of Commerce and administered by the American Society for Quality Control.

The award allows companies to make comparisons with other companies. This activity is called benchmarking. ***Benchmarking*** *is a continuous process of measuring products, services, and practices against competitors or industry leaders*. Covered in more detail in Chapter 14, benchmarking lets an organization know where they stand compared with others in their industry. Companies also use the award guidelines to determine a baseline. ***Baselining*** *is measuring the current level of quality in an organization.*

- ISO 14022
  Environmental Labels and Declarations—Symbols
- ISO 14023
  Environmental Labels and Declarations—Testing and Verifications
- ISO 14024
  Environmental Labels and Declarations—Environmental Labeling
  Type 1—Guiding Principles and Procedures
- ISO 1402X
  Type III Labeling
- ISO 14040
  Life Cycle Assessment—Principles and Framework
- ISO 14041
  Life Cycle Assessment—Life Cycle Inventory Analysis
- ISO 14042
  Life Cycle Assessment—Impact Assessment
- ISO 14043
  Life Cycle Assessment—Interpretation
- ISO 14050
  Terms and Definitions—Guide on the Principles of ISO/TC 207/SC6 Terminology Work
- ISO Guide 64
  Guide for the Inclusion of Environmental Aspects of Product Standards

Figure 13.5 ISO 14000 Product-Oriented Document Requirements

Baselines are used to show where a company is, so that it knows where it should concentrate its improvement efforts.

Standards to be used as baselines and benchmarks for total quality management are set in seven areas (Figure 13.6). The seven categories are leadership, information and analysis, strategic quality planning, human resource development and management, management of process quality, quality and operational results, and customer focus and satisfaction. The following descriptions are paraphrased from the Malcolm Baldrige National Quality Award criteria.

1.0 Leadership The criteria in Section 1.0 are used to examine senior-level management's commitment to, and involvement in the process of quality improvement. Company leaders are expected to develop and sustain a customer focus supported by visible quality values. Leaders must be responsible for managing quality throughout all aspects of an organization. Subcategories include senior executive leadership, leadership system and organization, and public responsibility and corporate citizenship.

- 1.0 Leadership
  - 1.1 Senior Executive Leadership
  - 1.2 Leadership System and Organization
  - 1.3 Public Responsibility and Corporate Citizenship
- 2.0 Information and Analysis
  - 2.1 Management of Information and Data
  - 2.2 Competitive Comparisons and Benchmarking
  - 2.3 Analysis and Uses of Company Data
- 3.0 Strategic Quality Planning
  - 3.1 Strategic Development
  - 3.2 Strategic Deployment
- 4.0 Human Resource Development and Management
  - 4.1 Human Resource Planning and Evaluation
  - 4.2 High Performance Work Systems
  - 4.3 Employee Education, Training and Development
  - 4.4 Empoyee Well-being and Satisfaction
- 5.0 Management of Process Quality
  - 5.1 Design and Introduction of Products and Services
  - 5.2 Process Management: Product and Service Production and Delivery
  - 5.3 Process Management: Support Services
  - 5.4 Management of Supplier Performance
- 6.0 Quality and Operational Results
  - 6.1 Product and Service Quality Results
  - 6.2 Company Operational and Financial Results
  - 6.3 Supplier Performance Results
- 7.0 Customer Focus and Satisfaction
  - 7.1 Customer and Market Knowledge
  - 7.2 Customer Relationship Management
  - 7.3 Customer Satisfaction Results
  - 7.4 Customer Satisfaction Comparison

Figure 13.6 Malcolm Baldrige Award Categories, 1996 Criteria (U.S. Department of Commerce)

2.0 Information and Analysis This category investigates the company's use of data and information to encourage quality excellence. Quality and performance information must be used to improve operational and competitive performance. The award recognizes that data are only useful when put to work to identify areas for improvement. Competitive comparisons and benchmarking are encouraged. Subcategories include management of information and data, competitive comparisons and benchmarking, and analysis and uses of company-level data.

3.0 Strategic Quality Planning To score well in this category, a company needs to have key quality requirements integrated into the overall business plan, both short and long term. The company must utilize strategic quality and performance data in the planning process. Subcategories are strategic development and strategic deployment.

4.0 Human Resource Development and Management Within human resource development and management, reviewers for the Baldrige Award are interested in a company's plans and actions to enable their workforce to develop to its fullest potential. The environment or culture created within a company to support quality excellence is reviewed. Employee involvement, education, training, and recognition are considered in this category. Subcategories include human resource planning and evaluation; high performance work systems; employee education, training, and development; and employee well-being and satisfaction.

5.0 Process Management Quality excellence is an ongoing process. Within this category, the company is judged on its plans supporting the continuous improvement of quality and operational performance goals. Reviewers assess the processes that support designing, developing, making, or providing a quality product or service. Subcategories are design and introduction of quality products and services, process management in product and service production and delivery, process management in support services, and management of supplier performance.

6.0 Business Results Category 6, business results, examines a company's quality performance, past and present. Examiners look for positive trends in quality improvement, supplier quality, and overall company operational performance. In this section, the company's financial performance and the effects of the quality system on that performance are reviewed. Subcategories include product and service quality results, company operational and financial results, and supplier performance results.

7.0 Customer Focus and Satisfaction The final category of the Baldrige Award criteria deals with the company's relationship with their customers. This category focuses on a company's knowledge of customer requirements and marketplace competitiveness. Reviewers also determine if the company has put this knowledge to work in the improvement of products, processes, and services. This category clarifies a company's commitment to their customers. The subcategories are customer and market knowledge, customer relationship management, customer satisfaction determination and results, and customer satisfaction comparison.

SUMMARY

ISO 9000, supplier certification requirements, and the Malcolm Baldrige Award criteria provide direction and support for companies interested in developing a total quality system. As discussed in Chapter 2, five key aspects of total quality are customer focus, teamwork, problem solving, measurement, and continuous improvement of the system that provides the product or service. Table 13.3 summarizes how ISO 9000 differs from the Baldrige Award criteria and how the two compare with a comprehensive quality system. Perhaps the key difference among the three is the purpose of each. The Malcolm Baldrige Award uses results-based performance assessments to improve competitiveness, whereas the ISO 9000 standards focus on functional requirements and record keeping to support international cooperation. Neither of these has the clear continuous improvement mandate that characterizes a total quality system. The Baldrige Award criteria, which set standards for leadership and customer-driven quality, most closely reflect the total quality concept.

Table 13.3 Comparison of ISO 9000, the Malcolm Baldrige Award Criteria, and Continuous Improvement/Quality Management

	ISO 9000	*Baldrige Award*	*CI/QM*
Scope	Quality of system structure	Quality of management	Quality management and corporate citizenship Continuous improvement
Basis for defining quality	Features and characteristics of product or service	Customer-driven	Customer-driven
Purpose	Clear quality system requirements for international cooperation Improved record keeping	Results-driven competitiveness through total quality management	Continuous improvement of customer service
Assessment	Requirements-based	Performance-based	Based on total organizational commitment to quality
Focus	International trade Quality links between suppliers and purchasers Record keeping	Customer satisfaction Competitive comparisons	Processes needed to satisfy internal and external customers

Lessons Learned

1. ISO 9000 is a requirements-based assessment that supports the development of a quality system structure.
2. ISO 9000 standards focus on functional requirements and record keeping to support international cooperation.
3. QS 9000 combines requirements from ISO 9000 with sector- and customer-specific requirements.
4. QS 9000 requires that suppliers create, document, and implement a quality system.
5. Companies competing for the Malcolm Baldrige National Quality Award must perform well in the following categories: leadership, information and analysis, strategic quality planning, human resource development and management, management of process quality, quality and operational results, and customer focus and satisfaction.

6. The Malcolm Baldrige Award uses results-based performance assessments to improve competitiveness.
7. Neither ISO 9000 nor the Malcolm Baldrige Award has the comprehensive emphasis on continuous improvement mandated by a total quality system.
8. The Baldrige Award criteria, with their standards for leadership and customer-driven quality, most closely reflect the total quality system. ■

Chapter Problems

Quality Systems

1. What is meant by the term "quality system"?
2. Why would a quality system be critical to providing a quality product or service?
3. What attributes would you expect to be present in a company that has a sound quality system?
4. Describe the quality system that existed at your most recent place of employment. How would you rate its effectiveness? Support your rating with examples.

ISO 9000/QS 9000

5. Describe which types of companies would use ISO 9001, ISO 9002, and ISO 9003.
6. Find an article about a company that is in the process of achieving or has achieved ISO 9000 certification. What were the steps that they had to take? What difficulties did they encounter?
7. Contact a local company that is in the process of achieving or has achieved ISO 9000 certification. What were the steps that they had to take? What difficulties did they encounter?
8. Describe the differences between the ISO 9000 series of requirements and the requirements for QS 9000.
9. Find an article discussing QS 9000. After reading the article, answer any or all of the following questions:
   a. Who needs to become certified?
   b. What steps does a company have to take to become certified?
   c. What are the reactions to QS 9000?

Malcolm Baldrige

10. Describe the main premise of each of the criteria for the Malcolm Baldrige Award.
11. Research a Malcolm Baldrige Award winner. What did they have to change about their quality system in order to become a winner? What did they consider the most important criterion? Why? How did they go about achieving that criterion?
12. Discuss the differences among ISO 9000, the Baldrige Award, and total quality.

CASE STUDY 13.1
Malcolm Baldrige Award Criteria

Many firms have implemented comprehensive quality-improvement programs. Recent articles in trade publications as well as nationally known magazines such as *Forbes*, *Business Week*, and *Newsweek* have reported on what these companies have been doing in the area of quality.

Assignment

Read one or more comprehensive articles from a current business or news magazine and on the basis of your reading assess the quality improvement activities of one company. Choose one of the seven areas found in the Malcolm Baldrige Award criteria and make your assessment along the same guidelines. You may wish to strengthen your assessment by contacting the company itself. Use the following questions to aid you in investigating the firm.

1. Leadership
   a. What is the attitude and involvement of top management? How is this visible? Check top management's understanding of quality, their investment of time and money in quality issues, their willingness to seek help on quality management, their support of each other and subordinates, the level of importance they place on quality, and their participation in quality process.
   b. What importance does the company management place on developing a quality culture? How is this visible? Check their understanding of quality, their investment of time and money in quality issues, their willingness to seek help on quality management, their support of each other and subordinates, the level of importance they place on quality, their participation in quality process, and their training of employees.
   c. What is your perception of importance of quality to this company? How is this visible? Is the company's position based on eliminating defects by inspection? judging cost of quality by scrap and rework? or preventing defects through design of process and product?
2. Information and Analysis
   a. What types of data and information does the company collect? Are these records on customer-related issues? on internal operations? on company performance? on cost and financial matters?
   b. How does the company ensure the reliability of the data throughout their company? Are their records consistent? standardized? timely? updated? Is there rapid access to data? What is the scope of the data?

c. How are their information and control systems used? Do they have key methods of data collection and analysis? systematic collection on paper? systematic collection with computers? How do they use the information collected to solve problems? Can problems be traced to their source?

3. Strategic Quality Planning

a. What importance does the company place on quality in its strategic planning? How is this visible? Check management's understanding of quality, their investment of time and money in quality issues, how their interest is reflected in the strategic plan, the level of importance they assign the strategic plan, and whether quality control is evident throughout the plan or in just one section.

b. How does the company develop their plans and strategies for the short term? for the long term?

c. Which benchmarks does the company use to measure quality? Are the benchmarks relative to the market leader or to competitors in general? What are management's projections about the market?

d. How does the company rank the following:

- Cost of manufacturing and product provision
- Volume of output
- Meeting schedules
- Quality

e. Does the company management feel that a certain level of defects is acceptable as a cost of doing business and a way of company life?

f. Is quality first incorporated into the process at the concept development/preliminary research level? the product development level? the production/operations level? at final inspection?

g. Does the company define quality according to performance? aesthetic issues? serviceability? durability? features? reliability? fit and finish? conformance to specifications?

h. Are the following measures used to evaluate overall quality?

- Zero defects
- Parts per million
- Reject or rework rate
- Cost of quality

i. The company's quality improvement program is best described as:

- There is no formal program.
- The program emphasizes short-range solutions.
- The program emphasizes motivational projects and slogans.
- A formal improvement program creates widespread awareness and involvement.
- The quality process is an integral part of ongoing company operations and strategy.

j. Perceived barriers to a better company are

- Top-management inattention
- Perception of program costs
- Inadequate organization of quality effort
- Inadequate training
- Costs of quality not computed
- Low awareness of need for quality emphasis
- Crisis management a way of life (leaves no time)
- No formal program/process for improvement
- The management system
- The workers

k. The following steps in a quality-improvement plan have (have not) been taken:

- Obtained top-management commitment to establish a formal policy on quality
- Begun implementation of a formal companywide policy on quality
- Organized cross-functional improvement teams
- Established measures of quality (departmentally and in employee evaluation system)
- Established the cost of quality
- Established and implemented a companywide training program
- Begun identifying and correcting quality problems
- Set and begun to move toward quality goals

l. What objectives has the company set?

- Less rework
- Less scrap
- Fewer defects
- Plant utilization
- Improved yield
- Design improvements
- Fewer engineering changes (material, labor, process changes)
- Workforce training
- Improved testing
- Lower energy use
- Better material usage
- Lower labor hours per unit

4. Human Resource Development and Management

a. Quality is incorporated into the human resources system through

- Job descriptions of the president, vice president, managers, etc.
- Performance appraisals

- Individual rewards for quality-improvement efforts
- Hiring practices
- Education and training programs

b. How does the company encourage employees to buy in to quality?

- Through incentives
- Through stressing value to customers
- Through surveys or other forms of customer feedback
- Through the example of management

c. Has the quality message been distributed throughout the organization so that employees can use it in their day-to-day job activities? How has it been distributed?

d. If the importance of quality has been communicated throughout the corporation, how is that emphasis made visible?

- All employees are aware of the importance of quality. How is this visible?
- All employees are aware of how quality is measured. How is this visible?
- All employees are aware of the role of quality in their job. How is this visible?
- All employees are aware of how to achieve quality in their job. How is this visible?

e. Have teams been established to encourage cross-functional quality efforts?

- Yes, multifunctional teams seek continuous improvement.
- Yes, teams have been formed in nonmanufacturing and staff departments to encourage quality companywide.

f. Which of the following topics are stressed in employee participation groups? Is this stress reflected in the minutes, in attitudes, in solutions to problems?

- Product quality
- Service quality
- Quality of work life
- Productivity
- Cost
- Safety
- Energy
- Schedules
- Specifications
- Long-term versus quick-fix solutions

g. What percentage of employees have been involved in training programs in the past three years?

- On the managerial/supervisory level?
- On the nonmanagerial level?
- Among clerical and shop workers?

h. Which of the following topics are covered in the training programs? What is the depth of coverage?

- Process control
- Problem solving
- Data gathering and analysis
- Quality tools (Pareto, cause/effect charts, etc.)

5. Management of Process Quality

a. How are designs of products, services, and processes developed so that customer requirements are translated into design and quality requirements?

b. How do people handle problem solving and decision making? How is this visible?

- Crisis management prevails. Quality problems arise and are fought on an ad hoc basis.
- Individuals or teams are set up to investigate major problems.
- Problem solving is institutionalized and operationalized between departments. Attempts are frequently made to blame others.
- Problems and potential problems are identified early in development. Data and history are used for problem prevention.

c. The following quality-related costs are compiled and analyzed on a regular basis:

- Scrap
- Product liability
- Product redesign
- Repair
- Warranty claims
- Consumer contacts/concerns
- Inspection
- Specification/documentation review
- Design review
- Engineering change orders
- Service after service
- Rework
- Quality audits
- Test and acceptance
- Supplier evaluation and surveillance

d. How are costs of quality (or nonquality) calculated?

- Costs are not computed; there is little or no awareness of total costs.
- Direct costs are computed for rework, scrap, and returns, but the total cost of nonquality is not.
- Total costs are computed and are related to percentage of sales or operations.
- The costs of quality (prevention, appraisal, failure) are computed and reduced to two to three percent of actual sales.

e. Are periodic audits conducted to determine if the system is meeting the goals?

f. How are supplier relationships studied?

- Suppliers are certified by requiring evidence of statistical process control (SPC).
- Defects in shipments are identified and suppliers have to pay for them.
- A close working relationship has been established, allowing the suppliers to participate in the design/manufacture of the products.
- A just-in-time system has been established.
- Suppliers are rated with a formal system based on quality levels, capacity, production facilities, and delivery on schedule.

6. Quality and Operational Results

a. How does the company track the key measures of product, service, and process quality?

b. How does the company benchmark itself against other companies?

c. How does the company justify expenditures on quality improvements?

- By traditional accounting procedures (expenditures justified only if lower than cost of product failure)
- By normal capital budgeting procedures that incorporate risk-adjusted net present value and discounted costs of capital
- By cash flow increase
- By the bottom line: return on investment, net profit percent, etc.

d. The finance and accounting system has the following roles in the quality management system:

- It calculates and tracks cost of quality.
- It measures and reports quality trends.
- It distributes quality data to appropriate persons.
- It designs performance measures.
- It compares performance to competition.
- It innovates new ways to evaluate quality investments.
- It turns the traditional cost-accounting system into a quality management system.

e. What tangible benefits has this company seen related to their quality program? How are the benefits visible? How are they measured?

- Increased sales
- Increased return on investment
- Customer satisfaction
- Lower cost
- Higher selling price
- More repeat business
- Higher market share
- Improved cash flow

7. Customer Focus and Satisfaction

a. How does the company provide information to the customer? How easy is that information to obtain?

b. How are service standards defined?

c. How are customer concerns handled? What is the follow-up process?

d. Which statement(s) describe company efforts to provide the best customer service:

- The process has unequivocal support of top management.
- Middle management is able to make significant changes.
- Employees assume the major responsibility for ensuring customer satisfaction.
- Formal training in customer satisfaction is provided to all employees.

d. How does the company gather quality feedback from their customers?

- Through customer surveys
- Through a telephone hot line
- Through customer focus groups
- Through sales force reports
- Through service rep reports

e. Customer complaints are received by

- The CEO or his/her office
- The marketing or sales department
- The quality-assurance staff
- Service support departments
- Customer complaint bureaus

14

Benchmarking and Auditing

Learning Opportunities:

1. To understand the concepts of benchmarking and auditing
2. To understand the basics of performing a benchmark assessment
3. To understand the basics of conducting an audit

"How Are We Doing?"

Throughout time companies have compared their products and services with those of their competitors. Sometimes this comparison is done formally, sometimes informally. Benchmarking and auditing allow a company to judge themselves against standards or competitors. Properly conducted, benchmarking against external standards or companies or auditing internal performance can provide a wealth of information about a company's competitive position.

BENCHMARKING

During a ***benchmarking*** *process, a company compares its performance against a set of standards or against the performance of best-in-its-class companies.* With the information provided by the comparison, a company can determine how to improve its own performance. Benchmarks serve as reference points. The measurements and information gathered are used to make conclusions about current performance and any necessary improvements. Companies may choose different aspects of their operations to benchmark. Typical areas to benchmark include procedures, operations, processes, quality-improvement efforts, and marketing and operational strategies.

Benchmarking can be done at several levels of complexity. Some companies choose to conduct a benchmarking assessment at the perception level. From a *perception benchmark assessment,* a company hopes to learn how they are currently performing. A perception assessment can focus on internal issues, seeking to answer questions related to what the people within the company think about themselves, the management, the company, or the quality-improvement process. Such an assessment reveals information about the company's current performance levels. This assessment may later serve as a baseline to compare with future benchmarking experiences.

Companies beginning the quest for ISO 9000 certification, qualified supplier certification, or a national quality award may choose to perform a *compliance benchmark assessment.* This more in-depth benchmarking experience verifies a company's compliance with stated requirements and standards. The information gathered will answer questions about how a company is currently performing against the published standards. This assessment will also help a company locate where compliance to standards is weak.

A third type of benchmarking assessment investigates the effectiveness of the quality system a company has designed and implemented. Complying with requirements, such as having a quality manual, does not guarantee that systems are in place to ensure the use of such a manual. An *effectiveness benchmark assessment* verifies that a company complies with the requirements and has effective systems in place to ensure that the requirements are being fulfilled.

A fourth type of benchmarking assessment deals with continuous improvement. A *continuous improvement benchmark assessment* verifies that quality is an integral and permanent facet of an organization. It judges if the company is providing lip service to quality issues or has embraced the quality philosophies and put systems into place that support continual improvement on a day-to-day basis.

Perceptions, compliance, effectiveness, and continuous improvement can be verified through conducting a thorough review of the existing business practices. This review should follow an organized format. Activities can be judged on the basis of visual observations by the reviewers, interviews with those directly involved, personal knowledge, and factual documentation.

Purpose of Benchmarking

Benchmarking is a time-consuming process. Companies planning a benchmarking assessment should carefully consider the motivating factors. Specifically, why is the company planning to do this? Benchmarking measures an organization against recognized standards or the best-performing companies in the industry. Those beginning a benchmarking assessment program should have plans in place to use the information generated by the comparison. A major pitfall of benchmarking is the failure to use the results of benchmarking to support a larger improvement strategy. Benchmarking will provide targets for improved performance.

The reasons for benchmarking are many and varied. A company may embark on a benchmarking assessment to determine if they are able to comply with quality standards set by their customers. Benchmarking will point out areas where improvements are needed before seeking certification. Benchmarking against standards verifies whether or not a company meets the certification standards and qualifications set by a customer. On a larger scale, benchmarking can be used to determine if a company's quality systems are able to meet the requirements appropriate to meet ISO 9000 or national quality award standards. Benchmarking answers such questions as: Is the company's quality system properly constructed and documented? Are systems in place to allocate resources and funding appropriately? Which areas have the greatest improvement needs? What are our internal and external customer needs? The information gathered in the assessment guides continuous improvement objectives, plans, and projects.

Benefits of Benchmarking

The primary benefit of benchmarking is the knowledge gained as to where a company stands when compared against standards set by their customers, themselves, or national certification or award requirements. With this knowledge, a company can develop strategies for meeting their own continuous improvement goals. The benchmarking experience will identify strong assets within the company as well as opportunities for improvement. Most quality-assurance certifications involve discovering how the company is currently performing, strengthening the weaknesses, and then verifying compliance with the certification standards. Since a benchmarking assessment provides an understanding of how the company is performing, it is a valuable tool throughout the certification process.

Standards for Comparison

Typically, when a company chooses to perform a benchmarking assessment, one of the following is chosen for comparison: the Malcolm Baldrige National Quality Award, the International Organization for Standardization's ISO 9000, the Deming Prize, supplier certification requirements, or other companies who are the best in their field.

Malcolm Baldrige National Quality Award

The Baldrige Award criteria are used by companies pursuing the United States's highest award for quality-management systems. The award criteria are also a popular and rigorous benchmarking tool for companies seeking a better understanding of their performance. Using the specific categories—leadership, information and analysis, strategic quality planning, human resource development and management, management of process quality, quality and operational results, and customer focus and satisfaction—companies can discover their capabilities and areas for improvement. The completeness and thoroughness of their coverage is a strong reason for using the criteria during benchmarking.

ISO 9000

Benchmarking assessments using ISO 9000 documentation are often employed when a company is seeking ISO 9000 certification. The ISO 9000 series of standards was developed to help companies effectively document the quality systems that need to be created and implemented to maintain an efficient total quality system. The standards cover areas like quality management and quality assurance. The documentation serves as a guide during the benchmarking experience. Comparisons can be made between the company's existing quality systems and those required or suggested by ISO 9000 standards.

Deming Prize

The Deming Prize, its guidelines, and its criteria are overseen by the Deming Prize Committee of the Union of Japanese Scientists and Engineers. The award guidelines are rigorous and can be used to judge whether or not an organization has successfully achieved organizationwide quality control.

Supplier Certification Requirements

Several larger manufacturers have supplier certification requirements, such as QS 9000, that a supplier must meet in order to maintain certified or preferred supplier status. A company wishing to remain a supplier must conform to the expectations set by their customers. These guidelines are often used to help companies assess where they currently stand against the requirements and determine where improvements need to be made. Once the improvements are in place, the requirements serve as a benchmark against which company performance will be measured by the customer.

Best-in-Field Companies

Comparing one's performance against those companies judged best in their field can be a powerful tool for companies wishing to improve their position in the marketplace. By comparing their own performance with that of the market leader, companies can better understand their own assets and capabilities as well as the areas needing improvement. It is important to realize that the companies to benchmark against are companies who perform well in the area under study; they may not necessarily be competitors. For instance, a corporation interested in benchmarking its packaging and shipping system may choose to gather information from an organization in an unrelated field that is known to possess an excellent system. The effectiveness of comparison company benchmarking is limited by the ability to obtain performance information from the comparison company.

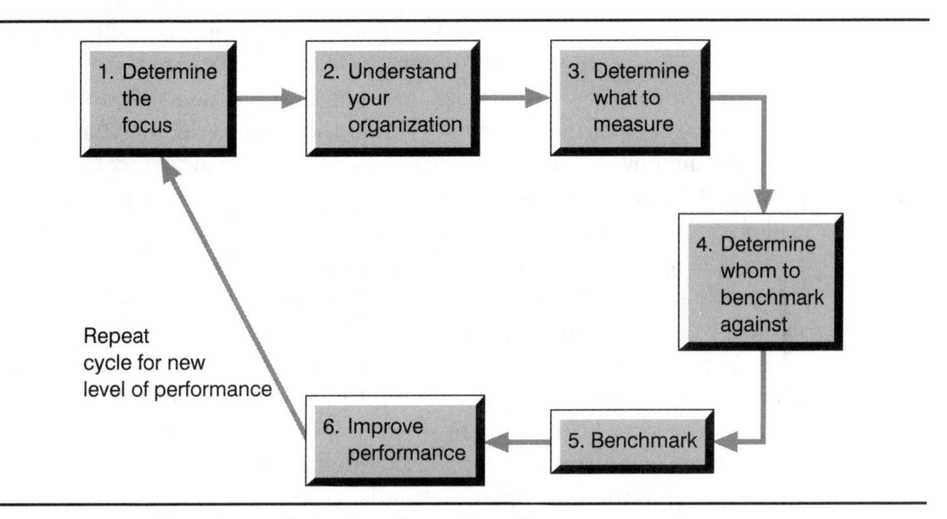

Figure 14.1 Flowchart of the Benchmarking Process

How to Benchmark

A variety of different plans and methods is available to aid interested companies in the benchmarking process. Certain steps and procedures will be part of every benchmarking experience. The following steps are usually facets of benchmarking (Figure 14.1).

1. *Determine the focus*. At the beginning of a benchmarking experience, those involved must determine what aspect of their company will be the focus of the study. The focus may be based on customer requirements, on standards, or on a general continuous improvement process. Information gathered during the benchmarking experience should support the organization's overall mission, goals, and objectives. Be aware that benchmarking and gathering information about processes is of greater value than focusing on metrics. The narrow focus on metrics can lead to stumbling on apples-and-oranges comparisons, while the focus on processes encourages improvement and adoption of new methods. To be of greater value, benchmarking should be a tool used to support a larger strategic objective.

2. *Understand your organization*. Critical to any process is defining and understanding all aspects of a situation. Individuals involved in the process need to develop an understanding of the company. To create a plan and conduct the process, information concerning the external customers, internal customers, and their major inputs and outputs is vital to achieve an understanding of the system under study. Often this step receives less attention than it should. Since people working for the company are performing the benchmarking assessment, the company is a known entity. Avoid the tendency to treat this step trivially. Use flowcharts to describe the processes involved. These will greatly enhance everyone's understanding of the system to be studied during benchmarking.

3. *Determine what to measure*. Once an understanding of the systems present in a company has been gained, it is time to determine the measures of performance. These measures will allow those conducting the benchmarking assessment to judge the performance of the company. This is the time to define what is truly critical for the company to remain competitive. These critical factors for success will be supported by standards for procedures, processes, and behaviors. Benchmarking will pinpoint questions to be answered and issues to be resolved, as well as processes and procedures to be improved. It is important to identify the questions pertinent to the company's particular operations. A well-developed list of items to be benchmarked will result in more consistent assessments and comparisons.

4. *Determine whom to benchmark against*. Choices for whom to benchmark against should be made by considering the activities and operations under investigation, the size of the company, the number and types of customers, the types of transactions, even the locations of facilities. Careful attention should be paid to selecting appropriate companies. Similarities in size and types of transactions or products may be more important in some instances than selecting a competitor. For instance, the bursar's office at a university may choose to benchmark against successful banking operations, not other universities' bursar's offices. A manufacturing company studying inventory control may be interested in the inventory control activities of a mail-order catalog operation.

5. *Benchmark*. The areas of the company that have been chosen for the benchmarking assessment should be notified prior to beginning the process. The authorization to proceed with the process should be obtained and notification should come from the highest levels of the company to ensure cooperation. During the benchmarking process, investigators collect and analyze data pertaining to the measures established in step 3. Performance measures and standards that are critical to the success of the company are used to study the company. Investigators are charged with the duty of verifying compliance to the performance measures and standards. The ability to perform to those measures and standards is judged. Compliance can be verified on the basis of interviews of those involved and direct observation of the processes.

6. *Improve performance*. Once the data and information have been gathered, a report summarizing the significant strengths and weaknesses of the area under study is created. In this report, the gap between the existing and the desired levels of performance is documented. A good report will focus on patterns of standards violations and elements missing from a strong system. The report should include recommendations for improving the processes. This report does not need to detail each of the observations made by the investigators. It should not be a list of all of the infractions seen.

In a successful benchmarking experience, the final report becomes a working document to aid the continuous improvement process. The information gathered in this report is used to investigate root causes, solve them, reduce process variation, and establish systems to prevent the occurrence of nonconformities. The benchmarking document is a powerful customer feedback tool and should be used accordingly.

EXAMPLE 14.1 The Benchmarking Process

The WP and Me Corporation, a pet food company, would like to determine whether or not it could perform well against the standards for the Malcolm Baldrige National Quality Award. While not applying for the award this year, WP and Me plans to apply for the award within the next few years. The purpose of this benchmarking experience is to see how they currently compare with the criteria and what areas they should focus on in the future. The company hopes to benefit from this type of compliance benchmarking experience before attempting to win the award.

The company undertakes the following steps:

Step 1. Determine the Focus. The Malcolm Baldrige Award has seven criteria: leadership, information and analysis, human resources development and management, management of process quality, quality and operational results, customer focus and satisfaction. WP and Me decides to focus on its ability to comply with all the award criteria areas.

Step 2. Understand the Organization. WP and Me Corporation now concentrates on understanding the current status of affairs at the company. Since their pet food has been developed on the basis of a careful study of pets' diet and health, WP and Me has always viewed a nutritional quality product as the cornerstone of their success. They encourage their people to understand the needs of their customers, two- and four-legged. They have drafted a mission statement to encourage the pursuit of continual improvement: "WP and Me Corporation strives to operate with integrity, providing a quality nutritional product for our customers. We challenge our employees to work together as teams to continuously improve our existing products and processes."

WP and Me operates on the assumption that decisions must be made to optimize the system that provides quality products to its customers. Over the past several years, information has been kept concerning customer satisfaction, operational effectiveness, continuous improvement efforts, employee development, and operational quality. This information will be used during the benchmarking process to determine the status of the WP and Me Corporation against the Malcolm Baldrige Award Criteria.

Step 3. Determine What to Measure. For each of the criteria categories, WP and Me must determine what measures are appropriate to determine compliance with the criteria. For this information they turn to the award criteria guidelines. Some examples of the areas WP and Me must consider include:

a. Customer surveys, product and service data
b. Audits
c. Amount of scrapped, reworked, and returned product
d. Customer satisfaction ratings

Step 4. Determine Whom to Benchmark Against. WP and Me are benchmarking themselves against the Baldrige Award criteria.

Step 5. Benchmark. During the actual benchmarking experience, participants search for indications that WP and Me is meeting the criteria based on the measures developed in step 3. For instance, in the category of leadership, benchmarking participants note that involvement in quality begins with the highest level of management, the founder of WP and Me. Dr. WP and her senior executives serve as role models. Their behavior exemplifies the activities and actions expected of someone practicing continuous improvement. They are involved in continuous improvement projects at their level, and they serve as resource people for projects underway throughout the firm. The decision-making approach adopted by senior-level managers encourages system optimization. When decisions are made, emphasis is placed on determining, as nearly as possible, the potential effects of such a decision on other areas of the organization.

In the area of human resources and development, WP and Me Corporation encourages employees to take part in further education and training. The reward and recognition system has been modified to reflect WP and Me's belief in the importance of continuous improvement and education.

Step 6. Improve Performance. While WP and Me is a company that has a strong quality philosophy and visible commitment from management, the benchmarking experience has revealed areas where improvements are needed. As stated earlier, this is one of the benefits that WP and Me wanted to get from the benchmarking experience. Now that certain areas have been identified for improvement, WP and Me can concentrate its efforts on these areas. For instance, in the area of information and analysis, investigators have found some room for improvement. Certain report formats and computer systems are outdated and do not best fulfill the needs of internal customers. Continuous improvement teams can now be formed to remedy the problems.

AUDITING

***Audits** are designed to appraise the activities, practices, records, or policies of an organization; they determine whether a company has the ability to meet or exceed a standard.* A variety of circumstances can initiate an audit. Audit programs may be part of customer contract requirements, or government regulations may require an audit. Audits do not have to have an outside instigator; it is not unusual to see a company create internal auditing systems to verify its own performance. Internal and supplier audits allow a company the opportunity to verify conformance to specifications and procedures. Audits may also examine aspects of equipment, software, documentation, and procedures. Whatever the reason, audits provide companies with information concerning their performance, the performance of their product, and areas for improvement.

Audits should be a positive experience used to improve the system. When deficiencies are uncovered, they should be seen as opportunities to look for solutions, not to fix blame. Audits enable a company to answer a variety of questions. Is the company achieving its quality objectives? Are procedures being followed? Are new and more efficient methods of performance documented and used where applicable? Are quality records being properly retained and used to solve production problems? Are preventive maintenance schedules being followed? Audits of systems such as material handling can reveal poor practices that need to be improved. Supplier quality and record-keeping practices can also be checked. Since audits identify opportunities for improvement, companies may perform a product or service integrity audit to verify that the process is performing in an optimal fashion. Using audits to identify process problems reduces opportunities for nonconformities.

The frequency of audits varies according to need. Areas having a significant effect directly on product creation, service provision, and product or service safety or quality are targeted for more frequent audits. When not based on customer or government requirements, the frequency of audits is usually related to the need to balance audit effectiveness with economics. To decrease the impact and disruption caused by a major companywide audit, an organization may choose to audit one or two areas or systems each month until all key areas have been evaluated.

An audit involves comparisons, checks of compliance, and discoveries of discrepancies. Because this news is not always positive, those conducting an audit may not be well received by the area being audited. To be successful, good auditors should be polite, objective, and professional. In some instances, a great deal of perseverance is necessary to find the information desired. Audits should be conducted in an objective and factual manner. The auditor's opinion should be unbiased, unfiltered, and undistorted. Audits are not subjective assessments against personal standards. The objectives, criteria, and measures against which an area is to be compared should be well defined before beginning the audit. Those involved in the audit should be notified as early as possible about the scope and breadth of the audit.

Types of Audits

Audits are designed to determine if deficiencies exist between actual performance and desired standards. They can cover the entire company, or a division of a company, or any portion of the processes that provide a product or service. Audits may be focused on product development and design, material procurement, or production. A customer may request an audit prior to awarding a contract to a supplier. Other types of audits include product design audits, preproduction audits, production audits, and supplier quality system audits.

Designing an Audit

Applying the Shewhart plan-do-study-act cycle, made popular by Dr. Deming, can help a company create an organized system for conducting an audit and using the information to make improvements. Typical auditing programs include a planning phase, the actual audit, reports recommending improvements, and follow-up action plans (Figure 14.2).

1. *Plan:* To begin, those planning the audit need to identify its purpose or objective. A statement of purpose clarifies the focus of the audit. Following this, planners will need to identify the who, what, where, when, why, and how related to the audit. Who is to be audited? Who is to perform the audit? What does this audit hope to accomplish? Is the audit to judge conformance to standards? If so, what are the critical standards? What are the performance measures? When and

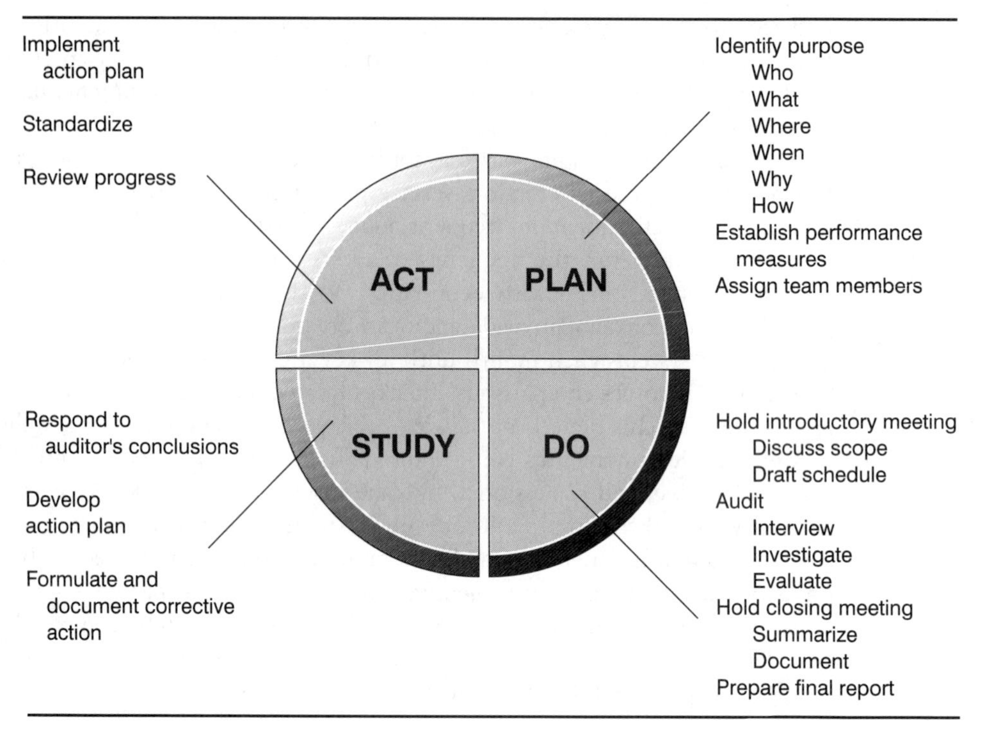

Figure 14.2 The Auditing Process

where the audit will be conducted must be set. Those about to be audited should be informed by an individual in a position of authority. Clear statements of the reasons behind the audit (why), the performance measures (what), and the procedures (how) should be given to those being audited. An audit is a valuable working document for improvement. It is important to determine how the results will be used and who will have access to the results before the auditing process begins.

2. *Do:* Using the information clarified in the planning phase, the audit is conducted. Often an introductory meeting is held by the participants to discuss the scope, objectives, schedule, and paperwork considerations. After the opening meeting, examiners begin the process of reviewing the process, product, or system under study. Auditors may require access to information concerning quality systems, equipment operation procedures, preventive maintenance records, inspection histories, or planning documents. Auditors may conduct interviews with those involved in the process of providing a product or service. Any and all information related to the area under study is critical for the success of the audit.

During the review process, auditors document their findings. These findings are presented in a general summary at a closing meeting of the participants. Within a short period of time, perhaps 10 to 20 days, the auditors will prepare a written report that documents their findings, conclusions, and recommendations.

3. *Study:* Audits provide information about the participant's strengths, weaknesses, and areas for improvement. Upon receipt, the auditor's report is read by the participants in the audit. During this phase of the audit cycle, they respond to the report and develop an action plan based on the recommendations of the auditors. This action plan should specify the steps and time frame involved in dealing with the issues raised by the audit.

4. *Act:* Once adopted, the action plan becomes the focus of the improvement activities related to the audited area. Auditors and company administrators should follow up at predetermined intervals to evaluate the status of the continuous improvement action plan. This ensures that the recommendations and conclusions reached by the auditors, and supported by an action plan, assist the company in reaching its continuous improvement goals.

EXAMPLE 14.2 Conducting an Audit

As part of checking its ability to meet the standards set by the Malcolm Baldrige National Quality Award, the WP and Me Corporation has decided to perform an internal audit on its main production line. To conduct an effective audit they followed the four steps presented in this chapter.

Step 1. Plan. To begin, those planning the audit identified the purpose or objective of the audit as being "to judge conformance to the following standards: cleanliness/lack of contamination, nutritional quality control, and compliance with written standard operating procedures (SOPs)." Following this step, the planners made a list:

Who: Who is to be audited? The main production line
Who is to perform the audit? W&T Consultants

What: What does this audit hope to accomplish? Assessment of conformance to the following standards: cleanliness/lack of contamination, nutritional quality control, and compliance with written SOPs
What are the performance measures?
Level of cleanliness and lack of contamination
Structure and performance of nutritional quality control system
Level of compliance with written SOPs

Where: Number 1 plant, main production line

When: February 1 through February 5
Report from auditors due February 28
Response from production line 1 due March 30
Follow-up by management every 30 days to check compliance

Why: To ensure that WP and Me produces the highest-quality nutritional dog food, under the cleanest possible conditions, at an optimum performance level

How: Auditors will conduct interviews, review quality documentation, review standard operating procedures, and observe line activities to create a comparison

Main production was informed by the vice president of operations about the audit and its purpose. At this meeting the performance measures and the procedures of conducting an audit were reviewed with those being audited. This made clear to the people on the line, as well as to their supervisors, manufacturing engineers, and others, the importance of the audit and the results generated by it. Since an audit is a valuable working document for improvement, the vice president of operations discussed how the audit results would be used to guide the improvement efforts within the department.

Step 2. Do. Using the information clarified in the planning phase, the audit was conducted by W&T Consultants. After the introductory meeting to discuss the scope, objectives, schedule, and paperwork considerations, the examiners began reviewing the process, the quality system, equipment operation procedures, preventive maintenance records, inspection histories, and planning documents. The auditors conducted interviews with line personnel, engineers, and supervisors.

During the review process, the auditors documented their findings. While the overall cleanliness and adherence to work standards were very good, at the closing meeting the following concerns were raised:

a. In the raw material storage area, several containers of additives had been contaminated by fumes from extraction ducts that were not airtight.
b. Also in the raw material storage area, several containers had lost their labels. There are written instructions requiring the involvement of quality assurance when the labels are replaced; however, the personnel responsible for replacing the labels were unaware of the authorized procedure. The procedure they described did not match the authorized standard operating procedure.
c. On the production line, because of the lack of space, a work schedule for the cleaning crew did not communicate the special cleaning instructions. The cleaning supervisor had been given no guidance concerning standard operating procedures.
d. While studying the procedures for ordering raw material and cleaning solutions, one of the auditors noticed that there were no purchase specifications for the critical cleaning solutions. There was also no established standard operating procedure for verifying incoming supplies.
e. The number 2 machine was not covered by a preventive maintenance program. There was no safety inspection record in the files. There was no information at all concerning the equipment's description, serial number, location, dates of inspection, condition, repairs, or standard operating procedures.
f. While the number 2 machine had almost no information available, other machines had incomplete records.

These are just a few of the findings presented in the general summary at a closing meeting of the participants. Within 20 days, the auditors prepared a written report that documented their findings, conclusions, and recommendations in detail.

Step 3. Study. The audit provided information about the participants' strengths, weaknesses, and areas for improvement. Number 1 line spent many hours developing an action plan based on the recommendations of the auditors. The action plan listed specific steps and time frames to deal with the issues raised by the audit. For example,

a. To correct the situation with incomplete equipment documentation (concerns 2e and 2f), number 1 line members will work with the maintenance department to create files containing the following information for all machines on number 1 line: equipment description, serial number, location, standard operating procedures, safety inspection records, dates of inspection, equipment condition and repairs. Completion date: June 1.
b. Number 1 line members will work with the maintenance department to create a preventive maintenance program suitable for all machines on number 1 line. Completion date: July 30.
c. To deal with concern 2c, number 1 line members will work with the cleaning department to develop standard operating procedures for all machines on number 1 line. Completion date: April 1.
d. A thorough training and review program will be established for all current and future employees on number 1 line. This will counteract situations evident in concern 2b. Completion date: April 30.
e. The raw material storage area will be redesigned to prevent future contamination. Completion date: August 30.

Step 4. Act. Management at WP and Me reviewed the action plan and adopted it as the focus of the improvement activities related to line 1. The vice president of operations will follow up every 30 days to determine the progress being made and to be available to remove any obstacles preventing implementation of the plan. This will ensure that the recommendations and conclusions reached by the auditors, and supported by an action plan, will be handled in a timely fashion.

SUMMARY

Benchmarking and auditing are both processes that enable a company to learn more about itself. A company that invests time and effort in either a benchmarking assessment or an audit will be able to verify its performance and uncover areas requiring improvement. In many cases better performance can be gained by understanding what a company is currently capable of doing versus where it needs to be performing to remain competitive.

■ *Lessons Learned*

1. Identification of performance measures enhances both benchmarking assessments and audits.
2. Benchmarking is the process of comparing a company's performance against a set of standards or against the performance of a best-in-its-class company.

3. Measurements and information gathered during both benchmarking and audits are used to make conclusions about current performance and any necessary improvements.
4. Audits are designed to appraise the activities, practices, records, and policies of an organization to determine the company's ability to meet or exceed a standard. ■

Chapter Problems

Benchmarking

1. Describe the steps involved in benchmarking.
2. Why would a company be interested in benchmarking? What would they hope to gain by benchmarking?
3. Select a company, either service- or manufacturing-related, and describe the steps and activities involved in benchmarking. Be sure to indicate which type of benchmarking would be most appropriate and why, as well as whom they should benchmark against and what they should measure.
4. A local bank considers itself a full-service bank. They offer a number of products and services to individual customers, businesses, trusts, and credit card users. Recently, the bank completed an internal quality-improvement process that included determining a mission statement and resolving issues that prevented employees from performing to the best of their ability. At this time they are evaluating external customer service. Competition in the banking industry is fierce, and it is critical that the bank discover and develop policies, procedures, and practices that will lead to customer delight. The bank's mission statement is simple: "All customers, regardless of the size of their accounts or the number of their transactions, are important!"

Auditing

5. Employees at the bank are already trained to be polite and eager to serve the customers. The bank has been able to identify several other features that are important to their customers: timeliness, accuracy, immediate service, knowledge, courtesy, convenience, and accessibility. The bank is considering benchmarking as a way to determine how they can improve. The following are some questions they must seek answers to:
   a. Which of the four types of benchmarking would be appropriate in this situation? Why?
   b. Describe the benchmarking steps that the bank should take. What should happen during each step?
   c. Describe at least three measures of performance or indicators of viability that would be appropriate for the bank to investigate. Why did you choose each of these?

   d. Against whom should the bank benchmark?
   e. After obtaining the benchmarked information, what should the bank do with it?
6. Describe the steps involved in auditing.
7. Why would a company be interested in auditing? What benefits would they hope to gain by auditing?
8. Select a company, either service- or manufacturing-related, and describe the steps and activities involved in auditing. Be sure to indicate which areas of the company would be most appropriate to audit and why, as well as who should be involved in the audit and what they should investigate.

CASE STUDY 14.1
Benchmarking

PART 1

TST, a local performing arts association, has been experiencing several years of financial prosperity. Their board of directors feels that it is time to update the formal strategic long-range plan (SLRP). The purpose of the SLRP is to focus and guide the thinking and activities of the board of directors and staff of the organization. The implementation of the SLRP is the responsibility of the board as well as of the staff.

Members of the board are very enlightened about total quality management and are determined to put their knowledge to use in creating the new SLRP. They begin their revision with the creation of mission and vision statements to guide the activities of the organization.

The vision:

> Our performing arts association will advance a national reputation of excellence to ever-expanding local and touring audiences. In addition to performing, the performing arts association will embrace vigorous and respected educational and community outreach programs, which will include a school and associated preprofessional performance troupes.

The mission:

> The purpose of the performing arts association is to educate, enlighten, and entertain the widest possible audience in the city and on tour by maintaining a professional company characterized by excellence. Its purpose is also to maintain a preprofessional training company and a school. The performing arts association will develop and maintain the artistic, administrative, technical, and financial resources necessary for those endeavors.

After developing the mission and vision statements, the board has decided to determine the viability of their organization when compared with other, similar performing arts associations. They have chosen to do benchmarking for this reason.

Assignment

1. Describe the type of benchmarking the performing arts association will be performing.
2. What benefits will the association see from benchmarking?
3. Describe the steps the association should take in the benchmarking process.

PART 2

Following the steps suggested by texts on benchmarking, the performing arts association begins its benchmarking process with a brainstorming session to determine the *focus* of the benchmarking activities. The three general areas the board feels are indications of viability are resource development, artistic excellence, and marketing effectiveness.

Resource development involves the financial aspects of the performing arts association. The association receives income from several sources, but all revenue can be classified as either unearned income—corporate and individual gifts, agency/government funding, and contributions—or earned income—ticket sales. Revenue is used to pay staff, purchase shows, employ performing artists, maintain the theater, market upcoming shows, and sponsor educational programs that are presented prior to the performance. Marketing efforts include advertising and promotional and discounted tickets. Staff members, as well as volunteers, devote a significant amount of time to efforts to solicit contributions from agencies, corporations, and individuals. These efforts are vital to the ability of the association to offer a wide variety of performances.

Artistic excellence involves such aspects as the program mix offered during the year, the performers themselves, their age and training, the number of community outreach programs, and preperformance lectures. Marketing effectiveness is based on the relationship of attendance, ticket sales, and dollars spent on marketing.

The results of the discussions are helpful in *understanding the company,* how it is viewed both internally and externally. Using their focus areas (resource development, artistic excellence, and marketing effectiveness) as a beginning, the board must *determine what to measure*.

Assignment

For each of the three areas—resource development, artistic excellence, and marketing effectiveness—brainstorm what you consider to be appropriate measures of performance (or indicators of viability).

PART 3

To identify the appropriate measures of performance, the board holds a brainstorming session. The session reveals the following:

Indications of viability:

1. Efficiency of programs
2. Marketing effectiveness
3. Educating the audience
4. Other measures of performance

Assignment

The board has listed a large number of potential measures of success or indicators of viability. To seek all of this information from the associations chosen as benchmarks would require a significant amount of time. Which would you select as the key benchmarks? Why?

PART 4

The board holds an additional session to determine the key benchmarks indicating the viability of the association. The following areas are chosen to investigate as key benchmarks:

Resource development

- Percentage of earned versus contributed income
- Percentage of contributed income from corporations, funding agencies, and individuals

Artistic excellence

- Number of performances on main stage
- Number of outreach programs
- Preperformance lecture attendance

Marketing effectiveness

- Attendance as percentage of house
- Percentage of total tickets that are complimentary or discounted tickets
- Percentage of earned income dedicated to marketing efforts

Now that the performance measures have been selected, before performing the actual benchmarking, the board needs to determine *whom to benchmark against*.

Assignment

On what factors should the selection of whom to benchmark against be based?

PART 5

The board decides to base their selection of performing arts associations to benchmark on several factors. From this they are able to select five performing arts associations to compare themselves with.

Their next step is to begin *benchmarking*. One of the committee members is chosen to contact each of the selected associations and gather the information. At the next board meeting, this individual will share the information with the other board members. The information from their benchmarking efforts will be used to *improve performance*.

Assignment

Study Figures 1 through 8 and discuss how well the performing arts association is doing when compared with other best-in-field associations.

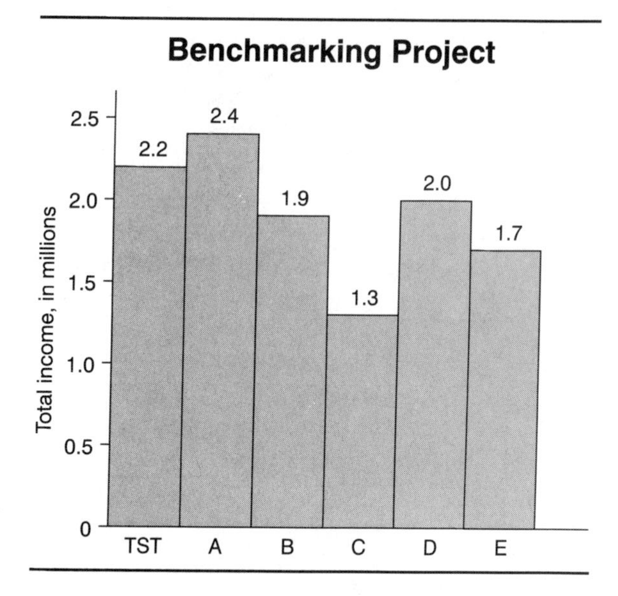

Figure 1 Marketing Expense as a Percent of Ticket Revenue

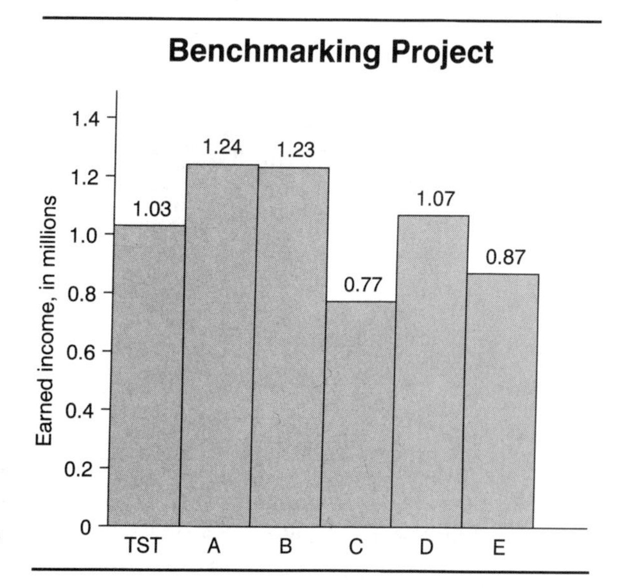

Figure 2 Marketing Expense as a Percent of Total Budget

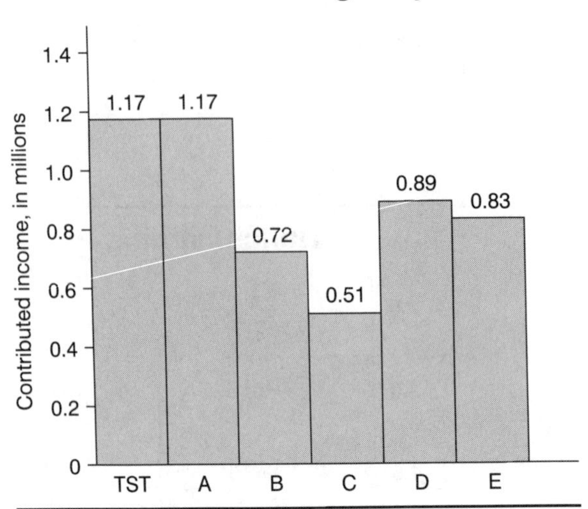

Figure 3 Total Tickets Available per Season

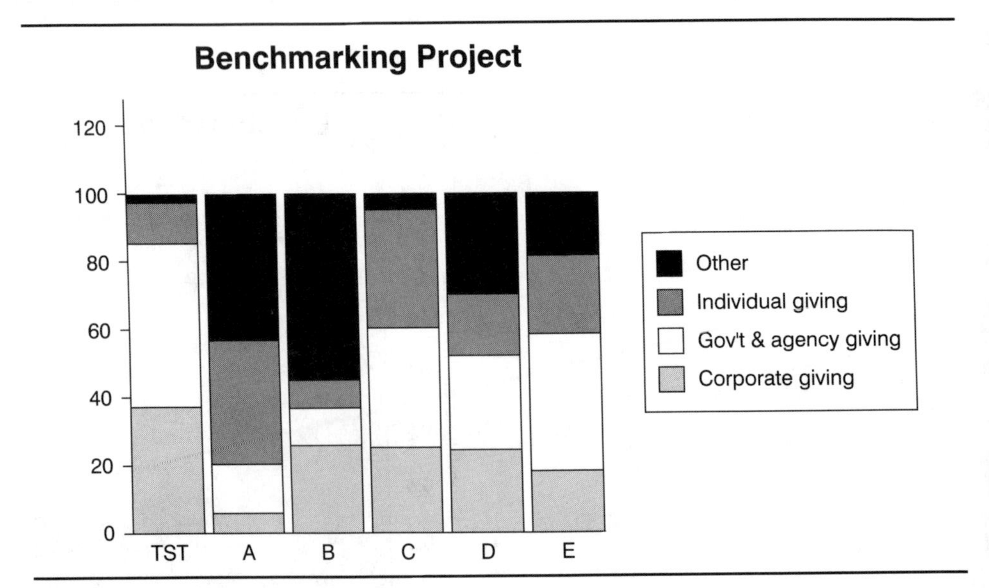

Figure 4 Performance Hall Size

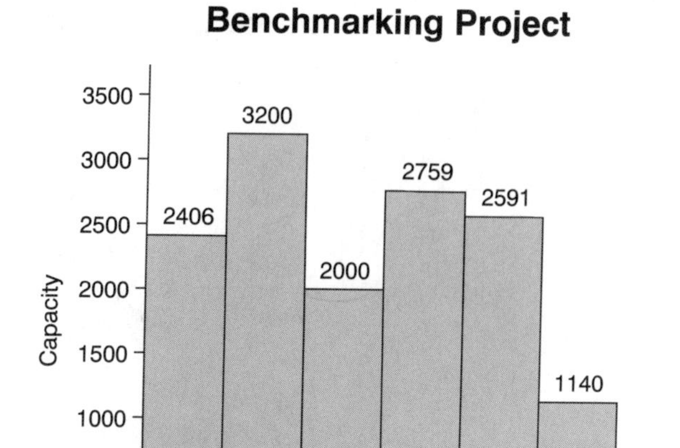

Figure 5 Composition of Contributed Income

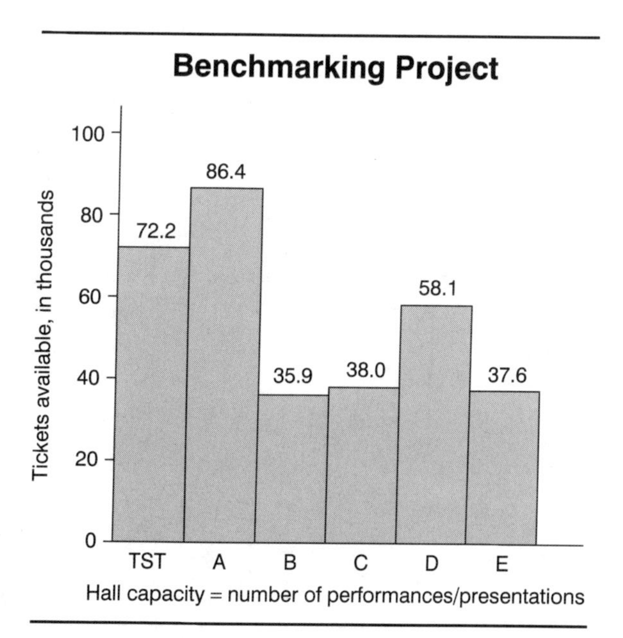

Figure 6 Total Contributed Income (in Millions)

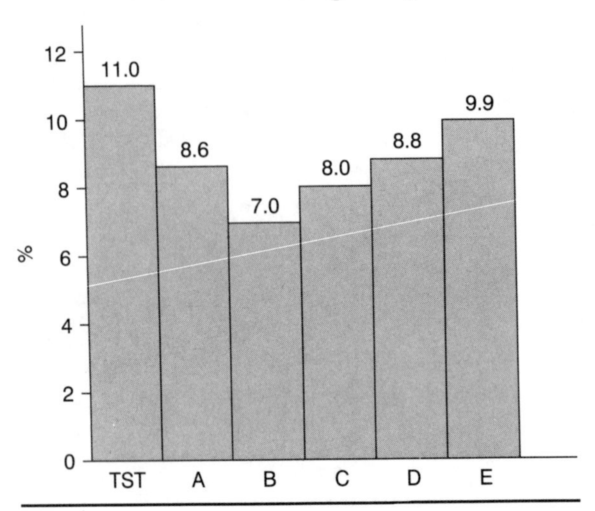

Figure 7 Total Earned Income (in Millions)

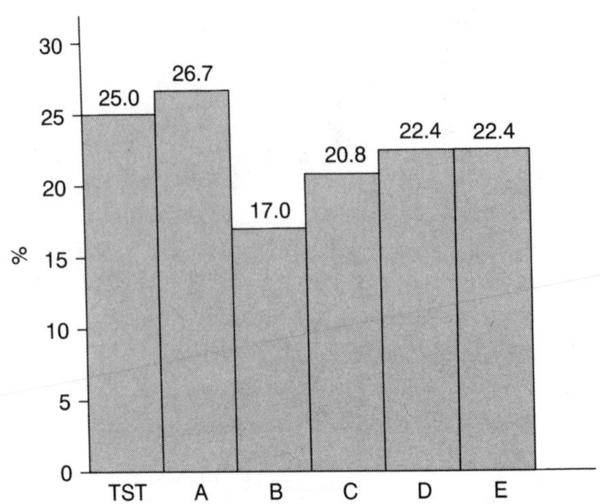

Figure 8 Total Income (in Millions)

APPENDIX 1

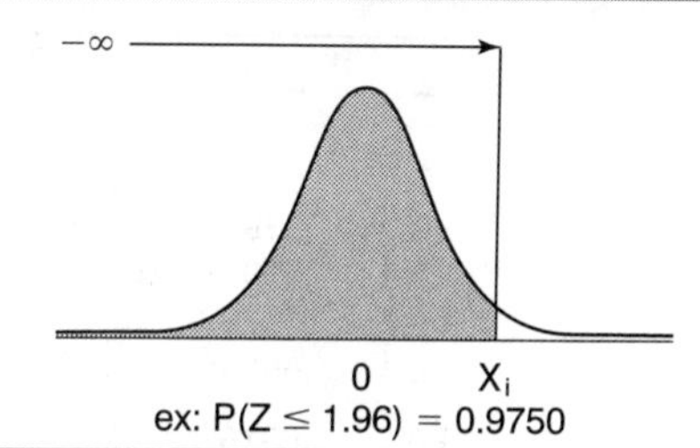

Normal Curve Areas $P(Z \leq Z_0)$

ex: $P(Z \leq 1.96) = 0.9750$

Z	*0.09*	*0.08*	*0.07*	*0.06*	*0.05*	*0.04*	*0.03*	*0.02*	*0.01*	*0.00*
−3.8	.0001	.0001	.0001	.0001	.0001	.0001	.0001	.0001	.0001	.0001
−3.7	.0001	.0001	.0001	.0001	.0001	.0001	.0001	.0001	.0001	.0001
−3.6	.0001	.0001	.0001	.0001	.0001	.0001	.0001	.0001	.0002	.0002
−3.5	.0002	.0002	.0002	.0002	.0002	.0002	.0002	.0002	.0002	.0002
−3.4	.0002	.0003	.0003	.0003	.0003	.0003	.0003	.0003	.0003	.0003
−3.3	.0003	.0004	.0004	.0004	.0004	.0004	.0004	.0005	.0005	.0005
−3.2	.0005	.0005	.0005	.0006	.0006	.0006	.0006	.0006	.0007	.0007
−3.1	.0007	.0007	.0008	.0008	.0008	.0008	.0009	.0009	.0009	.0010
−3.0	.0010	.0010	.0011	.0011	.0011	.0012	.0012	.0013	.0013	.0013
−2.9	.0014	.0014	.0015	.0015	.0016	.0016	.0017	.0018	.0018	.0019
−2.8	.0019	.0020	.0021	.0021	.0022	.0023	.0023	.0024	.0025	.0026
−2.7	.0026	.0027	.0028	.0029	.0030	.0031	.0032	.0033	.0034	.0035
−2.6	.0036	.0037	.0038	.0039	.0040	.0041	.0043	.0044	.0045	.0047
−2.5	.0048	.0049	.0051	.0052	.0054	.0055	.0057	.0059	.0060	.0062
−2.4	.0064	.0066	.0068	.0069	.0071	.0073	.0075	.0078	.0080	.0082
−2.3	.0084	.0087	.0089	.0091	.0094	.0096	.0099	.0102	.0104	.0107
−2.2	.0110	.0113	.0116	.0119	.0122	.0125	.0129	.0132	.0136	.0139
−2.1	.0143	.0146	.0150	.0154	.0158	.0162	.0166	.0170	.0174	.0179
−2.0	.0183	.0188	.0192	.0197	.0202	.0207	.0212	.0217	.0222	.0228
−1.9	.0233	.0239	.0244	.0250	.0256	.0262	.0268	.0274	.0281	.0287
−1.8	.0294	.0301	.0307	.0314	.0322	.0329	.0336	.0344	.0351	.0359
−1.7	.0367	.0375	.0384	.0392	.0401	.0409	.0418	.0427	.0436	.0446
−1.6	.0455	.0465	.0475	.0485	.0495	.0505	.0516	.0526	.0537	.0548
−1.5	.0559	.0571	.0582	.0594	.0606	.0618	.0630	.0643	.0655	.0668
−1.4	.0681	.0694	.0708	.0721	.0735	.0749	.0764	.0778	.0793	.0808
−1.3	.0823	.0838	.0853	.0869	.0885	.0901	.0918	.0934	.0951	.0968
−1.2	.0985	.1003	.1020	.1038	.1056	.1075	.1093	.1112	.1131	.1151
−1.1	.1170	.1190	.1210	.1230	.1251	.1271	.1292	.1314	.1335	.1357
−1.0	.1379	.1401	.1423	.1446	.1469	.1492	.1515	.1539	.1562	.1587
−0.9	.1611	.1635	.1660	.1685	.1711	.1736	.1762	.1788	.1814	.1841
−0.8	.1867	.1894	.1922	.1949	.1977	.2005	.2033	.2061	.2090	.2119
−0.7	.2148	.2177	.2206	.2236	.2266	.2296	.2327	.2358	.2389	.2420
−0.6	.2451	.2483	.2514	.2546	.2578	.2611	.2643	.2676	.2709	.2743
−0.5	.2776	.2810	.2843	.2877	.2912	.2946	.2981	.3015	.3050	.3085
−0.4	.3121	.3156	.3192	.3228	.3264	.3300	.3336	.3372	.3409	.3446
−0.3	.3483	.3520	.3557	.3594	.3632	.3669	.3707	.3745	.3783	.3821
−0.2	.3859	.3897	.3936	.3974	.4013	.4052	.4090	.4129	.4168	.4207
−0.1	.4247	.4286	.4325	.4364	.4404	.4443	.4483	.4522	.4562	.4602
−0.0	.4641	.4681	.4721	.4761	.4801	.4840	.4880	.4920	.4960	.5000

Z	0.00	0.01	0.02	0.03	0.04	0.05	0.06	0.07	0.08	0.09
+0.0	.5000	.5040	.5080	.5120	.5160	.5199	.5239	.5279	.5319	.5359
+0.1	.5398	.5438	.5478	.5517	.5557	.5596	.5636	.5675	.5714	.5753
+0.2	.5793	.5832	.5871	.5910	.5948	.5987	.6026	.6064	.6103	.6141
+0.3	.6179	.6217	.6255	.6293	.6331	.6368	.6406	.6443	.6480	.6517
+0.4	.6554	.6591	.6628	.6664	.6700	.6736	.6772	.6808	.6844	.6879
+0.5	.6915	.6950	.6985	.7019	.7054	.7088	.7123	.7157	.7190	.7224
+0.6	.7257	.7291	.7324	.7357	.7389	.7422	.7454	.7486	.7517	.7549
+0.7	.7580	.7611	.7642	.7673	.7704	.7734	.7764	.7794	.7823	.7852
+0.8	.7881	.7910	.7939	.7967	.7995	.8023	.8051	.8078	.8106	.8133
+0.9	.8159	.8186	.8212	.8238	.8264	.8289	.8315	.8340	.8365	.8389
+1.0	.8413	.8438	.8461	.8485	.8508	.8531	.8554	.8577	.8599	.8621
+1.1	.8643	.8665	.8686	.8708	.8729	.8749	.8770	.8790	.8810	.8830
+1.2	.8849	.8869	.8888	.8907	.8925	.8944	.8962	.8980	.8997	.9015
+1.3	.9032	.9049	.9066	.9082	.9099	.9115	.9131	.9147	.9162	.9177
+1.4	.9192	.9207	.9222	.9236	.9251	.9265	.9279	.9292	.9306	.9319
+1.5	.9332	.9345	.9357	.9370	.9382	.9394	.9406	.9418	.9429	.9441
+1.6	.9452	.9463	.9474	.9484	.9495	.9505	.9515	.9525	.9535	.9545
+1.7	.9554	.9564	.9573	.9582	.9591	.9599	.9608	.9616	.9625	.9633
+1.8	.9641	.9649	.9656	.9664	.9671	.9678	.9686	.9693	.9699	.9706
+1.9	.9713	.9719	.9726	.9732	.9738	.9744	.9750	.9756	.9761	.9767
+2.0	.9772	.9778	.9783	.9788	.9793	.9798	.9803	.9808	.9812	.9817
+2.1	.9821	.9826	.9830	.9834	.9838	.9842	.9846	.9850	.9854	.9857
+2.2	.9861	.9864	.9868	.9871	.9875	.9878	.9881	.9884	.9887	.9890
+2.3	.9893	.9896	.9898	.9901	.9904	.9906	.9909	.9911	.9913	.9916
+2.4	.9918	.9920	.9922	.9925	.9927	.9929	.9931	.9932	.9934	.9936
+2.5	.9938	.9940	.9941	.9943	.9945	.9946	.9948	.9949	.9951	.9952
+2.6	.9953	.9955	.9956	.9957	.9959	.9960	.9961	.9962	.9963	.9964
+2.7	.9965	.9966	.9967	.9968	.9969	.9970	.9971	.9972	.9973	.9974
+2.8	.9974	.9975	.9976	.9977	.9977	.9978	.9979	.9979	.9980	.9981
+2.9	.9981	.9982	.9982	.9983	.9984	.9984	.9985	.9985	.9986	.9986
+3.0	.9987	.9987	.9987	.9988	.9988	.9989	.9989	.9989	.9990	.9990
+3.1	.9990	.9991	.9991	.9991	.9992	.9992	.9992	.9992	.9993	.9993
+3.2	.9993	.9993	.9994	.9994	.9994	.9994	.9994	.9995	.9995	.9995
+3.3	.9995	.9995	.9995	.9996	.9996	.9996	.9996	.9996	.9996	.9997
+3.4	.9997	.9997	.9997	.9997	.9997	.9997	.9997	.9997	.9997	.9998
+3.5	.9998	.9998	.9998	.9998	.9998	.9998	.9998	.9998	.9998	.9998
+3.6	.9998	.9998	.9999	.9999	.9999	.9999	.9999	.9999	.9999	.9999
+3.7	.9999	.9999	.9999	.9999	.9999	.9999	.9999	.9999	.9999	.9999
+3.8	.9999	.9999	.9999	.9999	.9999	.9999	.9999	.9999	.9999	.9999

Appendix 2

Factors for Computing Central Lines and 3σ Control Limits for $\overline{X}$, s, and R Charts

Observations in Sample, n	Chart for Averages: Factors for Control Limits			Chart for Ranges: Factor for Central Line	Chart for Ranges: Factors for Control Limits					Chart for Standard Deviations: Factor for Central Line	Chart for Standard Deviations: Factors for Control Limits			
	A	A_2	A_3	d_2	d_1	D_1	D_2	D_3	D_4	c_4	B_3	B_4	B_5	B_6
2	2.121	1.880	2.659	1.128	0.853	0	3.686	0	3.267	0.7979	0	3.267	0	2.606
3	1.732	1.023	1.954	1.693	0.888	0	4.358	0	2.574	0.8862	0	2.568	0	2.276
4	1.500	0.729	1.628	2.059	0.880	0	4.698	0	2.282	0.9213	0	2.266	0	2.088
5	1.342	0.577	1.427	2.326	0.864	0	4.918	0	2.114	0.9400	0	2.089	0	1.964
6	1.225	0.483	1.287	2.534	0.848	0	5.078	0	2.004	0.9515	0.030	1.970	0.029	1.874
7	1.134	0.419	1.182	2.704	0.833	0.204	5.204	0.076	1.924	0.9594	0.118	1.882	0.113	1.806
8	1.061	0.373	1.099	2.847	0.820	0.388	5.306	0.136	1.864	0.9650	0.185	1.815	0.179	1.751
9	1.000	0.337	1.032	2.970	0.808	0.547	5.393	0.184	1.816	0.9693	0.239	1.761	0.232	1.707
10	0.949	0.308	0.975	3.078	0.797	0.687	5.469	0.223	1.777	0.9727	0.284	1.716	0.276	1.669
11	0.905	0.285	0.927	3.173	0.787	0.811	5.535	0.256	1.744	0.9754	0.321	1.679	0.313	1.637
12	0.866	0.266	0.886	3.258	0.778	0.922	5.594	0.283	1.717	0.9776	0.354	1.646	0.346	1.610
13	0.832	0.249	0.850	3.336	0.770	1.025	5.647	0.307	1.693	0.9794	0.382	1.618	0.374	1.585
14	0.802	0.235	0.817	3.407	0.763	1.118	5.696	0.328	1.672	0.9810	0.406	1.594	0.399	1.563
15	0.775	0.223	0.789	3.472	0.756	1.203	5.741	0.347	1.653	0.9823	0.428	1.572	0.421	1.544
16	0.750	0.212	0.763	3.532	0.750	1.282	5.782	0.363	1.637	0.9835	0.448	1.552	0.440	1.526
17	0.728	0.203	0.739	3.588	0.744	1.356	5.820	0.378	1.622	0.9845	0.466	1.534	0.458	1.511
18	0.707	0.194	0.718	3.640	0.739	1.424	5.856	0.391	1.608	0.9854	0.482	1.518	0.475	1.496
19	0.688	0.187	0.698	3.689	0.734	1.487	5.891	0.403	1.597	0.9862	0.497	1.503	0.490	1.483
20	0.671	0.180	0.680	3.735	0.729	1.549	5.921	0.415	1.585	0.9869	0.510	1.490	0.504	1.470

APPENDIX 3

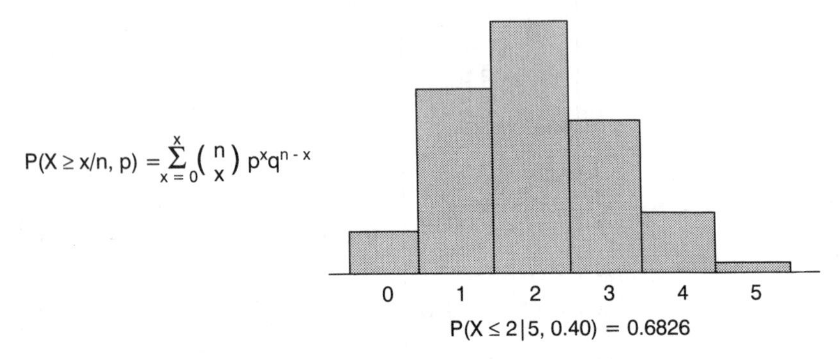

Cumulative Binomial Probability Distribution

n = 5

d \ p	*0.01*	*0.02*	*0.03*	*0.04*	*0.05*	*0.06*	*0.07*	*0.08*	*0.09*	*0.10*
0	0.9510	0.9039	0.8587	0.8154	0.7738	0.7339	0.6957	0.6591	0.6240	0.5905
1	0.9990	0.9962	0.9915	0.9852	0.9774	0.9681	0.9575	0.9456	0.9326	0.9185
2	1.0000	0.9999	0.9997	0.9994	0.9998	0.9980	0.9969	0.9955	0.9937	0.9914
3	1.0000	1.0000	1.0000	1.0000	1.0000	0.9999	0.9999	0.9998	0.9997	0.9995
4	1.0000	1.0000	1.0000	1.0000	1.0000	1.0000	1.0000	1.0000	1.0000	1.0000

d \ p	*0.11*	*0.12*	*0.13*	*0.14*	*0.15*	*0.16*	*0.17*	*0.18*	*0.19*	*0.20*
0	0.5584	0.5277	0.4984	0.4704	0.4437	0.4182	0.3939	0.3707	0.3487	0.3277
1	0.9035	0.8875	0.8708	0.8533	0.8352	0.8165	0.7973	0.7776	0.7576	0.7373
2	0.9888	0.9857	0.9821	0.9780	0.9734	0.9682	0.9625	0.9563	0.9495	0.9421
3	0.9993	0.9991	0.9987	0.9983	0.9978	0.9971	0.9964	0.9955	0.9945	0.9933
4	1.0000	1.0000	1.0000	0.9999	0.9999	0.9999	0.9999	0.9998	0.9998	0.9997
5	1.0000	1.0000	1.0000	1.0000	1.0000	1.0000	1.0000	1.0000	1.0000	1.0000

d \ p	*0.21*	*0.22*	*0.23*	*0.24*	*0.25*	*0.26*	*0.27*	*0.28*	*0.29*	*0.30*
0	0.3077	0.2887	0.2707	0.2536	0.2373	0.2219	0.2073	0.1935	0.1804	0.1681
1	0.7167	0.6959	0.6749	0.6539	0.6328	0.6117	0.5907	0.5697	0.5489	0.5282
2	0.9341	0.9256	0.9164	0.9067	0.8965	0.8857	0.8743	0.8624	0.8499	0.8369
3	0.9919	0.9903	0.9886	0.9866	0.9844	0.9819	0.9792	0.9762	0.9728	0.9692
4	0.9996	0.9995	0.9994	0.9992	0.9990	0.9988	0.9986	0.9983	0.9979	0.9976
5	1.0000	1.0000	1.0000	1.0000	1.0000	1.0000	1.0000	1.0000	1.0000	1.0000

d \ p	*0.31*	*0.32*	*0.33*	*0.34*	*0.35*	*0.36*	*0.37*	*0.38*	*0.39*	*0.40*
0	0.1564	0.1454	0.1350	0.1252	0.1160	0.1074	0.0992	0.0916	0.0845	0.0778
1	0.5077	0.4875	0.4675	0.4478	0.4284	0.4094	0.3907	0.3724	0.3545	0.3370

n = 5 (continued)

d \ p	0.31	0.32	0.33	0.34	0.35	0.36	0.37	0.38	0.39	0.40
2	0.8234	0.8095	0.7950	0.7801	0.7648	0.7491	0.7330	0.7165	0.6997	0.6826
3	0.9653	0.9610	0.9564	0.9514	0.9460	0.9402	0.9340	0.9274	0.9204	0.9130
4	0.9971	0.9966	0.9961	0.9955	0.9947	0.9940	0.9931	0.9921	0.9910	0.9898
5	1.0000	1.0000	1.0000	1.0000	1.0000	1.0000	1.0000	1.0000	1.0000	1.0000

d \ p	0.41	0.42	0.43	0.44	0.45	0.46	0.47	0.48	0.49	0.50
0	0.0715	0.0656	0.0602	0.0551	0.0503	0.0459	0.0418	0.0380	0.0345	0.0312
1	0.3199	0.3033	0.2871	0.2714	0.2562	0.2415	0.2272	0.2135	0.2002	0.1875
2	0.6651	0.6475	0.6295	0.6114	0.5931	0.5747	0.5561	0.5375	0.5187	0.5000
3	0.9051	0.8967	0.8879	0.8786	0.8688	0.8585	0.8478	0.8365	0.8247	0.8125
4	0.9884	0.9869	0.9853	0.9835	0.9815	0.9794	0.9771	0.9745	0.9718	0.9688
5	1.0000	1.0000	1.0000	1.0000	1.0000	1.0000	1.0000	1.0000	1.0000	1.0000

n = 6

d \ p	0.01	0.02	0.03	0.04	0.05	0.06	0.07	0.08	0.09	0.10
0	0.9415	0.8858	0.8330	0.7828	0.7351	0.6899	0.6470	0.6064	0.5679	0.5314
1	0.9985	0.9943	0.9875	0.9784	0.9672	0.9541	0.9392	0.9227	0.9048	0.8857
2	1.0000	0.9998	0.9995	0.9988	0.9978	0.9962	0.9942	0.9915	0.9882	0.9841
3	1.0000	1.0000	1.0000	1.0000	0.9999	0.9998	0.9997	0.9995	0.9992	0.9987
4	1.0000	1.0000	1.0000	1.0000	1.0000	1.0000	1.0000	1.0000	1.0000	0.9999
5	1.0000	1.0000	1.0000	1.0000	1.0000	1.0000	1.0000	1.0000	1.0000	1.0000

d \ p	0.11	0.12	0.13	0.14	0.15	0.16	0.17	0.18	0.19	0.20
0	0.4970	0.4644	0.4336	0.4046	0.3771	0.3513	0.3269	0.3040	0.2824	0.2621
1	0.8655	0.8444	0.8224	0.7997	0.7765	0.7528	0.7287	0.7044	0.6799	0.6554
2	0.9794	0.9739	0.9676	0.9605	0.9527	0.9440	0.9345	0.9241	0.9130	0.9011
3	0.9982	0.9975	0.9966	0.9955	0.9941	0.9925	0.9906	0.9884	0.9859	0.9830
4	0.9999	0.9999	0.9998	0.9997	0.9996	0.9995	0.9993	0.9990	0.9987	0.9984
5	1.0000	1.0000	1.0000	1.0000	1.0000	1.0000	1.0000	1.0000	1.0000	0.9999
6	1.0000	1.0000	1.0000	1.0000	1.0000	1.0000	1.0000	1.0000	1.0000	1.0000

d \ p	0.21	0.22	0.23	0.24	0.25	0.26	0.27	0.28	0.09	0.30
0	0.2431	0.2252	0.2084	0.1927	0.1780	0.1642	0.1513	0.1393	0.1281	0.1176
1	0.6308	0.6063	0.5820	0.5578	0.5339	0.5104	0.4872	0.4644	0.4420	0.4202
2	0.8885	0.8750	0.8609	0.8461	0.8306	0.8144	0.7977	0.7804	0.7626	0.7443
3	0.9798	0.9761	0.9720	0.9674	0.9624	0.9569	0.9508	0.9443	0.9372	0.9295
4	0.9980	0.9975	0.9969	0.9962	0.9954	0.9944	0.9933	0.9921	0.9907	0.9891
5	0.9999	0.9999	0.9999	0.9998	0.9998	0.9997	0.9996	0.9995	0.9994	0.9993
6	1.0000	1.0000	1.0000	1.0000	1.0000	1.0000	1.0000	1.0000	1.0000	1.0000

d \ p	0.31	0.32	0.33	0.34	0.35	0.36	0.37	0.38	0.39	.040
0	0.1079	0.0989	0.0905	0.0827	0.0754	0.0687	0.0625	0.0568	0.0515	0.0467
1	0.3988	0.3780	0.3578	0.3381	0.3191	0.3006	0.2828	0.2657	0.2492	0.2333
2	0.7256	0.7064	0.6870	0.6672	0.6471	0.6268	0.6063	0.5857	0.5650	0.5443
3	0.9213	0.9125	0.9031	0.8931	0.8826	0.8714	0.8596	0.8473	0.8343	0.8208
4	0.9873	0.9852	0.9830	0.9805	0.9777	0.9746	0.9712	0.9675	0.9635	0.9590
5	0.9991	0.9989	0.9987	0.9985	0.9982	0.9978	0.9974	0.9970	0.9965	0.9959
6	1.0000	1.0000	1.0000	1.0000	1.0000	1.0000	1.0000	1.0000	1.0000	1.0000

d \ p	0.41	0.42	0.43	0.44	0.45	0.46	0.47	0.48	0.49	0.50
0	0.0422	0.0381	0.0343	0.0308	0.0277	0.0248	0.0222	0.0198	0.0176	0.0156
1	0.2181	0.2035	0.1895	0.1762	0.1636	0.1515	0.1401	0.1293	0.1190	0.1094

n = *6* (continued)

d \ p	0.41	0.42	0.43	0.44	0.45	0.46	0.47	0.48	0.49	0.50
2	0.5236	0.5029	0.4823	0.4618	0.4415	0.4214	0.4015	0.3820	0.3627	0.3437
3	0.8067	0.7920	0.7768	0.7610	0.7447	0.7280	0.7107	0.6930	0.6748	0.6562
4	0.9542	0.9490	0.9434	0.9373	0.9308	0.9238	0.9163	0.9083	0.8997	0.8906
5	0.9952	0.9945	0.9937	0.9927	0.9917	0.9905	0.9892	0.9878	0.9862	0.9844
6	1.0000	1.0000	1.0000	1.0000	1.0000	1.0000	1.0000	1.0000	1.0000	1.0000

n = *7*

d \ p	0.01	0.02	0.03	0.04	0.05	0.06	0.07	0.08	0.09	0.10
0	0.9321	0.8681	0.8080	0.7514	0.6983	0.6485	0.6017	0.5578	0.5168	0.4783
1	0.9980	0.9921	0.9829	0.9706	0.9556	0.9382	0.9187	0.8974	0.8745	0.8503
2	1.0000	0.9997	0.9991	0.9980	0.9962	0.9937	0.9903	0.9860	0.9807	0.9743
3	1.0000	1.0000	1.0000	0.9999	0.9998	0.9996	0.9993	0.9998	0.9982	0.9973
4	1.0000	1.0000	1.0000	1.0000	1.0000	1.0000	1.0000	0.9999	0.9999	0.9998
5	1.0000	1.0000	1.0000	1.0000	1.0000	1.0000	1.0000	1.0000	1.0000	1.0000

d \ p	0.11	0.12	0.13	0.14	0.15	0.16	0.17	0.18	0.19	0.20
0	0.4423	0.4087	0.3773	0.3479	0.3206	0.2951	0.2714	0.2493	0.2288	0.2097
1	0.8250	0.7988	0.7719	0.7444	0.7166	0.6885	0.6604	0.6323	0.6044	0.5767
2	0.9669	0.9584	0.9487	0.9380	0.9262	0.9134	0.8995	0.8846	0.8687	0.8520
3	0.9961	0.9946	0.9928	0.9906	0.9879	0.9847	0.9811	0.9769	0.9721	0.9687
4	0.9997	0.9996	0.9994	0.9991	0.9988	0.9983	0.9978	0.9971	0.9963	0.9953
5	1.0000	1.0000	1.0000	1.0000	0.9999	0.9999	0.9999	0.9998	0.9997	0.9996
6	1.0000	1.0000	1.0000	1.0000	1.0000	1.0000	1.0000	1.0000	1.0000	1.0000

d \ p	0.21	0.22	0.23	0.24	0.25	0.26	0.27	0.28	0.29	0.30
0	0.1920	0.1757	0.1605	0.1465	0.1335	0.1215	0.1105	0.1003	0.0910	0.0824
1	0.5494	0.5225	0.4960	0.4702	0.4449	0.4204	0.3965	0.3734	0.3510	0.3294
2	0.8343	0.8159	0.7967	0.7769	0.7564	0.7354	0.7139	0.6919	0.6696	0.6471
3	0.9606	0.9539	0.9464	0.9383	0.9294	0.9198	0.9095	0.8984	0.8866	0.8740
4	0.9942	0.9928	0.9912	0.9893	0.9871	0.9847	0.9819	0.9787	0.9752	0.9712
5	0.9995	0.9994	0.9992	0.9989	0.9987	0.9983	0.9979	0.9974	0.9969	0.9962
6	1.0000	1.0000	1.0000	1.0000	0.9999	0.9999	0.9999	0.9999	0.9998	0.9998
7	1.0000	1.0000	1.0000	1.0000	1.0000	1.0000	1.0000	1.0000	1.0000	1.0000

d \ p	0.31	0.32	0.33	0.34	0.35	0.36	0.37	0.38	0.39	0.40
0	0.0745	0.0672	0.0606	0.0546	0.0490	0.0440	0.0394	0.0352	0.0314	0.0280
1	0.3086	0.2887	0.2696	0.2513	0.2338	0.2172	0.2013	0.1863	0.1721	0.1586
2	0.6243	0.6013	0.5783	0.5553	0.5323	0.5094	0.4868	0.4641	0.4419	0.4199
3	0.8606	0.8466	0.8318	0.8163	0.8002	0.7833	0.7659	0.7479	0.7293	0.7102
4	0.9668	0.9620	0.9566	0.9508	0.9444	0.9375	0.9299	0.9218	0.9131	0.9037
5	0.9954	0.9945	0.9935	0.9923	0.9910	0.9895	0.9877	0.9858	0.9836	0.9812
6	0.9997	0.9997	0.9996	0.9995	0.9994	0.9992	0.9991	0.9989	0.9986	0.9984
7	1.0000	1.0000	1.0000	1.0000	1.0000	1.0000	1.0000	1.0000	1.0000	1.0000

d \ p	0.41	0.42	0.43	0.44	0.45	0.46	0.47	0.48	0.49	0.50
0	0.0249	0.0221	0.0195	0.0173	0.0152	0.0134	0.0117	0.0103	0.0090	0.0078
1	0.1459	0.1340	0.1228	0.1123	0.1024	0.0932	0.0847	0.0767	0.0693	0.0625
2	0.3983	0.3771	0.3564	0.3362	0.3164	0.2973	0.2787	0.2607	0.2433	0.2266
3	0.6906	0.6706	0.6502	0.6294	0.6083	0.5869	0.5654	0.5437	0.5219	0.5000
4	0.8937	0.8831	0.8718	0.8598	0.8471	0.8337	0.8197	0.8049	0.7895	0.7734
5	0.9784	0.9754	0.9721	0.9684	0.9643	0.9598	0.9549	0.9496	0.9438	0.9375
6	0.9981	0.9977	0.9973	0.9968	0.9963	0.9956	0.9949	0.9941	0.9932	0.9922
7	1.0000	1.0000	1.0000	1.0000	1.0000	1.0000	1.0000	1.0000	1.0000	1.0000

n = *8*

d \ p	*0.01*	*0.02*	*0.03*	*0.04*	*0.05*	*0.06*	*0.07*	*0.08*	*0.09*	*0.10*
0	0.9227	0.8508	0.7837	0.7214	0.6634	0.6096	0.5596	0.5132	0.4703	0.4305
1	0.9973	0.9897	0.9777	0.9619	0.9428	0.9208	0.8965	0.8702	0.8423	0.8131
2	0.9999	0.9996	0.9987	0.9969	0.9942	0.9904	0.9853	0.9789	0.9711	0.9619
3	1.0000	1.0000	0.9999	0.9998	0.9996	0.9993	0.9987	0.9978	0.9966	0.9950
4	1.0000	1.0000	1.0000	1.0000	1.0000	1.0000	0.9999	0.9999	0.9997	0.9996
5	1.0000	1.0000	1.0000	1.0000	1.0000	1.0000	1.0000	1.0000	1.0000	1.0000

d \ p	*0.11*	*0.12*	*0.13*	*0.14*	*0.15*	*0.16*	*0.17*	*0.18*	*0.19*	*0.20*
0	0.3937	0.3596	0.3282	0.2992	0.2725	0.2479	0.2252	0.2044	0.1853	0.1678
1	0.7829	0.7520	0.7206	0.6889	0.6572	0.6256	0.5943	0.5634	0.5330	0.5033
2	0.9513	0.9392	0.9257	0.9109	0.8948	0.8774	0.8588	0.8392	0.8185	0.7969
3	0.9929	0.9903	0.9871	0.9832	0.9786	0.9733	0.9672	0.9603	0.9524	0.9437
4	0.9993	0.9990	0.9985	0.9979	0.9971	0.9962	0.9950	0.9935	0.9917	0.9896
5	1.0000	0.9999	0.9999	0.9998	0.9998	0.9997	0.9995	0.9993	0.9991	0.9988
6	1.0000	1.0000	1.0000	1.0000	1.0000	1.0000	1.0000	1.0000	0.9999	0.9999
7	1.0000	1.0000	1.0000	1.0000	1.0000	1.0000	1.0000	1.0000	1.0000	1.0000

d \ p	*0.21*	*0.22*	*0.23*	*0.24*	*0.25*	*0.26*	*0.27*	*0.28*	*0.29*	*0.30*
0	0.1517	0.1370	0.1236	0.1113	0.1001	0.8990	0.0806	0.0722	0.0646	0.0576
1	0.4743	0.4462	0.4189	0.3925	0.3671	0.3427	0.3193	0.2969	0.2756	0.2553
2	0.7745	0.7514	0.7276	0.7033	0.6785	0.6535	0.6282	0.6027	0.5772	0.5518
3	0.9341	0.9235	0.9120	0.8996	0.8862	0.8719	0.8567	0.8406	0.8237	0.8059
4	0.9871	0.9842	0.9809	0.9770	0.9727	0.9678	0.9623	0.9562	0.9495	0.9420
5	0.9984	0.9979	0.9973	0.9966	0.9958	0.9948	0.9936	0.9922	0.9906	0.9887
6	0.9999	0.9998	0.9998	0.9997	0.9996	0.9995	0.9994	0.9992	0.9990	0.9987
7	1.0000	1.0000	1.0000	1.0000	1.0000	1.0000	1.0000	1.0000	0.9999	0.9999
8	1.0000	1.0000	1.0000	1.0000	1.0000	1.0000	1.0000	1.0000	1.0000	1.0000

d \ p	*0.31*	*0.32*	*0.33*	*0.34*	*0.35*	*0.36*	*0.37*	*0.38*	*0.39*	*0.40*
0	0.0514	0.0457	0.0406	0.0360	0.0319	0.0281	0.0248	0.0218	0.0192	0.0168
1	0.2360	0.2178	0.2006	0.1844	0.1691	0.1548	0.1414	0.1289	0.1172	0.1064
2	0.5264	0.5013	0.4764	0.4519	0.4278	0.4042	0.3811	0.3585	0.3366	0.3154
3	0.7874	0.7681	0.7481	0.7276	0.7064	0.6847	0.6626	0.6401	0.6172	0.5941
4	0.9339	0.9250	0.9154	0.9051	0.8939	0.8820	0.8693	0.8557	0.8414	0.8263
5	0.9866	0.9841	0.9813	0.9782	0.9747	0.9707	0.9664	0.9615	0.9561	0.9502
6	0.9984	0.9980	0.9976	0.9970	0.9964	0.9957	0.9949	0.9939	0.9928	0.9915
7	0.9999	0.9999	0.9999	0.9998	0.9998	0.9997	0.9996	0.9996	0.9995	0.9993
8	1.0000	1.0000	1.0000	1.0000	1.0000	1.0000	1.0000	1.0000	1.0000	1.0000

d \ p	*0.41*	*0.42*	*0.43*	*0.44*	*0.45*	*0.46*	*0.47*	*0.48*	*0.49*	*0.50*
0	0.0147	0.0128	0.0111	0.0097	0.0084	0.0072	0.0062	0.0053	0.0046	0.0039
1	0.0963	0.0870	0.0784	0.0705	0.0632	0.0565	0.0504	0.0448	0.0398	0.0352
2	0.2948	0.2750	0.2560	0.2376	0.2201	0.2034	0.1875	0.1724	0.1581	0.1445
3	0.5708	0.5473	0.5238	0.5004	0.4770	0.4537	0.4306	0.4078	0.3854	0.3633
4	0.8105	0.7938	0.7765	0.7584	0.7396	0.7202	0.7001	0.6795	0.6584	0.6367
5	0.9437	0.9366	0.9289	0.9206	0.9115	0.9018	0.8914	0.8802	0.8682	0.8555
6	0.9900	0.9883	0.9864	0.9843	0.9819	0.9792	0.9761	0.9728	0.9690	0.9648
7	0.9992	0.9990	0.9988	0.9986	0.9983	0.9980	0.9976	0.9972	0.9967	0.9961
8	1.0000	1.0000	1.0000	1.0000	1.0000	1.0000	1.0000	1.0000	1.0000	1.0000

n = *9*

d \ p	*0.01*	*0.02*	*0.03*	*0.04*	*0.05*	*0.06*	*0.07*	*0.08*	*0.09*	*0.10*
0	0.9135	0.8337	0.7602	0.6925	0.6302	0.5730	0.5204	0.4722	0.4279	0.3874
1	0.9966	0.9869	0.9718	0.9522	0.9288	0.9022	0.8729	0.8417	0.8088	0.7748

n = *9* (continued)

d \ p	0.01	0.02	0.03	0.04	0.05	0.06	0.07	0.08	0.09	0.10
2	0.9999	0.9994	0.9980	0.9955	0.9916	0.9862	0.9791	0.9702	0.9595	0.9470
3	1.0000	1.0000	0.9999	0.9997	0.9994	0.9987	0.9977	0.9963	0.9943	0.9917
4	1.0000	1.0000	1.0000	1.0000	1.0000	0.9999	0.9998	0.9997	0.9995	0.9991
5	1.0000	1.0000	1.0000	1.0000	1.0000	1.0000	1.0000	1.0000	1.0000	0.9999
6	1.0000	1.0000	1.0000	1.0000	1.0000	1.0000	1.0000	1.0000	1.0000	1.0000

d \ p	0.11	0.12	0.13	0.14	0.15	0.16	0.17	0.18	0.19	0.20
0	0.3504	0.3165	0.2855	0.2573	0.2316	0.2082	0.1869	0.1676	0.1501	0.1342
1	0.7401	0.7049	0.6696	0.6343	0.5995	0.5652	0.5315	0.4988	0.4670	0.4362
2	0.9327	0.9167	0.8991	0.8798	0.8591	0.8371	0.8139	0.7895	0.7643	0.7382
3	0.9883	0.9842	0.9791	0.9731	0.9661	0.9580	0.9488	0.9385	0.9270	0.9144
4	0.9986	0.9979	0.9970	0.9959	0.9944	0.9925	0.9902	0.9875	0.9842	0.9804
5	0.9999	0.9998	0.9997	0.9996	0.9994	0.9991	0.9987	0.9983	0.9977	0.9969
6	1.0000	1.0000	1.0000	1.0000	1.0000	0.9999	0.9999	0.9998	0.9998	0.9997
7	1.0000	1.0000	1.0000	1.0000	1.0000	1.0000	1.0000	1.0000	1.0000	1.0000

d \ p	0.21	0.22	0.23	0.24	0.25	0.26	0.27	0.28	0.29	0.30
0	0.1199	0.1069	0.0952	0.0846	0.0751	0.0665	0.0589	0.0520	0.0458	0.0404
1	0.4066	0.3782	0.3509	0.3250	0.3003	0.2770	0.2548	0.2340	0.2144	0.1960
2	0.7115	0.6842	0.6566	0.6287	0.6007	0.5727	0.5448	0.5171	0.4898	0.4628
3	0.9006	0.8856	0.8696	0.8525	0.8343	0.8151	0.7950	0.7740	0.7522	0.7297
4	0.9760	0.9709	0.9650	0.9584	0.9511	0.9429	0.9338	0.9238	0.9130	0.9012
5	0.9960	0.9949	0.9935	0.9919	0.9900	0.9878	0.9851	0.9821	0.9787	0.9747
6	0.9996	0.9994	0.9992	0.9990	0.9987	0.9983	0.9978	0.9972	0.9965	0.9957
7	1.0000	1.0000	0.9999	0.9999	0.9999	0.9999	0.9998	0.9997	0.9997	0.9996
8	1.0000	1.0000	1.0000	1.0000	1.0000	1.0000	1.0000	1.0000	1.0000	1.0000

d \ p	0.31	0.32	0.33	0.34	0.35	0.36	0.37	0.38	0.39	0.40
0	0.0355	1.0311	0.0272	0.0238	0.0207	0.0180	0.0156	0.0135	0.0117	0.0101
1	0.1788	0.1628	0.1478	0.1339	0.1211	0.1092	0.0983	0.0882	0.0790	0.0705
2	0.4364	0.4106	0.3854	0.3610	0.3373	0.3144	0.2924	0.2713	0.2511	0.2318
3	0.7065	0.6827	0.6585	0.6338	0.6089	0.5837	0.5584	0.5331	0.5078	0.4826
4	0.8885	0.8747	0.8602	0.8447	0.8283	0.8110	0.7928	0.7738	0.7540	0.7334
5	0.9702	0.9652	0.9596	0.9533	0.9464	0.9388	0.9304	0.9213	0.9114	0.9006
6	0.9947	0.9936	0.9922	0.9906	0.9888	0.9867	0.9843	0.9816	0.9785	0.9750
7	0.9994	0.9993	0.9991	0.9989	0.9986	0.9983	0.9979	0.9974	0.9969	0.9962
8	1.0000	1.0000	1.0000	0.9999	0.9999	0.9999	0.9999	0.9998	0.9998	0.9997
9	1.0000	1.0000	1.0000	1.0000	1.0000	1.0000	1.0000	1.0000	1.0000	1.0000

d \ p	0.41	0.42	0.43	0.44	0.45	0.46	0.47	0.48	0.49	0.50
0	0.0087	0.0074	0.0064	0.0054	0.0046	0.0039	0.0033	0.0028	0.0023	0.0020
1	0.0628	0.0558	0.0495	0.0437	0.0385	0.0338	0.0296	0.0259	0.0225	0.0195
2	0.2134	0.1961	0.1796	0.1641	0.1495	0.1358	0.1231	0.1111	0.1001	0.0898
3	0.4576	0.4330	0.4087	0.3848	0.3614	0.3386	0.3164	0.2948	0.2740	0.2539
4	0.7122	0.6903	0.6678	0.6449	0.6214	0.5976	0.5735	0.5491	0.5246	0.5000
5	0.8891	0.8767	0.8634	0.8492	0.8342	0.8183	0.8015	0.7839	0.7654	0.7461
6	0.9710	0.9666	0.9617	0.9563	0.9502	0.9436	0.9363	0.9283	0.9196	0.9102
7	0.9954	0.9945	0.9935	0.9923	0.9909	0.9893	0.9875	0.9855	0.9831	0.9805
8	0.9997	0.9996	0.9995	0.9994	0.9992	0.9991	0.9989	0.9986	0.9984	0.9980
9	1.0000	1.0000	1.0000	1.0000	1.0000	1.0000	1.0000	1.0000	1.0000	1.0000

n = *10*

d \ p	0.01	0.02	0.03	0.04	0.05	0.06	0.07	0.08	0.09	0.10
0	0.9044	0.8171	0.7374	0.6648	0.5987	0.5386	0.4840	0.4344	0.3894	0.3487
1	0.9957	0.9838	0.9655	0.9418	0.9139	0.8824	0.8483	0.8121	0.7746	0.7361

n = *10* (continued)

d \ p	0.01	0.02	0.03	0.04	0.05	0.06	0.07	0.08	0.09	0.10
2	0.9999	0.9991	0.9972	0.9938	0.9885	0.9812	0.9717	0.9599	0.9460	0.9298
3	1.0000	1.0000	0.9999	0.9996	0.9990	0.9980	0.9964	0.9942	0.9912	0.9872
4	1.0000	1.0000	1.0000	1.0000	0.9999	0.9998	0.9997	0.9994	0.9990	0.9984
5	1.0000	1.0000	1.0000	1.0000	1.0000	1.0000	1.0000	1.0000	0.9999	0.9999
6	1.0000	1.0000	1.0000	1.0000	1.0000	1.0000	1.0000	1.0000	1.0000	1.0000

d \ p	0.11	0.12	0.13	0.14	0.15	0.16	0.17	0.18	0.19	0.20
0	0.3118	0.2785	0.2484	0.2213	0.1969	0.1749	0.1552	0.1374	0.1216	0.1074
1	0.6972	0.6583	0.6196	0.5816	0.5443	0.5080	0.4730	0.4392	0.4068	0.3758
2	0.9116	0.8913	0.8692	0.8455	0.8202	0.7936	0.7659	0.7372	0.7078	0.6778
3	0.9822	0.9761	0.9687	0.9600	0.9500	0.9386	0.9259	0.9117	0.8961	0.8791
4	0.9975	0.9963	0.9947	0.9927	0.9901	0.9870	0.9832	0.9787	0.9734	0.9672
5	0.9997	0.9996	0.9994	0.9990	0.9986	0.9980	0.9973	0.9963	0.9951	0.9936
6	1.0000	1.0000	0.9999	0.9999	0.9999	0.9998	0.9997	0.9996	0.9994	0.9991
7	1.0000	1.0000	1.0000	1.0000	1.0000	1.0000	1.0000	1.0000	0.9999	0.9999
8	1.0000	1.0000	1.0000	1.0000	1.0000	1.0000	1.0000	1.0000	1.0000	1.0000

d \ p	0.21	0.22	0.23	0.24	0.25	0.26	0.27	0.28	0.29	0.30
0	0.0947	0.0834	0.0733	0.0643	0.0563	0.0492	0.0430	0.0374	0.0326	0.0282
1	0.3464	0.3185	0.2921	0.2673	0.2440	0.2222	0.2019	0.1830	0.1655	0.1493
2	0.6474	0.6169	0.5863	0.5558	0.5256	0.4958	0.4665	0.4378	0.4099	0.3828
3	0.8609	0.8413	0.8206	0.7988	0.7759	0.7521	0.7274	0.7021	0.6761	0.6496
4	0.9601	0.9521	0.9431	0.9330	0.9219	0.9096	0.8963	0.8819	0.8663	0.8497
5	0.9918	0.9896	0.9870	0.9839	0.9803	0.9761	0.9713	0.9658	0.9596	0.9527
6	0.9988	0.9984	0.9979	0.9973	0.9965	0.9955	0.9944	0.9930	0.9913	0.9894
7	0.9999	0.9998	0.9998	0.9997	0.9996	0.9994	0.9993	0.9990	0.9988	0.9984
8	1.0000	1.0000	1.0000	1.0000	1.0000	1.0000	0.9999	0.9999	0.9999	0.9999
9	1.0000	1.0000	1.0000	1.0000	1.0000	1.0000	1.0000	1.0000	1.0000	1.0000

d \ p	0.31	0.32	0.33	0.34	0.35	0.36	0.37	0.38	0.39	0.40
0	0.0245	0.0211	0.0182	0.0157	0.0135	0.0115	0.0098	0.0084	0.0071	0.0060
1	0.1344	0.1206	0.1080	0.0965	0.0860	0.0764	0.0677	0.0598	0.0527	0.0464
2	0.3566	0.3313	0.3070	0.2838	0.2616	0.2405	0.2206	0.2017	0.1840	0.1673
3	0.6228	0.5956	0.5684	0.5411	0.5138	0.4868	0.4600	0.4336	0.4077	0.3823
4	0.8321	0.8133	0.7936	0.7730	0.7515	0.7292	0.7061	0.6823	0.6580	0.6331
5	0.9449	0.9363	0.9268	0.9164	0.9051	0.8928	0.8795	0.8652	0.8500	0.8338
6	0.9871	0.9845	0.9815	0.9780	0.9740	0.9695	0.9644	0.9587	0.9523	0.9452
7	0.9980	0.9975	0.9968	0.9961	0.9952	0.9941	0.9929	0.9914	0.9897	0.9877
8	0.9998	0.9997	0.9997	0.9996	0.9995	0.9993	0.9991	0.9989	0.9986	0.9983
9	1.0000	1.0000	1.0000	1.0000	1.0000	1.0000	1.0000	0.9999	0.9999	0.9999
10	1.0000	1.0000	1.0000	1.0000	1.0000	1.0000	1.0000	1.0000	1.0000	1.0000

d \ p	0.41	0.42	0.43	0.44	0.45	0.46	0.47	0.48	0.49	0.50
0	0.0051	0.0043	0.0036	0.0030	0.0025	0.0021	0.0017	0.0014	0.0012	0.0010
1	0.0406	0.0355	0.0309	0.0269	0.0233	0.0201	0.0173	0.0148	0.0126	0.0107
2	0.1517	0.1372	0.1236	0.1111	0.0996	0.0889	0.0791	0.0702	0.0621	0.0547
3	0.3575	0.3335	0.3102	0.2877	0.2660	0.2453	0.2255	0.2067	0.1888	0.1719
4	0.6078	0.5822	0.5564	0.5304	0.5044	0.4784	0.4526	0.4270	0.4018	0.3770
5	0.8166	0.7984	0.7793	0.7593	0.7384	0.7168	0.6943	0.6712	0.6474	0.6230
6	0.9374	0.9288	0.9194	0.9092	0.8980	0.8859	0.8729	0.8590	0.8440	0.8281
7	0.9854	0.9828	0.9798	0.9764	0.9726	0.9683	0.9634	0.9580	0.9520	0.9453
8	0.9979	0.9975	0.9969	0.9963	0.9955	0.9946	0.9935	0.9923	0.9909	0.9893
9	0.9999	0.9998	0.9998	0.9997	0.9997	0.9996	0.9995	0.9994	0.9992	0.9990
10	1.0000	1.0000	1.0000	1.0000	1.0000	1.0000	1.0000	1.0000	1.0000	1.0000

n = *11*

d \ p	*0.01*	*0.02*	*0.03*	*0.04*	*0.05*	*0.06*	*0.07*	*0.08*	*0.09*	*0.10*
0	0.8953	0.8007	0.7153	0.6382	0.5688	0.5063	0.4501	0.3996	0.3544	0.3138
1	0.9948	0.9805	0.9587	0.9308	0.8981	0.8618	0.8228	0.7819	0.7399	0.6974
2	0.9998	0.9988	0.9963	0.9917	0.9848	0.9752	0.9630	0.9481	0.9305	0.9104
3	1.0000	1.0000	0.9998	0.9993	0.9984	0.9970	0.9947	0.9915	0.9871	0.9815
4	1.0000	1.0000	1.0000	1.0000	0.9999	0.9997	0.9995	0.9990	0.9983	0.9972
5	1.0000	1.0000	1.0000	1.0000	1.0000	1.0000	1.0000	0.9999	0.9998	0.9997
6	1.0000	1.0000	1.0000	1.0000	1.0000	1.0000	1.0000	1.0000	1.0000	1.0000

d \ p	*0.11*	*0.12*	*0.13*	*0.14*	*0.15*	*0.16*	*0.17*	*0.18*	*0.19*	*0.20*
0	0.2775	0.2451	0.2161	0.1903	0.1673	0.1469	0.1288	0.1127	0.0985	0.0859
1	0.6548	0.6127	0.5714	0.5311	0.4922	0.4547	0.4189	0.3849	0.3526	0.3221
2	0.8880	0.8634	0.8368	0.8085	0.7788	0.7479	0.7161	0.6836	0.6506	0.6174
3	0.9744	0.9659	0.9558	0.9440	0.9306	0.9514	0.8987	0.8803	0.8603	0.8389
4	0.9958	0.9939	0.9913	0.9881	0.9841	0.9793	0.9734	0.9666	0.9587	0.9496
5	0.9995	0.9992	0.9988	0.9982	0.9973	0.9963	0.9949	0.9932	0.9910	0.9883
6	1.0000	0.9999	0.9999	0.9998	0.9997	0.9995	0.9993	0.9990	0.9986	0.9880
7	1.0000	1.0000	1.0000	1.0000	1.0000	1.0000	0.9999	0.9999	0.9998	0.9998
8	1.0000	1.0000	1.0000	1.0000	1.0000	1.0000	1.0000	1.0000	1.0000	1.0000

d \ p	*0.21*	*0.22*	*0.23*	*0.24*	*0.25*	*0.26*	*0.27*	*0.28*	*0.29*	*0.30*
0	0.0748	0.0650	0.0564	0.0489	0.0422	0.0364	0.0314	0.0270	0.0231	0.0198
1	0.2935	0.2667	0.2418	0.2186	0.1971	0.1773	0.1590	0.1423	0.1270	0.1130
2	0.5842	0.5512	0.5186	0.4866	0.4552	0.4247	0.3951	0.3665	0.3390	0.3127
3	0.8160	0.7919	0.7667	0.7404	0.7133	0.6854	0.6570	0.6281	0.5889	0.5696
4	0.9393	0.9277	0.9149	0.9008	0.8554	0.8687	0.8507	0.8315	0.8112	0.7897
5	0.9852	0.9814	0.9769	0.9717	0.9657	0.9588	0.9510	0.9423	0.9326	0.9218
6	0.9973	0.9965	0.9954	0.9941	0.9924	0.9905	0.9881	0.9854	0.9821	0.9784
7	0.9997	0.9995	0.9993	0.9991	0.9988	0.9984	0.9979	0.9973	0.9966	0.9957
8	1.0000	1.0000	0.9999	0.9999	0.9999	0.9998	0.9998	0.9997	0.9996	0.9994
9	1.0000	1.0000	1.0000	1.0000	1.0000	1.0000	1.0000	1.0000	1.0000	1.0000

d \ p	*0.31*	*0.32*	*0.33*	*0.34*	*0.35*	*0.36*	*0.37*	*0.38*	*0.39*	*0.40*
0	0.0169	0.0144	0.0122	0.0104	0.0088	0.0074	0.0062	0.0052	0.0044	0.0036
1	0.1003	0.0888	0.0784	0.0690	0.0606	0.0530	0.0463	0.0403	0.0350	0.0302
2	0.2877	0.2639	0.2413	0.2201	0.2001	0.1814	0.1640	0.1478	0.1328	0.1189
3	0.5402	0.5110	0.4821	0.4536	0.4256	0.3981	0.3714	0.3455	0.3204	0.2963
4	0.7672	0.7437	0.7193	0.6941	0.6683	0.6419	0.6150	0.5878	0.5603	0.5328
5	0.9099	0.8969	0.8829	0.8676	0.8513	0.8339	0.8153	0.7957	0.7751	0.7535
6	0.9740	0.9691	0.9634	0.9570	0.9499	0.9419	0.9330	0.9232	0.9124	0.9006
7	0.9946	0.9933	0.9918	0.9899	0.9878	0.9852	0.9823	0.9790	0.9751	0.9707
8	0.9992	0.9990	0.9987	0.9984	0.9980	0.9974	0.9968	0.9961	0.9952	0.9941
9	0.9999	0.9999	0.9999	0.9998	0.9998	0.9997	0.9996	0.9995	0.9994	0.9993
10	1.0000	1.0000	1.0000	1.0000	1.0000	1.0000	1.0000	1.0000	1.0000	1.0000

d \ p	*0.41*	*0.42*	*0.43*	*0.44*	*0.45*	*0.46*	*0.47*	*0.48*	*0.49*	*0.50*
0	0.0030	0.0025	0.0021	0.0017	0.0014	0.0011	0.0009	0.0008	0.0006	0.0005
1	0.0261	0.0224	0.0192	0.0164	0.0139	0.0118	0.0100	0.0084	0.0070	0.0059
2	0.1062	0.0945	0.0838	0.0740	0.0652	0.0572	0.0501	0.0436	0.0378	0.0327
3	0.2731	0.2510	0.2300	0.2100	0.1911	0.1734	0.1567	0.1412	0.1267	0.1133
4	0.5052	0.4777	0.4505	0.4236	0.3971	0.3712	0.3459	0.3213	0.2974	0.2744
5	0.7310	0.7076	0.6834	0.6586	0.6331	0.6071	0.5807	0.5540	0.5271	0.5000
6	0.8879	0.8740	0.8592	0.8432	0.8262	0.8081	0.7890	0.7688	0.7477	0.7256
7	0.9657	0.9601	0.9539	0.9468	0.9390	0.9304	0.9209	0.9105	0.8991	0.8867
8	0.9928	0.9913	0.9896	0.9875	0.9852	0.9825	0.9794	0.9759	0.9718	0.9673
9	0.9991	0.9988	0.9986	0.9982	0.9978	0.9973	0.9967	0.9960	0.9951	0.9941
10	0.9999	0.9999	0.9999	0.9999	0.9998	0.9998	0.9998	0.9997	0.9996	0.9995
11	1.0000	1.0000	1.0000	1.0000	1.0000	1.0000	1.0000	1.0000	1.0000	1.0000

n = *12*

d \ p	0.01	0.02	0.03	0.04	0.05	0.06	0.07	0.08	0.09	0.10
0	0.8864	0.7847	0.6938	0.6127	0.5404	0.4759	0.4186	0.3677	0.3225	0.2824
1	0.9938	0.9769	0.9514	0.9191	0.8816	0.8405	0.7967	0.7513	0.7052	0.6590
2	0.9998	0.9985	0.9952	0.9893	0.9804	0.9684	0.9532	0.9348	0.9134	0.8891
3	1.0000	0.9999	0.9997	0.9990	0.9978	0.9957	0.9925	0.9880	0.9820	0.9744
4	1.0000	1.0000	1.0000	0.9999	0.9998	0.9996	0.9991	0.9984	0.9973	0.9957
5	1.0000	1.0000	1.0000	1.0000	1.0000	1.0000	0.9999	0.9998	0.9997	0.9995
6	1.0000	1.0000	1.0000	1.0000	1.0000	1.0000	1.0000	1.0000	1.0000	0.9999
7	1.0000	1.0000	1.0000	1.0000	1.0000	1.0000	1.0000	1.0000	1.0000	1.0000

d \ p	0.11	0.12	0.13	0.14	0.15	0.16	0.17	0.18	0.19	0.20
0	0.2470	0.2157	0.1880	0.1637	0.1422	0.1234	0.1069	0.0924	0.0798	0.0687
1	0.6133	0.5686	0.5252	0.4834	0.4435	0.4055	0.3696	0.3359	0.3043	0.2749
2	0.8623	0.8333	0.8023	0.7697	0.7358	0.7010	0.6656	0.6298	0.5940	0.5583
3	0.9649	0.9536	0.9403	0.9250	0.9078	0.8886	0.8676	0.8448	0.8205	0.7946
4	0.9935	0.9905	0.9867	0.9819	0.9761	0.9690	0.9607	0.9511	0.9400	0.9274
5	0.9991	0.9986	0.9978	0.9967	0.9954	0.9935	0.9912	0.9884	0.9849	0.9806
6	0.9999	0.9998	0.9997	0.9996	0.9993	0.9990	0.9985	0.9979	0.9971	0.9961
7	1.0000	1.0000	1.0000	1.0000	0.9999	0.9999	0.9998	0.9997	0.9996	0.9994
8	1.0000	1.0000	1.0000	1.0000	1.0000	1.0000	1.0000	1.0000	1.0000	0.9999
9	1.0000	1.0000	1.0000	1.0000	1.0000	1.0000	1.0000	1.0000	1.0000	1.0000

d \ p	0.21	0.22	0.23	0.24	0.25	0.26	0.27	0.28	0.29	0.30
0	0.0591	0.0507	0.0434	0.0371	0.0317	0.0270	0.0229	0.0194	0.0164	0.0138
1	0.2476	0.2224	0.1991	0.1778	0.1584	0.1406	0.1245	0.1100	0.0968	0.0850
2	0.5232	0.4886	0.4550	0.4222	0.3907	0.3603	0.3313	0.3037	0.2775	0.2528
3	0.7674	0.7390	0.7096	0.6795	0.6488	0.6176	0.5863	0.5548	0.5235	0.4925
4	0.9134	0.8979	0.8808	0.8623	0.8424	0.8210	0.7984	0.7746	0.7496	0.7237
5	0.9755	0.9696	0.9626	0.9547	0.9456	0.9354	0.9240	0.9113	0.8974	0.8822
6	0.9948	0.9932	0.9911	0.9887	0.9857	0.9822	0.9781	0.9733	0.9678	0.9614
7	0.9992	0.9989	0.9984	0.9979	0.9972	0.9964	0.9953	0.9940	0.9924	0.9905
8	0.9999	0.9999	0.9998	0.9997	0.9996	0.9995	0.9993	0.9990	0.9987	0.9983
9	1.0000	1.0000	1.0000	1.0000	1.0000	0.9999	0.9999	0.9999	0.9998	0.9998
10	1.0000	1.0000	1.0000	1.0000	1.0000	1.0000	1.0000	1.0000	1.0000	1.0000

d \ p	0.31	0.32	0.33	0.34	0.35	0.36	0.37	0.38	0.39	0.40
0	0.0116	0.0098	0.0082	0.0068	0.0057	0.0047	0.0039	0.0032	0.0027	0.0022
1	0.0744	0.0650	0.0565	0.0491	0.0424	0.0366	0.0315	0.0270	0.0230	0.0196
2	0.2296	0.2078	0.1876	0.1687	0.1513	0.1352	0.1205	0.1069	0.0946	0.0834
3	0.4619	0.4319	0.4027	0.3742	0.3467	0.3201	0.2947	0.2704	0.2472	0.2253
4	0.6968	0.6692	0.6410	0.6124	0.5833	0.5541	0.5249	0.4957	0.4668	0.4382
5	0.8657	0.8479	0.8289	0.8087	0.7873	0.7648	0.7412	0.7167	0.6913	0.6652
6	0.9542	0.9460	0.9368	0.9266	0.9154	0.9030	0.8894	0.8747	0.8589	0.8418
7	0.9882	0.9856	0.9824	0.9787	0.9745	0.9696	0.9641	0.9578	0.9507	0.9427
8	0.9978	0.9972	0.9964	0.9955	0.9944	0.9930	0.9915	0.9896	0.9873	0.9847
9	0.9997	0.9996	0.9995	0.9993	0.9992	0.9989	0.9986	0.9982	0.9978	0.9972
10	1.0000	1.0000	1.0000	0.9999	0.9999	0.9999	0.9999	0.9998	0.9998	0.9997
11	1.0000	1.0000	1.0000	1.0000	1.0000	1.0000	1.0000	1.0000	1.0000	1.0000

d \ p	0.41	0.42	0.43	0.44	0.45	0.46	0.47	0.48	0.49	0.50
0	0.0018	0.0014	0.0012	0.0010	0.0008	0.0006	0.0005	0.0004	0.0003	0.0002
1	0.0166	0.0140	0.0118	0.0099	0.0083	0.0069	0.0057	0.0047	0.0039	0.0032
2	0.0733	0.0642	0.0560	0.0487	0.0421	0.0363	0.0312	0.0267	0.0227	0.0193
3	0.2047	0.1853	0.1671	0.1502	0.1345	0.1199	0.1066	0.0943	0.0832	0.0730
4	0.4101	0.3825	0.3557	0.3296	0.3044	0.2802	0.2570	0.2348	0.2138	0.1938
5	0.6384	0.6111	0.5833	0.5552	0.5269	0.4986	0.4703	0.4423	0.4145	0.3872
6	0.8235	0.8041	0.7836	0.7620	0.7393	0.7157	0.6911	0.6657	0.6396	0.6128
7	0.9338	0.9240	0.9131	0.9012	0.8883	0.8742	0.8589	0.8425	0.8249	0.8062

n = *12* (continued)

d \ p	0.41	0.42	0.43	0.44	0.45	0.46	0.47	0.48	0.49	0.50
8	0.9817	0.9782	0.9742	0.9696	0.9644	0.9585	0.9519	0.9445	0.9362	0.9270
9	0.9965	0.9957	0.9947	0.9935	0.9921	0.9905	0.9886	0.9883	0.9837	0.9807
10	0.9996	0.9995	0.9993	0.9991	0.9989	0.9986	0.9983	0.9979	0.9974	0.9968
11	1.0000	1.0000	1.0000	0.9999	0.9999	0.9999	0.9999	0.9999	0.9998	0.9998
12	1.0000	1.0000	1.0000	1.0000	1.0000	1.0000	1.0000	1.0000	1.0000	1.0000

n = *13*

d \ p	0.01	0.02	0.03	0.04	0.05	0.06	0.07	0.08	0.09	0.10
0	0.8775	0.7690	0.6730	0.5882	0.5133	0.4474	0.3893	0.3383	0.2935	0.2542
1	0.9928	0.9730	0.9436	0.9068	0.8646	0.8186	0.7702	0.7206	0.6707	0.6213
2	0.9997	0.9980	0.9938	0.9865	0.9755	0.9608	0.9422	0.9201	0.8946	0.8661
3	1.0000	0.9999	0.9995	0.9986	0.9969	0.9940	0.9897	0.9837	0.9758	0.9658
4	1.0000	1.0000	1.0000	0.9999	0.9997	0.9993	0.9987	0.9976	0.9959	0.9935
5	1.0000	1.0000	1.0000	1.0000	1.0000	0.9999	0.9999	0.9997	0.9995	0.9991
6	1.0000	1.0000	1.0000	1.0000	1.0000	1.0000	1.0000	1.0000	0.9999	0.9999
7	1.0000	1.0000	1.0000	1.0000	1.0000	1.0000	1.0000	1.0000	1.0000	1.0000

d \ p	0.11	0.12	0.13	0.14	0.15	0.16	0.17	0.18	0.19	0.20
0	0.2198	0.1898	0.1636	0.1408	0.1209	0.1037	0.0887	0.0758	0.0646	0.0550
1	0.5730	0.5262	0.4814	0.4386	0.3983	0.3604	0.3249	0.2920	0.2616	0.2336
2	0.8349	0.8015	0.7663	0.7296	0.6920	0.6537	0.6152	0.5769	0.5389	0.5017
3	0.9536	0.9391	0.9224	0.9033	0.8820	0.8586	0.8333	0.8061	0.7774	0.7473
4	0.9903	0.9861	0.9807	0.9740	0.9658	0.9562	0.9449	0.9319	0.9173	0.9009
5	0.9985	0.9976	0.9964	0.9947	0.9925	0.9896	0.9861	0.9817	0.9763	0.9700
6	0.9998	0.9997	0.9995	0.9992	0.9987	0.9981	0.9973	0.9962	0.9948	0.9930
7	1.0000	1.0000	0.9999	0.9999	0.9998	0.9997	0.9996	0.9994	0.9991	0.9988
8	1.0000	1.0000	1.0000	1.0000	1.0000	1.0000	1.0000	0.9999	0.9999	0.9998
9	1.0000	1.0000	1.0000	1.0000	1.0000	1.0000	1.0000	1.0000	1.0000	1.0000

d \ p	0.21	0.22	0.23	0.24	0.25	0.26	0.27	0.28	0.29	0.30
0	0.0467	0.0396	0.0334	0.0282	0.0238	0.0200	0.0167	0.0140	0.0117	0.0097
1	0.2080	0.1846	0.1633	0.1441	0.1267	0.1111	0.0971	0.0846	0.0735	0.0637
2	0.4653	0.4301	0.3961	0.3636	0.3326	0.3032	0.2755	0.2495	0.2251	0.2025
3	0.7161	0.6839	0.6511	0.6178	0.5843	0.5507	0.5174	0.4845	0.4522	0.4206
4	0.8827	0.8629	0.8415	0.8184	0.7940	0.7681	0.7411	0.7130	0.6840	0.6543
5	0.9625	0.9538	0.9438	0.9325	0.9198	0.9056	0.8901	0.8730	0.8545	0.8346
6	0.9907	0.9880	0.9846	0.9805	0.9757	0.9701	0.9635	0.9560	0.9473	0.9376
7	0.9983	0.9976	0.9968	0.9957	0.9944	0.9927	0.9907	0.9882	0.9853	0.9818
8	0.9998	0.9996	0.9995	0.9993	0.9990	0.9987	0.9982	0.9976	0.9969	0.9960
9	1.0000	1.0000	0.9999	0.9999	0.9999	0.9998	0.9997	0.9996	0.9995	0.9993
10	1.0000	1.0000	1.0000	1.0000	1.0000	1.0000	1.0000	1.0000	0.9999	0.9999
11	1.0000	1.0000	1.0000	1.0000	1.0000	1.0000	1.0000	1.0000	1.0000	1.0000

d \ p	0.31	0.32	0.33	0.34	0.35	0.36	0.37	0.38	0.39	0.40
0	0.0080	0.0066	0.0055	0.0045	0.0037	0.0030	0.0025	0.0020	0.0016	0.0013
1	0.0550	0.0473	0.0406	0.0347	0.0296	0.0251	0.0213	0.0179	0.0151	0.0126
2	0.1815	0.1621	0.1443	0.1280	0.1132	0.0997	0.0875	0.0765	0.0667	0.0579
3	0.3899	0.3602	0.3317	0.3043	0.2783	0.2536	0.2302	0.2083	0.1877	0.1686
4	0.6240	0.5933	0.5624	0.5314	0.5005	0.4699	0.4397	0.4101	0.3812	0.3530
5	0.8133	0.7907	0.7669	0.7419	0.7159	0.6889	0.6612	0.6327	0.6038	0.5744
6	0.9267	0.9146	0.9012	0.8865	0.8705	0.8532	0.8346	0.8147	0.7935	0.7712
7	0.9777	0.9729	0.9674	0.9610	0.9538	0.9456	0.9365	0.9262	0.9149	0.9023
8	0.9948	0.9935	0.9918	0.9898	0.9874	0.9846	0.9813	0.9775	0.9730	0.9679
9	0.9991	0.9988	0.9985	0.9980	0.9975	0.9968	0.9960	0.9949	0.9937	0.9922

n = 13 (continued)

d \ p	0.31	0.32	0.33	0.34	0.35	0.36	0.37	0.38	0.39	0.40
10	0.9999	0.9999	0.9998	0.9997	0.9997	0.9995	0.9994	0.9992	0.9990	0.9987
11	1.0000	1.0000	1.0000	1.0000	1.0000	1.0000	0.9999	0.9999	0.9999	0.9999
12	1.0000	1.0000	1.0000	1.0000	1.0000	1.0000	1.0000	1.0000	1.0000	1.0000

d \ p	0.41	0.42	0.43	0.44	0.45	0.46	0.47	0.48	0.49	0.50
0	0.0010	0.0008	0.0007	0.0005	0.0004	0.0003	0.0003	0.0002	0.0002	0.0001
1	0.0105	0.0088	0.0072	0.0060	0.0049	0.0040	0.0033	0.0026	0.0021	0.0017
2	0.0501	0.0431	0.0370	0.0316	0.0269	0.0228	0.0192	0.0162	0.0135	0.0112
3	0.1508	0.1344	0.1193	0.1055	0.0929	0.0815	0.0712	0.0619	0.0536	0.0461
4	0.3258	0.2997	0.2746	0.2507	0.2279	0.2065	0.1863	0.1674	0.1498	0.1334
5	0.5448	0.5151	0.4854	0.4559	0.4268	0.3981	0.3701	0.3427	0.3162	0.2905
6	0.7476	0.7230	0.6975	0.6710	0.6437	0.6158	0.5873	0.5585	0.5293	0.5000
7	0.8886	0.8736	0.8574	0.8400	0.8212	0.8012	0.7800	0.7576	0.7341	0.7095
8	0.9621	0.9554	0.9480	0.9395	0.9302	0.9197	0.9082	0.8955	0.8817	0.8666
9	0.9904	0.9883	0.9859	0.9830	0.9797	0.9758	0.9713	0.9662	0.9604	0.9539
10	0.9983	0.9979	0.9973	0.9967	0.9959	0.9949	0.9937	0.9923	0.9907	0.9888
11	0.9998	0.9998	0.9997	0.9996	0.9995	0.9993	0.9991	0.9989	0.9986	0.9983
12	1.0000	1.0000	1.0000	1.0000	1.0000	1.0000	0.9999	0.9999	0.9999	0.9999
13	1.0000	1.0000	1.0000	1.0000	1.0000	1.0000	1.0000	1.0000	1.0000	1.0000

n = 14

d \ p	0.01	0.02	0.03	0.04	0.05	0.06	0.07	0.08	0.09	0.10
0	0.8687	0.7536	0.6528	0.5647	0.4877	0.4205	0.3620	0.3112	0.2670	0.2288
1	0.9916	0.9690	0.9355	0.8941	0.8470	0.7963	0.7436	0.6900	0.6368	0.5846
2	0.9997	0.9975	0.9923	0.9833	0.9699	0.9522	0.9302	0.9042	0.8745	0.8416
3	1.0000	0.9999	0.9994	0.9981	0.9958	0.9920	0.9864	0.9786	0.9685	0.9559
4	1.0000	1.0000	1.0000	0.9998	0.9996	0.9990	0.9980	0.9965	0.9941	0.9908
5	1.0000	1.0000	1.0000	1.0000	1.0000	0.9999	0.9998	0.9996	0.9992	0.9985
6	1.0000	1.0000	1.0000	1.0000	1.0000	1.0000	1.0000	1.0000	0.9999	0.9998
7	1.0000	1.0000	1.0000	1.0000	1.0000	1.0000	1.0000	1.0000	1.0000	1.0000

d \ p	0.11	0.12	0.13	0.14	0.15	0.16	0.17	0.18	0.19	0.20
0	0.1956	0.1670	0.1423	0.1211	0.1028	0.0871	0.0736	0.0621	0.0523	0.0440
1	0.5342	0.4859	0.4401	0.3969	0.3567	0.3193	0.2848	0.2531	0.2242	0.1979
2	0.8061	0.7685	0.7292	0.6889	0.6479	0.6068	0.5659	0.5256	0.4862	0.4481
3	0.9406	0.9226	0.9021	0.8790	0.8535	0.8258	0.7962	0.7649	0.7321	0.6982
4	0.9863	0.9804	0.9731	0.9641	0.9533	0.9406	0.9259	0.9093	0.8907	0.8702
5	0.9976	0.9962	0.9943	0.9918	0.9885	0.9843	0.9791	0.9727	0.9651	0.9561
6	0.9997	0.9994	0.9991	0.9985	0.9978	0.9968	0.9954	0.9936	0.9913	0.9884
7	1.0000	0.9999	0.9999	0.9998	0.9997	0.9995	0.9992	0.9988	0.9983	0.9976
8	1.0000	1.0000	1.0000	1.0000	1.0000	0.9999	0.9999	0.9998	0.9997	0.9996
9	1.0000	1.0000	1.0000	1.0000	1.0000	1.0000	1.0000	1.0000	1.0000	1.0000

d \ p	0.21	0.22	0.23	0.24	0.25	0.26	0.27	0.28	0.29	0.30
0	0.0369	0.0309	0.0258	0.0214	0.0178	0.0148	0.0122	0.0101	0.0083	0.0068
1	0.1741	0.1527	0.1335	0.1163	0.1010	0.0874	0.0754	0.0648	0.0556	0.0475
2	0.4113	0.3761	0.3426	0.3109	0.2811	0.2533	0.2273	0.2033	0.1812	0.1608
3	0.6634	0.6281	0.5924	0.5568	0.5213	0.4864	0.4521	0.4187	0.3863	0.3552
4	0.8477	0.8235	0.7977	0.7703	0.7415	0.7116	0.6807	0.6490	0.6188	0.5842
5	0.9457	0.9338	0.9203	0.9051	0.8883	0.8699	0.8498	0.8282	0.8051	0.7805
6	0.9848	0.9804	0.9752	0.9890	0.9617	0.9533	0.9437	0.9327	0.9204	0.9067
7	0.9967	0.9955	0.9940	0.9921	0.9897	0.9868	0.9833	0.9792	0.9743	0.9685
8	0.9994	0.9992	0.9989	0.9984	0.9978	0.9971	0.9962	0.9950	0.9935	0.9917
9	0.9999	0.9999	0.9998	0.9998	0.9997	0.9995	0.9993	0.9991	0.9988	0.9983
10	1.0000	1.0000	1.0000	1.0000	1.0000	0.9999	0.9999	0.9999	0.9998	0.9998
11	1.0000	1.0000	1.0000	1.0000	1.0000	1.0000	1.0000	1.0000	1.0000	1.0000

n = ***14*** (continued)

d \ p	*0.31*	*0.32*	*0.33*	*0.34*	*0.35*	*0.36*	*0.37*	*0.38*	*0.39*	*0.40*
0	0.0055	0.0045	0.0037	0.0030	0.0024	0.0019	0.0016	0.0012	0.0010	0.0008
1	0.0404	0.0343	0.0290	0.0244	0.0205	0.0172	0.0143	0.0119	0.0098	0.0081
2	0.1423	0.1254	0.1101	0.0963	0.0839	0.0729	0.0630	0.0543	0.0466	0.0398
3	0.3253	0.2968	0.2699	0.2444	0.2205	0.1982	0.1774	0.1582	0.1405	0.1243
4	0.5514	0.5187	0.4862	0.4542	0.4227	0.3920	0.3622	0.3334	0.3057	0.2793
5	0.7546	0.7276	0.6994	0.6703	0.6405	0.6101	0.5792	0.5481	0.5169	0.4859
6	0.8916	0.8750	0.8569	0.8374	0.8164	0.7941	0.7704	0.7455	0.7195	0.6925
7	0.9619	0.9542	0.9455	0.9357	0.9247	0.9124	0.8988	0.8838	0.8675	0.8499
8	0.9895	0.9869	0.9837	0.9800	0.9757	0.9706	0.9647	0.9580	0.9503	0.9417
9	0.9978	0.9971	0.9963	0.9952	0.9940	0.9924	0.9905	0.9883	0.9856	0.9825
10	0.9997	0.9995	0.9994	0.9992	0.9989	0.9986	0.9981	0.9976	0.9969	0.9961
11	1.0000	0.9999	0.9999	0.9999	0.9999	0.9998	0.9997	0.9997	0.9995	0.9994
12	1.0000	1.0000	1.0000	1.0000	1.0000	1.0000	1.0000	1.0000	1.0000	0.9999
13	1.0000	1.0000	1.0000	1.0000	1.0000	1.0000	1.0000	1.0000	1.0000	1.0000

d \ p	*0.41*	*0.42*	*0.43*	*0.44*	*0.45*	*0.46*	*0.47*	*0.48*	*0.49*	*0.50*
0	0.0006	0.0005	0.0004	0.0003	0.0002	0.0002	0.0001	0.0001	0.0001	0.0001
1	0.0066	0.0054	0.0044	0.0036	0.0029	0.0023	0.0019	0.0015	0.0012	0.0009
2	0.0339	0.0287	0.0242	0.0203	0.0170	0.0142	0.0117	0.0097	0.0079	0.0065
3	0.1095	0.0961	0.0839	0.0730	0.0632	0.0545	0.0468	0.0399	0.0339	0.0287
4	0.2541	0.2303	0.2078	0.1868	0.1672	0.1490	0.1322	0.1167	0.1026	0.0898
5	0.4550	0.4246	0.3948	0.3656	0.3373	0.3100	0.2837	0.2585	0.2346	0.2120
6	0.6645	0.6357	0.6063	0.5764	0.5461	0.5157	0.4852	0.4549	0.4249	0.3953
7	0.8308	0.8104	0.7887	0.7656	0.7414	0.7160	0.6895	0.6620	0.6337	0.6047
8	0.9320	0.9211	0.9090	0.8957	0.8811	0.8652	0.8480	0.8293	0.8094	0.7880
9	0.9788	0.9745	0.9696	0.9639	0.9574	0.9500	0.9417	0.9323	0.9218	0.9102
10	0.9551	0.9939	0.9924	0.9907	0.9886	0.9861	0.9832	0.9798	0.9759	0.9713
11	0.9992	0.9990	0.9987	0.9983	0.9978	0.9973	0.9966	0.9958	0.9947	0.9935
12	0.9999	0.9999	0.9999	0.9998	0.9997	0.9997	0.9996	0.9994	0.9993	0.9991
13	1.0000	1.0000	1.0000	1.0000	1.0000	1.0000	1.0000	1.0000	1.0000	0.9999
14	1.0000	1.0000	1.0000	1.0000	1.0000	1.0000	1.0000	1.0000	1.0000	1.0000

APPENDIX 4

The Poisson Distribution

$$P(c) = \frac{(np)^c}{c!} e^{-np}$$ (Cumulative Values Are in Parentheses)

c \ np_0	0.1	0.2	0.3	0.4	0.5
0	0.905 (0.905)	0.819 (0.819)	0.741 (0.741)	0.670 (0.670)	0.607 (0.607)
1	0.091 (0.996)	0.164 (0.983)	0.222 (0.963)	0.268 (0.938)	0.303 (0.910)
2	0.004 (1.000)	0.016 (0.999)	0.033 (0.996)	0.054 (0.992)	0.076 (0.986)
3		0.010 (1.000)	0.004 (1.000)	0.007 (0.999)	0.013 (0.999)
4				0.001 (1.000)	0.001 (1.000)

c \ np_0	0.6	0.7	0.8	0.9	1.0
0	0.549 (0.549)	0.497 (0.497)	0.449 (0.449)	0.406 (0.406)	0.368 (0.368)
1	0.329 (0.878)	0.349 (0.845)	0.359 (0.808)	0.366 (0.772)	0.368 (0.736)
2	0.099 (0.977)	0.122 (0.967)	0.144 (0.952)	0.166 (0.938)	0.184 (0.920)
3	0.020 (0.997)	0.028 (0.995)	0.039 (0.991)	0.049 (0.987)	0.061 (0.981)
4	0.003 (1.000)	0.005 (1.000)	0.008 (0.999)	0.011 (0.998)	0.016 (0.997)
5			0.001 (1.000)	0.002 (1.000)	0.003 (1.000)

c \ np_0	1.1	1.2	1.3	1.4	1.5
0	0.333 (0.333)	0.301 (0.301)	0.273 (0.273)	0.247 (0.247)	0.223 (0.223)
1	0.366 (0.699)	0.361 (0.662)	0.354 (0.627)	0.345 (0.592)	0.335 (0.558)
2	0.201 (0.900)	0.217 (0.879)	0.230 (0.857)	0.242 (0.834)	0.251 (0.809)
3	0.074 (0.974)	0.087 (0.966)	0.100 (0.957)	0.113 (0.947)	0.126 (0.935)
4	0.021 (0.995)	0.026 (0.992)	0.032 (0.989)	0.039 (0.986)	0.047 (0.982)
5	0.004 (0.999)	0.007 (0.999)	0.009 (0.998)	0.011 (0.997)	0.014 (0.996)
6	0.001 (1.000)	0.001 (1.000)	0.002 (1.000)	0.003 (1.000)	0.004 (1.000)

c \ np_0	1.6	1.7	1.8	1.9	2.0
0	0.202 (0.202)	0.183 (0.183)	0.165 (0.165)	0.150 (0.150)	0.135 (0.135)
1	0.323 (0.525)	0.311 (0.494)	0.298 (0.463)	0.284 (0.434)	0.271 (0.406)
2	0.258 (0.783)	0.264 (0.758)	0.268 (0.731)	0.270 (0.704)	0.271 (0.677)
3	0.138 (0.921)	0.149 (0.907)	0.161 (0.892)	0.171 (0.875)	0.180 (0.857)
4	0.055 (0.976)	0.064 (0.971)	0.072 (0.964)	0.081 (0.956)	0.090 (0.947)
5	0.018 (0.994)	0.022 (0.993)	0.026 (0.990)	0.031 (0.987)	0.036 (0.983)
6	0.005 (0.999)	0.006 (0.999)	0.008 (0.998)	0.010 (0.997)	0.012 (0.995)
7	0.001 (1.000)	0.001 (1.000)	0.002 (1.000)	0.003 (1.000)	0.004 (0.999)
8					0.001 (1.000)

c \ np_0	2.1	2.2	2.3	2.4	2.5
0	0.123 (0.123)	0.111 (0.111)	0.100 (0.100)	0.091 (0.091)	0.082 (0.082)
1	0.257 (0.380)	0.244 (0.355)	0.231 (0.331)	0.218 (0.309)	0.205 (0.287)
2	0.270 (0.650)	0.268 (0.623)	0.265 (0.596)	0.261 (0.570)	0.256 (0.543)
3	0.189 (0.839)	0.197 (0.820)	0.203 (0.799)	0.209 (0.779)	0.214 (0.757)
4	0.099 (0.938)	0.108 (0.928)	0.117 (0.916)	0.125 (0.904)	0.134 (0.891)
5	0.042 (0.980)	0.048 (0.976)	0.054 (0.970)	0.060 (0.964)	0.067 (0.958)
6	0.015 (0.995)	0.017 (0.993)	0.021 (0.991)	0.024 (0.988)	0.028 (0.986)
7	0.004 (0.999)	0.005 (0.998)	0.007 (0.998)	0.008 (0.996)	0.010 (0.996)
8	0.001 (1.000)	0.002 (1.000)	0.002 (1.000)	0.003 (0.999)	0.003 (0.999)
9				0.001 (1.000)	0.001 (1.000)

c \ np_0	2.6	2.7	2.8	2.9	3.0
0	0.074 (0.074)	0.067 (0.067)	0.061 (0.061)	0.055 (0.055)	0.050 (0.050)
1	0.193 (0.267)	0.182 (0.249)	0.170 (0.231)	0.160 (0.215)	0.149 (0.199)
2	0.251 (0.518)	0.245 (0.494)	0.238 (0.469)	0.231 (0.446)	0.224 (0.423)

c \ np_0	2.6	2.7	2.8	2.9	3.0
3	0.218 (0.736)	0.221 (0.715)	0.223 (0.692)	0.224 (0.670)	0.224 (0.647)
4	0.141 (0.877)	0.149 (0.864)	0.156 (0.848)	0.162 (0.832)	0.168 (0.815)
5	0.074 (0.951)	0.080 (0.944)	0.087 (0.935)	0.094 (0.926)	0.101 (0.916)
6	0.032 (0.983)	0.036 (0.980)	0.041 (0.976)	0.045 (0.971)	0.050 (0.966)
7	0.012 (0.995)	0.014 (0.994)	0.016 (0.992)	0.019 (0.990)	0.022 (0.988)
8	0.004 (0.999)	0.005 (0.999)	0.006 (0.998)	0.007 (0.997)	0.008 (0.996)
9	0.001 (1.000)	0.001 (1.000)	0.002 (1.000)	0.002 (0.999)	0.003 (0.999)
10				0.001 (1.000)	0.001 (1.000)

c \ np_0	3.1	3.2	3.3	3.4	3.5
0	0.045 (0.045)	0.041 (0.041)	0.037 (0.037)	0.033 (0.033)	0.030 (0.030)
1	0.140 (0.185)	0.130 (0.171)	0.122 (0.159)	0.113 (0.146)	0.106 (0.136)
2	0.216 (0.401)	0.209 (0.380)	0.201 (0.360)	0.193 (0.339)	0.185 (0.321)
3	0.224 (0.625)	0.223 (0.603)	0.222 (0.582)	0.219 (0.558)	0.216 (0.537)
4	0.173 (0.798)	0.178 (0.781)	0.182 (0.764)	0.186 (0.744)	0.189 (0.726)
5	0.107 (0.905)	0.114 (0.895)	0.120 (0.884)	0.126 (0.870)	0.132 (0.858)
6	0.056 (0.961)	0.061 (0.956)	0.066 (0.950)	0.071 (0.941)	0.077 (0.935)
7	0.025 (0.986)	0.028 (0.984)	0.031 (0.981)	0.035 (0.976)	0.038 (0.973)
8	0.010 (0.996)	0.011 (0.995)	0.012 (0.993)	0.015 (0.991)	0.017 (0.990)
9	0.003 (0.999)	0.004 (0.999)	0.005 (0.998)	0.006 (0.997)	0.007 (0.997)
10	0.001 (1.000)	0.001 (1.000)	0.002 (1.000)	0.002 (0.999)	0.002 (0.999)
11				0.001 (1.000)	0.001 (1.000)

c \ np_0	3.6	3.7	3.8	3.9	4.0
0	0.027 (0.027)	0.025 (0.025)	0.022 (0.022)	0.020 (0.020)	0.018 (0.018)
1	0.098 (0.125)	0.091 (0.116)	0.085 (0.107)	0.079 (0.099)	0.073 (0.091)
2	0.177 (0.302)	0.169 (0.285)	0.161 (0.268)	0.154 (0.253)	0.147 (0.238)
3	0.213 (0.515)	0.209 (0.494)	0.205 (0.473)	0.200 (0.453)	0.195 (0.433)
4	0.191 (0.706)	0.193 (0.687)	0.194 (0.667)	0.195 (0.648)	0.195 (0.628)
5	0.138 (0.844)	0.143 (0.830)	0.148 (0.815)	0.152 (0.800)	0.157 (0.785)
6	0.083 (0.927)	0.088 (0.918)	0.094 (0.909)	0.099 (0.899)	0.104 (0.889)
7	0.042 (0.969)	0.047 (0.965)	0.051 (0.960)	0.055 (0.954)	0.060 (0.949)
8	0.019 (0.988)	0.022 (0.987)	0.024 (0.984)	0.027 (0.981)	0.030 (0.979)
9	0.008 (0.996)	0.009 (0.996)	0.010 (0.994)	0.012 (0.993)	0.013 (0.992)
10	0.003 (0.999)	0.003 (0.999)	0.004 (0.998)	0.004 (0.997)	0.005 (0.997)
11	0.001 (1.000)	0.001 (1.000)	0.001 (0.999)	0.002 (0.999)	0.002 (0.999)
12			0.001 (1.000)	0.001 (1.000)	0.001 (1.000)

c \ np_0	4.1	4.2	4.3	4.4	4.5
0	0.017 (0.017)	0.015 (0.015)	0.014 (0.014)	0.012 (0.012)	0.011 (0.011)
1	0.068 (0.085)	0.063 (0.078)	0.058 (0.072)	0.054 (0.066)	0.050 (0.061)
2	0.139 (0.224)	0.132 (0.210)	0.126 (0.198)	0.119 (0.185)	0.113 (0.174)
3	0.190 (0.414)	0.185 (0.395)	0.180 (0.378)	0.174 (0.359)	0.169 (0.343)
4	0.195 (0.609)	0.195 (0.590)	0.193 (0.571)	0.192 (0.551)	0.190 (0.533)
5	0.160 (0.769)	0.163 (0.753)	0.166 (0.737)	0.169 (0.720)	0.171 (0.704)
6	0.110 (0.879)	0.114 (0.867)	0.119 (0.856)	0.124 (0.844)	0.128 (0.832)
7	0.064 (0.943)	0.069 (0.936)	0.073 (0.929)	0.078 (0.922)	0.082 (0.914)
8	0.033 (0.976)	0.036 (0.972)	0.040 (0.969)	0.043 (0.965)	0.046 (0.960)
9	0.015 (0.991)	0.017 (0.989)	0.019 (0.988)	0.021 (0.986)	0.023 (0.983)
10	0.006 (0.997)	0.007 (0.996)	0.008 (0.996)	0.009 (0.995)	0.011 (0.994)
11	0.002 (0.999)	0.003 (0.999)	0.003 (0.999)	0.004 (0.999)	0.004 (0.998)
12	0.001 (1.000)	0.001 (1.000)	0.001 (1.000)	0.001 (1.000)	0.001 (0.999)
13					0.001 (1.000)

c \ np_0	4.6	4.7	4.8	4.9	5.0
0	0.010 (0.010)	0.009 (0.009)	0.008 (0.008)	0.008 (0.008)	0.007 (0.007)
1	0.046 (0.056)	0.043 (0.052)	0.039 (0.047)	0.037 (0.045)	0.034 (0.041)

c \ np_0	4.6	4.7	4.8	4.9	5.0
2	0.106 (0.162)	0.101 (0.153)	0.095 (0.142)	0.090 (0.135)	0.084 (0.125)
3	0.163 (0.325)	0.157 (0.310)	0.152 (0.294)	0.146 (0.281)	0.140 (0.265)
4	0.188 (0.513)	0.185 (0.495)	0.182 (0.476)	0.179 (0.460)	0.176 (0.441)
5	0.172 (0.685)	0.174 (0.669)	0.175 (0.651)	0.175 (0.635)	0.176 (0.617)
6	0.132 (0.817)	0.136 (0.805)	0.140 (0.791)	0.143 (0.778)	0.146 (0.763)
7	0.087 (0.904)	0.091 (0.896)	0.096 (0.887)	0.100 (0.878)	0.105 (0.868)
8	0.050 (0.954)	0.054 (0.950)	0.058 (0.945)	0.061 (0.939)	0.065 (0.933)
9	0.026 (0.980)	0.028 (0.978)	0.031 (0.976)	0.034 (0.973)	0.036 (0.969)
10	0.012 (0.992)	0.013 (0.991)	0.015 (0.991)	0.016 (0.989)	0.018 (0.987)
11	0.005 (0.997)	0.006 (0.997)	0.006 (0.997)	0.007 (0.996)	0.008 (0.995)
12	0.002 (0.999)	0.002 (0.999)	0.002 (0.999)	0.003 (0.999)	0.003 (0.998)
13	0.001 (1.000)	0.001 (1.000)	0.001 (1.000)	0.001 (1.000)	0.001 (0.999)
14					0.001 (1.000)

c \ np_0	6.0	7.0	8.0	9.0	10.0
0	0.002 (0.002)	0.001 (0.001)	0.000 (0.000)	0.000 (0.000)	0.000 (0.000)
1	0.015 (0.017)	0.006 (0.007)	0.003 (0.003)	0.001 (0.001)	0.000 (0.000)
2	0.045 (0.062)	0.022 (0.029)	0.011 (0.014)	0.005 (0.006)	0.002 (0.002)
3	0.089 (0.151)	0.052 (0.081)	0.029 (0.043)	0.015 (0.021)	0.007 (0.009)
4	0.134 (0.285)	0.091 (0.172)	0.057 (0.100)	0.034 (0.055)	0.019 (0.028)
5	0.161 (0.446)	0.128 (0.300)	0.092 (0.192)	0.061 (0.116)	0.038 (0.066)
6	0.161 (0.607)	0.149 (0.449)	0.122 (0.314)	0.091 (0.091)	0.063 (0.129)
7	0.138 (0.745)	0.149 (0.598)	0.140 (0.454)	0.117 (0.324)	0.090 (0.219)
8	0.103 (0.848)	0.131 (0.729)	0.140 (0.594)	0.132 (0.456)	0.113 (0.332)
9	0.069 (0.917)	0.102 (0.831)	0.124 (0.718)	0.132 (0.588)	0.125 (0.457)
10	0.041 (0.958)	0.071 (0.902)	0.099 (0.817)	0.119 (0.707)	0.125 (0.582)
11	0.023 (0.981)	0.045 (0.947)	0.072 (0.889)	0.097 (0.804)	0.114 (0.696)
12	0.011 (0.992)	0.026 (0.973)	0.048 (0.937)	0.073 (0.877)	0.095 (0.791)
13	0.005 (0.997)	0.014 (0.987)	0.030 (0.967)	0.050 (0.927)	0.073 (0.864)
14	0.002 (0.999)	0.007 (0.994)	0.017 (0.984)	0.032 (0.959)	0.052 (0.916)
15	0.001 (1.000)	0.003 (0.997)	0.009 (0.993)	0.019 (0.978)	0.035 (0.951)
16		0.002 (0.999)	0.004 (0.997)	0.011 (0.989)	0.022 (0.973)
17		0.001 (1.000)	0.002 (0.999)	0.006 (0.995)	0.013 (0.986)
18			0.001 (1.000)	0.003 (0.998)	0.007 (0.993)
19				0.001 (0.999)	0.004 (0.997)
20				0.001 (1.000)	0.002 (0.999)
21					0.001 (1.000)

c \ np_0	11.0	12.0	13.0	14.0	15.0
0	0.000 (0.000)	0.000 (0.000)	0.000 (0.000)	0.000 (0.000)	0.000 (0.000)
1	0.000 (0.000)	0.000 (0.000)	0.000 (0.000)	0.000 (0.000)	0.000 (0.000)
2	0.001 (0.001)	0.000 (0.000)	0.000 (0.000)	0.000 (0.000)	0.000 (0.000)
3	0.004 (0.005)	0.002 (0.002)	0.001 (0.001)	0.000 (0.000)	0.000 (0.000)
4	0.010 (0.015)	0.005 (0.007)	0.003 (0.004)	0.001 (0.001)	0.001 (0.001)
5	0.022 (0.037)	0.013 (0.020)	0.007 (0.011)	0.004 (0.005)	0.002 (0.003)
6	0.041 (0.078)	0.025 (0.045)	0.015 (0.026)	0.009 (0.014)	0.005 (0.008)
7	0.065 (0.143)	0.044 (0.089)	0.028 (0.054)	0.017 (0.031)	0.010 (0.018)
8	0.089 (0.232)	0.066 (0.155)	0.046 (0.100)	0.031 (0.062)	0.019 (0.037)
9	0.109 (0.341)	0.087 (0.242)	0.066 (0.166)	0.047 (0.109)	0.032 (0.069)
10	0.119 (0.460)	0.105 (0.347)	0.086 (0.252)	0.066 (0.175)	0.049 (0.118)
11	0.119 (0.579)	0.114 (0.461)	0.101 (0.353)	0.084 (0.259)	0.066 (0.184)
12	0.109 (0.688)	0.114 (0.575)	0.110 (0.463)	0.099 (0.358)	0.083 (0.267)
13	0.093 (0.781)	0.106 (0.681)	0.110 (0.573)	0.106 (0.464)	0.096 (0.363)
14	0.073 (0.854)	0.091 (0.772)	0.102 (0.675)	0.106 (0.570)	0.102 (0.465)
15	0.053 (0.907)	0.072 (0.844)	0.088 (0.763)	0.099 (0.669)	0.102 (0.567)
16	0.037 (0.944)	0.054 (0.898)	0.072 (0.835)	0.087 (0.756)	0.096 (0.663)
17	0.024 (0.968)	0.038 (0.936)	0.055 (0.890)	0.071 (0.827)	0.085 (0.748)
18	0.015 (0.983)	0.026 (0.962)	0.040 (0.930)	0.056 (0.883)	0.071 (0.819)
19	0.008 (0.991)	0.016 (0.978)	0.027 (0.957)	0.041 (0.924)	0.056 (0.875)

c \ np_0	11.0		12.0		13.0		14.0		15.0	
20	0.005	(0.996)	0.010	(0.988)	0.018	(0.975)	0.029	(0.953)	0.042	(0.917)
21	0.002	(0.998)	0.006	(0.994)	0.011	(0.986)	0.019	(0.972)	0.030	(0.947)
22	0.001	(0.999)	0.003	(0.997)	0.006	(0.992)	0.012	(0.984)	0.020	(0.967)
23	0.001	(1.000)	0.002	(0.999)	0.004	(0.996)	0.007	(0.991)	0.013	(0.980)
24			0.001	(1.000)	0.002	(0.998)	0.004	(0.995)	0.008	(0.988)
25					0.001	(0.999)	0.003	(0.998)	0.005	(0.993)
26					0.001	(1.000)	0.001	(0.999)	0.003	(0.996)
27							0.001	(1.000)	0.002	(0.998)
28									0.001	(0.999)
29									0.001	(1.000)

APPENDIX 5

About PQ Systems

Productivity-Quality Systems, Inc. is a full-service firm dedicated to helping customers continuously improve their organizations. We offer a comprehensive suite of products and services designed to improve quality, productivity, and competitive position for all industries. The full line of improvement products and services from PQ Systems includes:

- ***SQCpack***® for Microsoft® Windows™ combines powerful SPC techniques with flexibility and ease of use. In addition to variables, attributes, Pareto charting, and capability analysis, *SQCpack* for Windows features real-time charting, alarming, SQC Quality Advisor,™ ODBC, filtering, and an easy setup wizard. *SQCpack* is easily networked and the program comes with free technical support.
- ***PORTspy***™ ***Plus*** transfers data easily from a wide variety of measuring devices into *SQCpack* for Windows and other popular software programs. *PORTspy Plus* saves time, eliminates data input errors, guides users through setup with an easy-to-use wizard, and manages data from multiple devices simultaneously.
- ***MEASUREspy***™ bridges the communication gap between complex data from measurement devices and analysis software. It easily finds the important data in complicated, structured output, such as files from CMMs, and virtually eliminates keyboard or manual data entry.
- ***GAGEpack***® for Windows allows you to record all gage events in a database. This software will aid you in maintaining an unlimited number of gage records that record descriptions, calibration procedures and histories, and vendor information for all your gages.
- ***Quality* Workbench**© helps organizations with the daily control of their quality systems in order to comply with ISO 9000 and QS 9000. It features document control, audits and non-conformities, customer complaints, and an Internet viewer.
- ***R&Rpack***® use the power of a personal computer to perform tedious repeatability and reproducibility studies with minimum effort.
- ***DOEpack***® allows you to find the optimal conditions in a process by setting up and analyzing experimental designs for all stages of production.

- ***SPC Workout*** is an interactive, multimedia training course that provides effective step-by-step instruction on how to implement and use statistical process control.
- **FMEA *Investigator*** is an interactive, multimedia training course that provides effective instructions on how to conduct both design and process FMEAs.
- ***Total Quality Transformation®*** offers step-by-step help in facilitating the quality transformation in organizations. *TQT* is a part of the *Transformation of American Industry®* training project, which has been used in a variety of manufacturing and service organizations since 1984.
- **Consulting and Training Services** are offered by PQ Systems for companies at all stages of their quality management programs. A staff of highly-qualified consultants brings practical experience from both industrial and academic environments. Seminars and on-site training programs are available to help companies implement successful quality management programs.

PQ Systems invites your questions and comments about our products and services. Call PQ Systems at 1-937-885-2255. You can also send a fax to 937-885-2252 or an e-mail to sales@pqsystems.com. Visit PQ Systems on-line at http://www.pqsystems.com.

APPENDIX 6

SQCpack Student Version Software

The following series of figures provides information about the SQCpack student version software included with the text. The captions associated with the figures use example screens to show how to create data files and enter data.

Once the software has been accessed, the preliminary menu screen provides the user with several options. The most commonly used selections are Data, Control Charts, and Histograms & Capability.

Create a new data file by selecting Data. Once a data file has been created, this screen can be used to directly access control charts, histograms, or capability analyses for existing data.

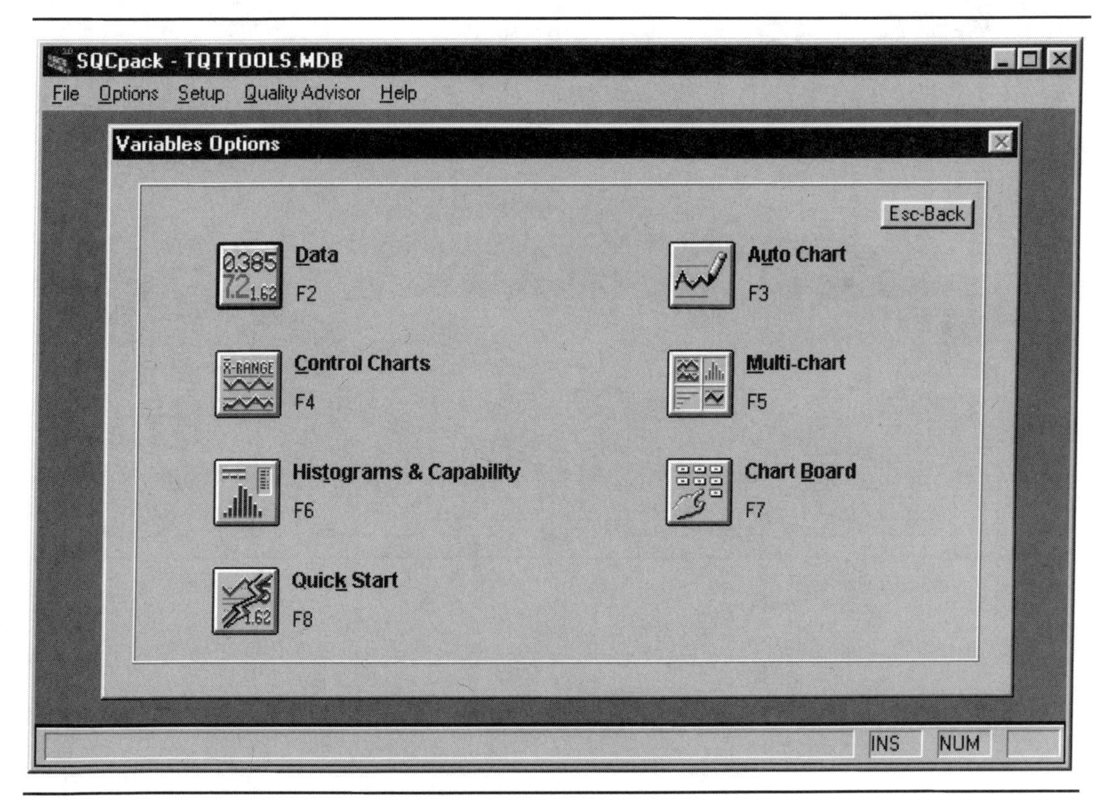

Figure 1 Variables Options Screen

To Create Data Files (Follow Figures 2, 3, 4, and 5.)

Having selected the Data option on the preliminary menu screen, a screen will appear which provides the user with the opportunity to access existing data files or create new data files.

To access an existing data file, click first on the desired file to highlight it, and then click on the desired activity (Enter Data, Modify Group, Copy Group, etc.).

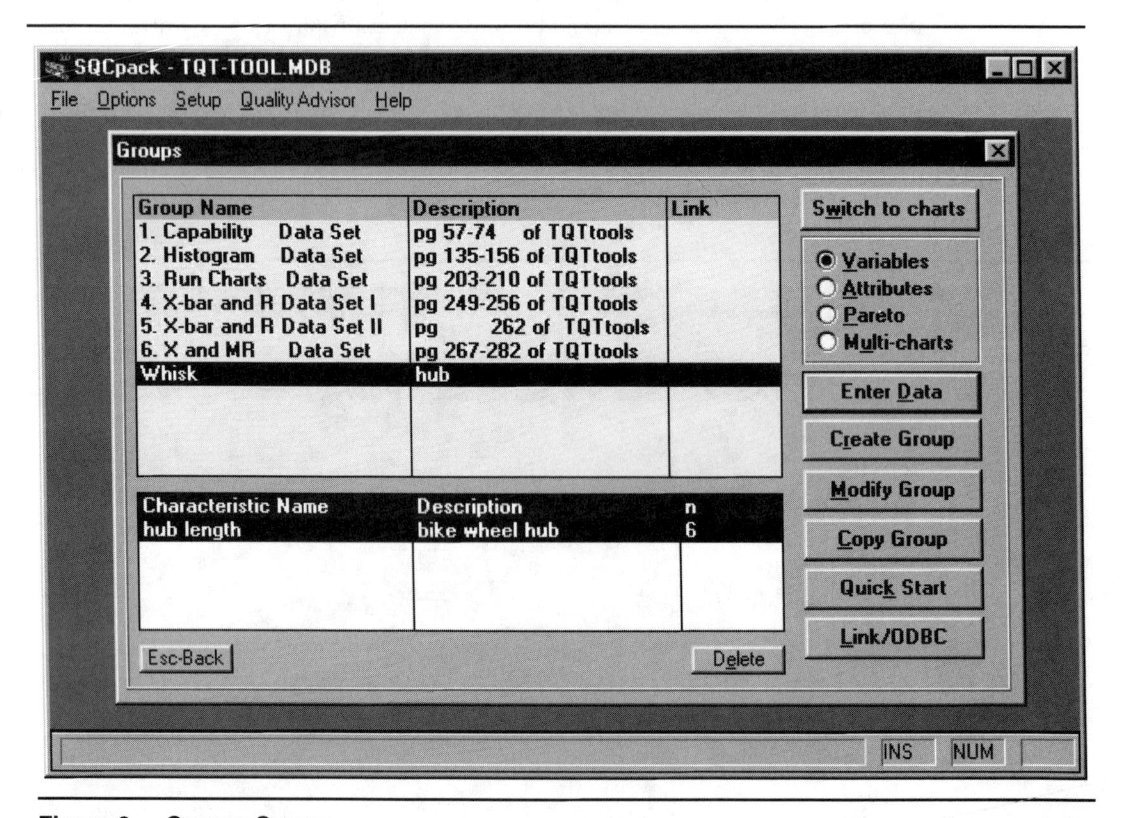

Figure 2 Groups Screen

Utilize the Create Group option to create new data files. Once these files have been established, the user may return to this screen to add data by selecting the file and clicking on Enter Data.

Having selecting the Create Group option, the screen in Figure 3 appears. Here information relevant to the characteristic under study is entered in order to create a data file.

The Variables Group screen also appears when the Modify Group option is selected on the Groups screen (Figure 2). Modifications to existing data file information can be made on this and the Variables Characteristics screen.

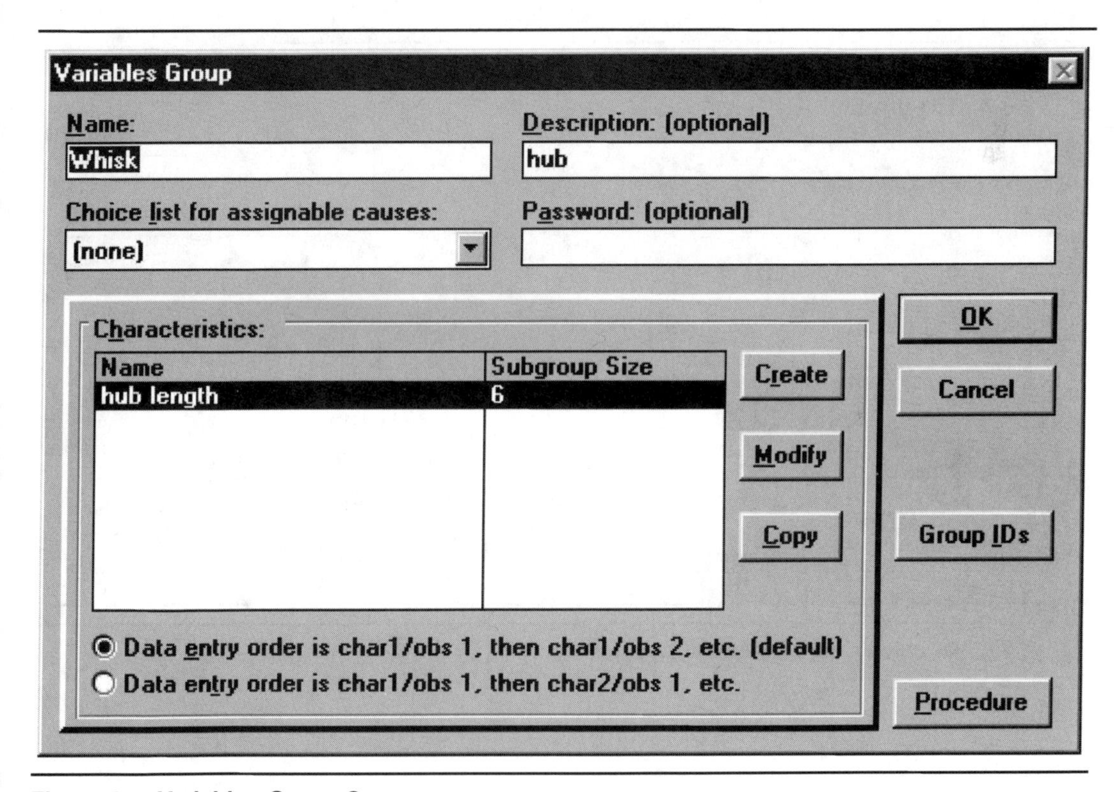

Figure 3 Variables Group Screen

To enter the description and pertinent details about the characteristic, click on Create to access the Variables Characteristics screen (Figure 4).

Having selected the Create option associated with Characteristics on the Variables Group screen, the user is able to enter information specific to the variable characteristic being studied. Once the necessary information has been added, to return to the Variables Group screen (Figure 3), click on OK. The information is automatically saved and a data file is created.

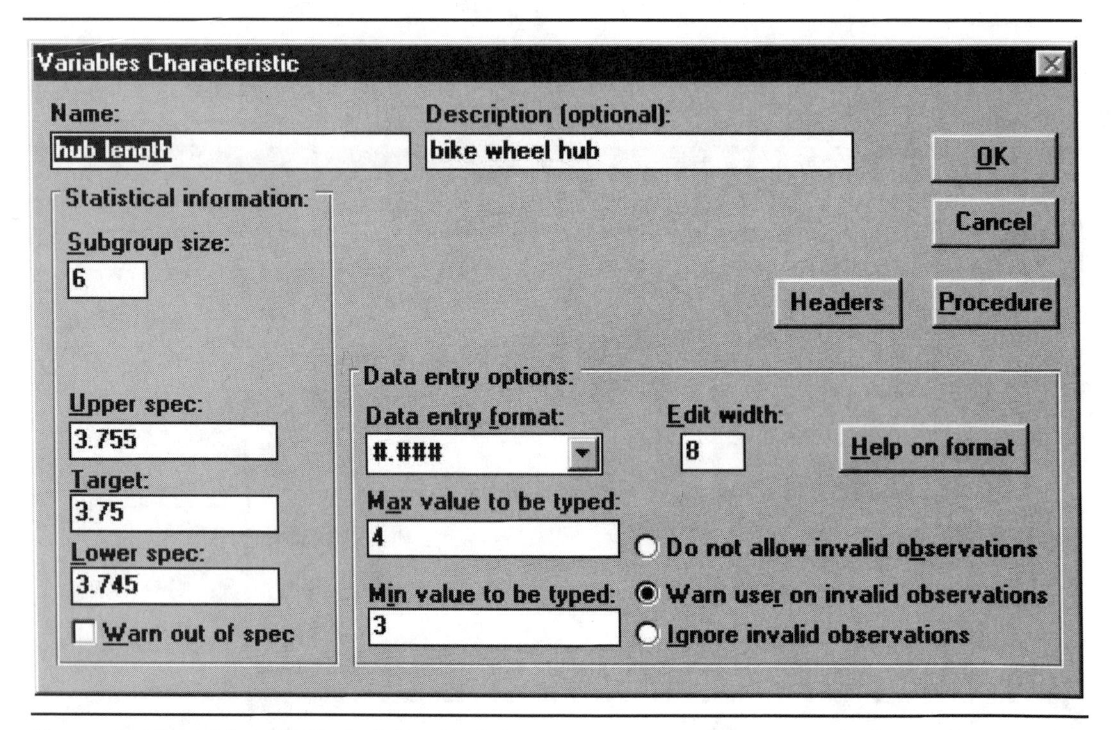

Figure 4 Variables Characteristics Screen

Upon returning to the Variables Group screen, to complete the creation of the data file, click on the option for Group ID (Figure 5).

Having selected the Group ID option on the Variables Group screen, the user can enter information which will help identify the data values when they are placed on control charts or in histograms. The most common selections are date and time. Other options include operator names and part identification numbers. Once the necessary information has been added, to return to the Variables Group screen, click OK.

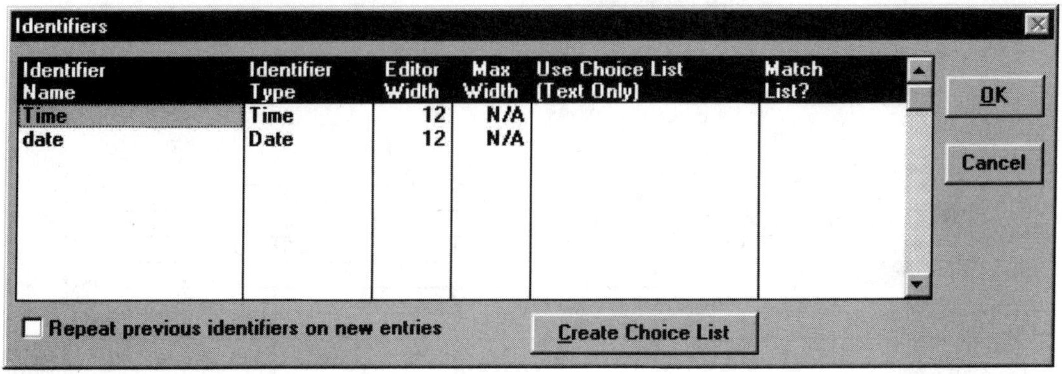

Identifiers

Identifier Name	Identifier Type	Editor Width	Max Width	Use Choice List (Text Only)	Match List?
Time	Time	12	N/A		
date	Date	12	N/A		

OK

Cancel

Repeat previous identifiers on new entries

Create Choice List

Figure 5 Identifiers Screen

To Complete Data Entry

Once the characteristics and group identifiers information has been added, click on OK on the Variables Group screen (Figure 3) to return to the Groups screen (Figure 2) to enter data. Click on Enter Data to access the Data Input screen (Figure 6).

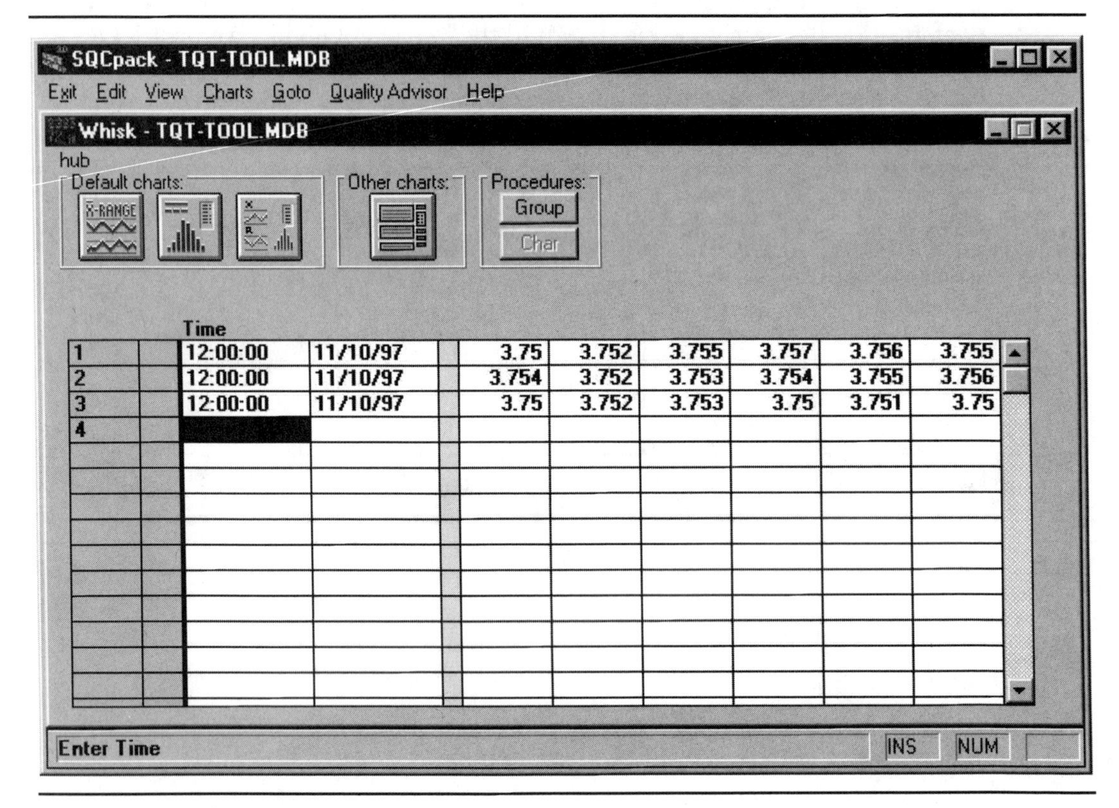

Figure 6 Data Entry

Data entry can also occur at any time from the Variables Options screen (Figure 1), once the Group data file has been created.

The system will prompt for data entry based on the information provided in the Groups, Characteristics and Group Identifiers screens. Control charts, histograms, and capability analysis can be accessed directly from this screen. If the data file already exists, control charts, histograms, and capability analysis can also be created directly from the Variables Option screen (Figure 1).

Figure 7 is an example of completed control charts, histogram, and capability analysis for data related to Whisk Wheel, the case study in Chapter 5.

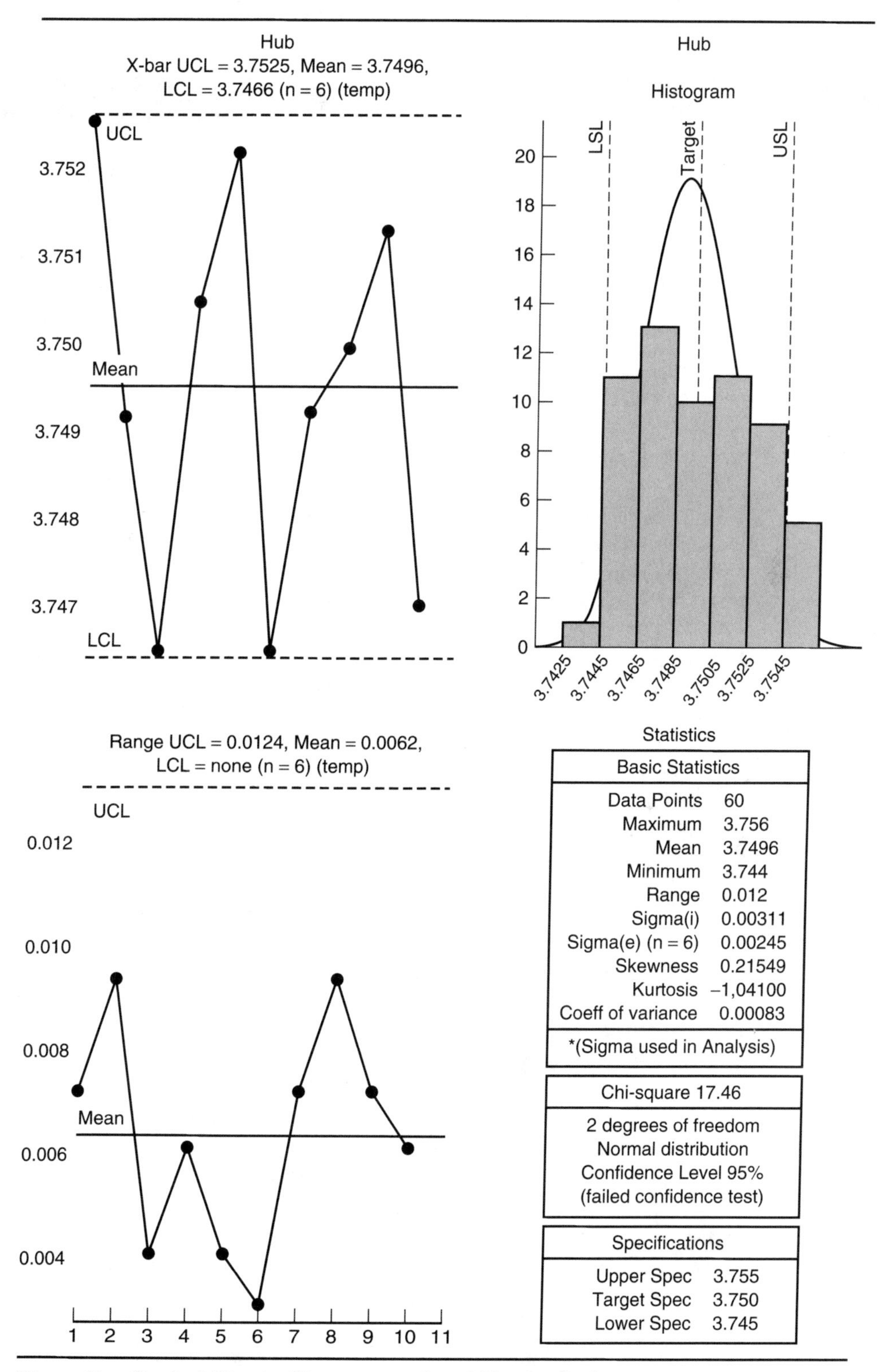

Figure 7 **Control Charts, Histograms, Capability Analysis**

Glossary

Acceptable Quality Level When a continuing series of lots is considered, a quality level that, for the purposes of sampling inspection, is the limit of a satisfactory process average.

Acceptance Sampling Inspection of a sample from a lot to decide whether to accept or not accept that lot. There are two types: attributes sampling and variables sampling. In attributes sampling, the presence or absence of a characteristic is noted in each of the units inspected. In variables sampling, the numerical magnitude of a characteristic is measured and recorded for each inspected unit; this involves reference to a continuous scale of some kind.

Acceptance Sampling Plan A specific plan that indicates the sampling sizes and the associated acceptance or non-acceptance criteria to be used. In attributes sampling, for example, there are single, double, multiple, sequential, chain, and skip-lot sampling plans. In variables sampling, there are single, double, and sequential sampling plans. (For detailed descriptions of these plans, see the standard ANS/ASQC A2-1987, Terms, Symbols, and Definitions for Acceptance Sampling.)

Accreditation Certification of a duly recognized body of the facilities, capability, objectivity, competence, and integrity of an agency, service or operational group or individual to provide the specific service or operation needed. For example, The Registrar Accreditation Board accredits those organizations that register companies to the ISO 9000 series standards.

***Accredited Registrars** Persons who perform audits for ISO 9000 standards. They are qualified organizations certified by a national body (e.g., the Registrar Accreditation Board in the U.S.).

American Society for Quality Control A professional, not-for-profit association that develops, promotes, and applies quality-related information and technology for the private sector, government, and academia. The Society serves more than 96,000 individual and 700 corporate members in the United States and 63 other countries.

***Assessment** An evaluation process that includes document reviews, audits, analysis, and report of findings.

Attribute data Go/no-go information. The control charts based on attribute data include percent chart, number of affected units chart, count chart, count-per-unit chart, quality score chart, and demerit chart.

*Definitions of asterisked terms have been provided by the author; all others come from "The Quality Glossary," *Quality Progress*, February 1992, pp. 20–29.

***Audit** A systematic appraisal procedure that examines, evaluates, and verifies that appropriate procedures, requirements, checklists, and programs are being followed effectively.

Average Chart A control chart in which the subgroup average, $\overline{X}$, is used to evaluate the stability of the process level.

Average Outgoing Quality The expected average quality level of outgoing product for a given value of incoming product quality.

Average Outgoing Quality Limit The maximum average outgoing quality over all possible levels of quality for a given acceptance sampling plan and disposal specification.

Benchmarking An improvement process in which a company measures its performance against that of best-in-class companies, determines how those companies achieved their performance levels, and uses the information to improve its own performance. The subjects that can be benchmarked include strategies, operations, processes, and procedures.

Blemish An imperfection that is severe enough to be noticed but should not cause any real impairment with respect to intended normal or reasonably foreseeable use. (See also Defect, Imperfection, Nonconformity.)

Brainstorming A technique that teams use to generate ideas on a particular subject. Each person in the team is asked to think creatively and write down as many ideas as possible. The ideas are not discussed or reviewed until after the brainstorming session.

c Chart or **Count of Nonconformities Chart** A control chart for evaluating the stability of a process in terms of the count of events of a given classification occurring in a sample of constant size.

Calibration The comparison of a measurement instrument or system of unverified accuracy to a measurement instrument or system of a known accuracy to detect any variation from the required performance specification.

***Capability** The amount of variation inherent in a stable process. Capability is determined using data from control charts and histograms from stable processes. When these indicate a stable process and a normal distribution, the indices C_p and C_{pk} can be calculated.

Cause-and-Effect Diagram A tool for analyzing process dispersion. It is also referred to as the Ishikawa diagram, because Kaoru Ishikawa developed it, and the fishbone diagram, because the complete diagram resembles a fish skeleton. The diagram illustrates the main causes and subcauses leading to an effect (symptom). The cause-and-effect diagram is one of the seven tools of quality.

Check Sheet A simple data-recording device. The check sheet is custom-designed by the user, which allows him or her to readily interpret the results. The check sheet is one of the seven tools of quality. Check sheets are often confused with data sheets and checklists.

Checklist A tool used to ensure that all important steps or actions in an operation have been taken. Checklists contain items that are important or relevant to an issue or situation. Checklists are often confused with check sheets and data sheets.

Common Causes Causes of variation that are inherent in a process over time. They affect every outcome of the process and everyone working in the process. (See also Special Causes.)

Company Culture A system of values, beliefs, and behaviors inherent in a company. To optimize business performance, top management must define and create the necessary culture.

Conformance An affirmative indication or judgment that a product or service has met the requirements of a relevant specification, contract, or regulation.

Continuous Improvement The ongoing improvement of products, services, or processes through incremental and breakthrough improvements.

Control Chart A chart with upper and lower control limits on which values of some statistical measure for a series of samples or subgroups are plotted. The chart frequently shows a central line to help detect a trend of plotted values toward either control limit.

***Control Limits** Statistically calculated limits placed on a control chart. The variability of the process is compared with the limits and special and common causes of variation are identified.

Corrective Action The implementation of solutions resulting in the reduction or elimination of an identified problem.

Cost of Poor Quality The costs associated with providing poor-quality products or services. There are four categories of costs: internal failure costs (costs associated with defects found before the customer receives the product or service), external failure costs (costs associated with defects found after the customer receives the product or service), appraisal costs (costs incurred to determine the degree of conformance to quality requirements), and prevention costs (costs incurred to keep failure and appraisal costs to a minimum).

Cost of Quality A term coined by Philip Crosby referring to the cost of poor quality.

C_p A widely used process capability index. It is expressed as

$$C_p = \frac{\text{upper specification limit} - \text{lower specification limit}}{6\sigma}$$

C_{pk} A widely used process capability index. It is expressed as

$$C_{pk} = \frac{|\mu - \text{nearer specification limit}|}{3\sigma}$$

***Crosby, Philip** Originated the zero defects concept. He has authored many books, including *Quality Is Free*, *Quality without Tears*, *Let's Talk Quality*, and *Leading: The Art of Becoming an Executive*.

Cumulative Sum Control Chart A control chart on which the plotted value is the cumulative sum of deviations of successive samples from a target value. The ordinate of each plotted point represents the algebraic sum of the previous ordinate and the most recent deviations from the target.

Customer Delight The result of delivering a product or service that exceeds customer expectations.

Customer Satisfaction The result of delivering a product or service that meets customer requirements.

Customer-Supplier Partnership A long-term relationship between a buyer and a supplier characterized by teamwork and mutual confidence. The supplier is considered an extension of the buyer's organization. The partnership is based on several commitments. The buyer provides long-term contracts and uses fewer suppliers. The supplier implements quality assurance processes so that incoming inspection can be minimized. The supplier also helps the buyer reduce costs and improve product and process designs.

Defect A product's or service's nonfulfillment of an intended requirement or reasonable expectation for use, including safety considerations. There are four classes of defects: Class 1, Very Serious, leads directly to severe injury or catastrophic economic loss; Class 2, Serious, leads directly to significant injury or significant economic loss; Class 3, Major, is related to major problems with respect to intended normal or reasonably foreseeable use; and Class 4, Minor, is related to minor problems with respect to intended normal or reasonably foreseeable use. (See also Blemish, Imperfection, Nonconformity.)

Deming, W. Edwards (deceased) A prominent consultant, teacher, and author on the subject of quality. After sharing his expertise in statistical quality control to help the U.S. war effort during World War II, the War Department sent Deming to Japan in 1946 to help that nation recover from its wartime losses. Deming published more than 200 works, including the well-known books *Quality, Productivity and Competitive Position*, and *Out of the Crisis*. Deming, who developed the 14 points for managing, was an ASQC Honorary member.

Deming Cycle See Plan-Do-Study-Act Cycle.

Deming Prize Award given annually to organizations that, according to the award guidelines, have successfully applied companywide quality control based on statistical quality control and will keep up with it in the future. Although the award is named in honor of W. Edwards Deming, its criteria are not specifically related to Deming's teachings. There are three separate divisions for the award: the Deming Application Prize, the Deming Prize for Individuals, and the Deming Prize for Overseas Companies. The award process is overseen by the Deming Prize Committee of the Union of Japanese Scientists and Engineers in Tokyo.

Dependability The degree to which a product is operable and capable of performing its required function at any randomly chosen time during its specified operating time, provided that the product is available at the start of that period. (Nonoperation-related influences are not included.) Dependability can be expressed by the ratio:

$$\frac{\text{Time available}}{\text{Time available} + \text{time required}}$$

Design of Experiments A branch of applied statistics dealing with planning, conducting, analyzing, and interpreting controlled tests to evaluate the factors that control the value of a parameter or group of parameters.

Diagnostic Journey and Remedial Journey A two-phase investigation used by teams to solve chronic quality problems. In the first phase—the diagnostic journey—the team journeys from the symptom of a chronic problem to its cause. In the second phase—the remedial journey—the team journeys from the cause to its remedy.

Documentation Written material describing the methods, procedures, or processes to be followed. Documentation may also provide supporting evidence that particular procedures are followed. Documentation may be in the form of a quality manual or process operation sheets.

Dodge-Romig Sampling Plans Plans for acceptance sampling developed by Harold F. Dodge and Harry G. Romig. Four sets of tables were published in 1940: single-sampling lot tolerance tables, double-sampling lot tolerance tables, single-sampling average outgoing quality limit tables, and double-sampling average outgoing quality limit tables.

80-20 Rule A term referring to the Pareto principle, which was first defined by J. M. Juran in 1950. The principle suggests that most effects come from relatively few causes; that is, 80% of the effects come from 20% of the possible causes.

Employee Involvement A practice within an organization whereby employees regularly participate in making decisions on how their work areas operate, including making suggestions for improvement, planning, goal setting, and monitoring performance.

Empowerment A condition whereby employees have the authority to make decisions and take action in their work areas without prior approval. For example, an operator can stop a production process if he detects a problem or a customer service representative can send out a replacement product if a customer calls with a problem.

Experimental Design A formal plan that details the specifics for conducting an experiment, such as which responses, factors, levels, blocks, treatments, and tools are to be used.

External Customer A person or organization who receives a product, a service, or information but is not part of the organization supplying it. (See also Internal Customer.)

Feigenbaum, Armand V. The founder and president of General Systems Co., an international engineering company that designs and implements total quality systems. Feigenbaum originated the concept of total quality control in his book *Total Quality Control,* which was published in 1951.

Fishbone Diagram See Cause-and-Effect Diagram.

Fitness for Use A term used to indicate that a product or service fits the customer's defined purpose for that product or service.

Flowchart A graphical representation of the steps in a process. Flowcharts are drawn to better understand processes. The flowchart is one of the seven tools of quality.

Funnel Experiment An experiment that demonstrates the effects of tampering. Marbles are dropped through a funnel in an attempt to hit a flat-surfaced target below. The experiment shows that adjusting a stable process to compensate for an undesirable result or an extraordinarily good result will produce output that is worse than if the process had been left alone.

Gauge Repeatability and Reproducibility The evaluation of a gauging instrument's accuracy by determining whether the measurements taken with it are repeatable (i.e., there is close agreement among a number of consecutive measurements of the output for the same value of the input under the same operating conditions) and reproducible (i.e., there is close agreement among repeated measurements of the output for the same value of input made under the same operating conditions over a period of time).

Go/No-Go State of a unit or product. The parameters are possible: go (conforms to specification) and no-go (does not conform to specification).

Histogram A graphic summary of variation in a set of data. The pictorial nature of the histogram lets people see patterns that are difficult to see in a simple table of numbers. The histogram is one of the seven tools of quality.

Imperfection A quality characteristic departure from its intended level or state without any association to conformance to specification requirements or to the usability of a product or service. (See also Blemish, Defect, Nonconformity.)

In-Control Process A process in which the statistical measure being evaluated is in a state of statistical control (i.e., the variations among the observed sampling results can be attributed to a constant system of chance causes). (See also Out-of-Control Process.)

Inspection Measuring, examining, testing, or gauging one or more characteristics of a product or service and comparing the results with specified requirements to determine whether conformity is achieved for each characteristic.

Instant Pudding A term used to illustrate an obstacle to achieving quality: the supposition that quality and productivity improvement is achieved quickly through an affirmation of faith rather than through sufficient effort and education. W. Edwards Deming used this term—which was initially coined by James Bakken of the Ford Motor Co—in his book *Out of the Crisis*.

Internal Customer The recipient (person or department) of another person's or department's output (product, service, or information) within an organization. (See also External Customer.)

Ishikawa, Kaoru (deceased) A pioneer in quality control activities in Japan. In 1943, he developed the cause-and-effect diagram. Ishikawa, an ASQC Honorary member, published many works, including *What Is Total Quality Control?*, *The Japanese Way*, *Quality Control Circles at Work*, and *Guide to Quality Control*. He was a member of the quality control research group of the Union of Japanese Scientists and Engineers while also working as an assistant professor at the University of Tokyo.

Ishikawa Diagram See Cause-and-Effect Diagram.

ISO 9000 Series Standards A set of five individual but related international standards on quality management and quality assurance developed to help companies effectively document the quality system elements to be implemented to maintain an efficient quality system. The standards were developed by the International Organization for Standardization (ISO), a special international agency for standardization composed of the national standards bodies of 91 countries.

Juran, Joseph M. The chairman emeritus of the Juran Institute and an ASQC Honorary member. Since 1924, Juran has pursued a varied career in management as an engineer, executive, government administrator, university professor, labor arbitrator, corporate director, and consultant. Specializing in managing for quality, he has authored hundreds of papers and 12 books, including *Juran's Quality Control Handbook*, *Quality Planning and Analysis* (with F. M. Gryna), and *Juran on Leadership for Quality*.

Just-in-Time Manufacturing An optimal material requirement planning system for a manufacturing process in which there is little or no manufacturing material inventory on hand at the manufacturing site and little or no incoming inspection.

Kaizen A Japanese term that means gradual unending improvement by doing little things better and setting and achieving increasingly higher standards. The term was made famous by Masaaki Imai in his book *Kaizen: The Key to Japan's Competitive Success*.

LCL See Lower Control Limit.

Leadership An essential part of a quality improvement effort. Organization leaders must establish a vision, communicate that vision to those in the organization, and provide the tools and knowledge necessary to accomplish the vision.

Lot A defined quantity of product accumulated under conditions that are considered uniform for sampling purposes.

Lower Control Limit Control limit for points below the central line in a control chart.

Maintainability The probability that a given maintenance action for an item under given usage conditions can be performed within a stated time interval when the maintenance is performed under stated conditions using stated procedures and resources. Maintainability has two categories: serviceability (the ease of conducting scheduled inspections and servicing) and repairability (the ease of restoring service after a failure).

Malcolm Baldrige National Quality Award An award established by Congress in 1987 to raise awareness of quality management and to recognize U.S. companies that have implemented successful quality management systems. Two awards may be given annually in each of three categories: manufacturing company, service company, and small business. The award is named after the late Secretary of Commerce Malcolm Baldrige, a proponent of quality management. The U.S. Commerce Department's National Institute of Standards and Technology manages the award and ASQC administers it.

MIL-Q-9858A A military standard that describes quality program requirements.

MIL-STD-105E A military standard that describes the sampling procedures and tables for inspection by attributes.

MIL-STD-45662A A military standard that describes the requirements for creating and maintaining a calibration system for measurement and test equipment.

***Nonconformance** State that exists when a product, service, or material does not conform to the customer requirements or specifications.

Nonconformity The nonfulfillment of a specified requirement. (See also Blemish, Defect, Imperfection.)

Nondestructive Testing and Evaluation Testing and evaluation methods that do not damage or destroy the product being tested.

np Chart Number of affected units chart.

Number of Affected Units Chart A control chart for evaluating the stability of a process in terms of the total number of units in a sample in which an event of a given classification occurs.

OC Curve Operating characteristic curve.

Operating Characteristic Curve A graph used to determine the probability of accepting lots as a function of the lots' or processes' quality level when using various sampling plans. There are three types: Type A curves, which give the probability of acceptance for an individual lot coming from finite production (will not continue in the future); Type B curves, which give the probability of acceptance for lots coming from a continuous process; and Type C curves, which for a continuous sampling plan, give the long-run percentage of product accepted during the sampling phase.

Out-of-Control Process A process in which the statistical measure being evaluated is not in a state of statistical control; the variations among the observed sampling results cannot be attributed to a constant system of chance causes. (See also In-Control Process.)

Out-of-Spec A term used to indicate that a unit does not meet a given specification.

p Chart The fraction nonconforming chart is a control chart used to monitor the proportion nonconforming in a lot of goods.

Pareto Chart A graphical tool for ranking causes from most significant to least significant. It is based on the Pareto principle, which was first defined by J. M. Juran in 1950. The principle, named after 19th century economist Vilfredo Pareto, suggests that most effects come from relatively few causes; that is, 80% of the effects come from 20% of the possible causes. The Pareto chart is one of the seven tools of quality.

***Parts per Million (PPM)** Describes the performance of a process in terms of either actual or projected defective material.

PDSA Cycle See Plan-Do-Study-Act Cycle.

Percent Chart A control chart for evaluating the stability of a process in terms of the percent of the total number of units in a sample in which an event of a given classification occurs. The percent chart is also referred to as a proportion chart.

***Plan-Do-Study-Act Cycle** A four-step process for quality improvement. In the first step (Plan), a plan to effect improvement is developed. In the second step (Do), the plan is carried out, preferably on a small scale. In the third step (Study), the effects of the plan are observed. In the last step (Act), the results are studied to determine what was learned and what can be predicted. The Plan-Do-Study-Act cycle was originally called the Plan-Do-Check-Act cycle but in the early 1990s was standardized as Plan-Do-Study-Act. It is still sometimes referred to as the Shewhart cycle because Walter A. Shewhart discussed the concept in his book *Statistical Method from the Viewpoint of Quality Control* or as the Deming cycle because W. Edwards Deming introduced the concept in Japan.

Prevention vs. Detection A term used to contrast two types of quality activities. Prevention refers to those activities designed to prevent nonconformances in products and services. Detection refers to those activities designed to detect nonconformances already in products and services. Another term used to describe this distinction is "designing in quality vs. inspecting in quality."

Process Capability A statistical measure of the inherent process variability for a given characteristic. The most widely accepted formula for process capability is 6σ.

Process Capability Index The value of the tolerance specified for the characteristic divided by the process capability. There are several types of process capability indexes, including the widely used C_{pk} and C_p.

***Process Control** Using statistical process control to measure and regulate a process.

Product or Service Liability The obligation of a company to make restitution for loss related to personal injury, property damage, or other harm caused by its product or service.

QFD See Quality Function Deployment.

Q90 Series Refers to the ANSI/ASQC Q90 series of standards, which is the Americanized version of the ISO 9000 series standards. The United States adopted the ISO 9000 series as the ANSI/ASQC Q90 series in 1987.

Quality A subjective term for which each person has his or her own definition. In technical usage, quality can have two meanings: (1) the characteristics of a product or service that bear on its ability to satisfy stated or implied needs and (2) a product or service free of deficiencies.

Quality Assurance/Quality Control Two terms that have many interpretations because of the multiple definitions for the words "assurance" and "control." For example, "assurance" can mean the act of giving confidence, the state of being certain, or the act of making certain; "control" can mean an evaluation to indicate needed corrective responses, the act of going, or the state of a process in which the variables are attributable to a constant system of chance causes. (For detailed discussion on the multiple definitions see ANSI/ASQC Standard A3-1987, "Definitions, Symbols, Formulas, and Tables for Control Charts.") One definition of quality assurance is: are all the planned and systematic activities implemented within the quality system that can be demonstrated to provide confidence that a product or service will fulfill requirements for quality? One definition for quality control is: the operational techniques and activities used to fulfill requirements for quality. Often, however, "quality assurance" and "quality control" are used interchangeably, referring to the actions performed to ensure the quality of a product, service, or process.

Quality Audit A systematic, independent examination and review to determine whether quality activities and related results comply with planned arrangements and whether these arrangements are implemented effectively and are suitable to achieve the objectives.

Quality Circles Quality improvement or self-improvement study groups composed of a small number of employees (10 or fewer) and their supervisor. Quality circles originated in Japan, where they are called quality control circles.

Quality Control See Quality Assurance/Quality Control.

Quality Costs See Cost of Poor Quality.

Quality Engineering The analysis of a manufacturing system at all stages to maximize the quality of the process itself and the products it produces.

Quality Function Deployment A structured method in which customer requirements are translated into appropriate technical requirements for each stage of product development and production. The QFD process is often referred to as listening to the voice of the customer.

Quality Loss Function A parabolic approximation of the quality loss that occurs when a quality characteristic deviates from its target value. The quality loss function is expressed in monetary units: the cost of deviating from the target increases quadratically the farther the quality characteristic moves from the target. The formula used to compute the quality loss function depends on the type of quality characteristic being used. The quality loss function was first introduced in this form by Genichi Taguchi.

***Quality Manual** The chief document for standard operating procedures, processes, and specifications that define a quality management system. The manual serves as a permanent reference guide for the implementation and maintenance of the quality management system described by the manual.

***Quality Plan** Integrates quality philosophies into an organization's environment. The plan will include specific continuous improvement strategies and actions. Plans are developed at the departmental, group, plant, division and company level. Lower-level plans should support the company's strategic objectives. A quality plan emphasizes defect prevention through continuous improvement rather than defect detection.

***Quality Records** Written verification that a company's methods, systems, and processes were performed according to the quality system documentation such as inspection or test results, internal audit results, and calibration data.

Quality Trilogy A three-pronged approach to managing for quality. The three legs are quality planning (developing the products and processes required to meet customer needs), quality control (meeting product and process goals), and quality improvement (achieving unprecedented levels of performance).

R Chart See Range Chart.

Random Sampling Sample units are selected in such a manner that all combinations of n units under consideration have an equal chance of being selected as the sample.

Range Chart A control chart in which the subject group, R, is used to evaluate the stability of the variability within a process.

Red Bead Experiment An experiment developed by W. Edwards Deming to illustrate that it is impossible to put employees in rank order of performance for the coming year based on their performance during the past year because performance differences must be attributed to the system, not to employees. Four thousand red and white beads (20% red) in a jar and six people are needed for the experiment. The participants' goal is to produce white beads, because the customer will not accept red beads. One person begins by stirring the beads and then, blindfolded, selects a sample of 50 beads. That person hands the jar to the next person, who repeats the process, and so on. When everyone has his or her sample, the number of red beads for each is counted. The limits of variation between employees that can be attributed to the system are calculated. Everyone will fall within the calculated limits or variation that could arise from the system. The calculations will show that there is no evidence one person will be a better performer than another in the future. The experiment shows that it would be a waste of management's time to try to find out why, say, John produced four red beads and Jane produced 15; instead, management should improve the system, making it possible for everyone to produce more white beads.

Registration to Standards A process in which an accredited, independent third-party organization conducts an on-site audit of a company's operations against the requirements of the standard to which the company wants to be registered. Upon successful completion of the audit, the company receives a certificate indicating that it has met the standard requirements.

Reliability The probability of a product performing its intended function under stated conditions without failure for a given period of time.

***Repair** Corrective action to a damaged product so that the product will fulfill the original specifications.

***Rework** Action taken on nonconforming products or services to allow them to meet the original specifications.

s Chart See Sample Standard Deviation Chart.

Sample Standard Deviation Chart A control chart in which the subgroup standard deviation, s, is used to evaluate the stability of the variability within a process.

Scatter Diagram A graphical technique to analyze the relationship between two variables. Two sets of data are plotted on a graph, with the y axis being used for the variable to be predicted and the x axis being used for the variable to make the prediction. The graph will show possible relationships (although two variables might appear to be related, they might not be—those who know most about the variables must make that evaluation). The scatter diagram is one of the seven tools of quality.

Seven Tools of Quality Tools that help organizations understand their processes in order to improve them. The tools are the cause-and-effect diagram, check sheet, control chart, flowchart, histogram, Pareto chart, and scatter diagram. (See individual entries.)

Shewhart, Walter A. (deceased) Referred to as the father of statistical quality control because he brought together the disciplines of statistics, engineering, and economics. He described the basic principles of this new discipline in his book *Economic Control of Quality of Manufactured Product*. Shewhart, ASQC's first Honorary member, was best known for creating the control chart. Shewhart worked for Western Electric and AT&T Bell Telephone Laboratories in addition to lecturing and consulting on quality control.

Shewhart Cycle See Plan-Do-Study-Act Cycle.

6σ Quality A term used generally to indicate that a process is well-controlled, i.e., ± 3 sigma from the centerline in a control chart. The term is usually associated with Motorola, which named one of its key operational initiatives "Six Sigma Quality."

SPC See Statistical Process Control.

Special Causes Causes of variation that arise because of special circumstances. They are not an inherent part of a process. Special causes are also referred to as assignable causes. (See also Common Causes.)

Specification A document that states the requirements to which a given product or service must conform.

SQC See Statistical Quality Control.

Statistical Process Control The application of statistical techniques to control a process. Often the term "statistical quality control" is used interchangeably with "statistical process control."

Statistical Quality Control The application of statistical techniques to control quality. Often the term "statistical process control" is used interchangeably with "statistical quality control," although statistical quality control includes acceptance sampling as well as statistical process control.

***Subcontractors** Persons who provide materials, parts, or services to the suppliers of original equipment manufacturers.

***Supplier Quality Assurance** Confidence that a supplier's product or service will fulfill its customers' needs. This confidence is achieved by creating a relationship between the customer and supplier that ensures the product will be fit for use with minimal corrective action and inspection. According to J. M, Juran, there are nine primary activities needed: (1) define product and program quality requirements, (2) evaluate alternative suppliers, (3) select suppliers, (4) conduct joint quality planning, (5) cooperate with the supplier during the execution of the contract, (6) obtain proof of conformance to requirements, (7) certify qualified suppliers, (8) conduct quality improvement programs as required, and (9) create and use supplier quality ratings.

***Suppliers** Persons who provide materials, parts, or services directly to manufacturers.

Taguchi, Genichi The executive director of the American Supplier Institute, the director of the Japan Industrial Technology Institute, and an honorary professor at Nanjing Institute of Technology in China. Taguchi is well known for developing a methodology to improve quality and reduce costs, which in the United States is referred to as the Taguchi Methods. He also developed the quality loss function.

Taguchi Methods The American Supplier Institute's trademarked term for the quality engineering methodology developed by Genichi Taguchi. In this engineering approach to quality control, Taguchi calls for off-line quality control, on-line quality control, and a system of experimental design to improve quality and reduce costs.

Tampering Action taken to compensate for variation within the control limits of a stable system. Tampering increases rather than decreases variation, as evidenced in the funnel experiment.

Top-Management Commitment Participation of the highest-level officials in their organization's quality improvement efforts. Their participation includes establishing and serving on a quality committee, establishing quality policies and goals, deploying those goals to lower levels of the organization, providing the resources and training that the lower levels need to achieve the goals, participating in quality improvement teams, reviewing progress organizationwide, recognizing those who have performed well, and revising the current reward system to reflect the importance of achieving the quality goals.

Total Quality Management A term initially coined in 1985 by the Naval Air Systems Command to describe its Japanese-style management approach to quality improvement. Since then, total quality management (TQM) has taken on many meanings. Simply put, TQM is a management approach to long-term success through customer satisfaction. TQM is based on the participation of all members of an organization in improving processes, products, services, and the culture they work in. TQM benefits all organization members and society. The methods for implementing this approach are found in the teachings of such quality control leaders as Philip B. Crosby, W. Edwards Deming, Armand V. Feigenbaum, Kaoru Ishikawa, and J. M. Juran.

TQM See Total Quality Management.

Trend Control Chart A control chart in which the deviation of the subgroup average, $\overline{X}$, from an expected trend in the process level is used to evaluate the stability of a process.

Type I Error An incorrect decision to reject something (such as a statistical hypothesis or a lot of products) when it is acceptable.

Type II Error An incorrect decision to accept something when it is unacceptable.

u Chart or **Nonconformities-per-Unit Chart** A control chart for evaluating the stability of a process in terms of the average count of events of a given classification per unit occurring in a sample; count per unit chart.

UCL See Upper Control Limit.

Upper Control Limit Control limit for points above the central line in a control chart.

Value-Adding Process Those activities that transform an input into a customer-usable output. The customer can be internal or external to the organization.

Variables Data Measurement information. Control charts based on variables data include average ($\overline{X}$) chart, range (R) chart, and sample standard deviation chart.

Variation A change in data, a characteristic, or a function that is caused by one of four factors: special causes, common causes, tampering, or structural variation.

Vision Statement: A statement summarizing an organization's values, mission, and future direction for its employees and customers.

Vital Few, Useful Many A term used by J. M. Juran to describe his use of the Pareto principle, which he first defined in 1950. (The principle was used much earlier in economics and inventory control methodologies.) The principle suggests that most effects come from relatively few causes; that is, 80% of the effects come from 20% of the possible causes. The 20% of the possible causes are referred to as the "vital few"; the remaining causes are referred to as the "useful many." When Juran first defined this principle, he referred to the remaining causes as the "trivial many," but realizing that no problems are trivial in quality assurance, he changed it to the "useful many."

$\overline{X}$ Chart Average chart.

Zero Defects A performance standard developed by Philip B. Crosby to address a dual attitude in the workplace: people are willing to accept imperfection in some areas, while, in other areas, they expect the number of defects to be zero. This dual attitude had developed because of the conditioning that people are human and humans make mistakes. However, the zero defects methodology states that, if people commit themselves to watching details and avoiding errors, they can move closer to the goal of zero defects. The performance standard that must be set is "zero defects," not "close enough."

Answers to Selected Problems

Chapter 4

4.19 Mean = 38.6
Mode = 34
Median = 38

4.20 Mean = 1.123
Mode = 1.122
Median = 1.123

4.21 Mean = 0.656
Mode = 0.654
Median = 0.655

4.22 Mean = 226.2
Mode = 227
Median = 227
$\sigma = 2.36$

4.26 $\sigma = 5.32$
R = 18

4.27 $\sigma = 0.0022$
R = 0.0065

4.28 Mean = 0.656
$\sigma = 0.0022$
Median = 0.655
Mode = 0.654

4.29 Mean = 0.077
$\sigma = 0.0032$

4.30 Mean = 24
Mode = 25
Median = 25
Range = 19
$\sigma = 5$

4.34 Area = 0.8413

4.35 Area = 0.0047 or 0.47% of the parts are above 0.93 mm.

4.36 Area = 0.9525

4.42 Area = 0.8186

4.43 X = 2871.25 foot-candles

4.44 2.27% of the batteries should survive longer than 1,000 days

4.45 37.07% will be below the LSL of 11.41 inches.

4.47 Mean = 0.0015
Mode = 0.0025, 0.0004
Median = 0.0014
$\sigma = 0.0008$
R = 0.0028
73.41% of the parts will meet spec.

Chapter 5

5.3 $\overline{\overline{X}} = 16$, $UCL_x = 20$, $LCL_x = 12$, R-bar = 7, $UCL_R = 15$, $LCL_R = 0$

5.4 $\overline{\overline{X}} = 50.2$, $UCL_x = 50.7$, $LCL_x = 49.7$, R-bar = 0.68, $UCL_R = 1.6$, $LCL_R = 0$

5.5 $\overline{\overline{X}} = 349$, $UCL_x = 357$, $LCL_x = 341$, R-bar = 14, $UCL_R = 8$, $LCL_R = 0$

5.10 $\overline{\overline{X}} = 0.0627$, $UCL_x = 0.0629$, $LCL_x = 0.0625$, R-bar = 0.0003, $UCL_R = 0.0006$, $LCL_R = 0$

5.14 $\overline{\overline{X}} = 0.0028$, $UCL_x = 0.0032$, $LCL_x = 0.0023$, R-bar = 0.0006, $UCL_R = 0.0014$, $LCL_R = 0$

5.15 $\overline{\overline{X}} = 0.0627$, $UCL_x = 0.0628$, $LCL_x = 0.0626$, s-bar = 0.0001, $UCL_s = 0.0002$, $LCL_s = 0$

5.16 $\overline{\overline{X}} = 0.0028$, $UCL_x = 0.0033$, $LCL_x = 0.0023$, s-bar = 0.0003, $UCL_s = 0.0007$, $LCL_s = 0$

Chapter 6

6.4 $\sigma = 3$, $6\sigma = 18$, $C_p = 0.44$, $C_{pk} = 0.33$

6.6 $\sigma = 0.0001$, $6\sigma = 0.0008$, $C_p = 1.667$, $C_{pk} = 1$

6.7 $\sigma = 0.33$, $6\sigma = 1.982$, $C_p = 0.505$, $C_{pk} = 0.303$
6.10 $\sigma = 0.00002$, $6\sigma = 0.00012$, $C_p = 0.33$, $C_{pk} = 0.167$
6.11 $\sigma = 0.0003$, $6\sigma = 0.0018$, $C_p = 0.55$, $C_{pk} = 0.55$
6.12 $\sigma = 0.0003$, $6\sigma = 0.0020$, $C_p = 0.5$, $C_{pk} = 0.55$

Chapter 7

7.1 X-bar = 24 mm, $UCL_x = 27$, $LCL_x = 21$, R-bar = 1, $UCL_r = 3$, $LCL_x = 0$
7.2 X-bar = 3790, $UCL_x = 4027$, $LCL_x = 3553$, R-bar = 90, $UCL_r = 292$, $LCL_x = 0$
7.3 X-bar = 34, $UCL_x = 39$, $LCL_x = 29$, R-bar = 5, $UCL_r = 13$, $LCL_x = 0$
7.4 X-bar = 0.812, $UCL_x = 0.817$, $LCL_x = 0.807$, R-bar = 0.005, $UCL_r = 0.013$, $LCL_x = 0$
7.5 X-bar = 3788, $UCL_x = 3951$, $LCL_x = 3625$, R-bar = 159, $UCL_r = 409$, $LCL_x = 0$
7.6 X-bar = 30.1, $UCL_x = 33$, $LCL_x = 28$, R-bar = 3.5, $UCL_r = 8$, $LCL_x = 0$
7.16 X-bar = 0.036, $UCL_x = 0.213$, $LCL_x = -0.141$, R-bar = 0.173, $UCL_R = 0.45$, $LCL_R = 0$

Chapter 8

8.1 0.50
8.2 0.008
8.3 P(orange & 5) = 1/24
P(orange) 6/24
P(five) = 3/24
8.4 60
8.5 Combination = 3003
Permutation = 360360
8.6 P(performs) = 0.91
8.7 P(H/P) = 0.78
8.8 P(0 of 5/male) = 0.25
8.9 P(Fabric 1) = 0.285
P(Fabric 2/Style 1) = 0.19
P(Style 4) = 0.255
P(Style 3/Fabric 3) = 0.23
8.10 a. 0.6
b. 0.4
c. independent
8.14 P(1) + P(0) = 0.262
P(1) = 0.238
P(0) = 0.024
8.15 0.419
8.17 P(0) = 0.545
8.18 P(2) = 0.3
8.19 P(1 or less) = 0.98
8.20 P(less than 2) = 0.98
8.21 P(2) = 0.10
8.22 P(more than 4) = 1 − P(4 or less) = 0.001
8.23 P(2) = 0.012
8.24 P(5) = 0.0015
8.25 P(more than 2) = 0.017
8.26 P(1 or less) = 0.84
8.28 P(at least 1) = 0.59
8.29 P(3 in five min) = 0.13
8.30 P(more than 1 call) = 0.22
8.31 P(more than 2) = 0.58
8.32 a. 2.71% of bikes will weigh below 8.3 kg
b. Area between 8.0 kg and 10.10 kg = 0.9950 − 0.00375 = 0.9913
8.33 P(2) = 0.014
8.34 P(more than 2) = 1 − 0.977 = 0.023

Chapter 9

9.3 p-bar = 0.068
UCL = 0.237
LCL = 0
9.4 p-bar = 0.004
UCL = 0.0174
LCL = 0
9.5 p-bar = 0.038
UCL = 0.085
LCL = 0
9.9 p-bar = 0.0131
UCL = 0.028
LCL = 0
9.10 p-bar = 0.0131
$UCL_{400} = 0.030$
LCL = 0
$UCL_{450} = 0.029$
LCL = 0
$UCL_{500} = 0.028$
LCL = 0
$UCL_{550} = 0.028$
LCL = 0
$UCL_{600} = 0.027$
LCL = 0
9.11 p-bar = 0.0185
$UCL_{400} = 0.0263$
LCL = 0.0107

Check points 2, 3, 4, 9, 11
point 2, check limits, out of control
point 3, do not check limits
point 4, do not check limits
point 9, do not check limits
point 11, check limits, out of control

9.13 np-bar = 4.5
UCL = 11
LCL = 0

9.14 p-bar = 9%
UCL = 21%
LCL = 0

9.16 p-bar = 0.0825
np-bar = 13
UCL = 23
LCL = 3

9.17 p-bar = 0.0405
np-bar = 12
UCL = 22
LCL = 2
process capability is np-bar = 12

9.18 p-bar = 0.013
np-bar = 5
UCL = 12
LCL = 0

9.19 c-bar = 8
UCL = 16
LCL = 0
process will not meet customer's specifications

9.21 c-bar = 5
UCL = 11
LCL = 0

9.22 c-bar = 6
UCL = 13
LCL = 0

9.23 u-bar = 2.345
UCL = 3
LCL = 1

9.25 u-bar = 1.18
n_{ave} = 101
UCL = 1.5
LCL = 0.86

Chapter 10

10.4 = 0.00035 Θ = 2886.25 hours
10.5 = 0.00197 Θ = 508 hours
10.8 Reliability of the system = 0.9768
10.9 Reliability of the system = 0.9561
10.10 Reliability of the system = 0.9859
10.12 Reliability of the system = 0.8996
10.13 Reliability of the system = 0.9950
10.14 Reliability of the system = 0.8569

Bibliography

Adam, P., and R. VandeWater. "Benchmarking and the Botton Line: Translating BPR into Bottom-Line Results." *Industrial Engineering*, February 1995, pp. 24–26.

Adcock, S. "FAA Orders Fix on Older 737s." *Newsday*, May 8, 1998.

Aft, L. *Quality Improvement Using Statistical Process Control*. New York: Harcourt Brace Jovanovich, 1988.

Alsup, F., and R. Watson. *Practical Statistical Process Control*. New York: Van Nostrand Reinhold, 1993.

AMA Management Briefing. *World Class Quality*. New York: AMA Publications Division, 1990.

American Society for Quality Control, P.O. Box 3005, Milwaukee, WI 53201-3005.

Bacus, H. "Liability: Trying Times." *Nation's Business*, February 1986, pp. 22–28.

Bamford, J. "Order in the Court." *Forbes*, January 27, 1986, pp. 46–47.

Bergamini, D. *Mathematics*. New York: Time, 1963.

Bernowski, K., and B. Stratton. "How Do People Use the Baldrige Award Criteria?" *Quality Progress*, May 1995, pp. 43–47.

Berry, Thomas. *Managing the Total Quality Transformation*. Milwaukee, WI: ASQC Quality Press, 1991.

Besterfield, D. *Quality Control*, 4th ed. Englewood Cliffs, NJ: Prentice Hall, 1994.

Biesada, A. "Strategic Benchmarking." *Financial World*, September 29, 1992, pp. 30–36.

Bishara, R., and M. Wyrick. "A Systematic Approach to Quality Assurance Auditing." *Quality Progress*, December 1994, pp. 67–69.

Bothe, D. "SPC for Short Production Runs." *Quality*, December 1988, pp. 58–59.

Bovet, S. F. "Use TQM, Benchmarking to Improve Productivity." *Public Relations Journal*, January 1994, p. 7.

Brocka, B., and M. Brocka. *Quality Management: Implementing the Best Ideas of the Masters*. Homewood, IL: Business One Irwin, 1992.

Brown, R. "Zero Defects the Easy Way with Target Area Control." *Modern Machine Shop*, July 1966, p. 19.

Bruder, K. "Public Benchmarking: A Practical Approach." *Public Press*, September 1994, pp. 9–14.

Brumm, E. "Managing Records for ISO 9000 Compliance." *Quality Progress*, January 1995, pp. 73–77.

Burr, Irving W. *Statistical Quality Control Methods*. New York: Marcel Dekker, 1976.

Butz, H. "Strategic Planning: The Missing Link in TQM." *Quality Progress*, May 1995, pp. 105–108.

Byrnes, Daniel. "Exploring the World of ISO 9000." *Quality*, October 1992, pp. 19–31.

Campanella, J. ed. *Principles of Quality Costs*. Milwaukee, WI: ASQC Quality Press, 1990.

Camperi, J. A. "Vendor Approval and Audits in Total Quality Management." *Food Technology*, September 1994, pp. 160–62.

Carson, P. P. "Deming Versus Traditional Management Theorists on Goal Setting: Can Both Be Right?" *Business Horizons*, September 1993, pp. 79–84.

Cook, B. M. "Quality: The Pioneers Survey the Landscape." *Industry Week*, October 21, 1991, pp. 68–73.

Crosby, P. B. *Cutting the Cost of Quality: The Defect Prevention Workbook for Managers*. Boston: Industrial Education Institute, 1967.

Crosby, P. B. *The Eternally Successful Organization: The Art of Corporate Wellness*. New York: New American Library, 1988.

Crosby, P. B. *Quality Is Free*. New York: Penguin Books, 1979.

Crosby, P. B. *Quality Is Free: The Art of Making Quality Certain*. New York: McGraw-Hill, 1979.

Crosby, P. B. *Quality without Tears: The Art of Hassle-Free Management*. New York: McGraw-Hill, 1979.

Cullen, C. "Short Run SPC Re-emerges." *Quality*, April 1995, p. 44.

Darden, W., W. Babin, M. Griffin, and R. Coulter. "Investigation of Products Liability Attitudes and Opinions: A Consumer Perspective," *Journal of Consumer Affairs*, June 22, 1994.

Davis, P. M. "New Emphasis on Product Warnings." *Design News*, August 6, 1990, p. 150.

Davis, P. M. "The Right Prescription for Product Tampering." *Design News*, January 23, 1989, p. 224.

Day, C. R. "Benchmarking's First Law: Know Thyself." *Industry Week*, February 17, 1992, p. 70.

Deming, W. Edwards. *The New Economics*. Cambridge, MA: MIT CAES, 1993.

Deming, W. Edwards. *Out of the Crisis*. Cambridge, MA: MIT Press, 1986.

Dentzer, S. "The Product Liability Debate." *Newsweek*, September 10, 1984, pp. 54–57.

DeToro, I. "The Ten Pitfalls of Benchmarking." *Quality Progress*, January 1995, pp. 61–63.

DeVor, R., T. Chang, and J. Sutherland. *Statistical Quality Design and Control*. New York: Macmillan, 1992.

Dobyns, L., and C. Crawford-Mason. *Quality or Else: The Revolution in World Business*. Boston: Houghton Mifflin, 1991.

Duncan, A. *Quality Control and Industrial Statistics*. Homewood, IL: Irwin, 1974.

Eaton, B. "Cessna's Approach to Internal Quality Audits." *IIE Solutions, Industrial Engineering*, June 1995, pp. 12–16.

Farahmand, K., R. Becerra, and J. Greene. "ISO 9000 Certification: Johnson Controls' Inside Story." *Industrial Engineering*, September 1994, pp. 22–23.

Feigenbaum, Armand V. "Changing Concepts and Management of Quality Worldwide." *Quality Progress*, December 1997, pp. 43–47.

Feigenbaum, Armand V. *Total Quality Control*. New York: McGraw-Hill, 1983.

Frum, D. "Crash!" *Forbes*, November 8, 1993, p. 62.

Galpin, D., R. Dooley, J. Parker, and R. Bell. "Assess Remaining Component Life with Three Level Approach." *Power*, August 1990, pp. 69–72.

Gerling, A. "How Jury Decided How Much the Coffee Spill Was Worth." *The Wall Street Journal*, September 4, 1994.

Gest, T. "Product Paranoia." *U.S. News & World Report*, February 24, 1992, pp. 67–69.

Geyelin, M. "Product Liability Suits Fare Worse Now." *The Wall Street Journal*, July 12, 1994.

Gitlow, H. S. *Planning for Quality, Productivity and Competitive Position*. Homewood, IL: Business One Irwin, 1990.

Gitlow, H. S., and S. J. Gitlow. *The Deming Guide to Quality and Competitive Position*. Englewood Cliffs, NJ: Prentice Hall, 1987.

Goodden, R. "Product Reliability Considerations Empower Quality Professionals." *Quality*, April 1995, p. 108.

Goodden, R. "Reduce the Impact of Product Liability on Your Organization." *Quality Progress*, January 1995, pp. 85–88.

Grahn, D. "The Five Drivers of Total Quality." *Quality Progress*, January 1995, pp. 65–70.

Grant, E., and T. Lang. "Why Product-Liability and Medical Malpractice Lawsuits Are So Numerous in the United States." *Quality Progress*, December 1994, pp. 63–65.

Grant, E., and R. Leavenworth. *Statistical Quality Control*. New York: McGraw-Hill, 1988.

Grant, P. "A Great Step Backwards." *Quality*, May 1985, p. 58.

Greene, R. "The Tort Reform Quagmire." *Forbes*, August 11, 1986, pp. 76–79.

Hare, L., R. Hoerl, J. Hromi, and R. Snee. "The Role of Statistical Thinking in Management." *Quality Progress*, February 1995, pp. 53–59.

Heldt, J. J. "Quality Pays." *Quality*, November 1988, pp. 26–27.

Heldt, J. J., and D. J. Costa. *Quality Pays*. Wheaton, IL: Hitchcock Publishing, 1988.

Himelstein, L. "Monkey See, Monkey Sue." *Business Week*, February 7, 1994, pp. 112–13.

Hockman, K., R. Grenville, and S. Jackson. "Road Map to ISO 9000 Registration." *Quality Progress*, May 1994, pp. 39–42.

Ireson, W., and C. Coombs. *Handbook of Reliability Engineering and Management*. New York: McGraw-Hill, 1988.

Ishikawa, K. *Guide to Quality Control*, rev. ed. White Plains, NY: Kraus International Publications, 1982.

Ishikawa, K. *What Is Total Quality Control? The Japanese Way*. Englewood Cliffs, NJ: Prentice Hall, 1985.

Juran, J. M. *Juran on Leadership for Quality: An Executive Handbook*. New York: Free Press, 1989.

Juran, J. M. *Juran on Planning for Quality*. New York: Free Press, 1988.

Juran, J. M. *Juran on Quality by Design: The New Steps for Planning Quality into Goods and Services*. New York: Free Press, 1992.

Juran, J. M. "The Quality Trilogy." *Quality Progress*, August 1986, pp. 19–24.

Juran, J. M., and F. M. Gryna. *Quality Planning and Analysis: From Product Development through Usage*. New York: McGraw-Hill, 1970.

Kackar, Raghu. "Taguchi's Quality Philosophy: Analysis and Commentary." *Quality Progress*, December 1986, pp. 21–29.

Kececioglu, D. *Reliability and Life Testing Handbook*. Englewood Cliffs, NJ: Prentice Hall, 1993.

King, R. "So What's the Law, Already?" *Forbes*, May 19, 1986, pp. 70–72.

Kolarik, W. J. *Creating Quality: Concepts, Systems, Strategies and Tools*. New York: McGraw-Hill, 1995.

Leibfried, K. H. *Benchmarking: A Tool for Continuous Improvement*. New York: Harper Business, 1992.

McElroy, A., and I. Fruchtman. "Use Statistical Analysis to Predict Equipment Reliability." *Power*, October 1992, pp. 39–46.

McGuire, P. "The Impact of Product Liability," Report 908. The Conference Board, 1988.

Malcolm Baldrige National Quality Award, U.S. Department of Commerce, Technology Administration, National Institute of Standards and Technology, Gaithersburg, MD.

Marcus, A. "Limits on Personal-Injury Suits Urged." *The Wall Street Journal*, April 23, 1991.

Martin, S. "Courting Disaster: A Huge Jury Award Sends the Wrong Message." *Bicycling*, March 1994, p. 136.

Mathews, J. "The Cost of Quality." *Newsweek*, September 7, 1992, pp. 48–49.
Mauro, T. "Frivolous or Not, Lawsuits Get Attention." *USA Today*, February 23, 1995.
Meier, B. "Court Rejects Coupon Settlement in Suit Over G.M. Pickup Trucks." *New York Times*, April 18, 1995.
Milas, G. "How to Develop a Meaningful Employee Recognition Program." *Quality Progress*, May 1995, pp. 139–42.
Miller, I., and J. Freund. *Probability and Statistics for Engineers*. Englewood Cliffs, NJ: Prentice Hall, 1977.
Moen, R., T. Nolan, and L. Provost. *Improving Quality through Planned Experimentation*. New York: McGraw-Hill, 1991.
Munoz, J., and C. Nielsen. "SPC: What Data Should I Collect? What Charts Should I Use?" *Quality Progress*, January 1991, pp. 50–52.
Nakhai, B., and J. Neves. "The Deming, Baldrige, and European Quality Awards." *Quality Progress*, April 1994, pp. 33–37.
Neave, H. *The Deming Dimension*. Knoxville, TN: SPC Press, 1990.
"Needed: A Backup for Ma Bell." *U.S. News & World Report*, September 30, 1991, p. 22.
Nesbitt, T. "Flowcharting Business Processes." *Quality*, March 1993, pp. 34–38.
Nutter, J. "Designing with Product Liability in Mind." *Machine Design*, May 24, 1984, pp. 57–60.
Orsini, J. "What's Up Down Under?" *Quality Progress*, January 1995, pp. 57–59.
Parsons, C. "The Big Spill: Hot Java and Life in the Fast Lane." *Gannett News Services*, October 25, 1994.
Petroski, H. "The Merits of Colossal Failure." *Discover*, May 1994, pp. 74–82.
Pittle, D. "Product Safety: There's No Substitute for Safer Design." *Trial*, October 1991, pp. 110–14.
Pond, R. *Fundamentals of Statistical Quality Control*. New York: Merrill, 1994.
Press, A., G. Carroll, and S. Waldman. "Are Lawyers Burning America?" *Newsweek*, March 20, 1995, pp. 30–35.
"Product Liability." *Business Insurance*, October 4, 1993, p. 36.
"Product Liability: The Consumer's Stake." *Consumer Reports*, June 1984, pp. 336–39.
"Product Liability Advice: Surpass Design Expectations." *Design News*, May 17, 1993, pp. 31–35.
Pyzdek, T. *What Every Engineer Should Know about Quality Control*. New York: Marcel Dekker, 1989.
Rice, C. M. "How to Conduct an Internal Audit and Still Have Friends." *Quality Progress*, June 1994, pp. 39–40.
Rienzo, T. F. "Planning Deming Management for Service Organizations." *Business Horizons*, May 1993, pp. 19–29.
Robinson, C., ed. *How to Plan an Audit*. Milwaukee, WI: ASQC Press, 1987.
Rowland, F. "Liability of Product Packaging." *Design News*, February 1, 1993, p. 120.
Rowland, F. "Product Warnings Updated." *Design News*, March 9, 1992, p. 168.
Rowland, J. R. "Should Congress Ease the Product Liability Law?" *American Legion*, April 1993, p. 10.
Russell, J. *Quality Management Benchmark Assessment*. Milwaukee, WI: 1991. ASQC Press, 1991.
Russell, J. "Quality Management Benchmark Assessment." *Quality Progress*, May 1995, pp. 57–61.
Savell, L. "Who's Liable When the Product Is Information?" *Editor and Publisher*, August 28, 1993, pp. 35–36.

"Short-Run SPC Re-emerges." QEI Speaker Interview. *Quality*, April 1995, p. 44.

Smith, G. *Statistical Process Control and Quality Improvement*. New York: Merrill, 1991.

Smith, R. "The Benchmarking Boom." *Human Resources Focus*, April 1994, pp. 1–6.

Spendolini, M. *The Benchmarking Book*. New York: Amcom, 1992.

Srikanth, M., and S. Robertson. *Measurements for Effective Decision Making*. Wallingford, CT: Spectrum Publishing Co., 1995.

Stevens, T. "Dr. Deming: Management Today Does Not Know What Its Job Is." *Industry Week*, January 17, 1994, pp. 20–28.

Taguchi, Genichi. *Introduction to Quality Engineering*. Dearborn, MI: American Supplier Institute, 1986.

Thorpe, J., and W. Middendorf. *What Every Engineer Should Know about Product Liability*. New York: Marcel Dekker, 1979.

Tobias, R. *Applied Reliability*. New York: Van Nostrand Reinhold, 1986.

Traver, R. "Nine-Step Process Solves Product Variability Problems." *Quality*, April 1995, p. 94.

Traver, R. "Pre-Control: A Good Alternative to X-bar and R Charts." *Quality Progress*, September 1985, pp. 11–13.

Tsuda, Y., and M. Tribus. "Planning the Quality Visit." *Quality Progress*, April 1991, pp. 30–34.

"U.S. FAA: FAA Orders Immediate Inspection for High-time Boeing 737s, Extends Inspection Order." *M2 Press Wire*, May 11, 1998.

Vardeman, S. *Statistics for Engineering Problem Solving*. Boston: PWS Publishing, 1994.

Walpole, R., and R. Myers. *Probability and Statistics for Engineers and Scientists*. New York: Macmillan, 1989.

Walters, J. "The Benchmarking Craze." *Governing*, April 1994, pp. 33–37.

Walton, M. *Deming Management at Work*. New York: Pedigree Books, 1991.

Walton, M. *The Deming Management Method*. New York: Putnam, 1986.

Watson, G. "Digital Hammers and Electronic Nails—Tools of the Next Generation." *Quality Progress*, July 1998, pp. 21–26.

Watson, R. "Modified Pre-Control." *Quality*, October 1992, p. 61.

Weimer, G. A. "Benchmarking Maps the Route to Quality." *Industry Week*, July 20, 1992, pp. 54–55.

Wheeler, D. *Advanced Topics in Statistical Process Control*. Knoxville, TN: SPC Press, 1995.

Wheeler, D. *Short Run SPC*. Knoxville, TN: SPC Press, 1991.

Wheeler, D. *Understanding Variation: The Key to Managing Chaos*. Knoxville, TN: SPC Press, 1993.

Wheeler, D., and D. Chambers. *Understanding Statistical Process Control*. Knoxville, TN: SPC Press, 1992.

Wiesendanger, B. "Benchmarking for Beginners." *Sales and Marketing Management*, November 1992, pp. 59–64.

Wilson, L. *Eight-step Process to Successful ISO 9000 Implementation*. Milwaukee, WI: ASQC Press, 1996.

Winslow, R. "Hospitals' Weak Systems Hurt Patients, Study Says." *The Wall Street Journal*, July 5, 1995.

Zaciewski, R. "Attribute Charts Are Alive and Kicking." *Quality*, March 1992, pp. 8–10.

Zollers, F. E. "Product Liability Reform: What Happened to the Crisis?" *Business Horizons*, September 1990, pp. 47–52.

Index